The Nuaulu World of Plants

Royal Anthropological Institute

The RAI Series

Series Editor: Jeremy MacClancy

The RAI Book Series is devoted to the publishing of major works in anthropology, both monographs and edited volumes. Enquiries as to future projects are welcome, and may be addressed to publications@therai.org.uk. Further details may be found at www.therai.org.uk/publications/rai-book-series.

Volume 1 *Built in Niugini: Constructions in the Highlands of Papua New Guinea*
Paul Sillitoe

Volume 2 *Made in Niugini: Technology in the Highlands of Papua New Guinea*
Paul Sillitoe

Volume 3 *The Nuaulu World of Plants: Ethnobotanical Cognition, Knowledge and Practice Among a People of Seram, Eastern Indonesia*
Roy Ellen

The Nuaulu World of Plants

Ethnobotanical Cognition, Knowledge and Practice Among a People of Seram, Eastern Indonesia

Roy Ellen

The RAI Series, Volume Three

Sean Kingston Publishing
www.seankingston.co.uk
Canon Pyon

First published in 2020 by
Sean Kingston Publishing
www.seankingston.co.uk
Canon Pyon

British Library Cataloguing in Publication Data

A catalogue record for this book is available from the British Library.

ISBN 978-1-912385-25-6

For Guy and Aurelia as they begin to discover the world of plants

Frontispiece *Tuisa Neipani-tomoien of Rouhua village lighting a fire as he prepares medicinal plants in his swidden shelter, August 1973. In the foreground are at least six plant species attributed with medicinal properties.*

Contents

FIGURES

TABLES

PREFACE

This book has been a long time gestating. In a sense it has been with me for my entire professional life, from the time when, as a research student, I first encountered the work pioneered by Harold Conklin, and began to understand its significance during my period of doctoral fieldwork amongst the Nuaulu. Not only did I begin to accumulate data on Nuaulu ethnobotany from my early days in the field, but there is evidence of my planning for articles and analyses. In due course, some of this work has been published in various papers, and I suppose most of the key themes that have become subsequently important in my work – folk classification, the formation of concepts of nature, the management of biological resources, forest knowledge and knowledge transmission – have been reflected in these. But the priorities of an academic career tend to be with short outputs that make an impact, and ethnobotanical research is something that requires commitment over the long term. Its interdisciplinarity character means that a particular combination of skills and collaborative links have to be acquired and experienced; comprehensiveness of data needs serial field trips to expand the corpus and check queries, while international herbaria often operate on a time scale more reminiscent of the Vatican than the urgent deadlines of contemporary academia.

The scale of this work, its complexities and the time period to which it relates, inevitably mean that I have acquired many debts along the way. My research amongst the Nuaulu was sponsored in its early years by LIPI, Lembaga Ilmu Pengetahuan Indonesia (the Indonesian Academy of Sciences), and subsequently by Pattimura University in Ambon. This monograph draws on nine field trips, totalling some 33 months. In 1970–1 (over 18 months) I was supported by a Social Science Research Council (SSRC) studentship, the London-Cornell Project for South-East Asia and the Central Research Fund of the University of London; in 1973 (three months) by the Central Research Fund and a Hayter Travel Award; in 1975 (three months) by an SSRC grant for work mainly on Nuaulu ethnozoology; in 1981 (one month) by the British Academy through the British Institute in South-East Asia and a South-East Asia Travel Grant from the University of Kent at Canterbury, and in 1986 (one month) by a Nuffield Foundation Social Science Research Fellowship, while working on a multi-sited project on inter-island trade; and in 1990 (one month) by the Centre for South-East Asian Studies at the University of Kent. Three major Economic and Social Research Council (ESRC) awards allowed me to expand the data corpus on Nuaulu plant knowledge significantly, supporting fieldwork in 1996 (four months) and again in 2003 (one month): Grant R000–236082, 1995–8, on 'Deforestation and forest knowledge in south central Seram'; Grant R000–239310, 2001–4 on 'Frequency and periodicity in Nuaulu ritual reproduction' and grant R000–221106, 2005–7, on 'The ethnography, ethnobotany and dispersal of palm starch extraction'. A British Academy-ASEASUK Research Committee award (2009–10), for 'A comparative study of the sociocultural concomitants of cassava diversity in four eastern Indonesian populations in relation to environmental security', involved further fieldwork in 2009, and British Academy SG 131590, 2014–16, 'Demography, kinship and ritual reproduction: Nuaulu cultural resilience in the new Indonesia', funded fieldwork in 2015. Finally, a Leverhulme Trust Emeritus Fellowship (EM-2018-057-6, 2018–20) for 'Nuaulu ethnobotanical cognition and knowledge' has permitted the extensive checking of data and has facilitated the writing-up stage and publication of this volume. To all these bodies I am truly indebted.

During this time I have benefitted from the expertise of staff associated with a number of institutions. In connection with the 1969–71 field trip, I am grateful to Dr Chang Kiaw Lan at the Singapore Botanic Gardens. At Kew, I have received expert support and expertise since 1969, particularly from numerous staff attached to the herbarium: including Sir George Taylor, and thereafter, C.M. Wilmot-Dear, Aaron Davis, S. Knapp, Bill Baker, Tom Cope, David Middleton, Paul Wilkin, Sally Dawson, Abdulrokhman Kartonegoro, Ana Claudia Araujo, Ray Harley, Eve Lucas, Maroa Vorontsova, Peter Wilkie, S. Sengum, Alison Moore, Martin Xan, Gwilym Lewis, Marie Briggs, B Mackinder, Leonard Forman, John Dransfield, J.B.T. Blewett, Emily Brennan, Rogier de Kok, Mark Nesbitt, Tim Utteridge and the South-East Asia Regional Team; and particularly for the 1996 field trip, Mark Coode. In Leiden, I would like to thank Christel Schollaardt and Roxali Bijmoer at the herbarium of the Naturalis Biodiversity Center, and Tinde van Andel. In Bogor, I am indebted to Johannis Mogea at the Herbarium Bogoriense, and his successor as head of the Kepala Bidang Botani Herbarium, Dr Joeni Setijo Rahajoe, Mohammad Fathi Royyani and other staff, and Irham Muhammad at LIPI Cibinong. Johan Iskandar and Wati Supangkat-Iskandar at Padjadjaran University Bandung have provided continuous assistance and friendship for this and other projects. In Ambon, Hermien Soselisa has been my main contact in recent years, and I would like to thank her, Dr H.B. Telelepta, Professor Ir.J.W. Mosse and other colleagues and friends at Pattimura University.

In and around Seram, and in working with the Nuaulu, my stays have been made much easier by the assistance of Rosemary Bolton, Hunanatu Matoke and Niksen Sopacua. Komisi Soumori, Saite Soumori, Kaisa Neipani-tomoien and Sonohue Matoke, as successive 'kepala dusun' in Rouhua, were not only my official hosts while undertaking fieldwork, but also became generous research colleagues and good friends. In addition, I need to thank Raja Desa Nuanea, Sahune Matoke-hanaie and Merpati Sounaue-ainakahata for permissions with respect to work in their area of jurisdiction. It would be impossible to mention all those individual Nuaulu who have assisted me in the field directly on matters relating to plant knowledge, but Retaone Soumori and Heunaka Neipani-tomoien have over many years played a prominent and proactive role as research associates, and deserve special mention. To these and others I also apologize for the inevitable errors that will have crept in to my attempts to describe their knowledge, despite my best attempts to avoid them.

At Kent, in connection with the grants listed above, I am grateful to former research students and colleagues who have helped prepare mounted vouchers, produced databases, analysed plot data, organized and curated material-culture collections, and assisted with computing issues: Alan Bicker, Alison Hoare, Amy Warren, Susi Soemarwoto and Simon Platten. Christine Eagle, at the time a departmental IT coordinator, was responsible for setting up the original database and transferring the plot data to digital diagrams. Most recently, Mercy Morris has performed small miracles in organizing the appendices, completing the work on the Nuaulu herbarium voucher specimens housed in the Ethnobiology Laboratory, and producing some of the artwork. For advice on the identification of particular plant groups I am grateful to the following: the late D.G. Coursey (yams), Jean Kennedy (Musaceae), Dave Roberts, Kent (Nepenthes), Sadao Sakamoto (millets and sorghum) and Asri Peni Wulandari (cassava).

Some parts of what follows have been previously published or presented as seminar or conference papers. Sections of Chapter 9 dealing with vegetative change and how people respond to this through classification, was first presented as part of a European COST network meeting on 'Stability and adaptation in classification systems in a cross-cultural perspective', hosted by Raheli Shalomi-Hen and Orly Goldenweisser, in Jerusalem in 2007. I am grateful to the Wenner-Gren Foundation and the editor of *Current Anthropology* for permission to reproduce passages (in Chapter 10) which previously appeared in Ellen 2018b; to the editors of the *Journal of Material Culture* to reproduce other passages, including part of Chapter 10, which appeared in Ellen 2009a; to Tanja Pommerening, Walter Bisang and De Gruyter for permission to reproduce passages from *Classification From Antiquity to Modern Times* in Chapter 11 (Ellen 2017b); and to Paul Sillitoe and CABI International for permission to reproduce extracts from *Indigenous Knowledge: Enhancing its Contribution to Natural Resources Management* (Ellen 2017c) which appear in Chapter 7.

The final shape of this book has been much influenced by my struggle over many years, together with my Nuaulu co-researchers and a multitude of field botanists and herbarium specialists, to connect the individual plants encountered with the names (scientific and vernacular) and cultural frameworks that people share within what might be described as distinct but overlapping 'communities of practice'. In the situations where this process takes place, problems have often been compounded by having to deal with specimens and other data that are fragmentary (sometimes just descriptions and representations), unfamiliar or ecologically decontextualized. I am pleased to say that these problems have been resolved in a way that allows us all to make sense 'sufficiently' of the differences we experience between plants of various kinds.

Roy Ellen,
Crockshard Farmhouse, Wingham

A NOTE ON ORTHOGRAPHY

The orthography of Nuaulu language adopted here follows that developed by Rosemary Bolton and used by her and Hunanatu Matoke in their 2005 edition of the *Kamus Sou Naunue–Sou Manai; Bahasa Indonesia–Nuaulu*. Observant readers will note slight changes in the way spoken Nuaulu has been rendered in this volume compared with my *Cultural Relations of Classification* (1993). The letters used to represent Nuaulu speech sounds are phonemic. They include five vowels and eleven consonants composed of the following phonetic features (Bolton 1990):

Vowels

front		unrounded		back rounded
high	close	i		u
mid	half open	e		o
	open		a	

Consonants

glottal				
voiceless stops	p	t	k	
fricatives		s		h
nasals	m	n		
flaps	r			
laterals	l			
semi-vowels	w	y		

The Indonesian alphabet includes all Nuaulu phonemes and is used here without modification. Stress is unmarked in most regular cases and normally occurs on the penultimate syllable. Nuaulu words appear in italicized ***boldface***, Indonesian words and words in other Indonesian languages in regular Roman but placed in inverted commas. Scientific names are placed in *italics*. Because of the large number of Nuaulu plant names referred to, I have sometimes abbreviated folk binomials (especially in tables) following the practice adopted in scientific nomenclature. For example, ***karatupa menesane*** (a type of chilli) becomes ***k. manesane***.

> ... if you know these substructures in the lower levels of the ontic hierarchy (beyond the plant, which is an organism) and go into the physical, chemical, molecular, and atomic structures, ever further down, the greater becomes the miracle how all that thing is a plant.... If you deform your experience by trying to explain what you experience by the things that you don't experience but that you know only by science, you get a perverted imagination of reality.
>
> (Voegelin 2004:290–1)

> Sorting and classifying are fundamental human activities. Young children can be completely absorbed in sorting fallen leaves, accurately, into piles according to species. They are quick to pick up the 'feel' of the species and sort them precisely, without worrying about either the names of the trees or the purpose of the sorting. A label or name remembered guarantees a slot in our mind reserved for the plant and its associations, and makes the noun concrete; for this reason, of course, names are inevitable whenever we become serious about managing or understanding a resource.
>
> (Lawrence and Hawthorne 2006:2)

> What is less often emphasized ... is that scientific names suffer from all of these same types of problem –impermanence, incompleteness, synonymy and inconsistency of application ... This is in spite of efforts made by scientists to place their names beyond the vagaries of normal language.
>
> (Hawthorne and Harris 2006:63)

> The only good classification is a living classification.
>
> (Bowker and Star 2000:326)

1

Introduction

Objectives and organization

The aim of this work is twofold. Firstly, it seeks to provide a coherent and systematic account of how plants feature as living organisms and phenomena in the world-view and lifeways of the Nuaulu, a people of the island of Seram in eastern Indonesia. Ethnobotanical ethnographies based on long-term fieldwork are still uncommon enough that they can be justified both in their own right and as a resource for understanding other aspects of regional ethnography. Secondly, the work seeks to provide a body of data for evaluating the principles and theories underpinning the way knowledge of the natural world is culturally organized more generally. In 1993 I published the *Cultural Relations of Classification*, which looked at the ways Nuaulu construct and arrange categories of the animal world. Since that time much more has been published on ethnobiological classification and so now, using a different corpus, I seek to examine how some of these ideas have stood the test of time.

The work is organized into eleven chapters. The present chapter (Chapter 1) provides some background necessary to understand what follows: geographic context, accounts of the vegetation, and of the history of human occupation on Seram and its ecological impact. I provide a brief introductory description of the Nuaulu way of life, and outline my methodology, beginning with the history of ethnobotany in the Moluccas. Chapter 2 describes how Nuaulu experience and cognize the plant world. It addresses the question 'what is a plant?' and covers the language of plant description, the structure and semantics of plant nomenclature, and life form concepts as a way of grouping plants at the most inclusive level. Three chapters that discuss the plants Nuaulu experience in terms of different spatial and ecological orientations then follow. But although the work seeks to provide an overview of the main issues in Nuaulu ethnobotanical classification, it does not systematically or uniformly detail the intricacies of every vascular plant group. Neither should it be understood as mainly a gazeteer of plant uses, although I argue that how plants are used is intrinsic to how Nuaulu think about them.

Chapter 3 sets out the received framework for describing ethnobotanical classification, as derived from Brent Berlin, and introduces the systematic descriptions of plant classification by focusing on the example of ferns and their allies, pteridophytes. Chapter 4 looks at swiddens[1] and the plants associated most closely with them: starchy roots and tubers, bananas and plantains, grains, leafy vegetables, pulses, fruits and other garden crops, and those grown in interstitial spaces on village land. It also looks at managed fallow, village spaces, path-sides and grassy biotopes. It reflects on how Nuaulu experience the world of cultivated plants, and engages with the problem of how plant diversity is conceptualized below the species level. Chapter 5 focuses on trees and forest as part of a semi-managed landscape, examining 'forest' as an ecological, conceptual and experiential entity. It analyses the categories ***ai*** (tree), ***wane*** (vine) and ***meute*** (rattan), and looks at the extent to which complexes of resemblance can be detected within the life form 'tree'. Concluding that there is remarkably little attempt to divide up the large tree life form, it asks how, therefore, tree diversity can be handled above the basic level.

Chapter 6 looks at bamboos, pandans and palms, three important groups of plants mainly known to Nuaulu from near forest and the garden fringes, whose physical saliences are matched by their usefulness and symbolic values. Utility is addressed critically as a concept in Chapter 7, by looking at plants as food, healing agents, materials for making objects and as fuel, and at their role in ritual. The emphasis in this chapter is on those plants we describe as shrubs and herbs. Chapter 8 focuses on the correspondence between Nuaulu plant categories and scientifically defined taxa, examining – progressively – one-to-one correspondence at the species level, the notions of under- and over-differentiation, the placing of different genera from the same family in the same basic Nuaulu category, and genera from different biological

1 Fields created by clearing vegetation (usually trees) and burning the trash. These are then cultivated for a few years before allowing secondary growth to recolonize the area and so establish a fallow of indeterminate length.

families in the same category, and the relationship of life forms and other 'higher taxa' to biological systematics. The problems are addressed in the light of recent practice in scientific taxonomy, and the chapter revisits the issue of what we might mean by scientific taxa in the first place. Chapter 9 looks at plant knowledge over time: introducing the historical ethnobotany of the Nuaulu, exploring evidence for proto-domestication and incipient cultivation, reviewing the impact of the great Columbian exchange on species composition and use, on folk classification and the ways plants as a domain are ontologized. The chapter also looks at other exotic plants introduced from the late colonial period onwards, and the impact of colonial and post-colonial ways of seeing the plant world.

Chapter 10 is the first of two inviting us to rethink some aspects of the framework of the 'new normal' consensus deriving from the work of Brent Berlin and Scott Atran. Rather than treating plants and animals in formal terms as part of hermetically sealed domains, I suggest that we should consider ethnobotanical classification as a process within a wider dynamic system. Particular attention is paid to place and the localization of knowledge, and to the concept of life form as it relates to the major plant groupings. The chapter rehearses some limitations of the Berlinian scheme and the problems of comparative ethnobotany in the context of a new interest in 'convergence metaphysics'. It asks how we might reconcile general and comparative features of classification given that plant knowledge is socially embedded in plural local contexts. The chapter concludes by reflecting on parallel issues in Nuaulu plant classification and scientific taxonomy, which it suggests might also be viewed as a community of folk practice. I show how the work of the Germano-Dutch naturalist George Rumphius, including his insights into seventeenth-century Ambonese plant nomenclature and classification, might be seen to be in dialogue with contemporary Nuaulu, in a sense mediating between these two worlds. Finally, Chapter 11 looks at the relevance of Nuaulu data for theory in ethnobotany and cultural cognition more generally, in the light of some recent trends. We return to the idea of plants as a domain of thought and practice, exploring the mutuality of sociality, body imagery and plant knowledge, and plant classification and the meanings of life, before finishing with some reflections on methodological reflexivity and ontological friction.

Geographic context

The island of Seram lies three degrees south of the Equator, between Sulawesi to the west and New Guinea to the east, in a part of the world now known, because of its characteristically transitional biodiversity, as Wallacea (Figure 1.1). Wallace's line marks the western boundary, running north from between Bali and Lombok and east of Borneo. The line is also at the interface of two tectonic plates, the Asian and the Pacific. By the end of the last ice age it also reflected the boundary between two continental masses, Sunda (mainland South-East Asia and the larger western islands of the archipelago) and Sahul (Australasia, Papuasia) (Whitmore 1981). Lying within the South-East Asia monsoon area, south Seram has a rainy season between May and August. In the Amahai subdistrict there was in 2013 an average yearly rainfall of 302.5 mm and 18.8 rain days a year, with the rainy season falling between June and July. There was an average temperature of 26.5 degrees Celsius, with the hottest months being December – March, during which the average rose to 31 (Badan Pusat Statistik 2014:11–12). Since 1970 there has been an increase in average annual rain of 19.3 mm, from 283.2, and of average annual temperature of 1.3 degrees, from 25.2 (Ellen 1978a:212–4).

The macro-geomorphology of the Moluccas focuses on small islands and short fast-flowing rivers, in contrast to the large islands, lowlands and larger rivers of the western archipelago. The underlying geology of Seram is largely coralline, but with evidence of previous volcanic activity around its periphery. The coastal area of south-central Seram is essentially Quarternary, comprising primarily alluvial deposits containing conglomerates and the characteristic outcrops of coralline limestones ('karang'). The phyllite formations further inland, together with areas of elevated reefs and conglomerates, complete the petrographical structure of the area from which Nuaulu extract resources although these latter deposits tend not to encroach on the cultivated area of the villages (Ellen 1978a:215–19, map 14; Germeraad 1946:20–2, 34, 105 and map). Soil patterns tend to be closely related to variations in forest type, stage of soil evolution and composition of parent material. Based on work in the Manusela National Park, Payton (1993) has shown measurable differences between the soils of Holocene alluvial land that support more species-rich forest, and older severely leached and weathered soils characteristic of dissected Pleistocene surfaces that support species-poor dipterocarp forest. There is marked altitudinal zonation of soils reflecting rainfall and temperature, and distinguishing lower and upper montane, and alpine and subalpine zones. These latter zones are rarely frequented by Nuaulu hunters and foragers. This picture is basically consistent with the little we know of the soils underlying Nuaulu cultivated areas in the region of Sepa on the south coast, where the initially rich soils of recently cleared gardens suffer gradual nutrient depletion during subsequent phases of the cultivation cycle, followed by improvement during fallow, the 'semi-fallow' of grove land and regeneration to mature forest. Soils associated with ***nokase*** limestone are regarded as superior due to high lime content, but as less easy to cultivate due to greater stoniness than the more easily worked soils on ***tapune*** sandstones and schists (Ellen 1978a:217–19).

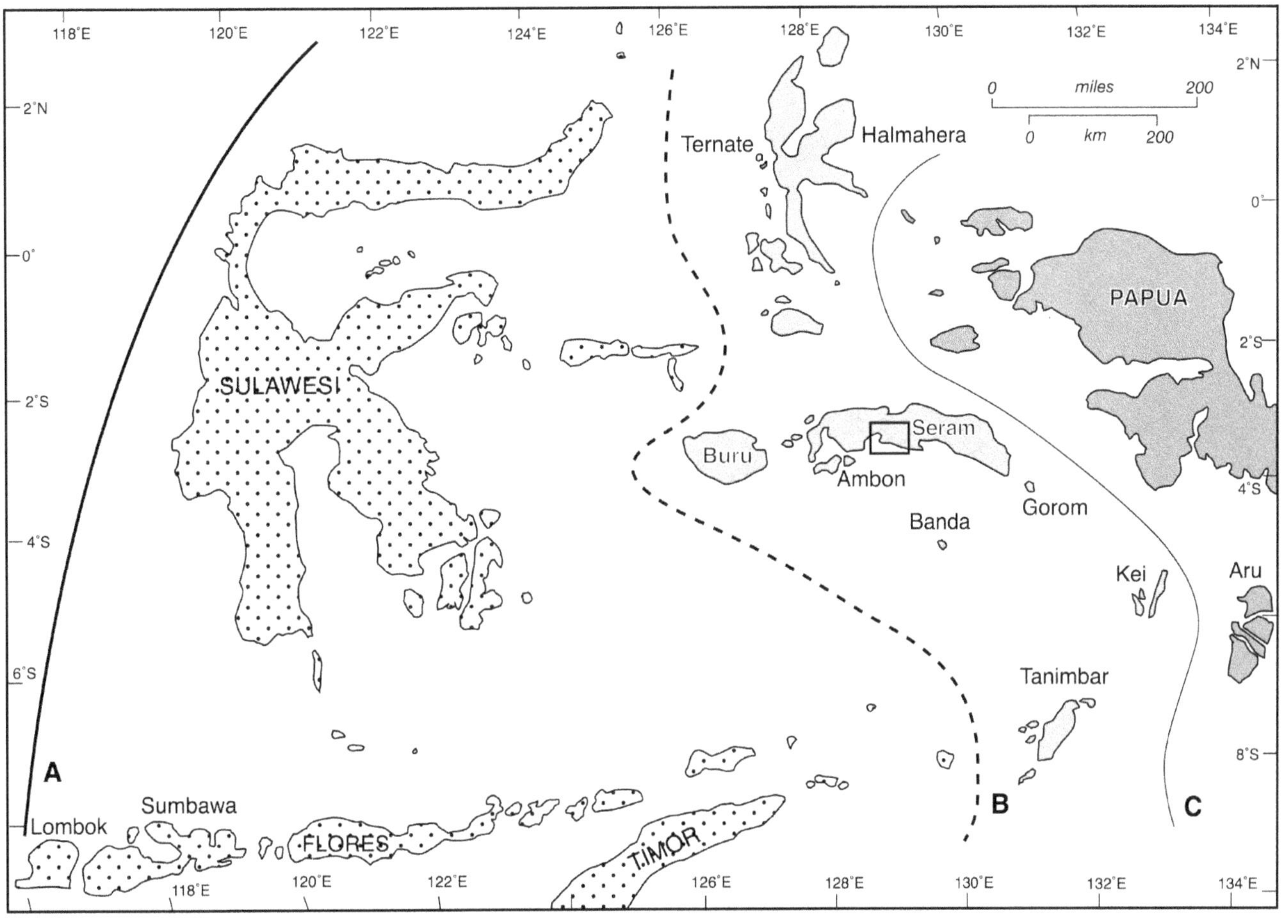

Figure 1.1 Eastern Indonesia, showing the location of Seram, Wallace's line (A), Weber's line (B), and the western boundary of the Australasian biogeographic area (C), which is Lydekker's line marking the edge of the Sahul shelf. Modified from Ellen 1978b, figure 1, after Darlington 1957:462. The box indicates the area expanded in Figure 1.4 (b).

The vegetation of Seram

Seram lies in the Malesian phytogeographic area, in a 'district' that Van Steenis (1950a:lxxv) first circumscribed – based on generic taxa of phanerogams – as including Buru, Halmahera, Seram and the chain of islands down to Kei, but excluding Aru and the Raja Ampat islands that show greater affinities with New Guinea. To the south, Tanimbar is regarded as having greater botanical affinities with Nusa Tenggara Timor. In compositional terms, the vegetation of Seram reflects, therefore, its position in the Wallacean transitional area (Figure 1.1), although plant distributions are less easily explained through the Wallace hypothesis than are animals (van Welzen, Parnell and Slik 2011; see also Michaux 2010). Wallace (1863) first theorized the biogeographic cline that runs from mainland South-East Asia to the Pacific, marked by what we now call the Wallace line, which approximately describes the median point where the biota of Sunda becomes that of Sahul. Of course, we now see this as a zone of gradation, forming different patterns for different groups of organisms, and named Wallacea in honour of Wallace. It was this transitional status that much intrigued me, both as an anthropologist and as an ethnobiologist, when I first embarked on ethnographic research in the Moluccas at the end of 1969. For example, the flora of Seram combines Asiatic and Australasian species in different ways, with decreasing numbers of dipterocarps as we move eastwards (about 300 species from Borneo and just one on Seram, *Shorea selanica*), matched by increasing numbers of LAURACEAE, and especially eucalypts (one species on Seram: *Eucalyptus deglupta*) and their near relatives (such as *Melaleuca cajuputi*). In terms of mammals, there is a comparable cline between eutherian megafaunas to the west and marsupials to the east. Since Papuan languages overlap with Austronesian languages in the Moluccas, and Melanesian features of human social organization and culture intermingle with Indonesian features, this is all expressed in ethnobotanically interesting ways. But although for the groups of organisms that most interested Wallace – birds and insects – Moluccan species are plentiful, for most macro-organisms, including for vascular plants, these islands are highly depauperate. Thus, whereas Borneo supports some 1000 recognized

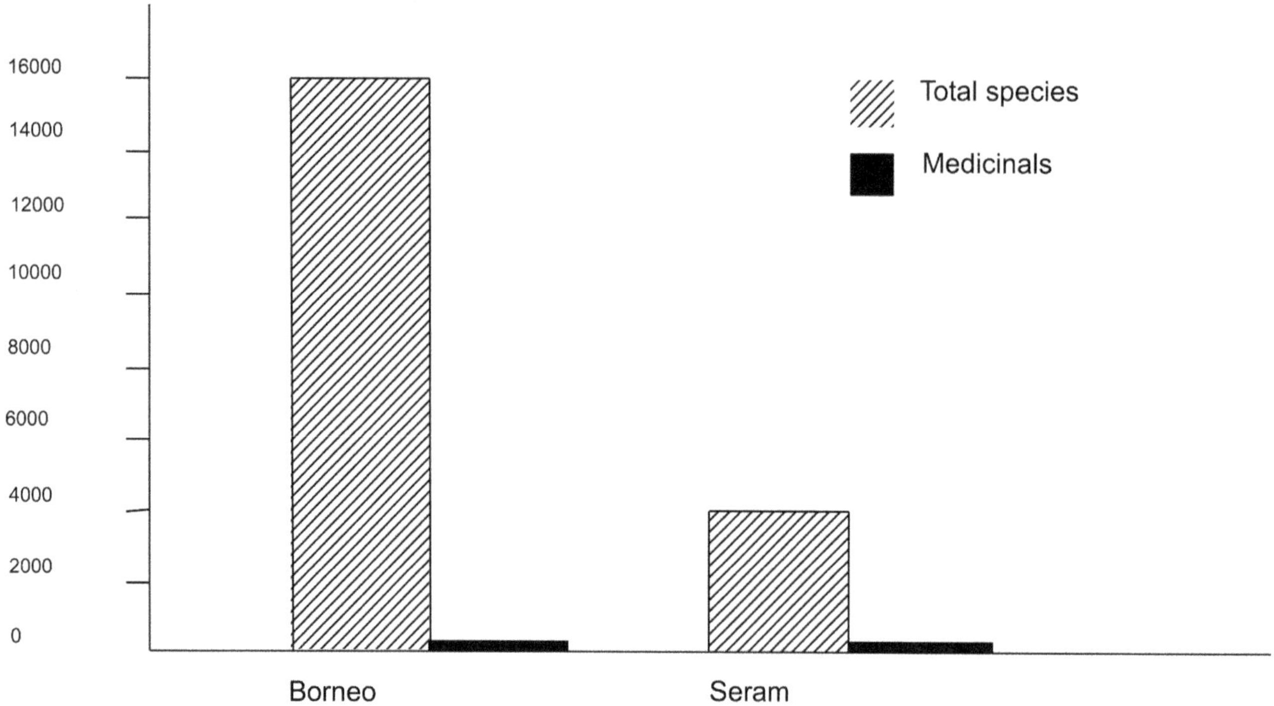

Figure 1.2 Total vascular plant species in relation to those reported as medicinals, Borneo and Seram compared (Ellen and Puri 2016).

fern species, Seram supports barely 534 (Edwards 1993; MacKinnon, Hatta, Halim and Mangalik 1997; Monk, de Fretes and Reksodiharjo-Lilley 1997). Of course, this must to some extent reflect the comparative land masses of these islands and the amount of scientific attention they have received, but these proportions are not reflected, for example, in the numbers of plants used as medicinals, with a ratio of 1:71 for Borneo and 1:12 for Seram, as can be seen from Figure 1.2 (Ellen and Puri 2016). At the same time, the relatively low levels of biodiversity found in the terrestrial areas of the Moluccas have been historically offset by the movement of many species into the Moluccas via human agency, from both South-East Asia and the Pacific, and including organisms as varied as the cassowary (*Casuarius casuarius*) and the teak tree (*Tectona grandis*), in addition to all those other useful plants that arrived in the region as a result of what Alfred Crosby (1972) called the great 'Columbian exchange', following the European discovery of the Americas. In other words, the eastern limit of Van Steenis' Moluccan 'district' conforms quite closely with general biogeographic discontinuities and language and cultural differences. So, as we would now say, Wallacea is a biocultural (Maffi 2001) and not just a biological phenomenon.

Much of the surface of Seram is still covered with tropical rainforest, which varies in composition and structure with altitude, with proximity to freshwater and the coast, and, most importantly, with the long-term impacts of human occupation. The variation and composition of forest in the area from which Nuaulu extract resources will be discussed in greater detail in Chapter 5, and also in relation to forest regeneration through swiddening. In general terms though, for the island as a whole, we can distinguish between montane, lowland and swamp forests. In terms of altitudinal zonation, the highest areas are alpine (Edwards, Payton, Proctor and Riswan 1990). In addition, we can distinguish grassland areas, bamboo forests, plantation land (mainly along the coast, but also at higher altitude, *Agathis dammara* plantations were established by the Dutch during the nineteenth century), and areas of permanent agriculture. Mangrove is characteristic of some coastal areas, but in those frequented by the Nuaulu it is mainly found around the mouth of the Ruatan and around the eastern part of Elpaputih Bay (where it is now much depleted through pollution and extraction), and at the mouths of some of the larger rivers flowing south into the Banda sea. The littoral is in most places similar: sandy coasts giving way to stony scrubby foreshores and quickly followed by plantation or grove land, as described by Warburg (1893) for Seram-Laut.

Human occupation

The deep human history of Seram is still poorly understood. There has been some excavation and surface archaeology conducted in western Seram (Latinis 2002) and surface archaeology in the Nuaulu area (Ellen and Glover 1979; Glover and Ellen 1975). On the basis of this, incidental finds over many years, and extrapolation from work conducted elsewhere in Sahul, the archaeological traces suggest palaeolithic dates of between 55,000 and 31,000 BP for first occupation (Spriggs 1998:53–4). There is as yet no direct evidence (such as pollen or charcoal) in the subfossil record of Seram that might provide clues for the first appearance of farming. We have possible traces of human activity for some parts of Indonesia, radiocarbon dated to about 50,000 BP, from palynology, charcoal and archaeology, suggesting repeated biomass burning interrupting ecological successions (Hunt and Rabett 2014:26; Maloney 1998). Recent work on El Niño Southern Oscillation variability over the long term suggests that it might account for the fire evidence in the record (Donohue and Denham 2010). Although rainforest archaeology is notoriously difficult, comparative evidence suggests that arboriculture was certainly present in the central Moluccas by the early Holocene (Stark and Latinis 1996:15). If, as implied by data from New Guinea and the western Indonesian archipelago, the appearance of neolithic culture is linked to the spread of Austronesian languages (Blust 1993), the earliest we can expect it in the Moluccas is around 4,500–3,500 BP, though archaeologicial dates for the Moluccas generally do not yet go back earlier than 3300 BP (Spriggs 1998:57).

The ecological and plant geographic evidence, however, suggests considerable human agency. Domesticated banana (*Musa acuminata* hybrids), sugar cane (*Saccharum officinarum*), the greater yam (*Dioscorea alata*), taro (*Colocasia esculenta*) and sago (*Metroxylon sagu*) all appear to have moved fairly early from the New Guinea mainland or have been partly domesticated in eastern Indonesia (e.g. Donohue and Denham 2010). There is now not only evidence for the introduction into the Moluccas of the cassowary (*Casuarius casuarius*) (White 1975), but of two species of phalanger (*Phalanger orientalis* and *Spilocuscus maculatus*) from New Guinea (Heinsohn 2009), and of the warty pig (*Sus celebensis*) by 7,000 BP, and of the dog before 4,000 BP (Oskarsson *et al.* 2012). With the early introduction of both animal and plant species from elsewhere, and forest extraction that is best explained through a 'hunter-horticulture' model, we might place the origins of low-intensity gardening and arboriculture (Latinis 2000) to as early as 28,000–13,000 BP, this latter being the earliest date for *Colocasia* established for the Bismark archipelago (Spriggs 1998:54). Such a complex is most likely to have included extensive use of *Canarium* from before 13,000, and the present distribution of *Canarium* in the lowland forests of Seram can only be explained through long-term anthropic influence (Edwards 1994; Ellen 2019c). Similarly, it is likely that the establishment of more extensive swidden cultivation was underway during this period, altering forest dynamics through managed long fallows. Rice appears to have been a relatively late arrival in the Moluccas, and only obtained a foothold as a cultivated staple in parts of Halmahera and Seram before the historical period, certainly no earlier than 5,000 BP (Glover and Higham 1996).

During the historical period, the main agents transforming land use and plant ecology have been, in addition to swiddening, the management of non-timber forest products for trade, timber extraction (Ellen 1985), human population movement, and the growth in spice production determined by outside forces (Ellen 1979b). There is contested evidence for the earliest international trade of clove and nutmeg, but this has undeniably been an important factor relevant to changes in flora and human-plant interaction for at least 2,000 years, as has been interisland trade connecting Seram with Melanesia for much longer (Ellen 2003b). Clove was an unimportant trade product on Seram at the time Europeans first arrived in the central Moluccas, but established itself as a major export crop as a result of Dutch attempts to control its trade in the northern Moluccas. Subsequent extirpation by the Dutch of clove gardens established along the coast of Seram (and particularly on the Hoamoal peninsula), and an attempt to confine production to the Ambonese islands, significantly altered the distribution of production. As the Dutch ability to maintain their monopology declined, clove planting increased along the coast of Seram as a private trade. Moreover, as we have already noted above, the period 1,500–1,900 CE saw the arrival of a large number of cultigens of Amerindian origin that made farming much more efficient, and to some extent replaced existing cultivated species (especially *Manihot esculenta*, *Xanthosoma*, sweet potato, papaya, and chilli).

By the late nineteenth century the importance of the clove trade had declined, and other cash-crops were becoming important, such as coconut (for copra), *Agathis* (for damar), and hevea rubber. Although the Dutch had established a sawmill in Amahai early on, logging was not to become a significant transforming agent in the Nuaulu area until the 1980s, at about the same time as the Indonesian government established transmigration zones along the Ruatan valley. The nature reserves at Wae Mual and Wae Nua were established by Ministerial Decree in 1972 (Amir and Wind 1978), and the Manusela National Park, opened in 1997, has now largely incorporated these reserves.

The anthropogenic forest ecology of Seram

Seram has a dynamic ecological system altered and maintained through human interventions, in which different subsistence strands – swiddening, hunting, plant gathering – all interconnect in complex ways. This system, while it can be understood as involving some interlocking sustainable cycles, has at the same time been subject to progressive change over many centuries (Figure 1.3). Its present state is, therefore, the outcome of patterns of human-resource interaction at different periods in the past, and we need to approach our investigation into it by using the tools of historical ecology. At the risk of oversimplification, we might imagine this system to consist of (1) forest landscapes, (2)

BCA	
26,000 or earlier	sago (*Metroxylon sagu*), possibly from New Guinea
5000 or earlier	pig and possibly cassowary, from New Guinea
2000 or earlier	taro and Eumusa section of *Musa* bananas, from the western archipelago
2000 or earlier	dog
CA	
1500	introduction of *Cervus timorensis* (deer), from the western archipelago
1600	introduction of first New World species: *Xanthosoma*, sweet potato
1600	beginning of main clove plantations on Seram
1654-1702	Rumphius in Ambon
1733-1750	publication of *Herbarium Amboinense*
1850	introduction of *Manihot esculenta* (cassava); expansion of coastal copra trade and production of coconut oil
1880	introduction of *Hevea* rubber
1980	creation of large scale transmigration zones; expansion of logging activity

Figure 1.3 Tentative historical timeline for anthropogenic impacts on the vegetation of Seram.

swamp landscapes, and (3) landscapes of exchange. Early European and colonial perceptions of Seramese as 'lazy' and of their subsistence as merely 'gardening' and foraging (Ellen 1988b:119) were partly a misunderstanding based on not appreciating the complexity of this system, and partly a result of observations based on only the most visible of their subsistence activities, the tending of small plots ('kebun') containing a diverse mixture of crops not recognizable as agriculture at the time of first European contact, and which contrasted with the wet-field rice agriculture more typical of the western archipelago (Hunt and Rabett 2014).

Historical studies of the environment (e.g. Crumley 1994), particularly those inforned by quantitative ethnobotany, have not been well represented hitherto in work on the forests of South-East Asia. Although some formative and pioneering work on ethnobotany and ethnoecology is associated with the work of Conklin (e.g. 1954) amongst the Hanunóo and Ifugao, the core influential body of work was undertaken in the Amazon, both in historical ecology (Balée 1998; Balée and Erickson 2006) and in quantitative ethnobotany (e.g. Phillips and Gentry 1993). In relying on Amazonian parallels and methodologies to interpret landscape change in island South-East Asia, it is understandable that we first turn to the 'intact' rainforest masses of Sunda. In Borneo, for example, there has been some research on the ecological consequences of settlement abandonment (Puri 2005), and on the likely occurrence of dark earths as evidence of a sustained process of soil enrichment through a management regime of swiddening and long fallow (Sheil *et al.* 2012). However, eastern Indonesia and Wallacea present us with a special case, with historic human patterns of settlement and movement reflecting in broad terms the biogeographic disjuncture with, and the influence of, the Melanesian world.

Despite the shortage of modern studies, we are now beginning to understand the ecology of lowland rainforest on Seram. We have already noted the location of Seram in a zone of transition (Wallacea) between the dipterocarp forests of Sunda and the Australo-Pacific tree flora of Sahul. The characteristic families (in lowland areas) include MYRTACEAE (particularly the distinctive eucalypts), MYRISTICACEAE, LAURACEAE and GUTTIFERAE. This pattern is confirmed in data I collected in the Nuaulu area in the 1990s (Ellen 2007, 2010). Data from plots yielded 42 families, with few clear dominants, confirming the patchiness of species composition of forest below 1,000 metres reported by Edwards *et al.* (1990, 1993).

Historically and conceptually, Nuaulu themselves perceive their subsistence and general cultural orientation as one of forest and mountain, rather than of coast and sea. All forest is seen as being in a state of flux, and Nuaulu recognize that this is in large part due to interaction with humans. This is represented in their social practice and everyday interaction through a detailed toponymic grid, the main components of which are named rivers and even smaller creeks, supplemented by names of peaks, hills, prominent rock outcrops, stones, waterfalls, lakes, swamps, caves and such like. In addition, there are more transient features: large trees, paths, log bridges, burned patches and patches of grassland; plus the recorded evidence of human activity, such as gardens belonging to particular individuals, old gardens, fences, traps, warning signs, bivouacs and huts, sago-processing equipment along streams, abandoned gardens and – most importantly – old village sites (Appendix 4: plots 5 and 11; Ellen 2018a:26–7). As a reference system, these toponyms begin with the names of particular mountains on the one hand, and of large rivers on the other. A unique set of toponyms therefore serves to identify particular patches of forest, which to some extent bypasses the need to identify forest in terms of its distinctive floristic composition or habitat structure. Toponyms, moreover, demonstrate the investment of history and reflect the overall 'cultural density' of the forest. This vision of the forest shows how – as far as Nuaulu are concerned – their extractive environment is continuous, with no sharp breaks between settlements, gardens and different kinds of forest and swamp, and with strategies of management and genetic material moving freely throughout the area. These strategies have encouraged high species richness through the establishment of ***sin wesie*** (protected sacred forest) and the continued use of ***nia monae*** (old village sites), and ***nisi ahue*** (long-term fallow). ***Sin wesie*** areas (e.g. Appendix 4: plot 7) are owned by individual clans, are resource rich, and their composition may reflect the age and successional stage of long fallows. They are designated as part of the preparation for various ritual cycles, often several decades in advance, and provide both construction materials for sacred houses and food species – mainly sago and *Canarium* – for the feasts that punctuate these cycles (Ellen 2007, 2010, 2016). Recent long fallow (***ahue***) provides the greatest concentration of resources in areas most influenced by humans by whatever measure (Appendix 4: plots 4 and 6), while ***nia monae*** refers to old village sites at different stages following abandonment, and flags an ecologically distinctive habitat through potently encoded history and culturally lived landscape. Some of these sites contain visible sacred stones and yield ceramic and lithic surface finds, but none have been excavated.

Evidence from each of these different kinds of site show how patterns of use have impacted on the distribution of species. As we have noted, arboriculture has been practised over many hundreds of years (probably much longer), and has focused on a small number of crucial nut- and seed-yielding trees, most notably *Canarium, Aleurites* and *Pandanus.* Such resources are the product of many generations of selective human interaction and modification (deliberate and inadvertent), optimizing usefulness and enhancing local diversity. Low tree diversity is influenced by disturbance, with abundant evidence that human agency has had consequences for ecology. In particular, lowland forest gives rise to 'a higher than expected proportion of *Canarium*' (Edwards 1994:7), and 'mosaic forest with *Canarium*' (Matthews and Gosden 1997) is typical of many places with long histories of *Canarium* extraction (e.g. Matthews and Gosden 1997), especially where *C. indicum* constitutes part of a traditional agroforestry system (Nevenimo *et al.* 2007:123). Selective logging and collection for exchange (resins, rattan, birds) has also been significant, e.g. *Agathis dammara* (for its resin) at sub-montane altitudes and *Melaleuca cajuputi* (for medicinal oil) in drier and more open areas.

However, for several millennia the extraction of non-domesticated forest resources and the life histories of settlement sites have been inseparable from the long-term impact of small-scale forest-fallow swiddening. The Nuaulu term ***nisi ahue*** designates secondary growth of various kinds. The status of ***ahue*** is ambiguous, ranging from plots that have been cleared within the last three years to those which are 25 years old or more. Not only is it a temporally wide category, it is spatially extensive. Of all land within a 4 km radius of the village of Rouhua in 1971, totalling some 2430 hectares, 15 per cent was ***ahue*** and 58 per cent ***wesie*** (mature forest). Of the 102 hectares designated garden land and cultivated by Nuaulu from Rouhua in 1970–1, 23 per cent was classified as ***ahue***. We can therefore conclude that long-fallow swiddening as a process has relentlessly changed lowland forest landscapes at a rate of 1 hectare a year per household for several millennia, with regrowth modified to increase its utility, for example by planting useful trees. Wherever fire is used to manage forest, as in clearance for swiddens, it enhances distribution of seed-bearing trees (Harris 1996). Thus, as we go back in time, it becomes increasingly reasonable to see swiddens as a means of producing economically productive managed fallow, rather than fallow being simply a means of restoring forest to make it available for swiddens (Belharte 2011).

By comparison with the dominant arboreal landscapes just described, the swamp landscapes of Seram might seem primeval, but here too first appearances are deceptive. Freshwater swamps on Seram are dominated by *Metroxylon sagu,* which has been the main source of dietary starch for at least 1,000 years. Swamp ecosystems are carefully managed through selective harvesting of palms before sexual maturation, and the transplantation of clonal suckers (Ellen 2006a); while sago management and reliance has influenced other aspects of the wider landscape, for example by mitigating forest conversion (Sasaoka, Laumonier and Sugimura 2014). Genetic recombination is, however, possible in large swamp areas where palms flower, fruit, seed and germinate without human intervention. Likely, *Metroxylon sagu* has spread westwards from an area of endemism in mainland New Guinea, for which there is genetic, lexical, ethnographic and archaeological evidence (Ellen 2004c; Hunt and Premathilake 2012). Only in the twentieth century have serious inroads been made to swamp landscapes, through neglect, dietary change or removal through deliberate draining.

But both forest and swamp landscapes have also – over the long term – been elements in what we might describe as 'landscapes of exchange', probably for as long as there have been Austronesian speakers in Maluku. Humans have selectively dispersed plant matter within Sahul for as long as they have lived there, and even low human population densities can have significant impacts, as the examples of *Canarium* and *Metroxylon* show. The development of characteristic maritime exchange networks focused on small islands, and surplus sago production was encouraged in mainland areas by trade with sago-deficient islands (Ellen 1979b, 2003b, 2011). Such landscapes of exchange entirely transformed the ecology of small islands through clove and nutmeg production in the pre-colonial period, a process that accelerated in the early colonial period. The rise of regional and global trade also changed the physical habitats of some small islands, through overextraction (e.g. of mangrove), coral destruction and land reclamation and defence. Similarly, the late colonial period saw significant landscape changes through agroforestry (*Agathis dammara*) and estate crops, such as coconut and *Hevea* rubber. Logging and deforestation ecologically transformed some islands (e.g. Kei Kecil) turning them into savanna environments, which in turn led to the decline in sago habitats and the rise in the importance of cassava (Ellen and Soselisa 2012b). Since the 1980s the main human impacts on forest have been through logging and transmigration resettlement, and, since 2008, through oil-palm plantations in north Seram (Liswanti *et al.* 2013:4).

The Nuaulu way of life

The Nuaulu presently (2015) comprise a group of 2,000 or so Austronesian speakers who have occupied central Seram for at least 500 years. The language, and in particular those aspects relevant to ethnobotanical knowledge, are dealt with in more detail in Chapter 2. Until around 1880 Nuaulu lived in dispersed clan and house-based settlements in the central highlands, from the river Sapalewa in the west to Manusela in the east. Under pressure for more effective colonial administration they moved to the coast after 1880, settling in a concentrated fashion near to the important Muslim domain of Sepa on the south coast, with which they had an historic bond (Ellen 2018a:14–34), and where they eventually formed three hamlets: Watane (Nuaulu Lama), Aihisuru and Hahualan. By the Second World War there were additional coastal settlements to the west (Bunara) and east (Rouhua) of Sepa. This was the situation that existed at the time of my first fieldwork on Seram in 1970.

The Nuaulu currently comprise some fourteen exogamous patrilineal clans that continue to exercize a considerable degree of autonomy. Each is normally composed of two ***numa*** (houses) that are in a complementary ritual relationship, and clans themselves are in an exchange relationship with other clans for major life-cycle ceremonies. Kinship terminology points to an ideal marriage between classificatory bilateral cross cousins, though in practice this is only achieved by a small percentage of the population, usually involving the heads of houses and their eldest sons. Although clans are robustly autonomous, precedence is recognized, with the clan Matoke being *primus inter pares* and its head being deferred to in certain ritual matters. Under the Dutch colonial dispensation three ranked officials were recognized: the ***ia onate iaia*** (head of the clan Matoke-hanaie), the ***ia onate ankarua*** and the ***ia onate ankatiga***. During the period 1970–2015 there was some mobility in the location of Nuaulu clans, but the main changes came with the opening of the government transmigration zone along the Ruatan valley in the early 1980s, and the creation of metalled roads to service them. This had a major impact on the landscape, and also attracted Nuaulu households and clans, who moved to an area that was regarded as their traditional heartland, was richer in resources, for example sago, and was less under the effective control of the 'desa' of Sepa and its raja. By 2012, the intial settlements of Simalouw and Tahena Ukuna had been absorbed into a new 'desa' – Nuanea – under the control of the Nuaulu themselves, and with their own government head styled locally 'raja'. One settlement on the south coast (Aihisuru) was abandoned in 1980 and the other settlements close to Sepa are much depleted, but the settlement of Rouhua has continued to expand, giving rise to a satellite settlement at Moni further to the east.

Most Nuaulu settlements are currently administratively located in the Amahai subdistrict, presently comprising 1149,07 square kilometres, within the Regency of the Central Moluccas (Kabupaten Maluku Tengah). This has its capital at Masohi, and comprises the central part of Seram, the Ambon-Lease islands outside of the city of Ambon (the provincial capital) and the Banda islands. Though there have been changes in the boundaries since 1970, these have not been in those areas principally inhabited and used by Nuaulu. The only exception is that the most recent boundary changes have left the Christian Nuaulu settlement of Rouhua Baru (near Waraka) in the new subdistrict of Elpaputih Bay. The boundaries of 'desa' (lowest level administrative units) have, however, much changed, with Nuanea at 125.00 km² (14 m a.s.l., 3.02 S, 129.06 E), Sepa at 223.60 km² (9 m a.s.l., 3.35 S, 129.11 E) and at Tamilouw 596.00 km² (6 m a.s.l., 3.39 S, 129.20 E) (Badan Pusat Statistik 2014:3–4, 9). The approximate boundaries of what I designate the 'Nuaulu extractive environment' are shown superimposed on Figure 1.4. This area encompasses the historic area inhabited by people we can describe as 'Nuaulu' as determined by toponymic evidence and their own claims to ownership and oral history. It also encompasses the furthest extent of their extractive activities and contemporary movements for subsistence, and therefore the area that determines, for the most part, their environmental and ethnobotanical knowledge. All plants (including all voucher specimens) referred to in this monograph are from this area, apart from traded plant products, or unless otherwise indicated.

The subsistence system of the Nuaulu population of south Seram is best described as 'interstitial' (Ellen 1978a, 1988). Nuaulu subsist largely through sago extraction, hunting, swidden cultivation and forest gathering, with some fishing. Today they are more than ever farmers, tending swiddens, growing a variety of vegetable crops, and tending increasingly large groves of cash-crops. However, this is partly a consequence of political interventions and movements at the end of the nineteenth century. In addition to taro and yams (the main historic staple of swiddens), the principle source of starch on Seram was and continues to be sago from the palm *Metroxylon sagu*. We now know a lot about sago management: that it is complex and involves organization that goes beyond what is crudely described as extraction (Ellen 2004b, 2006a, 2011). Other plant products have long been extracted from forests and managed to various degrees. Hunting has always been important, and its efficiency has increased with the introduction of new species: first cassowary and pig from New Guinea during prehistory, and in proto-historic times deer from the western archipelago (Figure 1.3). The main Nuaulu starch staple comes from the stems of the sago palm *Metroxylon sagu*, and the management and extraction of this palm features centrally in cultural life. Nuaulu also rely on swidden cultivation, with each household usually creating a new swidden of around one hectare each year. The main starch crops are yams and taro, but also

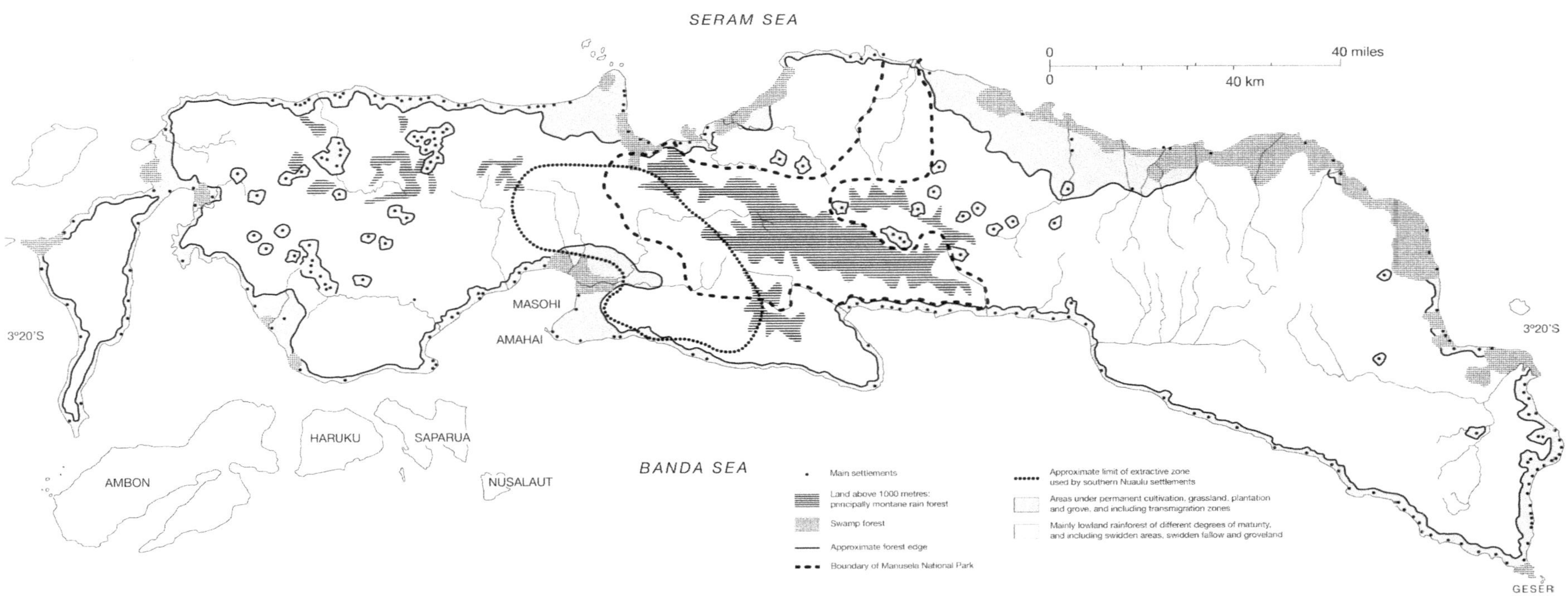

Figure 1.4a *The island of Seram, showing broad land use and vegetation zones, and the limits of the Nuaulu extractive environment.*

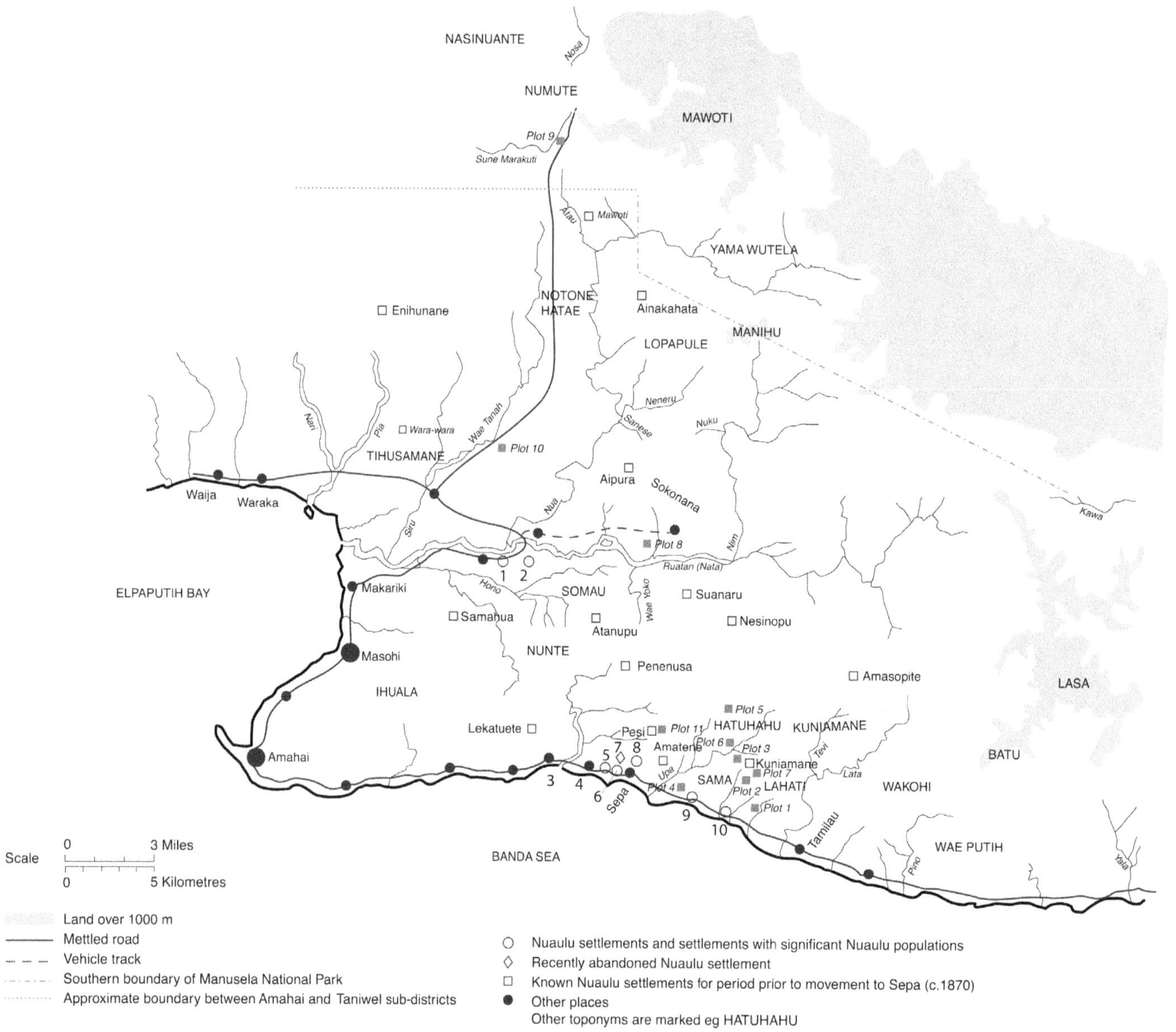

Figure 14b South Seram, showing former and current Nuaulu settlements, roads and boundaries as of 2015, and locations of numbered 1996 study plots discussed in text and described in Appendix 4. The numbered settlements are as follows: 1, Simalouw (Kilo 9); 2, Tahena Ukuna (Kilo 12, after 2012 absorbed into Nuanea); 3, Hatuheno; 4, Nuelitetu; 5, Bunara; 6, Watane; 7, Aihisuru; 8. Hahualan; 9, Rouhua; 10, Moni.

increasingly sweet potato, *Xanthosoma* and particularly cassava. Swiddens are also planted with a wide range of other leafy vegetables, pulses, fruits, spices and other useful plants. As swiddens grow older they contain many kinds of fruit trees and other long-term tree resources. Increasingly, land is devoted to cash-crops: for more than 100 years mainly clove and coconut (for copra) and some nutmeg, though since 1980 there has been diversification into coffee, cacao, green beans and similar. Hunting (principally deer, pig, cassowary and cuscus) is the most important source of animal protein, along with freshwater fishing. A few Nuaulu engage in marine fishing, but a great deal of saltwater fish is purchased for consumption from local non-Nuaulu traders. The collection of a wide variety of forest products, both food (e.g. honey) and non-food (timber, rattan) is important for the Nuaulu economy.

A history of Moluccan ethnobotany

The earliest descriptions of Seramese plants are necessarily ethnobotanical, in the sense that they are embedded in accounts of their cultural significance, and written by people who were not professional botanists. Although there

Figure 1.5 Portrait of G.E. Rumphius by his son Paulus Augustus dated to 1695/6 which appeared in the first edition of his D'Amboinsche Rariteitkamer, published in 1705. Photography and copyright courtesy of the Linnean Society of London.

are descriptions of the plants of Seram and surrounding islands from the earliest European encounters (e.g. Galvão 1971 [1544]), our historical knowledge is dominated by the towering work of Georgius Everhardus Rumphius (2011 [1741–50]) conducted between 1654 and 1702, and all subsequent work has been effectively undertaken in his shadow (Figure 1.5). Certainly in my own work, Rumphius provides a kind of comparative ethnographic lens through which to interpret Nuaulu knowledge and classification of plants, full of clues that help us penetrate particular Nuaulu ways of thinking. *The Ambonese Herbal*, in particular, has been a constant fund of empirical data and inspiration, partly because Rumphius's work confirms a remarkable continuity in patterns of plant knowledge and use; because he provides a benchmark of Moluccan plant life at a moment when it was on the brink of significant change; and partly because Rumphius was not only a shrewd observer himself, but also celebrated the knowledge of the local Ambonese who provided him with his data. In some significant respects Rumphius shows us an understanding of the plant world that reflects many features of the European folk knowledge of his time, before that knowledge became separated from scientific botany through the imposition of post-Linnaean and Darwinian orthodoxies, and the professionalization and institutionlization of the subject (Ellen 2004a). For these reasons, I shall frequently return to the classificatory observations and practices of Rumphius in coming to grips with how we might best understand Nuaulu plant-knowledge practices.

The nineteenth century saw the rise of European (and most relevant to this account, Dutch) colonial botany, the development of botanic gardens and the deliberate use of local knowledge for economic purposes. Much of this work attempted to inventory the whole of the Dutch East Indies systematically, and is typified by compendia of useful plants such as those produced by Heyne (1913–17) and Ochse (Ochse and Bakhuizen van den Brink 1931, Ochse *et al.* 1931). By this time the Moluccas had been largely displaced as a region of scientific scrutiny or of economic botany motivated by commercial interest. It had become an imperial backwater.

Few of the scientific expeditions to the Moluccas listed by Nolthenius (1935:418–23) for the period 1800–1930 were primarily concerned with plants, though some were interested in general natural history, such as those of C.B.H. von Rosenberg (1859–66) and S. Hoedt (1862–63), who collected specifically for the Leiden Natural History Museum. The British collectors A.R. Wallace (1857–61) and H.O. Forbes (1878–83), and the Italian O. Beccari (1872–5), were primarily concerned with zoology, though they devoted some attention to plants in their reports of their travels; while still others were specifically preoccupied with marine biology. Of the botanists, O. Warburg (1888–9) of Berlin focused mainly on nutmeg, producing a comprehensive work on the subject, including its ethnobotany (Warburg 1902). The Second Freiburg Moluccas Expedition, 1911–12, led by K. Deninger, had as part of its team E. Stressemann, who made linguistic and botanical collections, and O. Tauern, who focused on ethnography. Rubenkoning (1959) has listed additional significant scientific expeditions related to the interests of Rumphius, some of which are not mentioned by Nolthenius. Of these, those particularly relevant to the botany and ethnobotany of Seram are: the geological expedition (1917–18) of L.M.R. Rutten and W. Hotz, which also secured important botanical material, and the fieldwork of A. Rant (1929, 1931; see Rant 1934), and P.J. Eyma (1937–8). While by no means confined to Seram, it is also important to mention the work of E.D. Merrill and Charles Robinson (Merrill 1913, 1917), who through their interpretation of the work of Rumphius made a critically important contribution to our understanding of the botany of Seram.

As far as actual botanical voucher specimens are concerned, the most extensive collections on Seram are currently held by the National Herbarium of the Netherlands (the old Rijksherbarium), now part of Naturalis in Leiden. These have material dating to around 1800, and one firm date of 1841 for a specimen of *Crotalaria retusa*, though the collections do not become significantly large until 1857–9, and then again in the period 1914–17, with the names of J.E. Teijsmann and G.H. de Vriesse (1859–1860) being prominent (Figure 1.6; see also Teijsmann 1877). For recent botanical expeditions, a useful list is provided by Monk de Fretes and Reksodiharjo-Lilley (1997:881–6), who note expeditions by the Rijksmuseum Leiden and the Herbarium Bogoriense (1978), a study of the botany and ethnobotany of the mangrove system (1981–2) by the Herbarium Bogoriense, the Indonesia-Japanese botanical expeditions of 1983, 1984–5 and 1986, the 1985 expedition by L.B.N. Tehore, and the 1985 preliminary survey of the east Indonesian mangrove expedition. Much research during this period was supervised by Johannis Mogea of the Herbarium Bogoriense, who took a felicitous interest in my investigations from 1996 onwards. In addition, there is the work with Pattimura University leading to the report on forest ecology by Glatzel (1992), and that of various scientists associated with the Operation Raleigh expedition of 1987 (Edwards, MacDonald and Proctor 1993). A chronology of plant collectors as individuals is provided in the first two volumes of the *Flora Malesiana* Series 1 (Van Steenis 1950b). Apart from my own work, beginning in 1970, other recent ethnobotanical work includes the doctoral theses of Suharno (1997) on Alune agriculture, the collaborative work of botanist Wolff with linguist Florey (Florey and Wolff 1998; Wolff and Florey 1998), also on Alune, the doctoral thesis of Emily Brennan (2018) on Nuaulu bark-cloth, and the reports and dissertations of Ranlund (2011), Cosiaux (2012), Purwanto and Cosiaux (2013) on the Saleman area of north Seram.

The Nuaulu ethnobotany project and methodology

The ethnographic and ethnobotanical fieldwork that forms the basis of this monograph was conducted between 1970 and 2015, mainly with the Nuaulu population of Rouhua. Ethnobotany has been a feature of my Nuaulu ethnographic work since the beginning. My doctoral work on the ecology of settlement, and in particular on swidden cultivation, involved plant collection, and enquiries related to the Nuaulu plant world. During 1970–1 I developed the Nuaulu Ethnobotanical Index, which subsequently logged all data on plant names and uses that I accumulated between 1970 and 1996. In 1970–1 I had the support of the Botanic Gardens in Singapore and Kew. This work continued intermittently as fieldwork permitted up until 1996, and can be characterized as working with particular informants on particular plants and plant groups fairly opportunistically. During this period I was also conducting work on Nuaulu ethnozoological classification, which was more systematic

Figure 1.6 Herbarium specimen of the fern Diplazium ceramicum collected by G.H. de Vriesse and J.E. Teijsmann on Seram (1859–60), who described it as Asplenium ceramicum.

Figure 1.7 Retaone Soumori drying specimens over a fire of coconut husks, Rouhua February 1996.

and deliberately focused on developing field methods – such as sorting tests – which enabled me to get a much more dynamic view of Nuaulu animal knowledge, and its intracultural variability (Ellen 1979a, 1993). My experience in this programme of work encouraged me to develop the various protocols that I was to use in the 1996 Nuaulu ethnobotany project funded by the British Economic and Social Research Council (ESRC).

The backbone of the 1996 project was the transformation of the Nuaulu Ethnobotanical Index into the Nuaulu Ethnobotanical Database (hereafter, NED), to hold all systematic data relating to Nuaulu plant knowledge, organized according to local name. At the same time fieldwork involved a major attempt to collect voucher specimens in support of the anthropological parts of the analysis. Working with Kew, and within the terms of an agreement with the Indonesian National Herbarium in Bogor, voucher specimens were obtained in triplicate where possible, through deliberate collecting trips in the forest and in the settlement area, mainly in the vicinity of Rouhua, but also in forest plots in the Ruatan valley and near the trans-Seram highway at higher altitudes (Appendix 4). Once identified in terms of local names, and discussed with informants, specimens were dried or preserved wet using alcohol (the Schweinfurth method), depending on the prevailing weather conditions. Where specimens could not be properly sun-dried they were dried above open fires (Figure 1.7) in a specially made local press constructed from sago leafstalks. Fleshy fruits and other plant material that could not be reliably dried, were kept separately in sealed plastic bags in alcohol. In general terms, the guidelines followed were those recommended by Bridson and Forman (1998). Specimens were shipped to Bogor, where they were subsequently divided, with the second and third sets sent to Kew and the University of Kent. Each voucher specimen was accompanied by a label (Figure 1.8) that logged data on the

NUAULU ETHNOBOTANY PROJECT
SOUTH SERAM, CENTRAL MALUKU
R.F. ELLEN COLLECTION

No.	96-0009	Date	3/2/96
N.E.D.	223	Alt. m.	10
Locality	Rouhua		
Habitat	edge of path just outside village		
Descrip.	paired pinnate lvs which close-up on touching		
Engl. name	Touch-me-not		
Nuaulu name 1	***pitiri malu***		
Nuaulu name 2	***pitiri malu***		
Nuaulu name 3	***pitiri malu***		
Family	FABACEAE		
Scientific name	Mimosa pudica		
Author	L.		
Uses	Secondary food		
Notes	Known for the sensitivity of its leaves		
Coll. by	R.F.Ellen		
Dist.to	UKC		
Photo ref.	96-03-11		

NUAULU ETHNOBOTANY PROJECT
SOUTH SERAM, CENTRAL MALUKU
R.F.ELLEN COLLECTION

Number 96-943 Date 1/3/96

Altitude 500 m

Locality Tanaa valley, Wae Pia (the Tanaa is a trib. of the Pia, which flows S to Elpaputih Bay), map ref. approx 126 555′ E, 3 10 S; subdistrict Teun, Nila and Serua

Habitat Forest plot 10: low altitude forest, isolated tree of its species, no other examples seen; record on 60 m line transect which included several species of *Myristica*] and *Syzygium*; also *Glochidion*, *Calophyllum*, *Melastoma*, *Ficus*, *Heritiera* and *Calamus*

English name unknown

Description k.o. pachycual treelet Base of the leaf blade very gradually tapering into a short petiole. Infructescence approximating or exceeding the leaves, composed of numerous ochreous seeds

Nuaulu name 1 ***anakoine*** (Heunaka Sounaue-ainakahata)
Nuaulu name 2 ***anakoine*** (Sonohue Matoke-pina)

Family PRIMULACEAE [MYRSINACEAE] Scientific name Tapeinosperma Hook f.

Uses unknown

Collected by Heunaka Sounaue-ainakahata, Waenisa Sounaue-ainakahata, Ketane Sounaue-ainakahata, Sonohue Matoke-pina, Naunesi Sounaue-ainakahata

Distributed to KEW Photograph 96-11-23

Notes fr. and lvs. M. Coode (Kew) reports that this genus is new to Seram. Specimen and photo sent to P. Stevens at Arnold Arboretum

*Figure 1.8 Two examples of voucher labels: specimens of (a) Mimosa pudica (**pitiri malu**), (b) Tapeinosperma sp. (**anakoine**).*

context of the specimen collected, its description and notes on use. Data were later added on botanical identification. In addition, the label allowed for up to three Nuaulu name identifications from three independent research subjects, in order to capture any significant variation in naming practices at the individual specimen level. All specimens were checked and sorted in Bogor and forwarded to Kew. The same procedures and arrangements were adopted on subsequent collecting trips.

Between 1970 and 2015, a total of 1,761 specimens of vascular plants were collected in the Nuaulu area and delivered to herbaria in Kew (KEW) – 695, Bogor (BOGOR) – 587, Singapore (SING) – 8, and Canterbury (UKC) – 460; and to the biological laboratory of Padjadjaran University in Bandung (UNPAD) – 11 (Table 1.1). Compared with Berlin, Breedlove and Raven (1974b:xix), who report having collected some 100,000 specimens, this is small, but by general comparative standards for ethnobotany it is reasonably substantial (cf. Hunn 2008; Taylor 1990). The specimens were mostly in the form of dried plants suitable for herbarium voucher sheets (Figures 1.9 and 1.10), though there were also separate collections of wood, dry carpological specimens and larger fleshy material (usually fruits) in spirit, in support of the herbarium sheets. The cassava sent to Bandung was in the form of leaf specimens for DNA analysis. Of all of these, 1,175 were identified to species level or below, 437 to genus level and 88 only to family level. A further three were identified but have subsequently gone missing with unknown locations. Sixty-two specimens could not be identified for various reasons, either because they were lost in transit, or more usually because the specimens were infertile (had no fruit or flowers).

All voucher specimens have a unique identification number (Appendix 1): a reference to the field trip by year followed by the voucher number (e.g.71-56, 96-245, 03-04). Where voucher specimens have been subsequently divided to form separate specimens, this is indicated by adding a lower-case letter to an existing number (e.g. 71-56, 71-56a, 71-56b). In the few cases where numbers have become detached and lost, but the herbarium and field trip is known, the number is rendered, for example, BOGOR 96-00. Where specimens themselves have gone astray, the relevant entries in Appendix 1 are marked 'labels only'. Voucher specimens 15-99 to 15-134 were collected by Emily Brennan in and around Nuanea and incorporated into the 2015 set for distribution and identification. The sixty vouchers deposited in Kew in 2015 are still in a queue awaiting identification.

In addition to identification from voucher specimens in herbaria, some physical material was identified with the aid of other specialists (the late Sadao Sakamoto for

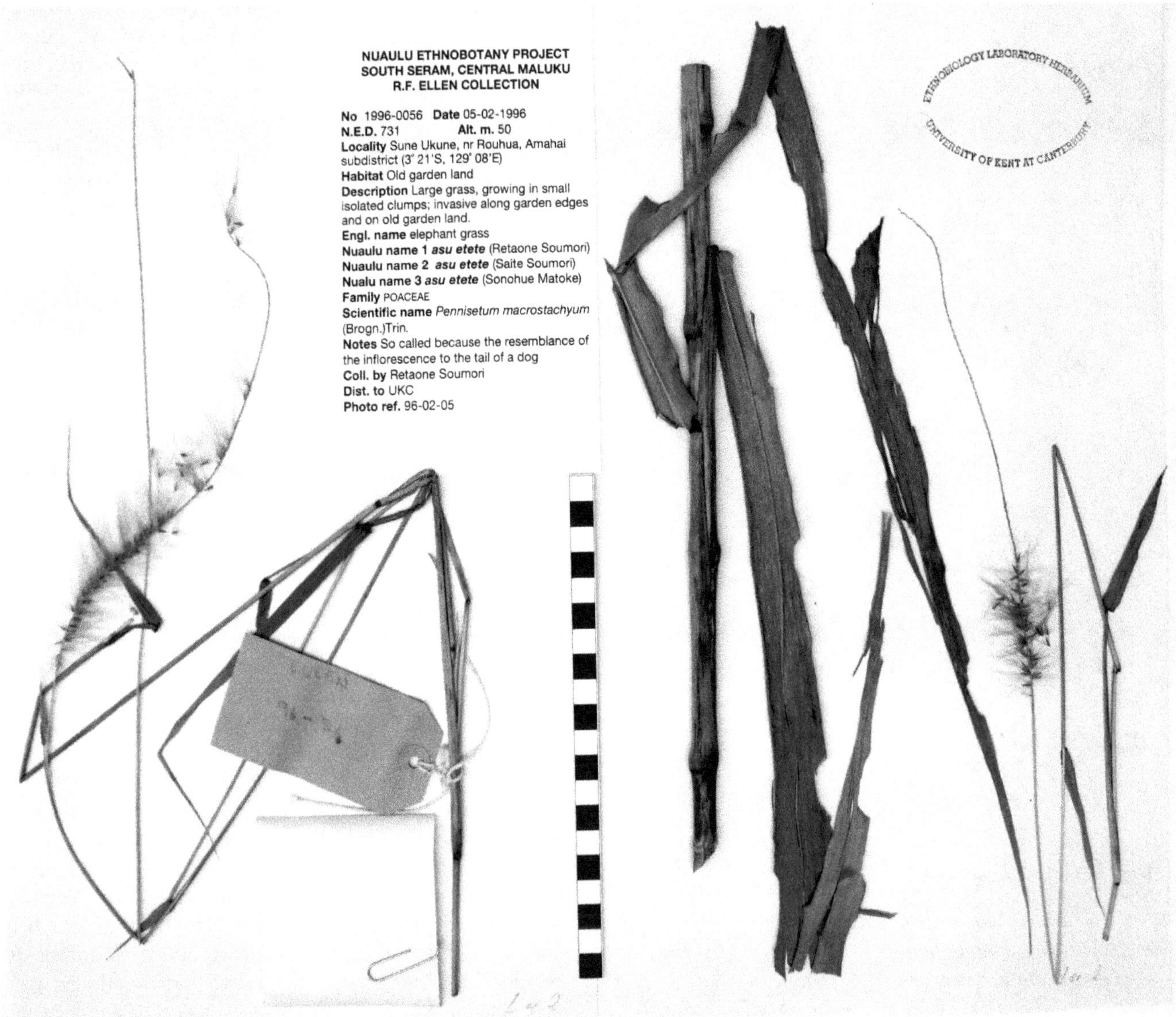

*Figure 1.9 Herbarium sheet for UKC 96-56, a voucher for Pennistetum macrostachyum (Brongn.) Trin. (**asu etute**) in the University of Kent Ethnobiology collection.*

millets and sorghum, Dave Roberts for pitcher plants and Asri Peni Wulandari for the DNA bar-coding of cassava). Well-known (and some not so well-known) species (usually domesticates) were identified by Ellen in the field or from photographs, using specialist literature (Ochse and Bakhuizen van den Brink 1931; Ochse *et al.* 1931; PROSEA 1992–2003; Vaughan and Geissler 1997). More specialist publications were used for particular groups of plants (e.g. Coursey 1967 for yams). Where numbered voucher specimens were not available, all taxa identified to genus or species level were confirmed as such by checking against descriptions, in the first instance, in the PROSEA volumes, and where that proved insufficient, against other specialist literature as indicated in the penultimate column of Appendix 1.

Scientific names obtained as a result of all determinations (both from vouchers and by other means) have been checked against THE PLANT LIST,[2] and are listed in Appendices 1–3. Where THE PLANT LIST name has superseded the one originally provided by the identifying authority, both have been provided in the Appendices, but only the most recent is used in the text. All vascular plant types and Nuaulu folk categories confidently identified represent 121 families, 391 genera and at least 485 named species and hybrids; 82 have been identified to life form level only, 38 have only been been identified only to family level, 138 only to genus level, while 11 have no identification at all (6 of which it is doubtful are plant terms at all). There remain, therefore,

2 www.theplantlist.org (static since 2013).

Figure 1.10 Herbarium sheet for KEW 96-836, a voucher for Solanum torvum Sw. ***(tonne hanaie****, [syn]* ***tonne wesie****) (specimens. kew.org/herbarium/K000610976).*

around 131 vascular plant types named in Nuaulu, and with a separate record in the NED, that we have been unable to identify to generic level (12.9 per cent of all those named categories), and 93 that we could not identify to family level (9.16 per cent of all named categories).

The aggregate data gathered on earlier field trips, data already entered into the NEI and data collected in 1996, were used to develop the NED using Filemaker Pro 6. In the NED there is one record for every Nuaulu plant-term report, regardless of taxonomic level of application. Thus, while most records are for terms that gloss the species level of scientific classification, many are for various superordinate groups, up to and including terms for life forms and folk kingdoms, and for subordinate groups in the case of varieties, cultivars or landraces of cultigens, such as cassava or banana. Originally, the database included all plant types (including algae), lichens and fungi. Subsequently, the NED has been divided into separate databases for vascular plants, cryptogams and fungi. Although reference is made to these other databases in this work, for example in Chapter 2, the main reference here is to the vascular plant database. There are currently 1,012 records in the database, reflecting the number of plants 'recognized' by Nuaulu. However, of these, eight were recognized as distinct types but are not accompanied by names in the NED. These names may have disappeared during various stages of data processing and storage, or may exist but I have been unable to retreive them.

The basic NED record is illustrated in Figure 1.11 and consists of sixteen fields grouped into five sections as follows: (A) basic – 1) record number, 2) names (including synonyms and lexical variants), 3) description in English, 4) photograph or photographic cross-reference (at bottom of record); (B) linguistic and folk classificatory data – 5) names in other languages, especially Ambonese Malay; 6) nomenclature type (uninomial, binomial, trinomial, and whether reduplicated); 7) category type (primary, secondary, terminal; specific, generic, varietal *sensu* Berlin); (C) botanical data – 8) vegetal habitat type (e.g. primary forest, secondary forest, swidden); 9) domestication status (domesticate or non-domesticate); (D) cultural data – 10) code for use type; and 11) a notes field. This latter is potentially large, listing all other information available, including semantic and etymological data on plant names, references to fieldnotes and other sources, and where appropriate, to records in the earlier NEI. The left-hand side of the record is devoted to (E) taxonomic identification – 12) voucher specimen number, 13) biological family, 14) scientific name, including author and 15) herbarium of deposit. The template allows for up to 10 specimens to be identified for each record. Finally, there is a field (16) for a correlation index, which is an expression of the relationship between scientific taxon and Nuaulu category (see Chapter 8).

The objective of the 1996 fieldwork phase was to achieve a minimum of three identifications (ideally based on voucher collections) for each record in the NED. As of 2015 (1,012 records), there are 551 records with just one botanical identification, 125 with two identifications, and 244 records with three or more identifications. As we go to press, there remain 95 records without any firm identification. The largest number of collections and identifications for one folk category is 13 for ***ai osi*** (*Aglaia sapindina* and *Chisocheton ceramicus*, both Meliaceae) and for ***kaote*** (Congo Jute, *Urena lobata*). Where there is only one botanical determination does not necessarily mean identifications are any less secure, as these include the most common and easily recognizable cultivated species (e.g. ***nione***, the coconut *Cocos nucifera*). In some cases, despite having secured many collections for a particular folk category, botanical identifications were not forthcoming, and larger numbers of collections were subsequently made

	1970–1	1973	1975	1996	2003	2009	2015	totals	ID to family level only	ID to genus level only	ID to species level or below	no ID
Singapore	8	0	0	0	0	0	0	8	0	3	5	0
Bogor	0	0	0	518	0	0	69	587	27	108	434	18
Kew	81	7	0	547	0	0	60	695	37	198	436	25
Bandung	0	0	0	0	0	11	0	11	0	0	11	0
UKC	6	4	2	432	13	0	3	460	24	128	289	19
Total	95	11	2	1,497	13	11	132	1,761	88	437	1,775	62

Table 1.1 Vascular plant specimens (mainly herbarium vouchers) collected between 1970 and 2015, and institutions of deposit.

NUAULU ETHNOBOTANICAL DATABASE
VASCULAR PLANTS

A1 File number	**668**
A2 Nuaulu name	**(ai) supa**
A3 English description	**k.o. tree: large waxy lvs and hard green fruit singly or in pairs interspersed amongst lvs.**
A4 Photographs	**96-13-14, 96-13-15, 96-18-14, 96-18-1**
B5 AM or other local term	**AM. 'waringin'**
B6 Type of Nuaulu term	**uninom binom**
B7 Category	**prim term generic**
C8 Vegetation type	**F1**
C9 Domestication status	**N** 16 Corr index **08**
D10 Use abbreviation	**D**
D11 Notes	**1970-1. Ethnobot. index 595; source of white bark cloth (ai unte putie); inner bark removed and taken back to village where it is boiled for 2-3 hours to retain its pliabilty and to make subsequent work easier.**

E12 Specimen no.	E13 Biological family	E14 Scientific name and Author	E15 Herbarium or other authority for identification
96-1305	**Moraceae**	**Ficus tinctoria G. Forst.**	**BOGOR**
96-490	**Moraceae**	**Ficus L. sp.**	**KEW**
96-491	**Moraceae**	**Ficus L. sp.**	**UKC**
96-686	**Moraceae**	**Ficus tinctoria G. Forst.**	**BOGOR**
96-687	**Moraceae**	**Ficus tinctoria G. Forst.**	**UKC**
96-688	**Moraceae**	**Ficus ?callosa Willd.**	**KEW**
96-489	**Moraceae**		**BOGOR**
96-1306	**Moraceae**	**Ficus L. sp.**	**KEW**
96-1307	**Moraceae**	**Ficus L. sp.**	**UKC**
B15-120	**Moraceae**	**Ficus tinctoria G. Forst.**	**BOGOR**
B15-121			**KEW**
B15-122			**KEW**

*Figure 1.11 Example of a record in the Nuaulu Ethnobotanical Database: **ai supa** (the humped fig, Ficus tinctoria). Key: F1, primary forest; D (use), attire; N, non-domesticated.*

where previously identifications had not been secured. As the NED is updated, so numbers alter, and those provided here can only ever be approximate.

Another problem encountered was the occurrence of different identifications for the same Nuaulu named category. In some cases this can be accounted for by the Nuaulu using the same term for a number of closely related forms. For example, for voucher specimens consistently described using the term ***mono nuetetue*** we have identifications from Kew for various Compositae: *Synedrella nodiflora, Crassocephalum crepidioides* and *Ageratum conyzoides*. In other cases different Nuaulu may have provided the same term to refer to more distantly related plants, this reflecting intracultural variations in classificatory practices. In still other cases we may be dealing with cross-contamination between voucher specimens that has arisen in transit. However, sometimes it is simply that plant specialists working in different herbaria at different times have provided different identifications for material that we know to be from the same source. In this situation the anomaly occasionally reflects use of different synonyms or spellings (in which case the problem is resolvable). Two examples of conflicting determinations that cannot be resolved this way, or be plausibly explained as a result of a corrupted data source, are *Centotheca lappacea* (KEW 1971-50) and *Setaria palmifolia* (BOGOR 2015-17) for the grass ***ekene***; and *Cyperus diffusus* (KEW 1971-47) and *Scleria* sp. (KEW 1996-47) for the sedge ***hahuseite***. In fifteen cases (1.47 per cent) taxonomists in different herbaria provided different identifications at the family level, and in fifty-five cases at the genus level, based on the material we were able to provide them with. It would be easy to dismiss these kinds of anomalies as 'errors' made by either Nuaulu experts, by the ethnographer (for example in providing only poor sterile material for identification) or by herbarium taxonomists, and to assume that if we searched hard enough we would find a correct and consistent identification. This may or may not be the case, but the problem does open up the issue of the flexibility with which names might be used in unwritten folk classifications, and also comparable hazards in the identification practices of both local experts and professional botanists. I shall return to the theoretical implications of all this in Chapters 8 and 10.

As a major part of the argument of this work is that how people classify plants is embedded in, amongst other things, how they use them, reference is made systematically to the cultural representations and applications of different taxa. However, only enough data are provided here to support the general argument – usually a brief list of uses as recorded in the NED, referring where possible to the Nuaulu material-culture collection in the British Museum (e.g. BM 1972 As 1.46 – 15), in the Rijksmusem voor Volkenkunde (now the Nationaal Museum van Wereldculturen) in Leiden (e.g. RMV 1971-81), and in the Ethnobiology Laboratory of the University of Kent (e.g. UKC Ellen 1970.52), and to published work. Where necessary, reference is made to processing, use and management, but for more detail readers will have to refer to other sources.

It can be seen from Appendix 1 that the voucher specimens collected are heavily biased in favour of particular taxa, often those that are especially frequently encountered or used, though in some cases the bias reflects a deliberate collecting strategy to obtain taxa that proved difficult to identify, or which I wished to collect to exhaustion in order to satisfy myself that the data were as complete as possible.

Most of the systematic data reported here have been acquired through the collection of plants, either by myself or by Nuaulu co-researchers, and through informal conversations with people at the point and place at which the collections were made, and during subsequent follow-up discussions in the village as specimens were being processed. However, I have in addition used a number of other methods to acquire complementary data. In 1996 I conducted eleven plot surveys, in which as wide a range as possible of mixed forest vegetation with which Nuaulu were actively interacting was sampled (Ellen 2007, 2010). The main purpose of these surveys was to measure the extent to which Nuaulu had knowledge of particular kinds of forest and landscape, but it also served to provide data on particular plants. Trees above 10 cm dbh (diameter at breast height) were recorded, and a minimum of three members of the all-male teams asked to identify them and provide information about them and their growth patterns, ecology and uses. One advantage of this approach was to appreciate the extent that *in situ* context of a living specimen was a significant determinant of knowledge about individual plants, and how knowledge about one plant was dependent on noting its relationship with other plants. A similar approach was adopted in the participatory mapping of swidden plots, and with similar consequences, athough voucher specimens were not systematically taken from swiddens. Large plastic sacks of representative specimens of trees from the forest plot surveys were taken back to the village, and, before drying, were the subject of focus-group discussion. Such discussions were unstructured, but would involve groups of upwards of twelve persons, both male and female, and of various ages, who would either confirm the identifications provided by team members during the plot surveys, or dispute them. These discussions proved to be a powerful tool for assessing the dynamic character of knowledge, and the extent and significance of disagreement. Finally, in addition to techniques that focused on the plant specimens themseves, data about plants and their folk classification was acquired through discussion of photographs, and in the course of general ethnographic enquiries around subsistence techniques, food, healing practices and material culture. The use of picture cards to elicit (not so much identification as) classificatory resemblance, a technique that had proved useful in my earlier study of Nuaulu animal classification, was reckoned to be less workable for plants due to differences, when compared with animals, in the perceptual qualities used when identifying them.

The language of elicitation has at all times been a combination of (south) Nuaulu and the local variant of Ambonese Malay, supplemented by occasional use of standard Indonesian. This latter is now widely understood, though this was not the case in 1970 when I first arrived in the field. The classificatory scope and meanings of ethnobotanical lexemes were primarily established using the usual range of methods of structural semantics, such as locating how particular lexemic terms contrasted with or were incorporated within or linked to a range of other lexemes, plus explicit queries about meaning. In this regard I have benefitted in particular from the discussions of Conklin (1962), Frake (1964) and Samarin (1967: especially 205–17) and the advice contained therein. The use of Ambonese Malay proved to be of analytic assistance in determining the meanings of some key 'intermediate', unaffiliated and life form categories, for as Nuaulu research subjects and co-researchers moved backwards and forwards between Nuaulu and Ambonese Malay, so they revealed classificatory arrangements and linkages that were otherwise linguistically hidden. Thus, an elaboration of the technique of 'back-translation' (e.g. Werner and Campbell 1970) proved to be productive in a fieldwork context. Subsequent work on the *Ambonese Herbal* allowed me to compare Rumphius's documentation of local language terms and his interpretations of plant groupings with contemporary Nuaulu practice, as a means of confirming and extending these patterns, and also to shed further light on Nuaulu plant classification in general.

Conventions adopted in the organization of this work

Since the publication of Rosemary Bolton's orthography (1990) and the appearance of the Nuaulu dictionary of Bolton and Matoke (2005) I have, on the whole, adopted the conventions that these authorities use for writing the language (see 'A Note on Orthography' included in the preliminary matter). Thus, /b/ in ***borune*** becomes /p/ in ***porune*** (a young cuscus), ***bobote*** and ***asabobote*** (jambu or water apples) become ***popote*** and ***asapopote***, ***tobako*** becomes ***topako*** (tobacco), ***bahane*** becomes ***pahane*** (various *Allium*), and so on. Only in the case of toponyms, such as 'Bunara' have I retained /b/, as these terms are now widely adopted in maps and official records, and to change them would be confusing.

As the reader will have already noted, there is variation in the terms that Nuaulu use to describe particular plants, as well as – in some cases – in the scientific identifications for living and preserved specimens. As Nuaulu plant names and scientific taxa are referred to repeatedly in the descriptive and analytic chapters, I have sometimes had to form a judgment as to which terms and scientific names to use. Except where there are specific reasons for doing otherwise, I have prioritized the identification most frequently obtained, and in cases where there are only two identifications, I have formed a judgment as to which to use based on all circumstantial evidence available to me. In the case of Nuaulu names, I have used the full name (even where in everyday speech abbreviated forms predominate), and the most common lexical variant and synonym. Many synonyms and the most frequent lexical variants are given in Appendices 1–3, and the full range of language data and scientific identifications, together with much unpublished and unanalysed data, can be found in the NED, which will eventually be available through the website of the Kent Centre for Biocultural Diversity (CBCD) at research.kent.ac.uk/nuaulu/about.

In the text and appendices I have indicated optional lexical variants by the use of parentheses: thus ***hata(ne)*** indicates that in a binomial formed from the basic term ***hatane***, the element ***ne*** is optional, as in ***hatane msinae*** or ***hata msinae***, a sago (*Metroxylon sagu*) folk variety with a red patch at the base of each spine on the leaf stalk. The complete list of scientific identifications (together with plant name authors) is provided in Appendix 1. In the main body of the text full Linnaean binomials, including authors, only appear where a name is not listed in Appendex 1. Botanical terminology generally follows Beentje (2010), plant identification procedures follow Lawrence and Hawthorne (2006), and general ethnobotanical methods, Martin (1995). Common English names are on the whole those most frequently encountered in British English usage, and are taken from the lists provided in the PROSEA volumes. My models in matters presentational and analytical has always been the botanical ethnographies of Harold Conklin and Brent Berlin. While my analysis is substantial in empirical scope, it has had to leave out much of the systematic detail provided in some other studies, for want of space and in order to achieve better analytical focus. I have had to be selective, and provide indicative examples only. Most Nuaulu plant categories, their nomenclatural properties and cognitive organization, the plants they refer to and their uses, are described here only briefly, in most cases only sufficiently to shed light on the wider issues of classification and understanding. For more detail the reader is referred to published papers on the ethnobotany of individual plants, such as those for sago (e.g. Ellen 2006a) and *Canarium* (e.g. Ellen 2019c), and to the NED.

When indicating (especially) higher order categories that do not have a clear lexical equivalent, or no lexical equivalent at all in Nuaulu language, or when referring in the abstract to categories that are imputed to exist more widely, I have employed full capitalization, as in PLANT, BAMBOO and FERN. I also use full capitals when referring to Brent Berlin's ethnobotanical ranks (Berlin, Breedlove and Raven 1974b; Berlin 1992), namely FOLK KINGDOM, LIFE FORM, INTERMEDIATE, GENERIC, SPECIFIC and VARIETAL, though use non-captitalized text where the technical Berlinian sense is not intended (e.g. 'life form'). Scientific families are distinguished by using small capitals, as in Arecaceae and Euphorbiaceae. Scientific binomials (e.g. *Canarium indicum*) are italicized, while other scientific taxonomic nomenclature is written in regular lowercase lettering with initial capitalization (e.g. Bambusoideae).

Given my earlier remarks on terminological variation, both as this applies to the Nuaulu language and scentific taxonomy, I need to add a cautionary note on numbers. This monograph relies extensively on quantification, though without much statistical manipulation. The reader must expect that there are sometimes inconsistences and occasions when the numbers do not add up due to the character of the original data. With respect to calculations involving scientific taxa we have to contend with disputed and changing identifications and changes in taxonomic groupings and boundary problems, for example those posed by hybridization and what constitutes a species rather than a variety. Changes in family allocations can be particularly vexing. Thus, during the time that the data incorporated into Figure 8.7 has been acquired, various revisions in the family allocation of genera have led to the disappearance of some family names altogether and the expansion of the content of other families, with consequences for the way scientifc taxa correspond to Nuaulu categories. One family significantly impacted by recent revision is CLUSIACEAE (GUTTIFERAE), which has lost the genus *Calophyllum* to the new family CALOPHYLLACEAE. The fact that many specimens could not be adequately identified, or were identified using 'unresolved' names, has also influenced what is possible by way of quantitative inference. Similarly, Nuaulu terms for plants are subject to much variation between individuals or by the same individual on different occasions, with consequences for the calculations presented here. The use of synonyms, lexical variants, context of elicitation, material available and the knowledge of particular individuals all influence the degree of unanimity that can be achieved. Indeed, we should be suspicious of too perfect a fit. And sometimes despite careful checking, my arithmetic will have failed. I have done my best to update all numbers to the best approximations possible as new information has come in, but it is indeed part of my conclusion that we must expect that the numbers will be constantly shifting and that therefore we can never be completely certain.

Finally, we come to the question of diagrams. Researchers have used diagrams as explanatory devices since the beginnings of the serious examination of folk classifications for plants and animals (e.g. Conklin 1954, 1962). In large part, the way ethnobiologists have used diagramatic representations has followed the use of similar diagrams by biological taxonomists. Rumphius did not use diagrams to reflect either his or native Ambonese classification in his *Ambonese Herbal*, but he did arrange text in particular ways (including as lists) that might be thought to have diagrammatic properties that reflect how he thought classifications should be organized. As we now know, organizing knowledge as text can have a significant impact on how we think about the relations between elements in a classification (e.g. Goody 1977; and Chapters 9 and 10 here). But if Rumphius avoided diagrams, Linneaus (1738) certainly did not, and it was Linneaus who set the standard for subsequent work on biological taxonomy. Taxonomic diagrams (sometimes quite complex) have become de rigeur in works on biological taxonomy and systematics ever since, as an examination of any random paper (e.g. Christenhusz and Chase 2014:581) or textbook (Jeffrey 1982) will indicate.

In my own work I have used diagrams too, and indeed celebrate them as Gell (1999) has done. But on the whole I have come to distrust them in published studies of folk classification, and so for the most part I use them only as transient working devices to ' think things through'. One of the main problems with tree diagrams (dendrograms), and their elaboration in taxonomic science and in studies of folk classification in particular, is that they are prone to suggest a misleading degree of certainty about an arrangement, and can be adapted to display alternative cognitive possibilities only awkwardly. Frequently they imply false notions of hierarchy and simplicity of connection, most relying on elementary two-dimensional digital distinctions. In some analyses such diagrams have been modified to take into account these problems, but on the whole the main issues have nevertheless been perpetuated and entrenched through subsequent developments in biological taxonomy, including cladistics, the introduction of computer-generated representations and most recently molecular (DNA) based approaches. This entrenchment has foreclosed certain options on how best to model the process by which classificatory decisions are made, and the form in which the relevant data are stored in our minds in cultures where knowledge is orally transmitted. Moreover, these kinds of diagram tend to playdown the significance of intracultural variation and flexibility in the way classifications are used in practice. To get round these issues some have experimented with Venn diagrams (Berlin 1992: e.g. 40–51; Ellen 1993:146), and – to express the way in which categories are linked through multiple connections – network diagrams. The danger is that these latter can sometimes become so complex (e.g. Friedberg 1970: opp. 1150, i–vi) that they are not easy to read, and do not plausibly reflect what is going on in any one individual mind, or indeed some putative collective cultural 'mind'. There is no reason to assume that classificatory space is two-dimensional, and recognition of this has led to experimentation with three-dimensional 'exploded' diagrams (e.g. Conklin 1962). Issues of diagramming are thus not purely technical, but raise major issues about how we might truthfully represent classificatory space in those cognitive and representational systems that we study. I shall return to these matters, including dimensionality, in Chapters 10 and 11, but need to announce here that in this monograph I have decided to limit my use of diagrams to illustrate Nuaulu classificatory relationships overall. Where I do adopt simple drawings, continuous lines indicate strong evidence for the perception of shared affinities, while broken lines indicate tentative or alternative linkages.

2

How the Nuaulu experience and talk about their plant world

Nuaulu language, in relation to plants and research methodology

Nuaulu speak an Austronesian language placed in the Ambon-Timor group. A summary of recent attempts to locate Nuaulu in relation to other languages was provided in *The Cultural Relations of Classification* (Ellen 1993:35–7), following in particular the work of Collins (1980) emphasizing genetic rather than lexicostatistical relations, and also Bolton (1990). Collins placed Nuaulu in his Patakai/Manusela grouping of Nunusaku languages with the languages of Piliana and Huaulu. Further recent work (e.g. *Atlas Bahasa Tanah Maluku* 1996) has adjusted the genetic relationship between languages, and the position occupied by the Nuaulu speech community in relation to other languages of Seram on present evidence is shown in Figure 2.1. Within the Nuaulu language grouping, we can distinguish between semi-intelligible northern Nuaulu and southern Nuaulu.

Although many of the surrounding indigenous languages are under threat, being in process of being replaced by Ambonese Malay (AM) or varieties of Indonesian (see, e.g. *Atlas Bahasa Tanah Maluku* 1996; Collins 1980; Florey 1991, 1993, 1997, 2001, 2005a, 2005b; Florey and Engelshoven 2001), Nuaulu language, though spoken by a small number of people, is robust in areas where a traditional way of life is still practised. In fact, the number of speakers has increased over the last fifty years. This can be attributed to population growth, and the fact that there has been relatively little conversion to other religions, by comparison with the demographic trend. This has in turn sustained wider cultural confidence in actions to protect their way of life, including by political means. In addition, the programme by the Summer Institute of Linguistics to document Nuaulu language, encouragement to use the language in liturgical and more mundane contexts, including the production of language materials, some specifically bearing on ethnobotanical knowledge (e.g. Bolton n.d.; Oh Swee Cheng 2005; Yesomkor 2005), has surely had an impact. There is no particular evidence of changes in language use that would suggest that those aspects of language, including special registers that encode ecological concepts and plant knowledge, are shifting in ways that we might not otherwise expect (Florey, Healey, Wolff and Manakane n.d.). Compared with other languages on Seram that are endangered (Florey 2001), Nuaulu language vitality seems strongly associated with the preservation of not only ethnobiological knowledge, but also esoteric knowledge associated with ritual practice. This remarkable resilience of Nuaulu language must also be understood against a backdrop of high rates of immigration, an almost 100 per cent rate of early years schooling by 2015 (where Indonesian is the only language of instruction), and the other myriad factors of globalization that threaten small local languages elsewhere.[1]

Nuaulu continue to live in a linguistically diverse environment, and the patterns of multiple-language use are broadly the same as I reported in 1993 (Ellen 1993:37). I noted then that most of the adult population (those over 15) were bilingual in Nuaulu and Sepa, and current evidence suggests that this remains unaltered. Sepa-Tamilouw remains an important vehicular language in south Seram, though immigration from south-east Sulawesi and other parts of the Moluccas since 1980 is beginning to erode its status. Proficiency in other local languages has declined since 1970, though knowledge of AM remains general, and most significantly competence in spoken Indonesian has risen substantially amongst the younger generation as a result of schooling. I would say that most Nuaulu (and certainly those on the south coast in the vicinity of Sepa) are effectively trilingual. Understanding the dynamics of linguistic pluralism is important in the context of this study, as other languages are a source of new nomenclature and even new classificatory structures, as we shall see in Chapter 9.

1 The Summer Institute of Linguistics (hereafter SIL) and its successor organization in the Moluccas, the Wycliffe Bible Translators, have been involved for twenty years in producing language materials for the purpose of aiding language retention, and these include texts concerning knowledge of natural history.

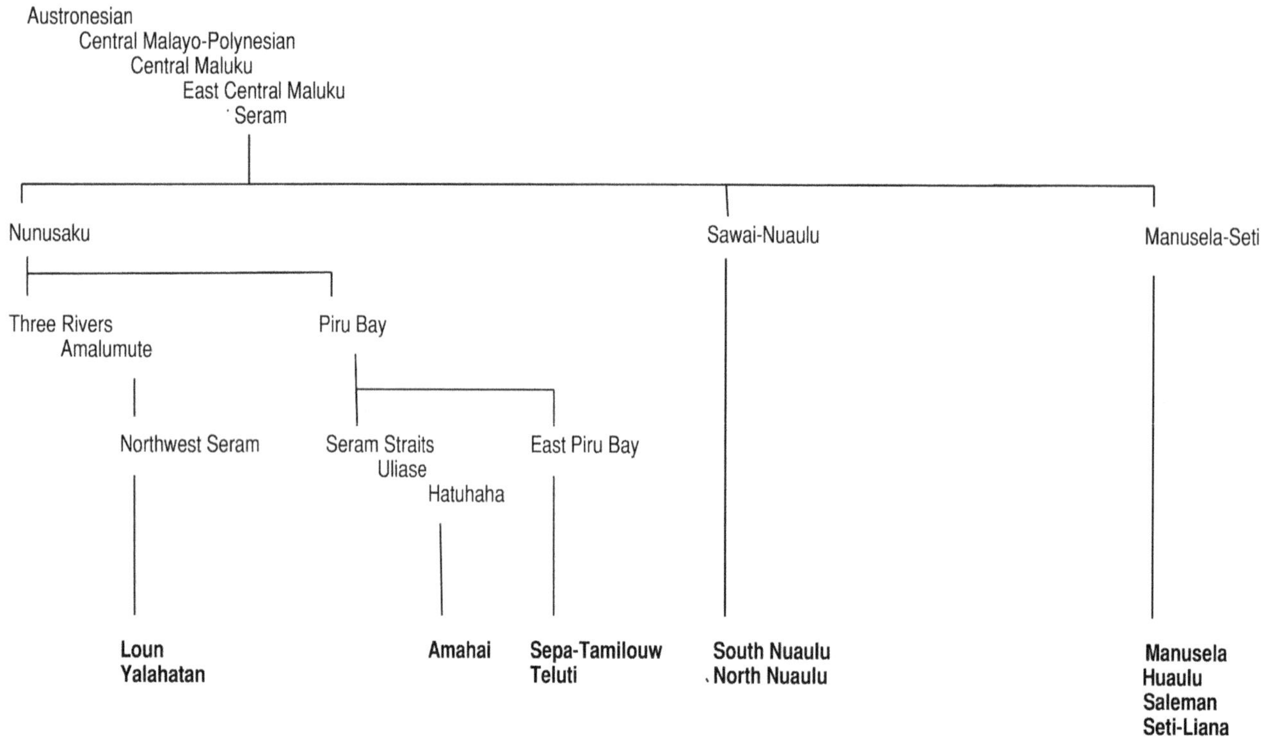

Figure 2.1 Language phylogeny showing relationship of Nuaulu to other languages mentioned in the text (Atlas Bahasa Tanah 1996).

What is a plant?

There is no word in Nuaulu clearly equivalent to that used in English to designate all plants in contrast to all animals, or non-plants, and yet it is equally clear that Nuaulu conceptually distinguish the domain of plants. The nearest we have is ***atuhu***, which, as in Indonesian 'tumbuhan', refers to things which grow (***ruhu*** = 'to grow'), wild plants and things that are planted, but it is also used more specifically to refer to crops. The existence of an ethnobotanical concept of plant, a folk kingdom concept coterminus with scientific Plantae, is evident from the syllogistic logic of folk classification, from the terms that people use to describe plant parts and processes (see below), and from the various sorting experiments and formal elicitation techniques that I employed fairly early on in my ethnobiological work amongst the Nuaulu (Ellen 1993). We might infer from direct questioning and language behaviour that the Nuaulu definition of a plant is of a living organism that is sessile, capable of reproducing (by seed, spores or vegetative diseminule), grows towards sunlight, has green fibrous parts and which obtains its sustenance (as Nuaulu see it) from the soil or a similar substrate. What might a 'generic' plant look like? Figure 2.3 illustrates one possibility, but it is clear, for example from Figures 2.4–2.7, and 6.4 (palm) and 6.8 (bamboo), that many plants will not really reflect this stereoptype, which tends to be for what we call 'herbs' – a 'plant without a persistent woody stem above ground' (Beentje 2010:56). It is difficult to envision what cognitive prototype Nuaulu might have for a 'plant', and to see how this might be conveyed simply in visual form. Quite apart from variation in the physical habit that plants take, both between species and within a species at different stages of maturity, other perceptual factors can be important: smell, taste, touch, even sound. No wonder that the use of picture cards as an aid to studying the folk classification of plants is not very satisfactory. It is problematic in animals, but even more so with plants.

By a process of polythetic aggregation or 'criteria clustering', whereby the characteristic cognitive profiles of different entities are perceived to overlap with each other to varying degrees, Nuaulu recognize that a number of scientific taxa share sufficient family resemblance to be grouped together in contrast to other organisms as PLANTS. These organisms are all Plantae in the phylogenetic sense,[2] and include both Gymnospermae and Angiospermae among the Spermatophyta, although Nuaulu make no nomenclatural or classificatory distinction between non-flowering seed-producing plants and flowering plants. Of the cryptogams,

2 For convenience, the nomenclature employed for macro taxonomic categories here follows the usage adopted by PROSEA (e.g. Prud'homme van Reine and Trono 2002:15–16), unless otherwise stated. With new DNA techniques plant systematics is a fast-changing field.

some mosses are sometimes included, and certainly all club-mosses, horsetails and ferns. For the purpose of this book and its analysis, and despite the fuzzy edges, these groups define the Nuaulu world of plants. All quantitative data on nomenclature and categories provided refer, unless otherwise indicated, to vascular plants within the Nuaulu Ethnobotanical Database (NED). In this respect, the folk kingdom category PLANT conforms to similar categories found in many other cultural traditions and languages. There are, however, in addition plant-like organisms that exist in an ambiguous classificatory zone that we should briefly review, partly because their examination helps clarify what Nuaulu mean by a plant, but also because scientific convention expects them to be treated together. In most folk biological systems, fungi and algae are classic liminal forms in this sense (Ellen 2008). In the classifying and naming strategies that Nuaulu employ there is a tension between placing fungi with plants and according them their own separate 'kingdom'; and in addition, seaweed is nomenclaturally aligned with fungi.

The most salient group of liminal organisms are, therefore, fungi. These Nuaulu designate as a separate domain through the term ***unate***. ***Unate*** refers to all visible fungi, with the exception of lichen-forming Ascomycetes; and is also applied to sea sponges (***unate nau moti***, literally, 'mushrooms of the reef'), and to all algal seaweed (***una nuae***), despite the first being phylogenetically Animalia and the latter either Plantae or Protista. The Nuaulu concept that ***unate*** labels is undeniably a FOLK KINGDOM in Berlinian terms, and since it has been described elsewhere (Ellen 2008) I shall treat it only briefly here. All ***unate*** are marked lexically by the contracted prefix ***una***- followed by an adjectival or other qualifier: for example, ***una msinae*** ('red mushroom') for *Pycnoporus sanguineus*, or ***una pate*** ('fig-tree mushroom') for *Trametes corrugata*. The broadly inclusive character of ***unate*** is in itself interesting, given that mushrooms as a phylogenetic grouping are extremely varied in shape, structure, colour, habit and reproduction. As there can be considerable morphological difference, say between rigid shelf mushrooms or bracket fungi, fragile fleshy mushrooms and the highly salient coral mushrooms, we must assume that placing them together in a single clearly labelled category (whether 'mushrooms' or ***unate***) must reflect some combination of cognitive prototype and common distinctive features. Though not regarded by Nuaulu as plants, fungi have similarities with them, and in some respects their lives and nomenclature intertwine. For example, many fungi are named through the tree host on which they grow.

Algae occupy a similar though more complex position. Seaweed – meaning mostly algal groups that include Phaeophyta (brown algae), Rhodophyta (red algae) and Chlorophyta (green algae) – are as a group described as ***una nuae*** (fungi of the sea), and are little differentiated by Nuaulu, who as an interior and highland people have had no particular reason to do so. Where they do describe seaweed types it is on an ad hoc basis, or using widely understood folk terms in Ambonese Malay or the languages of Sepa and Tamilouw, for example when species have some commercial significance, such as the 'agar-agar' genus *Gracilaria*. The term ***uruane*** is used to refer to filementous marine algae such as *Chaetomorpha javanica* Kütze (AM 'lumu-lumu') said to resemble veins and arteries in the human body. The term ***una nuae*** is also used for sponges. Freshwater and terrestrial algae are are usually Cynaphyta and Chlorophyta (e.g. Spyrogyra) and described as ***mapunue*** or ***kawanae*** by Nuaulu, related forms also found commonly growing on trees. The same terms are also applied to dry algae on rocks or stream edges.

Not subsumed under ***unate*** are forms of lichen forming Ascomycetes, such as the fruticose form which Nuaulu call ***ahane***$_1$,[3] and molds and mildew (which they call ***rekune***). Slime molds (no longer classified as fungi in Western science) are also given their own category, ***sona***, literally 'sago jelly', which they are said to resemble (Ellen 2008:488). ***Ahane***$_1$ appears as a branching filamentous growth on host trees that are often sick and dying, and is described as 'lani' in AM, and includes all species of *Usnea* known from Seram (*barbata, dasypoga, longisima*, cf. *hirta* and *misaminensis*). These are now placed (following fungal taxonomy) in the family PARMELIACEAE, a combination of a fungus from the Ascomycota division and an alga from the Chlorophyta. In the recent past it is said to have been used as a soft cushioning by Nuaulu when sleeping in the forest (Figure 2.2), and was described to me by Hotena as the hair of the ***setane***, a kind of spirit, which places it with other vernacular names for this lichen (such as 'beard lichen') from other parts of the world.

Some other lichens are named, such as the distinctive yellow form ***napate ere kunai***. Other lichens, like fungi, may be described in terms of their host, e.g. ***aha onie*** (lichen growing on the dipterocarp *Shorea selanica*), or ***matamaine*** or ***matamaine kihene*** (lit. 'wings of a cockroach'), a grey lichen growing on trees. Mosses may be described as ***ekote hanaie***, ***kawanae*** or ***nunte***, terms also applied to some lichens, liverworts, true mosses and the club-moss *Lycopodium cernuum*. ***Nunte*** is homonymous with the word for the semi-permeable membrane (***nunte***) stretched across the upper trough of a sago-processing apparatus and usually made from the fibre of old coconut-leaf bracts (***nione ne nunte***) (Ellen 2004b).

But not only are plants in Nuaulu terms living 'things' from a perceptual point of view, empirical entities in a cognized field of experience, they are also agentive 'lively

3 Note that ***ahane***$_1$ is the homophone of the term ***ahane***$_2$ used for several species of fern (see Chapter 3, e.g. Table 3.2 and Figure 3.5).

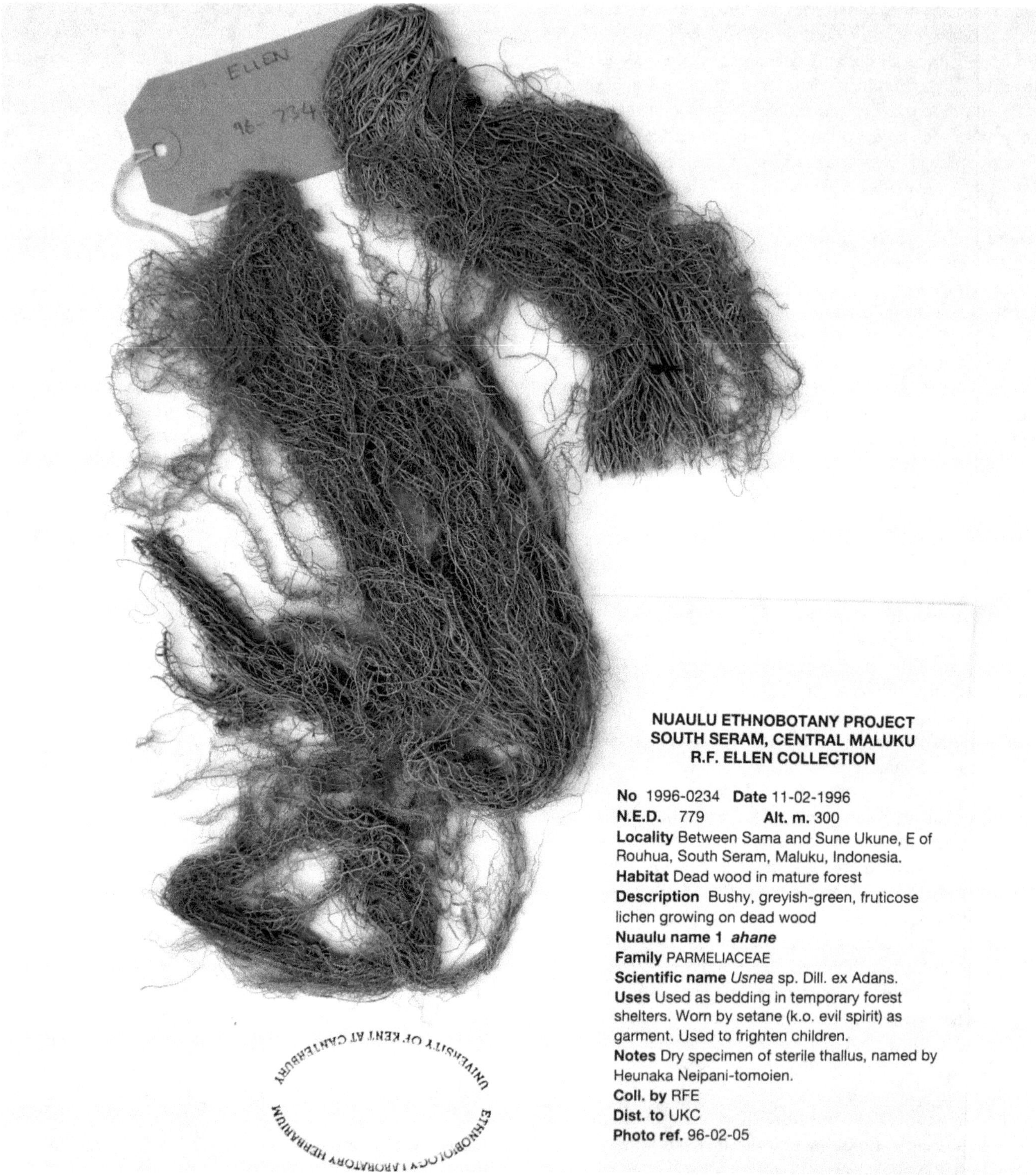

Figure 2.2 *Specimen of dried* ***ahane1****, the arborescent lichen Usnea, used as temporary bedding in the forest.*

materialities' as well. Moreover, the vegetal is a dimension of all things that live and of life as a phenomenon, no less than is the animalistic or indeed the humanistic dimension (Ellen 2017b); constituting also the context, the sensed landscape in which particular things (as various as stones, trees, waterways, animals, past and present human settlements and residues) are apprehended and understood. There is no possible construction of the Nuaulu environment that is not simultaneously physical and vegetal, potentially animated, and with a human dimension. The use of male and female terms (***hanaie***, ***pina***), and core genealogical terms such to ***anai*** (offspring), ***inae*** (mother) and ***ama*** (father), to describe and talk about plants, implies certain relational characteristics that plants share with animals and humans. They are all connected one to another, they all grow and they all reproduce. As a trope, the sago palm, with its numerous suckers and phenotypic continuity over generations, is widely compared to descent lines, the removal of suckers from a parent tree to clan segmentation, and the relationship between suckers from the same palm to siblinghood, a figurative language widely found amongst the sago peoples of lowland Melanesia (e.g. Gell 1975:144). But while the spirit of certain plants, such as ***kawasa***, the *Archidendron clypearia* tree used for making sacred shields, is specifically addressed when being harvested, and while the process of transformation is accompanied by reverence and a mode of engagement that emphasizes anthropomorphic and animistic properties, such treatment of plants is highly selective by species. You do not have to ask permission from either plant spirits or ancestors to cut plants in general, only certain plants, such as sago. We can see the same process at work when Carrithers, Bracken and Emery (2011) refer to 'The person-like character of a species' in conservation and taxonomic discourse, and when they show how the 'axiom of amity' finds an afterlife in it. The concern of some polemicists of the ontological turn with deconstructing the hegemony of the plant/animal distinction, and a special interest in 'planimals', reflects a degree of conceptual distress amongst observers of a biological world as revealed through modern science that takes a particular interest in the anomalies, whether the result of natural selection or bioengineering. Most folk, including the Nuaulu, have no problem here. There is likely an evolved human proclivity to distinguish things that most people label as plants from the things most people label as animals. The fact that there are fuzzy areas – fungi, lichens (fungal-algal symbiants), sessile animals that appear like plants (bryozoa, sea-lilies and sea-anenomes) or plants that that behave like animals, such as pitcher plants – intrigues, but does not fundamentally phase or concern Nuaulu, who have perfectly adequate ways of handling what for them are peripheral representational problems emerging from the classifications they invent.

The language of plant description

Most of what Nuaulu know about plants is not readily expressed in words, but comes rather from intimate and long-term physical interaction and experience with the plants themselves. Moreover, Nuaulu know many more plant taxa than they have names for, and sometimes several names for the same taxa. They know, for example, that certain plants, which they do not use, are eaten by other peoples, and may even know the names that those people employ. For example, some people in nearby Sepa are known to eat a species of the fleshy herb *Pentaphragma*. Nuaulu may also distinguish plant species, which they do not have names for, that are eaten by animals. However, language provides a means of codifying differences, and is useful in the transmission of knowledge. Indeed, words and expressions relating to plants constitute a remarkably high proportion of all words in the languages of people who subsist largely from their own environment. Thus, Bolton and Matoke (2005) list 267 names for Nuaulu plant taxa, and 92 other terms related specifically to plant related activities or descriptions, though this latter number is probably a conservative estimate. Together, these terms constitute 7.16 per cent of an approximate total of 5,014 words reported in their dictionary. But if we replace the terms for vascular plant taxa with the more complete number reported in the NED up until 2019 (1,015), then the percentage increases to 22 per cent. Either way, these numbers are remarkable and reflect the significance of plants in the lives of people living in small-scale subsistence populations. Through the work of Atran and Medin (2008) on dictionary evidence more generally, we can infer that once people are disconnected from the world of plants the percentage decreases dramatically, with a shift from specific species-level linguistic knowledge to more abstract and general terminologies. Moreover, the words used to describe plant parts and technical procedures involving plants – what Berlin, Breedlove and Raven (1974b:69) dub 'ethnophytography' – also provide us with indirect evidence for shared knowledge of different plant groupings, and evidence that plants as a whole are perceived as a domain of knowledge, in contrast to animals and other organisms.

The basic vocabulary of plant partonyms is shown in Table 2.1 and Figure 2.3, but even these may apply in different ways to different plant types, and therefore may require qualification. Thus, how ***tinaie***, usually meaning 'main stem', is employed with respect to a palm or a vine will vary. The way in which generic partonyms map on to various kinds of whole plant is illustrated in Figure 2.4, which compares ginger, banana, taro and maize; and in Figure 2.5, which compares sago, papaya, clove, lime and coconut with regard to inflorescences and fruits, in particular. Specialized terms for parts of the flower are shown in Figure 2.6, and for a tree in Figure 2.7. Sometimes, the relevant plant uninomial will be incorporated as the

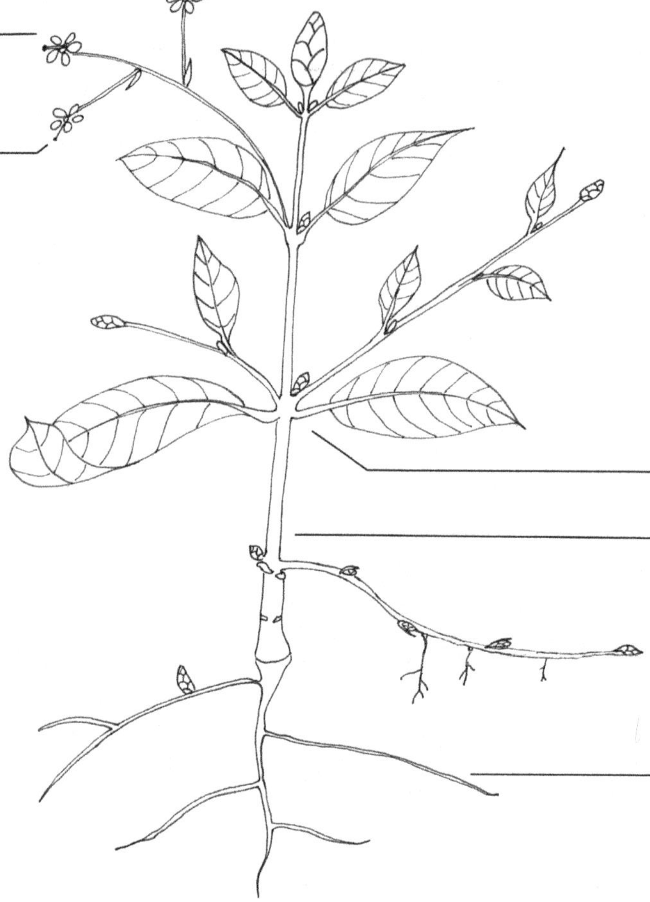

Figure 2.3 Basic Nuaulu plant partonyms in relation to a generalized plant.

first element in the partonymic expression used to describe a feature, as in ***nusi unte*** (lime peel) or ***uri kaie*** (banana leaf blade), though whether or not this is the case may depend on the morphosyntactic constraints of ordinary speech and the pragmatics of the issue at stake. In other cases, whole terms are plant type specific, as in ***kin-kinie*** for the flower of the durian, ***uri tatae*** for a 'hand' or bunch of bananas, ***makapotae*** for a young 'kenari' nut, ***mnaune*** for a hard, unripe *Areca* nut, ***kilau*** for the layers of an onion or bulb, and ***kokohahau*** for a pandan leaf. ***Msupate*** is a coconut that falls before it is ripe, ***kakenae*** the sheath around coconut-palm flowers, ***sopae*** a coconut palm from which the fruit has been taken, and ***wekete isie*** coconut juice. Moreover, on a 'hand' of bananas the fruits are distinguished by their positioning. The fruit at the highest point on the stalk (usually the largest fruit) is ***uri katahahai***, while the fruit at the end of the stalk, usually the smallest, is ***uri kisuhe***, while that possessing the **usue** ('jantung' in AM, or heart) is ***uri katatihue***. All those on one stalk are ***uri hunne***. These precise distinctions illustrate the potential dangers of confusing descriptive terms with actual plant taxa, and remind us that the 'art of describing' the constituent parts of plants evolves organically and pragmatically as people interact with the plant types individually, and not through the simple application of a stock set of terms.

Another feature of Nuaulu plant partonymy is that in many cases where a partonym includes the name of a plant type as an element, the plant name is intrinsic to the term. Thus, in a palm frond the leaflet is ***hatan totue*** (sago, *Metroxylon sagu*), ***kanai totue*** (betelnut, *Areca catechu*) or ***nione totue*** (coconut, *Cocos nucifera*) and the leaf stem, ***hatanasite***, ***kanai asite*** and ***nione asite*** respectively. Similarly, in banana ***uri anai*** is used for the growing plantlet arising from the base, and ***uri usue*** for the 'flower', and ***uri kaie*** for the leaf midrib, while ***soi isie*** is 'ginger root' (*Zingiber officianlis*), and ***yakon nene hunua***, the hair on top of maize cob (*Zea mays*). So, we should not always assume that ***soi isie*** refers to the same plant part as ***uen isie*** (yam tuber). Where parts of plants are especially important for commercial reasons, particular partonyms may be emphasized. Thus, in clove (***kupanahane***, *Syzgium aromaticum*), unopened buds or

ai anai	small tree, seedling
ai arine	buttress root
ai atue	crown or canopy branch (if already fallen)
aikune	main stem, trunk of tree; base of palm
ai rimaia	'all branches of a tree'
ai tinaie (sometimes ***ai hatae***)	main branch, stem
ai unte	bark
ai utone	interior (of wood, trunk, branch or twig)
anai	tree seedling, baby tuber
anai ikine	small baby tuber offshoot
ananae	seedling, planting material
ani mui	sapwood, tree rings
apane anoi	heartwood
asite	frond of banana plant or palm
atue	the top of a tree, canopy
haki nohae	old inflorescence or fruiting branches, for example of *Areca* palm
hana	fruits of *Areca* palm
hanae	a main branch
hanai apue	'hand' (bunch) of fruit
hatae	trunk, stick
honoe	fruit of bamboo or sugar cane, internode
huae	seed, fruit, tuber
hukune	knot, as in wood; node, branching point
hunne	flower, fruit, fruitescence (as in separate flowering/fruiting body of ginger which appears separately through the earth); maize head, single banana fruit, pineapple
isie	inside, contents (of starch storage organ); may sometimes refer to starch root or tuber as whole, as in ***kasipi isie***; tubers, corms, bulbs, rhizomes, as in ginger (***soi isie***), or taro
kaie	leaf midrib, flower stalk, leaf stalk, stem (e.g. of taro)
kainuni	thorn, spine
kakane	flower bud stalk (of clove), stalk of chilli fruit
kakaue	flower
kamatae	bunch, or stalk of bunch (as for banana); also trap, snare or lasso
kamatuane	an old leaf, as when no longer green
kaminae	waist of taro or other plant, point at which shoot becomes root, at soil surface; base of leaf stalk on *Areca* or coconut palm; stalk, single stalk of taro appearing through soil
kanai asite	frond of *Areca* palm
kanai tinaie	main stem, trunk of *Areca* palm
kanie	seed, kernal, stone, pip
kapanue	stone or pip and surrounding pulp found in durian, jackfruit and similar
kinaie	stem (e.g. taro)
kina(n)te	resin, sap
kin-kinie	flower of durian
kisae	segments (as in *Citrus* fruits)
kupako	flower
kupako nene noene	calyx, petals, sepals
mata ukune	top of tuber, where shoots emerge; inner parts of flower (style, stamens and pistils), stalk, stem
minete	young leaf, tip of leaf, bud
mpiae	sago pith
namte	root
namte nene hunua	hairy adventitious roots of yam tuber, or similar; hairy roots around top of taro or yam tuber; hairy roots between soil surface and tubers on taro
nante	tree canopy (also 'sky')
noene	flower (on tree)
ntone	sap, latex, resin, gum, exudate
rahue	young planted seedling
rimae	branch
sanae	branch, twig
sanai ikina	twiglets
sineune	bud/shoot
suaei	stalk, peduncle (as in banana infructescence)
tinaie	stem, main branch, trunk
tinete	outer (hard) epidermis of bamboo
tonote	small adventitious roots at base of palms
totue	leaf
tuhue	bud, shoot, sprout
uhune	unopened/young leaf shoot (taro), bud, young leaf, bamboo shoot; /t/ inserted when following vowel, as in ***wana tuhune***
ukune	crown or canopy of a tree
unie	thorn, spine
unte	skin or epidermis, as around tubers or in husk wrapping a maize cob
unte makae	hard outer bark
unte mane	soft inner bark
upone	tree-top
utone	heartwood, inside of bamboo (used as cordage)
wane	vine, as in yam; liana; trailer
wanie	young fruit

Table 2.1 Basic plant partonyms (ordered alphabetically).

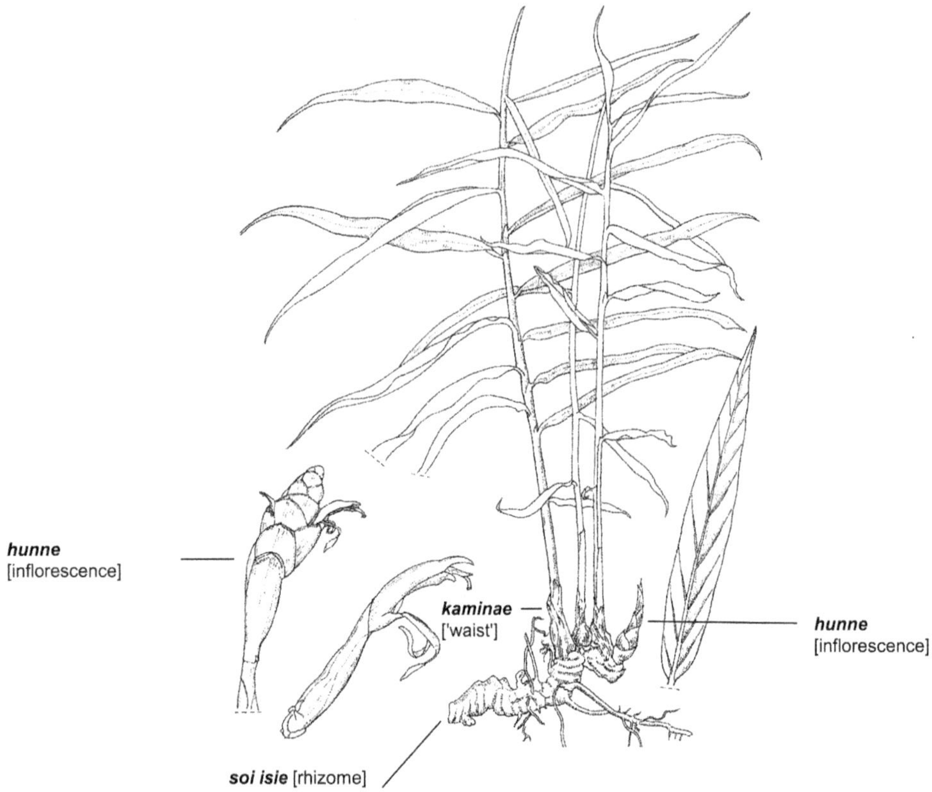

Figure 2.4a Nuaulu plant partonyms as applied to ginger (Zinziber officinale). Illustration courtesy of Backhuys-PROSEA, with permission.

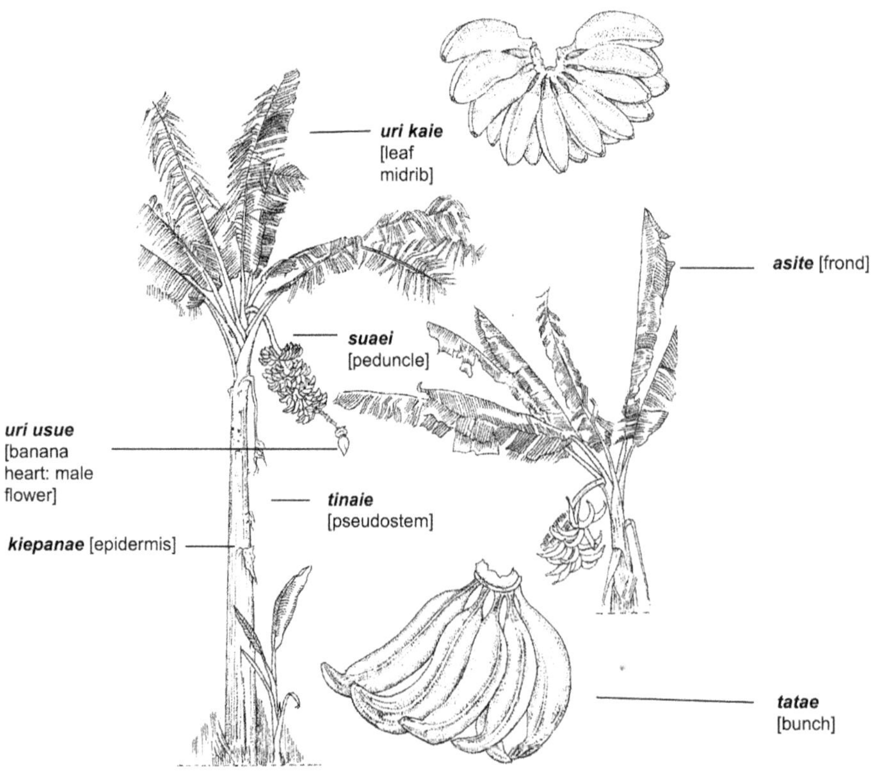

Figure 2.4b Nuaulu plant partonyms as applied to plantain (Musa x). Illustration courtesy of Backhuys-PROSEA, with permission.

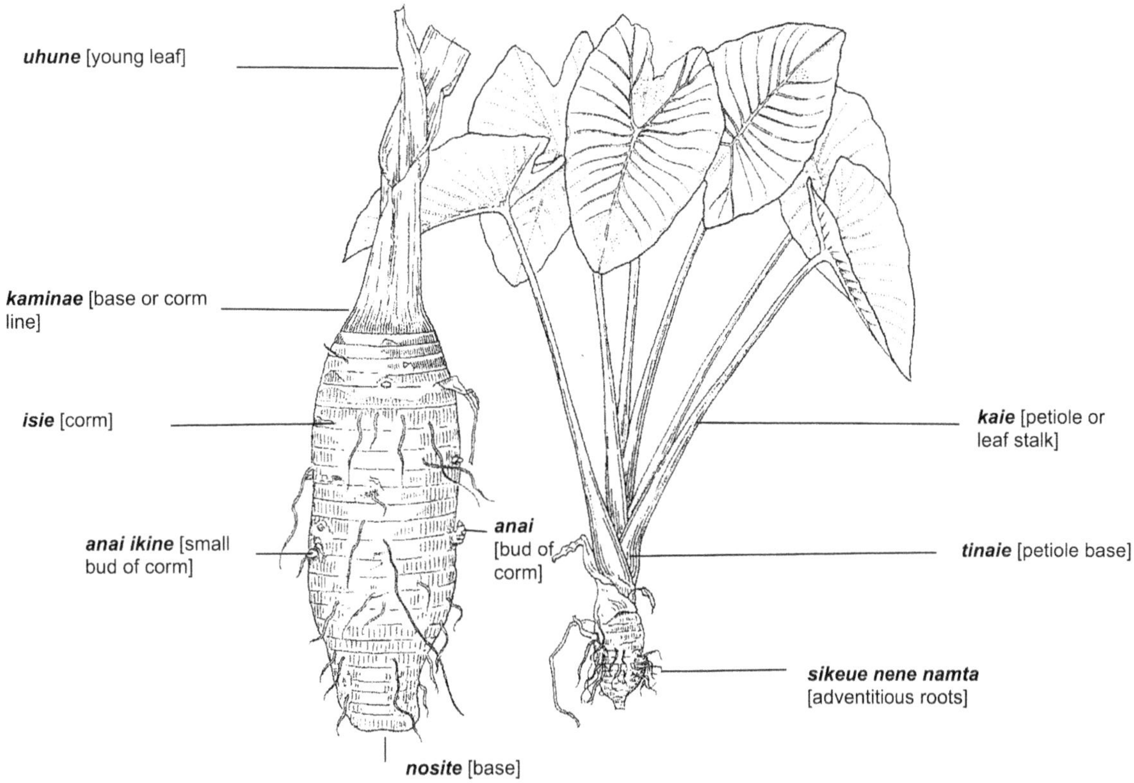

Figure 2.4c *Nuaulu plant partonyms as applied to taro (Colocasia esculena). Illustration courtesy of Backhuys-PROSEA, with permission.*

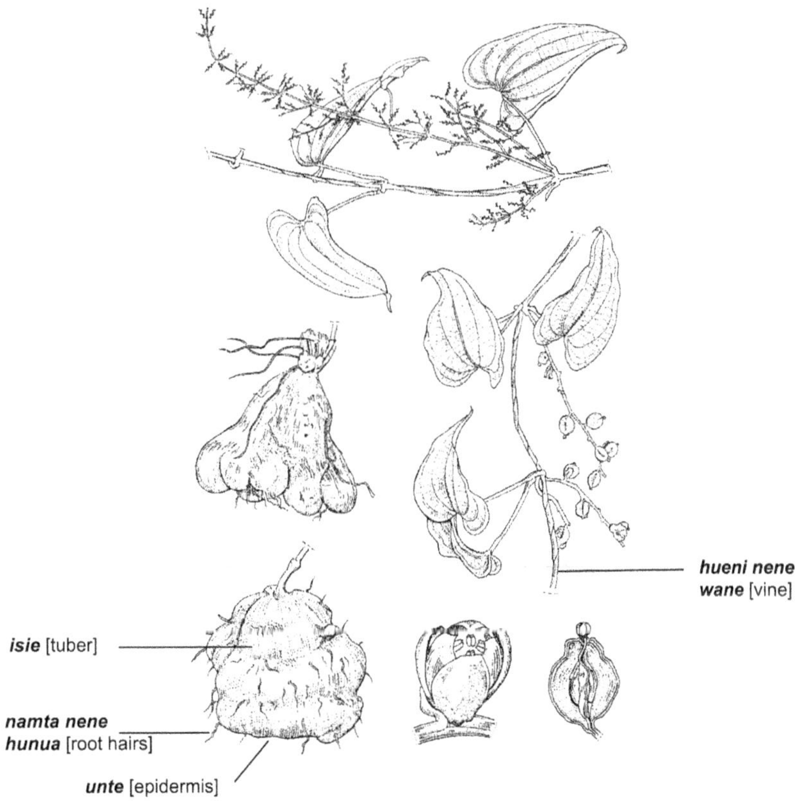

Figure 2.4d *Nuaulu plant partonyms as applied to yam (Dioscorea alata). Illustration courtesy of Backhuys-PROSEA, with permission.*

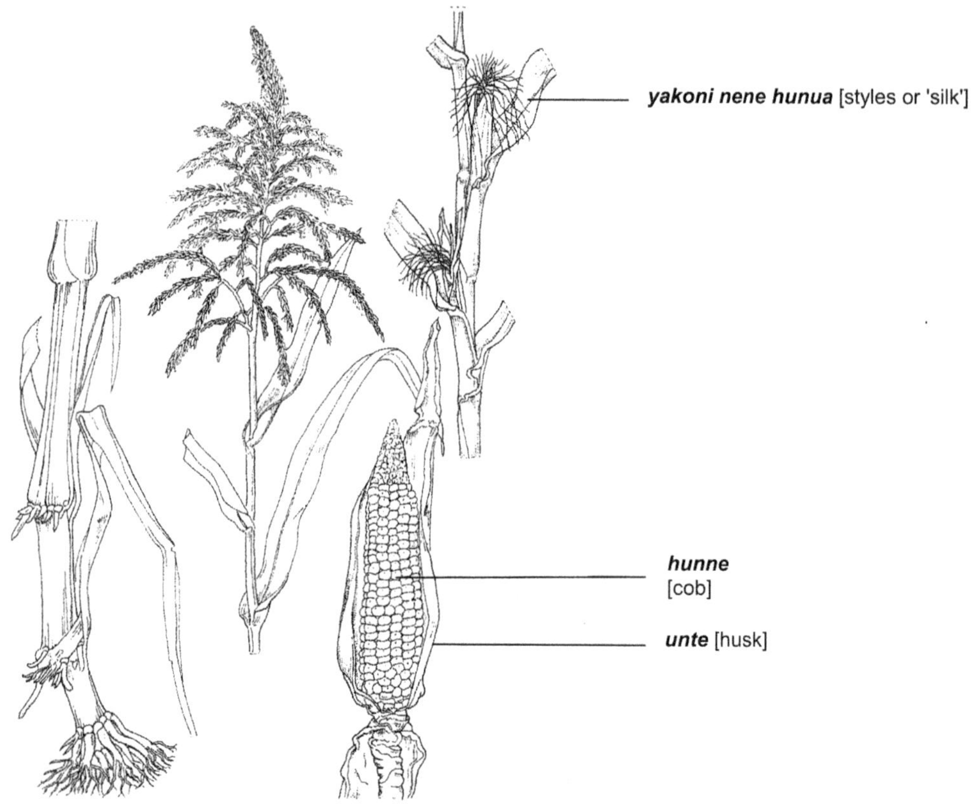

Figure 2.4e *Nuaulu plant partonyms as applied to maize (Zea mays). Illustration courtesy of Backhuys-PROSEA, with permission.*

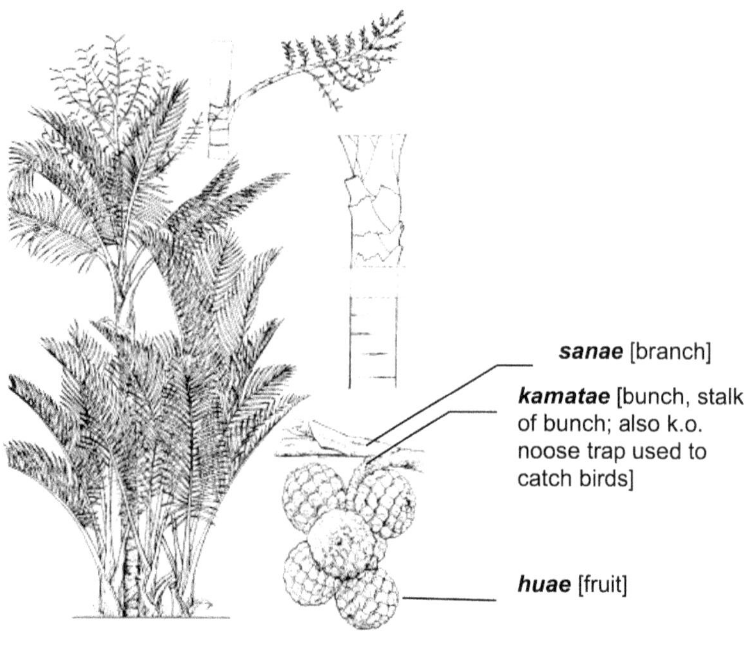

Figure 2.5a *Partonyms for selected fruits and flowers to show variation in application of terminology: (a) sago (Metroxylon sagu). A more complete account of sago partonyms can be found in Ellen (2006a). Illustration courtesy of Backhuys-PROSEA, with permission.*

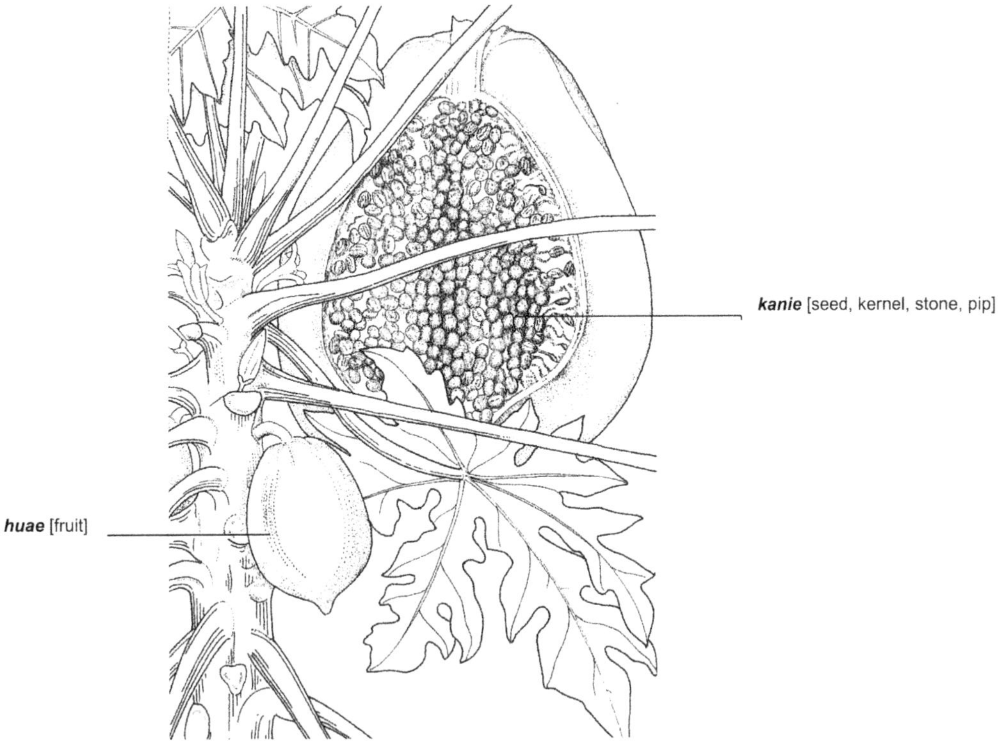

Figure 2.5b *Partonyms for selected fruits and flowers to show variation in application of terminology: papaya (Carica papaya). Illustration courtesy of Backhuys-PROSEA, with permission.*

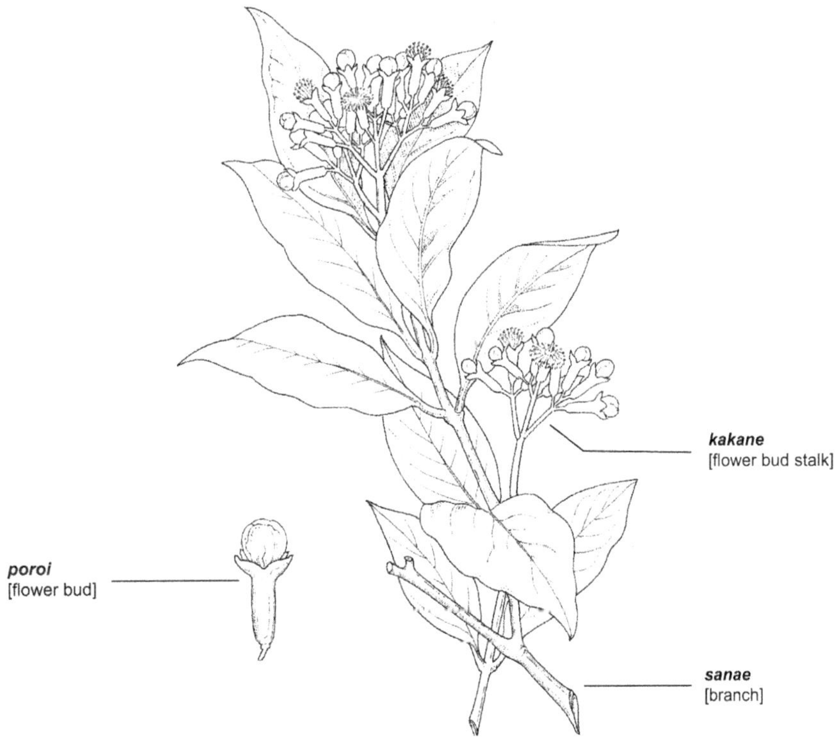

Figure 2.5c *Partonyms for selected fruits and flowers to show variation in application of terminology: clove (Syzgium aromaticum). Illustration courtesy of Backhuys-PROSEA, with permission.*

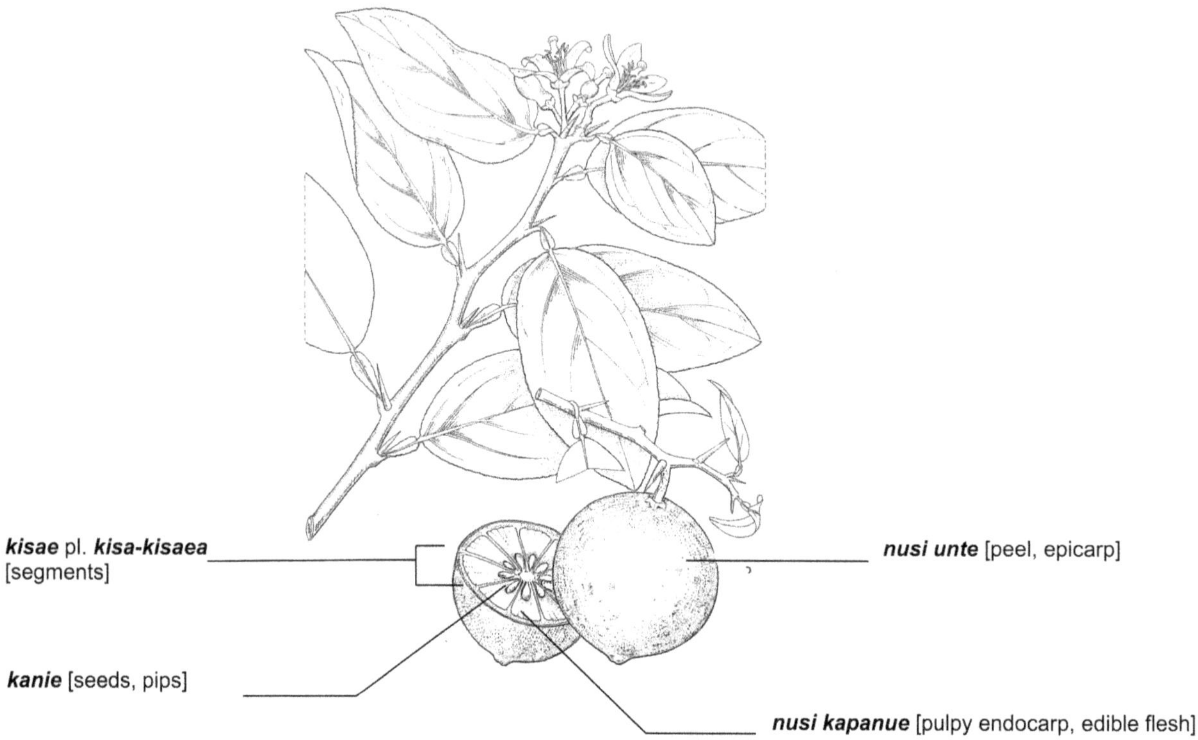

Figure 2.5d Partonyms for selected fruits and flowers to show variation in application of terminology: lime (Citrus aurantifolia). Illustration courtesy of Backhuys-PROSEA, with permission.

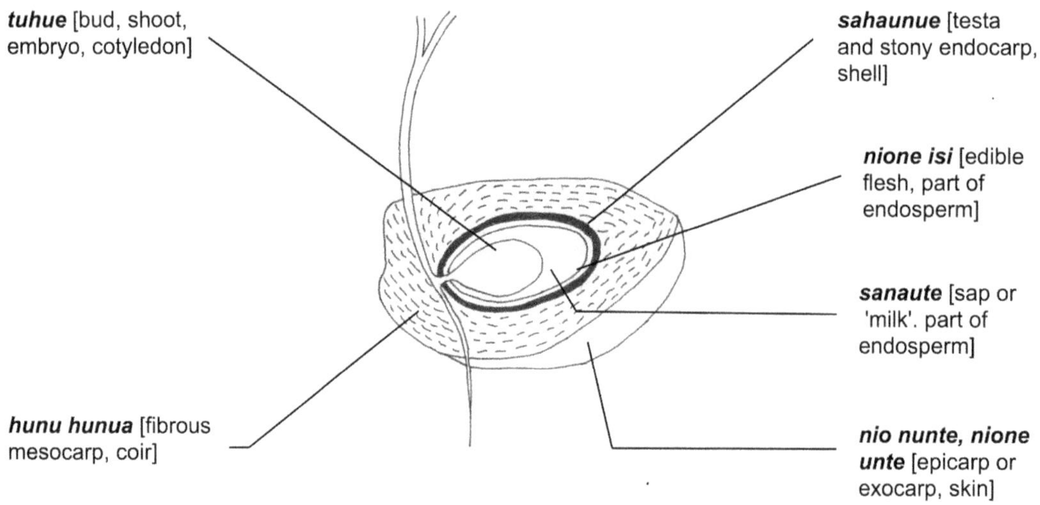

Figure 2.5e Partonyms for selected fruits and flowers to show variation in application of terminology: diagrammatic section of coconut (Cocos nucifera).

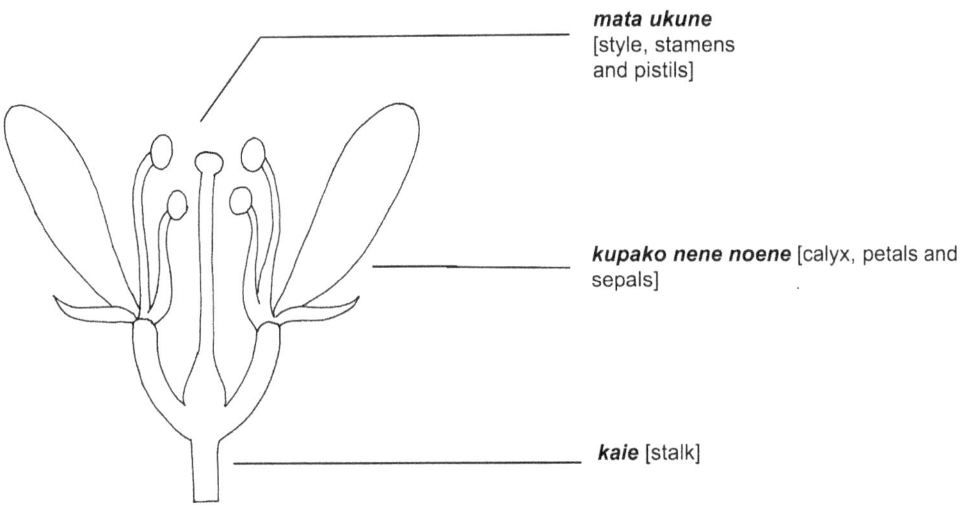

Figure 2.6 *Nuaulu partonyms applied to basic parts of a generic flower.*

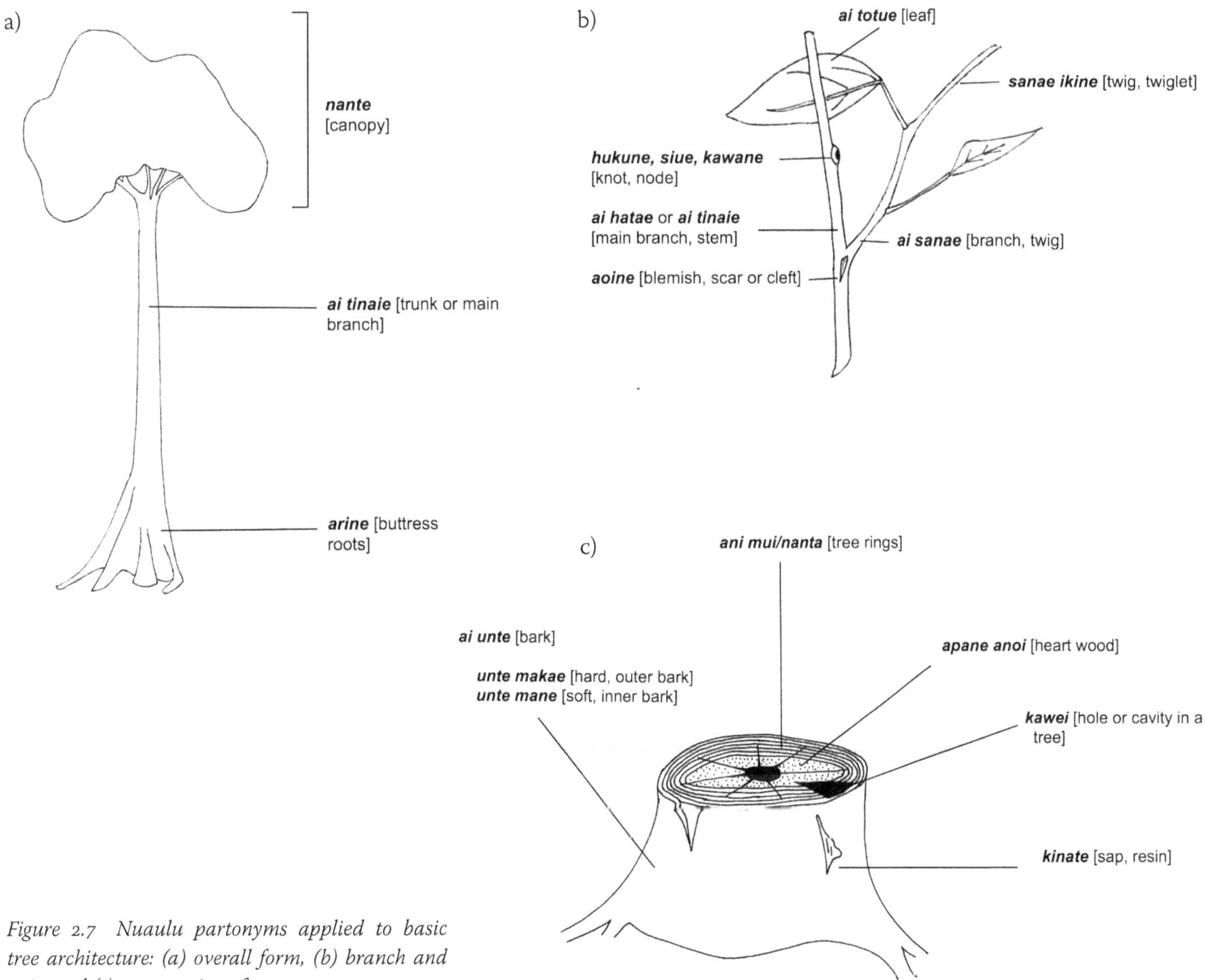

Figure 2.7 *Nuaulu partonyms applied to basic tree architecture: (a) overall form, (b) branch and twig, and (c) cross section of stump.*

pods (***poroi***) are distinguished from ***kakane*** (the flower stalk), which are in turn distinguished from ***sanae*** (the stalk of two or more flower stalks; elsewhere 'branch'), as dried buds and stalks command different market prices. The term ***kakane*** is used to describe the flower or fruit stalk of some other species (e.g. chilli, *Capsicum anuum*), but not all.

Other terminology distinguishes the growth stage and maturation, and again may sometimes be plant type specific. For example, ***kam honue*** distinguishes young green betel pepper from old, brown, tough betel pepper (***kam lea***), while ***kanai ikine*** or ***kanai wanie*** distinguishes young, usually green, soft *Areca* nut from old, hard *Areca* nut, requiring pulverizing before chewing (***kanai menai***). ***Munane*** distinguishes any leaf that has turned yellow from an old one that is already brown (***kamatuane***).

For some species, the amount of specialist vocabulary is very extensive, and plant partonyms connect to terminologies for equipment, processing and consumption. The largest and most complex Nuaulu example of this is for sago (Ellen 2004b, 2006a). These vocabularies reflect autecological knowledge, knowledge focused on and around a particular species. Note also that the names for growth or processing stages are often embedded in verbs describing the actual task, and we have to remember that this is how people experience the acquisition of knowledge. Finally, there are many terms and phrases that relate to general features of many plant-related activities, or that refer to processed parts of many vegetable products:

ahuneu	to grow a smaller tree into a bigger one
asikuku	to remove leaves from a branch or frond
aupono	to cut a tree so that the trunk or stalk does not break
ausona	to slide down a tree or palm trunk
hera	to strip and process a sago leaf, bamboo or coconut leaf, for example when making thatch or in preparation for basketry
sema	to flatten grass
sihu	to cut a banana from a tree
sinuhue	to fruit plentifully
aukoni	a stripped, prepared and cleaned coconut leaf or rattan cane

No wonder then, that with plants being such an all-pervading part of Nuaulu life experiences, kinds of plant, parts of plants and plant processes occupy a central place in the descriptive, interpretative and metaphorical repertoire of language. The most symbolically salient example of this are tip (***atue***) and trunk (***aikune***) analogies. In Nuaulu ***,ukune*** or ***ai-kune*** also has the extended meaning of base, beginning or cause, but does not extend to the concept of 'root', for which we find the term ***namte***. The figurative use of this pairing is particularly evident in ritual and mythological contexts, especially in sung verse, and is well illustrated through the work of other ethnographers of eastern Indonesia, such as Fox (1971) and Traube (1986). Indeed, so all pervading is plant terminology that whether Nuaulu refer primarily (exogenically) to plants or humans or houses seems to hardly matter, metaphorical language being consubstantial and dynamic. As in many languages, words used for plant parts and vegetal processes are also words that have other meanings. Some of these, such as words for stem and trunk, have near universal semantic reference. However, there are variations. In Nuaulu, while the junction of the ear and head is ***tinai ukune***, 'ear root', the outer ear as a whole and the earlobe is ***tinai totue***, 'ear leaf'. So, as the pattern of differences in the correspondence between names of plant parts varies, this gives rise to different semantic and figurative possibilities. Some special features of the partonymy of other plant groups (palms and bamboos) are dealt with in Chapter 6 (Figures 6.4 and 6.8).

Quantifying lexical knowledge

Just how many Nuaulu terms are there for different types of plant? This is an important question because plant names are used as a proxy for plant knowledge in cross-cultural studies, for example in comparing human populations according to subsistence regime. As we shall see, discerning the answer is not as easy as it might at first seem. Documenting plant names is not the same as documenting local plant categories for taxa. Some categories are unnamed, and others, some seventy-five in my accounting, are named using a number of 'true' synonyms; though the number must surely be much higher, given the many plant names on which taboos are placed in certain contexts, such as in mixed-gender situations and between people in particular kinship positions, and between clans. Moreover, some scientific taxa are divided into folk categories that in fact comprise sexual types of dioecious species.

One measure of the number of terms for vascular plants takes into account all those used by Nuaulu to refer to any type of plant at any degree of inclusion or level, and including all synonyms and lexical variants. We might call this the total aggregate number. My most recent estimate of this for the Nuaulu is 1,177.[4] This is an aggregate number in the sense that it comprises the total number of different words that I was able to record being used by anyone, from the many individuals who speak the Nuaulu language; it is not what any one individual knows. It includes synonyms and lexical variants, but determining what is a synomym and what is simply a lexical variant is not always straightforward. As ethnobotanists have learned to their cost – even those

4 This is made up as follows: 1,012 records in the NED including 8 unknown but presumed names, plus 75 true synonyms and a minimum of 90 additional lexical variants for the same plants, giving a total of 1,177.

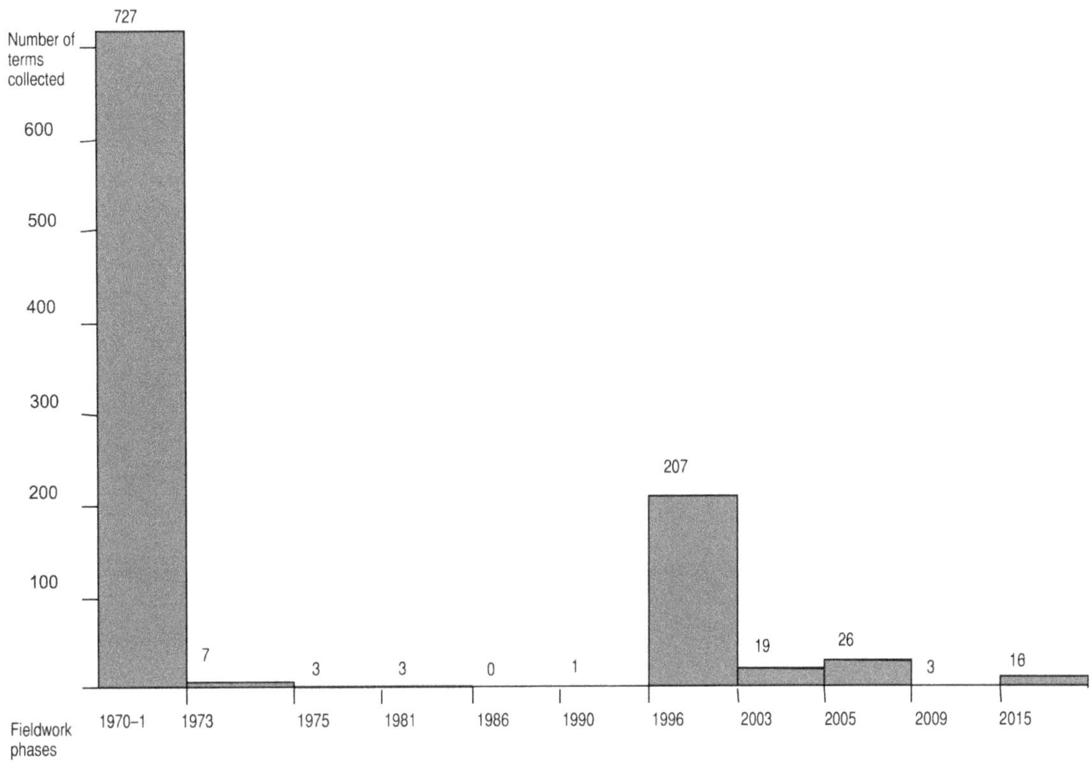

Figure 2.8 Number of new Nuaulu terms for vascular plants obtained during successive fieldwork phases (excluding synonyms).

fluent in a local language – it is rather easy to acquire a large number of erroneous answers to the question 'What is this plant called?' Informants will helpfully provide a name even if one is not in regular use, or may agree a name even if they have never heard it before. They may mishear questions, or may provide a descriptive term rather than a name. In long-term field studies it is usual to expect the number of discrete folk terms for plants to begin by rising steeply, but then to plateau as 'new' terms turn out to be already recorded. By 1971 I had recorded 784 names in the 'Nuaulu Ethnobotanical Index'. By 1996, after the collection of more data supported by voucher specimens, and after the dataset had been rejigged as the computerized NED, the total number of records had increased to 1,070. But this dataset had not been systematically edited, and despite the addition of new names between 1996 and 2019, a thorough systematic editing of the list in that year reduced the number to 1,012 records for named plants, as non-names, duplicates, synonyms and lexical variants were eliminated, and as a result of restricting the corpus to vascular plants. If we wish to include the total number of lexemes, including true synonyms but ignoring lexical variation and optional constructions (such as where uninomials and binomials are used for the same plant), the number is 1,090.[5]

5 See also Taylor (1990:29) on ways of establishing the lexemic status of words (nouns and phrasal nouns) used to describe plants.

Table 2.2 shows numbers for the acquisition of names for vascular plants during successive fieldwork phases. Note, however, that 'all terms' listed in row 10 are not the sum of all those listed in the individual columns for rows 1 to 9. This is because the vegetal types listed in the first column are not always mutually exclusive. Note also that almost all Nuaulu plant terms were acquired from people living in Rouhua. While Rouhua was in 2015 the largest Nuaulu settlement, and although we should expect a large overlap in shared knowledge of plant names between Nuaulu villages given population size, proximity and shared ecological diversity, there are nevertheless bound to be some differences, and therefore plant names that I have yet to come across. As with Balick's (1994) multiple-use-curve technique for determining the exhaustiveness of numbers of uses for a single plant species, diminishing returns in the number of new names indicates exhaustion of total number of terms, and the fact that most terms gathered before 1996 were secured without benefit of linked voucher specimens, suggests that some confidence can be placed in the ability to generate plant terms without benefit of voucher specimens. During successive fieldwork phases I continued to be presented with the same terms and vouchers for the same species, and only occasionally would encounter something new. We can illustrate this with a graph (Figure 2.8). Nevertheless, successive fieldwork phases (especially from 1996 onwards when ethnobotany became a more important

	1970–1	1973	1975	1981	1986	1990	1996	2003	2005	2009	2015	Totals
Ferns and their allies (incl. Cycads)	30	0	0	0	0	0	12	1	1	0	0	44
Swidden crops	243	2	1	2	0	0	24	5	6	3	8	294
Trees	276	5	2	0	0	1	118	5	10	0	6	423
Vines	48	0	0	0	0	0	19	0	1	0	1	69
Rattans	9	0	0	0	0	0	1	0	0	0	0	10
Pandans	10	0	0	0	0	0	1	2	0	0	0	13
Palms	41	0	0	0	0	0	6	0	5	0	0	52
Bamboos and their allies	13	0	0	0	0	0	3	0	1	0	0	17
Herbs	131	0	0	1	0	0	42	7	5	0	5	191
All terms	727	7	3	3	0	1	207	19	26	3	16	1,012

Table 2.2 Total number of new Nuaulu terms for vascular plants obtained during successive fieldwork phases (excluding synonyms), organized into selected morphological and functional categories. The order of plant groupings in column 1 follows the order of treatment in this book.

focus of research) presented opportunities to increase the number of terms. These opportunities derived from new plants and words entering the Nuaulu biocultural system during the intervening period (Chapter 9), the fact that collections of voucher specimens provided better physical material to prompt the elicitation of names, the ability to visit new areas (such as in 1996 the Nua river valley, and in 2003 the border of the Manusela National Park), and a deliberate focus on particular species (e.g. 2002–7, *Metroxylon sagu*, and in 2009, *Manihot esculenta*). In this way we can be fairly confident that over forty-five years I have accessed the greater part of Nuaulu terminology for kinds of plant. Moreover, my post-1971 experience of securing firm identifications through collecting voucher specimens, and a steady diminishing in the aquisition of new terms, suggests that early attempts (1970–1) to list comprehensively the majority of Nuaulu plant terms (1970–2015) had been suprisingly successful.

As in my work on Nuaulu animal categories (Ellen 1993), I define an ethnobiological category term as one that has lexemic status, is know by ten or more people, is used repeatedly to refer to particular taxonomic content, and which excludes what are obviously free variants of the same term, unincorporated borrowed terms, ad hoc descriptions, synonyms and terms for developmental stages and sexual types, where these are not also regarded as distinct natural kinds. In other words, it has to be a lexeme with a history and not simply one generated on the spot to satisfy a researcher.

There are many species for which there is no generally accepted name. In some cases, a term from another language may be used, which may then later be incorporated (e.g. ***pitiri malu*** < AM 'putri malu'). ***Pitiri malu*** (Figure 1.8s) is a common term for Nuaulu, but at what point does a rarely used borrowed term become a Nuaulu term, and at what point should we count it when quantifying numbers of terms for plants? The same problem arises with the use of ad hoc descriptors, say calling a yam ***siahu metene*** ('black lesser yam') not because that is a widely agreed term, but because it is a ***siahu*** that happens to be black or with dark flesh. Distinguishing names from expressions is widely discussed in the linguistic ethnobiology literature, and I have reviewed some of this in my account of the problem with respect to Nuaulu animal names (Ellen 1993:45–7).

There are Indonesian or Malay dialect terms for plants that are known to many Nuaulu, but which are not planted or harvested by them and for this reason did not get into my various inventories. One such significant absentee is *Glycine max* (L.) Merr., soya bean, in Indonesian 'kedelai'. Rumphius reports soya bean in the *Ambonese Herbal*, and it is certainly known to many Nuaulu as a product if not a plant, though it was never mentioned by them. The same is true for the pigeon pea, *Cajanus cajan* (L.) Millsp., 'kacang bali', and for the chickpea, *Cicer arietinum* L., 'kacang arab' or 'kacang kuda', and for the Lima bean *Phaseolus lunatus* L., 'kratok' or 'kacang cina'.

The Nuaulu population to which these remarks apply is that of Rouhua village, where I have undertaken most of my work. The variation of terms and synonyms discussed applies therefore to the corpus acquired here, though it is possible that there are some socio-geographic differences between Rouhua Nuaulu, and those living in other Nuaulu villages, and certainly between speakers of northern and southern Nuaulu. Verheijen (1984, 1993) has shown how plant terms can vary geographically for the Manggarai language area of Flores, though I have made no systematic attempt to undertake a similar survey here.

Free lexical variation and synonymy

Nuaulu, like members of all speech communities, vary in the precise way in which they use or pronounce the same plant name, and in the ways that they accept that it is the same name that is being used. Sometimes these slightly different names look like synonyms (either apparent uninomials or binomials), but mostly they are recognizable as free variants of the same word. For example, both ***kokune hanaie*** and ***koku hanaie*** are used to refer to the shrubby tree *Callicarpa longifolia*, and this variation is encoded in the written convention ***koku(ne) hanaie***. Similarly, tree-ferns of the genus *Cyathea* are flexibly referred to by either ***mukune*** or ***ai*** (tree) ***mukune***. This is reflected in the written convention ***(ai) mukune***. Sometimes one phoneme is replaced with another, or simply dropped, as in the free variants of the term for the tree *Antidesma stipulare*: ***ahau unu***, ***ahau un*** and ***ahau uni***. It can be inferred from these examples that a common source of free variation is the contraction of a term to create a prefix, by eliminating a noun marker, as in many uninomials labelling polytypic categories:

ahutaune > ***ahutau***: *Calophyllum*
asapopute > ***asapopo***: *Syzgium* with edible fruits ('jambu')
hatane > ***hata***: sago (*Metroxylon sagu*)
iane > ***ia***: mainly *Canarium indicum*
meute > ***meu***: rattan
monote > ***mono***: weed/grass
mukune > ***muku***: tree-fern
sinsinte > ***sinsin***: *Codiaeum variegatum*

These prefixes are generated when the term is attached to various qualifiers to form binomial noun phrases such as:

hata tamno
ia mkauke
meu onate
mono marosu
sinsin nawe

The specific shape of the prefix may alter or be elided. Some uninomials, however, cannot be further contracted and in many cases the use of the contracted form is optional. This is a general feature of Nuaulu language and exemplifies how classificatory practice can be encoded in word structure. A common lexical variant is where a prefix in a binomial is optional, as in ***nusi peka-peka***, or alternatively simply ***peka-peka*** (the pomelo: *Citrus maxima*). In some binomials, this may result in usages that seem odd and confusing. Thus, both ***wan(ate) apane*** (k.o. bamboo) and ***kan(ai) apane*** (*Areca* sp.) may in both cases be shortened to just ***apane***. This word is usually used as an adjective to describe a plant of the forest, but here it is converted to a noun, which inadvertently forms a homonym.

Lexical variation is not random in speech, and in addition to differences in spoken language between individuals, variation often follows linguistic context, where form of a noun phrase depends on where stress is placed. So, if stress is placed on the root ***hueni*** as the first element in a binomial, the adjectival qualifier will generally be shorted, as in ***hueni puti*** (where ***putie*** > ***puti***) (*Dioscorea alata* var.), but where stress is placed on the second element the adjectival qualifier of the binomial becomes ***huen putie*** (where ***hueni*** > ***huen***). Thus, speech rhythm seems to require some kind of acoustical balance in rendering the alternatives. Other examples include ***tunene rata*** or ***tune warata*** (*Annona muricata* var.), ***kam wapane*** or ***kamu apane*** (*Piper* sp.) and ***paha putie*** or ***pahane puti*** (*Allium sativum* var.). All lexical variants that I reliably heard being used and which I recorded are entered in the final column of Appendix 1. In Appendices 2 and 3 and usually in general text I use the variant I judge to be the most common. True synonyms are listed in the first three appendices.

Sometimes it is difficult to differentiate names that are simply acceptable idiolectal alternative labels for a category from 'true synonyms'. Some true synonyms are generated when individuals are not permitted to use a term for ritual reasons. For example, the term may be prohibited in mixed-sex company, or in particular clans, or for particular individuals or for all persons in particular circumstances (say in the forest rather than in the village), or, more pragmatically, in a particular linguistic context. Thus ***yapono***, ***haruku***, and ***pukune*** are all synonyms for the smaller variety of *Canarium indicum* (***iane mkauke***), and ***nawe*** and ***kahiaue*** synonyms for the larger variety (***iane hanate***). Sometimes the synonym can reflect classificatory relationships. For example, the tree-fern ***muku metene*** is also known as ***sonone***. While ***muku(ne)*** is the generic (primary) term used for all tree-ferns, use of ***sonone*** as an optional term reflects a real phylogenetic difference, in that while all other tree-ferns known to Nuaulu are species of the family CYATHACEAE, ***sonone*** is a member of MARATTIACEAE (*Angiopteris evecta*).

Gender terms and sexuality in plant nomenclature

There are few taxa where separate Nuaulu terms for developmental stages seem to be acting as if they were 'natural kinds' (cf. Ellen 1993:48–9), unlike Nuaulu terms for animals – e.g. ***teke tam niane*** (juvenile) > ***teke konomete*** (mature) in the case of the elapid snake *Dendralaphis pictus*). The only plant example I have come across refers to the palm *Caryota rumphiana*, where ***katue*** is used for the younger palm and ***panuke*** for the mature (starch-bearing) palm. What is interesting in this case is that both terms are uninomials. Much more common is the distinction between plant natural kinds using the adjectical sex qualifiers ***pina*** (female) and ***hanaie*** (male), some examples of which are given in Table 2.3. The terms ***hanaie*** and ***pina*** appear in about thirty-three Nuaulu plant binomials. However, it is never quite clear how these are being used until we examine individual cases. On the face of it, we might think that these refer to the male and female plants of the same species, and in some cases sex terms do equate with the male and female types of dioecious species (e.g. ***ahutaune pina*** and ***ahutaune hanaie*** for *Calophyllum inophyllum*, or ***kawasa pina*** and ***kawasa hanaie*** for *Pithecellobium clypearia*). Dioecious species are mainly found in the coastal vegetation systems of the tropics (e.g. Matallana *et al.* 2005) and even here are infrequent, representing, except in special situations, perhaps 14 per cent overall. They are a higher percentage of dominant woody species in these habitats. Of the more obvious dioecious species, none of those reported in the Nuaulu corpus (e.g. *Claoxylon*, *Carica*, the cycads, *Diospyros*, *Gnetum*, *Litsea*, *Nepenthes*, *Podocarpus* and *Urtica*) display folk names that reflect this.

One genus divided in the folk classification using gender terms is *Clerodendrum* (***unuhutu***), a conspicuous plant of the outskirts of villages that is tabooed for the clan Soumori, but for everyone else is considered an effective antiseptic treatment for minor cuts and bruises and head lice. It is also divided into two folk categories (***unuhutu hanaie*** and ***unuhutu pina***), based on the gender contrast. ***Hanaie*** is *C. rumphianum* (a short and robust plant characterized by a tall, dense, spike-like inflorescence and sharply dentate leaf margins), while ***pina*** (also known as ***msinae***, meaning red) is usually *C. speciosissimum* (a tall and less robust plant, with similar leaves to *rumphianum* but with less sharp leaf margin dentition and with a much shorter, more open, inflorescence). Neither of these species of *Clerodendrum* is technically bisexual, rather they avoid self-pollination by staggering the maturity of male and female reproductive parts. Nuaulu also place plants identified as *C. blumeanum* var. *blumeanum* and *C. colebrookianum* in ***unuhutu hanaie***.

The biology of some plants can lead, in some cases, to curious expressions in folk nomenclature. For example, things become complicated where species are optionally dioecious (male and female flowers on different plants) and monoecious (all plants bisexual), and where several types of female plant may be distinguished as contrasting natural kinds. This is so with ***iane*** (the 'kenari', *Canarium*), where we have ***iane hanaie*** distinguished from ***iane hanate*** (larger fruit) and ***iane mkauke*** (smaller fruit), neither term incidentally referring literally to size but rather to concepts that are often used to distinguish larger from smaller, of many natural kinds. In addition, there are folk synonyms of each of these associated with prohibitions on some clans using particular names, a common Nuaulu cultural practice. Moreover, the term ***iane*** can be used at greater levels of inclusiveness to include several species of *Canarium*, rather than simply varieties of *Canarium indicum* (Ellen 2019c). In effect, all other kinds of *Canarium* are either monoecious types or trees that carry both male and female flowers. This phenomenon is discussed further in Chapter 5 as a special case study.

So, as we can see, while male and female terms are rarely used by Nuaulu to specifically indicate sex types of dioecious species, they can be used to describe quite different paired species, usually either varieties of the same species, species of the same genus, genera in the same family, or sometimes genera in different families. Where a genus is locally bitypic, the terms may, for example, distinguish smaller from larger forms (e.g ***kokune pina*** for *Timonius sericeus* and ***kokune hanaie*** for *Callicarpa longifolia*, or ***nesa-nesa pina*** for *Digitoria setigera* and ***nesa-nesa hanaie*** for *Chloris barbata*, two otherwise unrelated weedy grasses).

These apparently confusing usages of the concepts male and female, mirror wider usages of the male-female opposition in Austronesian languages, and were found by Alice Peeters (1979) in Rumphius's *Ambonese Herbal*, reflecting wider Malay World conceptions. Rumphius, as it happens, does use the terms to refer to sex morphotypes of the same species – without recognizing that the male has a role in fertilizing the female flowers. Indeed, the sexuality of plants was being speculated about in Europe when Rumphius was writing, but the issue was quite unsettled (Taiz and Taiz 2017). In Malay conceptions plants described as male ('laki') are those producing fruits that are smaller than female ('perempuan') plants, and whose leaves, flowers and overall size are smaller than in the female form. Thus, Rumphius writes (2011, vol. 5:99) that Ambonese 'bungo-bungo laki-laki' is what we might now call *Lepidagathis rumphii* (an unresolved name in the ACANTHACEAE) and 'bungo bungo parampuan' *Eranthemum pulchellum*, a kind of amaranth.

Revel (1990:134–91) reports a much more comprehensive conceptual division of plants into male and female for Palawan. And while, if pressed on the matter, Nuaulu claim to employ male and female more generally to express pairing and difference between species perceived to be related, and of course between male and female plants of the same species where dioecy prevails, it is more of an implicit organizing device, relevant in certain circumstances and not in others (see also Verheijen 1984:32–4). Although some plants are additionally assigned gender affiliations through the customary management roles of males (e.g. sago) and females (e.g. leafy greens), this is not a hard-and-fast division of labour and is seemingly motivated mainly by practical rather than symbolic considerations, unlike the widespread classification of crops by gender-related

Female term	Scientific taxa	Male term	Scientific taxa
ahutaune pina	*Calophyllum inophyllum*	***ahutaune hanaie***	*Calophyllum inophyllum*
ai numa pina	LAURACEAE sp.	***ai numa hanaie***	LAURACEAE sp.
akunin pina	*Pithecellobium clypearia*	***akunin hanaie***	LAURACEAE: *Actinadaphne*
auhaune pina	*Leea indica*	***auhaune hanaie***	*Leea* sp.
hono-hono pina	*Pollia secundiflora*	***hono-hono hanaie***	*Aneilema* sp.
kamruku pina	*Asystasia australasia*		
kawasa pina	*Pithecellobium clypearia*	***kawasa hanaie***	*Pithecellobium clypearia*
kokune pina	*Timonius sericeus*	***kokune hanaie***	*Callicarpa longifolia*
makakoii pina	*Alpinia* sp.		
nesa-nesa pina	*Digitaria setigera*	***nesa-nesa hanaie***	*Chloris barbata*
pupu pina	*Nephrolepsis biserrata*	***pupu hanaie***	*Sphaerostephanos unitus*
saumahu pinaie	*Chamaecrista* sp.		
tonne pina	*Solanum melongena* var.	***tonne hanaie***	*Solanum torvum*
unuhutu pina	*Clerodendrum rumphianum*	***unuhutu hanaie***	*Clerodendrum speciosissimum*
		alalasie hanaie	poss. *Aquilaria* sp.
		ia hanaie	*Canarium indicum* ♂
		kananuku hanaie	*Hyptis capitata*
		muku hanaie	*Parkia speciosa*
		sipane hanaie	*Robiquetia gracilistipes*

Table 2.3 Use of gender terms in plant binomials.

practices reported for some New Guinea peoples (Sillitoe 1981:1). As we have seen, the pairing of some male and female plant categories is certainly related to features of morphology (e.g. large-small), but there is no symbolic distinction based on consistently selected morphological features, such as the erect-recumbant contrast described by Sillitoe for Wola. Rather we must conclude that in the Nuaulu case the male-female distinction is serving the portmanteau function of expressing relatedness through complementarity and contrast.

The structure and semantics of plant names

The set of distinctions introduced by Berlin (Berlin, Breedlove and Raven 1973, drawing on Conklin 1962), to describe plant nomenclature has stood the test of time. As in my 1993 book, however, I prefer to use the terms innovated by Bulmer (1974a, 1974b), and speak of uninomials, binomials and polynomials. Nuaulu plant uninomials (simple unitary lexemes) include plants as diverse as the fern ***oune*** (*Microsorum commutatum*), the palm ***nione*** (*Cocos nucifera*) and the banana ***uri*** (*Musa* x *paradisiaca* and related species and hybrids), mostly semantically unanalysable primary lexemes, often indicating their status as generic plurals by adding a suffix, /*e*/, /***ne***/, /***te***/ or /***ke***/, as in ***hata***/***ne***/ and ***mono***/***te***/. An example of a unitary complex or analysable primary lexeme would be ***une-une taia***, 'goat shit' (the common path-side weed *Synedrella nodiflora*), which though bi-segmental is clearly not a binomial. Similarly, ***hahu sina*** is not a kind of pig (***hahu***) but rather the nettle *Laportea interrupta*, while ***asu etute*** (literally 'dog's tail') is not a kind of dog but the invasive grass *Pennisetum macrostachyum* (Figure 1.9). By comparison, ***kane-kane wane unie***, 'thorny ***kane-kane*** vine' (the prickly stemmed *Smilax zeylanica*), would be an example of a productive complex lexeme. Juxtaposed uninomials, such as ***hahu-maianane*** (pig-deer) for pigs and deer collectively, seem to be virtually absent from plant nomenclature, though I have encountered ***ana-hotone-manipa*** used descriptively to refer collectively to the grain crops rice, sorghum and millet.

Reduplication is common (Table 2.4), with 46 terms containing reduplicates at all degrees of inclusiveness, most (33) being either uninomials or the first nominal element in a binomial, e.g. ***nori-nori*** (*Macaranga trichocarpa*) *or* ***hono-hono nuae*** (*Andrographis paniculata*), and just 9 appearing as adjectival qualifiers in binomials, such as ***uri abu-abu*** (*Musa* x *paradisiaca* var.). This is five per cent of all plant terms, compared with eight per cent of all animal terms. There seem to be no simple general rules to explain reduplication in plant names. In the case of certain groups of Nuaulu animals (birds and frogs especially) vocalization expressed through onomatopoeia would appear to be a motivating factor (Ellen 1993:50). Neither do reduplicates serve as diminutives of the non-reduplicated form, nor distingush exotics from a standard form, nor accentuate diagnostic differences between closely related forms of a type, as Kocher-Schmid describes for Nokopo (1991:264–6) or, as among the Maenge (Panoff 1969:23), to distinguish inedible types of otherwise edible forms, nor the indigenous type of a form that also contains exotics.

In looking at composite (secondary) lexemes, most are binomials, exemplified by ***hata msinae*** (*Metroxylon sagu*

Nuaulu name	Scientific name	English description
akahon tonu-tonu	unident.	k.o. tree
awane kin-kin	*Melothria indica*	vine with small white flower and triangular-ovate leaves
hono-hono hanaie	*Aneilema* sp.	herbaceous plant with striking white flower
hono-hono hanaie msinae	unident. ORCHIDACEAE	epiphytic orchid with red flower
hono-hono hanaie putie, hono-hono pina	*Pollia secundiflora*	erect perennial herb with white flower
hono-hono nuae	*Andrographis paniculata*	erect annual herb with very bitter taste
hono-hono wesie	*Amischotolype hispida*	a perennial monocot
kane-kane wane (***unie***)	*Smilax zeylanica*	climbing shrub with prickly stem
kasipi sin-sinte	*Manihot esculenta* var.	variegated-leaf cassava
kau-kau	*Pteris*	k.o. fern
kin-kinie	*Durio zebethinus*	durian flower
(***mono***) ***panu-panu***	*Desmodium sequax, Leptaspis urceolata*	term used for a number of common weeds
(***mono***) ***panu-panu totu onate***	*Cyathula prostrata*	k.o. amaranth
(***mono***) ***une-une taie***	*Synedrella nodiflora*	k.o. forb, generally treated as a weed
(***ai***) ***naka-naka***	*Lawsonia inermis*	henna
naka-naka nuae	unident. FABACEAE	shrub or small tree
namu-namu	*Cynometra cauliflora*	small cauliferous tree with distinctive pods
nesa-nesa hanaie	*Chloris barbata*	k.o. grass
nesa-nesa pina	*Digitaria setigera*	k.o. grass
nio weke hatu-hatu	*Cocos nucifera* var.	var. coconut
nori-nori	*Macaranga trichocarpa*	common tree of secondary growth; syn. ***hunori***
(***nusi***) ***peka-peka***	*Citrus maxima*	pomelo
sakalate kau-kau	*Theobroma cacao*	var. cacao tree
sin-sinte	*Codiaeum variegatum*	ornamental with variegated leaves
sina-sina wane	*Pityrogramma calomelanos*	k.o. fern
soka-sokae	*Selaginella, Psilotum, Lycopodiella*	various club-mosses, spike-mosses, and whisk ferns
tom-tom(***e***)	*Flacourtia rukam*	tree with small edible red fruit
tom-tom wesie	*Phyllanthus borneensis*	
tapako sin-sinte	*Nicotiana tabacum*	k.o. tobacco with variegated leaf
topi-topi	unident. RUTACEAE	
uri abu-abu	*Musa* x *paradisiaca*	var. banana
(***uri***) ***moe-moe***	*Musa* x *paradisiaca*	var. banana

Table 2.4 Reduplication in Nuaulu plant names.

var.), ***uri punau*** (*Musa* x *paradisiaca* var.) and ***iane mkauke*** (*Canarium indicum* var.). Most binomials are bi-segmental, as in the preceeding examples, but some are tri-segmental, as in ***kikun*** (***hua***) ***pukune*** (***hua*** = fruit, ***pukune*** = short: *Amomum compactum* var.), where – as in this case – the middle segment may often be optional. An example of a trinomial, or any other term implying three of more degrees of classificatory inclusiveness (that is three or more 'taxonomic levels' in the standard Berlinian model), would be ***hono-hono hanaie msinae*** (unident. ORCHIDACEAE), ***hono-hono hanaie putie*** (*Pollia secundiflora*) or ***nio weke hatu-hatu*** (*Cocos nucifera* var.).

Distinguishing uninomials from binomials is not always easy, as binomials may elide into uninomials over time, as in ***ahau unu*** > ***ahauni*** (*Antidesma stipulare, A. montanum*), ***ana upua*** > ***anaupua*** (the grain-like grass *Pennisetum macrostachyum*), ***kam natine*** > ***kamnatine*** (*Melastoma polyanthum*) and ***tohu pikane*** > ***topikane*** (*Saccharum*). Despite these features and other complicating factors making it difficult to be too precise when calculating numbers, on the basis of the NED, it is possible to identify 446 plant uninomials (44 per cent), 597 binomials (59 per cent) and 32 trinomials (3 per cent). Of these, 63 are optionally uninomial or binomial, meaning that 534 are exclusively binomial and 383 exclusively uninomial. Of the trinomial usages, 10 are optionally binomial and 1 optionally binomial-uninomial. Thus, the proportions are in line with the predictions of Berlin (Berlin 1992:116–17; Berlin, Breedlove and Raven 1974b:40–1) concerning the preponderance of binomials over uninomials and the rarity of trinomials. However, if we compare lexical structure for some selected morphological and functional groupings (Table 2.5), we can see some interesting variations, mainly in terms of the ratio of uninomials to binomials. Some of

	All terms	Uninomial	Binomial	Optional binomial/uninomial	Trinomial
Ferns and their allies (incl. cycads)	30	21	24	5	0
Swidden crops	243	87	211	15	12
Trees	423	225	217	28	7
Cultivated fruit trees	63	33	39	7	0
Vines	67	24	47	3	2
Rattans	9	1	9	0	0
Pandans	13	7	8	2	0
Palms	41	19	35	6	10
Bamboos and their allies	13	9	12	4	1
Herbs	191	84	104	9	5

Table 2.5 The lexical structure of Nuaulu plant terms plotted against some selected morphological and functional groupings. Note that the separate columns for uninomials, binomials and trinomials incorporate terms that are optionally one or the other. The order of plant groupings in column 1 follows the order of treatment in this book.

the differences reflect morphological discreteness. Thus, rattans for Nuaulu have a clearly defined family resemblance, where one uninomial defines the entire group. The high ratio, however, for swidden crops reflects management practices and the encouragement of diversity in particular cultigens. By contrast, many trees are less frequently encountered, and where they are managed, diversity is less easy to produce. This becomes even clearer when we compare cultivated fruit trees with trees as a whole.

Calculating the number of different adjectival qualifiers found in local plant binomials is neither accurate nor straightforward, but for the population of Rouhua in 1996 (410 individuals) I have recorded a minimum of 297 of all kinds, reduced from an initial count of 417 in the NED once all possible lexical variants and doubtful cases have been eliminated. If I had been able to include all Nuaulu settlements, no doubt the number would increase. Of the 600 or so binomials that are reported in the NED, a small number (33, or 11 per cent of all qualifiers) comprise two terms that also exist as independent uninomials, indicating that the second segment is a part of a category labelled by the uninomial specified in the first segment. An example in English would be 'oak' + 'tree', where both terms can exist independently, and where 'oak' is a kind of 'tree'. Nuaulu examples include (***wana***) ***ona*** (the bamboo *Dendrocalamus asper*) and ***tumno pateka*** (where ***tumno*** references cucurbits generally and ***pateka*** specifically the watermelon *Citrullus lanatus*). Mostly, binomials place the less inclusive item or adjectival qualifier as the second segment, as in the preceeding cases, but there are some examples where the qualifier is placed first, as in ***hunane wane*** ('moon + vine', *Merremia peltata*). More complexly, there being no overarching term for grain crops, ***ana manipa*** (rice-maize) will sometimes be used as a synonym for *Zea mays*, and ***hotone manipa*** for the millet *Setaria italica* (a classificatory conundrum discussed further in Chapter 4).

The remaining binomials comprise those with a uninomial plus a 'true' adjectival qualifier. Again, deciding what is 'distinct' can be problematic, as there are sometimes several lexical variants of basically the same qualifier, e.g. ***tun/tunne*** (as in ***kunie tun/tunne***: var. *Curcuma longa*) or ***umena/muena*** (as in ***hatane umena/hata muena***: *Metroxylon sagu* var.), while different lexemes that appear separately may also combine to form more complex qualifiers. Over time such qualifiers may become more separated, habituated and therefore distinct. Qualifers such as ***onate*** ('large') may refer to fruits, flowers, leaves or the whole plant, where ***onate*** and ***totu onate*** ('large leaf') may be optional names for the same plant (e.g. ***sinsin (totu) onate***: *Codiaeum variegatum* var.). In deciding what is a separate term I have tended to be cautiously conservative, but where the semantic evidence is poor, I have treated qualifiers as separate. The numbers provided are for the times a lexical segment appears in the database in a binomial. So, for example, the number for ***onate*** will not include instances where a term is bisegmental, but not a binomial, as in the uninomial ***asu etute*** (dog+tail) given earlier.

Many binomial adjectival qualifiers refer to appearance (colour, size, shape), habitat, use, gender, or animals; but others are semantically more complex and may allude to material culture, origin, social relationships, features of plant morphology and behaviour, use (as in ritual) and other aspects of sensation. Classifying these by type is not easy, as meanings combine different types and can be grouped in alternative ways. However, the frequency of some different kinds of adjectival qualifier in Nuaulu plant polynomials is shown in Table 2.6. Gratifyingly, the groupings I have used are remarkably similar to those adopted by Verheijen (1984:29) for Manggarai plant names, and might therefore plausibly form a basis for wider cross-language comparison. Gender qualifiers have already been discussed (Table 2.3). Qualifiers specifically referring to animals are indicated in Table 2.7, those referring to other plants in Table 2.8 and to toponyms in Table 2.9. Animal names sometimes refer to the preferred food or nesting sites of birds, but names involving other plants work in a more complex way and can be deceptive. Some are true adjectival qualifiers indicating some perceived resemblance or association, such as ***apalam hua papae***, a variety of *Mangifera indica* with a fruit shaped like a papaya. However, some binomials with a plant qualifier are simply synonyms for the same plant also known by a uninomial. For example, ***mukune sonone*** is the optional binomial for the tree-fern *Angiopteris evecta*, also simply known as ***sonone***. In some cases, binomials with a plant qualifier are an allusion to colour, as when ***kunie*** (turmeric) is added to indicate 'yellow'. Also, use of a plant qualifier may be an optional way of indicating a resemblance to a plant that is otherwise contrasted with it. Thus ***hotone***, the millet *Setaria italica*, is also known as ***ana hotone***, placing it with rice, which it resembles as a grain crop. In this case, the usage reflects a quasi-covert category covering grain crops, optionally labelled using the juxtaposed uninomial ***ana-hotone-manipa*** (see above). Toponyms occur as adjectival qualifiers but mainly for cultivars and often indicate the place from which they were first acquired, such as ***kasipii makuana***, a variety of cassava, *Manihot esculenta*, said to have originated from the long-abandoned Nuaulu village of Makuana in north Seram.

The remaining adjectival qualifiers refer to a wide variety of features of – and allusions to – the plants concerned, including in languages other than Nuaulu, references to material culture, and many qualifiers (49, or 16 per cent) with a meaning and etymology unknown, not only to me, but to many of my interlocuters. There is, however, one final binomial form that deserves special mention, as its existence provides a clue to a wider feature of the process of plant classification. In a number of cases there is a term that can best be translated as meaning 'the first, the best, or the true' (***tunne***) kind. Thus we have, for example, the pepper vine ***kam tunne*** (the betel pepper *Piper corylistachyon*), ***kunie tunne*** (the turmeric *Curcuma longa*), ***apalam tunne*** (the mango *Mangifera indica* var.), and (***awane***) ***munu tunne*** (*Derris eliptica*), a vine used as a fish poison. In each case, the term suggests that it is the most cognitively salient or main reference variant of a type of which there are also other less salient variants, thereby pointing to a family resemblance. In a few cases the form suggests also that it is 'best' in terms of the uses to which it is put: the best bamboo for construction purposes, the best turmeric for dyeing and for food, the best reed for arrow shafts and so on. Rosch and others (e.g. Rosch *et al.* 1976) have suggested that similar usages point to 'typicality effects', where one kind is somehow truer of its type than others placed in the same category, thus lending credence to the idea of cognitive prototype with core kinds and peripheral kinds, and the concept of 'type' used in taxonomic botany, and 'category extension' as a means of incorporating new and unfamiliar plants. I shall return to these ideas in Chapters 9 and 10.

Semantic domain (total number of different kinds of qualifier)	Example of Nuaulu qualifier	English gloss	Instances Recorded	Example of full Nuaulu name and scientific gloss
Colour (8)	***metene***	black/dark	16	***tohu tona metene***: *Saccharum officinarum* var.
	putie	white	46	***pahane putie***: *Allium sativum*
	msinae	red	41	***popote msinae***: *Eugenia aquea*
	masikunne	yellow	8	***kasipii masikunne***: *Manihot esculenta* var.
	marae	green-blue	11	***manahute marae***: *Amaranthus tricolor*
	sakalate	chocolate	1	***koitipi sakalate***: *Nepenthes maxima* var.
Gender (2)			34	see Table 2.3
Habitat (11) [see also Table 4.1]	***numa***	house	4	***ai numa***: k.o. forest tree
	niane	village	4	***sune niane***: *Artocarpus altilis* var.
	wesie	forest	11	(***awane***) ***munu wesie***: *Derris trifoliate*
	nuae	sea (near to)	13	***kasitena nuae***: *Ipomoea pes-caprae*
	wae, waene (anoi)	freshwater (in)	8	***manahute waene anoi***: *Cyathula argentica*
	apane	far forest	9	(***awane***) ***munu apane***: *Derris trifoliata*
Dimensions (6)	***onate, maiianae***	large	28	***kupanaha hua maiinae***: *Syzgium aromatica* var.
	ikine	small	11	***tunene hua ikine***: *Durio zibethinus* var.
	pukune	short	10	***kikun*** (***hua***) ***pukune***: *Amomum compactum*
	nawe	long	8	***sinsinte totu nawe***: *Codiaeum variegatum* var.

Semantic domain (total number of different kinds of qualifier)	Example of Nuaulu qualifier	English gloss	Instances Recorded	Example of full Nuaulu name and scientific gloss
Social relationships and ritual (14)	***ama***	father	1	***ai ama***: *Shorea selanica*
	anai	child	2	***neune anae***: *Casuarina* sp.
	hatasia	'nine' grouping of West Seram people, in contrast with Nuaulu, who are ***hatanima*** ('five' grouping)	1	***hueni huae hatasia***: *Dioscorea alata* var.
	inae	mother	7	***kinoke wane*** (***inae***): *Poikilospermum suaveolens*
	pinamo	unmarried menstruating girl	1	***ai pinamo***: *Annona reticulata*
	Numanaita	clan name	1	***uri numanaita***: *Musa* var.
	warata	Dutch/foreign/white person; generally indicating outside/other	6	***karatupa warata***: *Capsicum annuum* var.
Animals			33	see Table 2.7
Toponyms			25	see Table 2.9
Plant morphology and behaviour	***unie***	spiney	3	***awane unie***: *Rubus fraxinifolius*
	kainun, nuni	thorny	11	***hata nuni nasinana***: *Metroxylon sagu* var.
	wane	liana-like	25	***kikane wane***: *Bauhinia semibifida*
	mkauke	fine, grasshopper-like	9	***iane mkauke***: *Canarium indicum* var.
	hanate	west wind: reference to season when mature fruits are harvested	2	***kam hanate***: *Piper caninum*
	hunu (***hue***)	hairy	1	***aka hunuhue***: *Dioscorea pentaphylla*
	ai arine	root-like	1	***hueni ai arine***: *Dioscorea alata* var.
	marinanate	(fruit) pointing to the sky	1	***uri marinanate***: *Musa troglodytarum*
Other plant species or products			29	see Table 2.8
Sensation	***pinate***	bitter	1	***uri pinate***: *Musa acuminata* x *balbisiana* var.
	hau rihue	aromatic smell	1	***sikeue hau rihue***: *Colocasia esculenta*
	notute	fart	1	***awane notute***: *Paederia verticillata*
	hatu, makae	stone, hardness	9	***wana hatu***: *Dendrocalamus* sp.
	nikate	patterned	2	***sikeue nikate***: *Colocasia esculenta* var.
Material culture (27)	***matopae***	k.o. box made from sago leafstalk	1	***sinsin matopae:*** *Codiaeum variegatum var.*
	tona	female carrying basket	3	***tohu tona***: *Saccharum officinarum*
Other (10)	***honue, paru***	new		***kasipii paru***: *Manihot esculenta* var.

Table 2.6 Frequency of different kinds of adjectival qualifier in Nuaulu plant polynomials.

Adjectival qualifier	Animal taxa referred to	Number	Example
asu	dog: *Canis familiaris*	2	***asuhune***: *Dysoxylon cauliflorum*
asuan matahunue	cassowary eyelash: *Casuarius casuarius*	1	***asuan matahunue***: k.o. tree
hahu	pig: *Sus scrofa*	3	***saune hahu***: *Pandanus tectorius*
hehue	***mara hehue***: refers to coloration of certain kinds of cuscus during breeding season	3	***meu hehue***: *Calamus* sp.
inane	pearl (oyster): *Ostrea* sp.	1	***nunu inane***: *Ficus* sp.
kau(ke)	cricket, grasshopper	9	***ai kau***: *Rhodamnia cinerea*
kihoke	k.o. bird: *Lorius domicella*	1	***ai kihoke***: *Breynia cernua*
kinoe	anthribid fungus weevil: *Xenocerus*	1	***koae kinoe***: *Pandanus amboinensis*
kori [***-kori***]	butterfly	1	***nione kori***: *Cocos nucifera* var.
kunini	Moluccan Red Lory: *Eos bornea*	1	***nana kunini***: *Pterocarpus* sp.
kunte	mosquito	1	***meu kunte***: *Daemonorops* sp.
maianane	deer: *Cervus timorensis*	1	***mono maianane***: k.o. leguminous herb
manu	bird	2	***wainite manu aia***: *Phrynium* sp.
marane	marsupial cuscus: *Phalanger* and *Spilocuscus*	1	***utatam marane***: *Gnetum gnemon* var.
mara osu	marsupial cuscus: ♀ *Phalanger orientalis*	1	***mono mara osu***: *Ageratum conyzoides*
matamaine	cockroach: *Periplaneta australasiae*	2	***matamaine kihene***
mau	cat: *Felix domesticus*; ***manapesi mau***: cat's whiskers	1	***manapesi mau***: *Orthosiphon stamineus*
mitane	prawns and shrimps	1	***tohu mitane***: *Saccharum officinarum* var.
mkoue	quail: *Coturnix chinensis*	2	***ai mkoue***: k.o. tree (Araliaceae)
mnaha	common mouse: *Mus musculus*	3	***karatupa mnaha etute***: *Capsicum annuum* var.
naka-naka	sacred name for a python	1	***ai naka-naka***: k.o. tree
nakatua	*Cacatua moluccensis*	2	***sikeue nakatua***: *Colocasia esculenta* var.
nopane	k.o. skink	1	***kanai nopane***: *Areca catechu*
notane	bat	1	***kamruku notane***: probl. *Helicia serrata*
patona	*Python reticulatus*	1	***meute patona***: *Calamus* sp.
porane	ref. to young marsupial cuscus	1	***ai porane***: k.o. tree
puane	cuckoo-dove: *Macropygia amboinensis*	1	***puane*** (***ni***)***musie***: *Actephila excelsa* var. *javanica*
sopite	hornbill: *Rhyticeros plicatus*	1	***kanai sopite***: var. *Areca catechu*
sote	heron: *Dupetor flavicollis*	1	***ai sote***: k.o. tree
suto	crab: *Portunus pelagicus*	1	***ai suto***: *Premna serratifolia*
tekene	snake	3	***sikeue tekene***: *Colocasia esculenta* var.
une	caterpiller	1	***nusi une***: *Citrus reticulata*
une-une taia	goat (*Capra hircus*) faeces	1	***mono une-une taia***: *Synedrella nodiflora*

Table 2.7 Plant names referencing animals.

Adjectival qualifier	Plant referred to	Number	Example
ana	rice: *Oryza sativa*	1	***ana manipa***: *Zea mays*
hotone	millet: *Setaria italica*	1	***ana hotone***: *Setaria italica*
kanapua	cotton: *Gossypium acuminatum*	1	***kasipii kanapua***: *Manihot esculenta* var.
kasupa	*Bixa orellana*	1	***uri kasupa***: *Musa acuminate* var.
kau-kau	*Pteris* sp.	1	***sakalate kau-kau:*** *Theobroma cacao*
kunie	turmeric: *Curcuma*	2	***awane kunie***: *Arcangelisia flava*
loloeno	yam: *Dioscorea pentaphylla*	1	***siahue loloeno***: *Dioscorea esculenta* var.
namte	*Pongomia pinnata*	1	***ai namte***: *Polyathia laterifolia*
nasinana	*Caryota rumphiana*	1	***hata nuni nasinana***: *Metroxylon sagu* var.
nione	*Cocos nucifera*	1	***nione nione***: *Cocos nuciferas* var.
panu-panu	*Cyathula prostrata*	1	***mono panu-panu***: *Desmodium sequax*
papae	papaya: *Carica papaya*	2	***apalam hua papae***: *Mangifera indica*
rapa	k.o. tree, used for bark-cloth	1	***munu rapa***: unident. poisonous plant
sakalate	*Theobroma cacao*	1	***koitipi sakalate***: *Nepenthes maxima*
sinsinte	*Codiaeum variegatum*	2	***tapako sinsinte***: *Nicotiana tabacum* var.
wanate	*Dendrocalamus* sp.	1	***uri wanate***: *Musa* x var.

Table 2.8 Plant binomials referencing other kinds of plant (17 out of 29 kinds reported).

Adjectival qualifier	Place referred to	Number	Example
bali	Bali	1	***jeruk bali***: the pomelo, *Citrus maxima*
geser/pesa	Geser	1	***uri geser***: *Musa* x var.
haruku	Haruku: island in Lease group	1	***ia haruku***: *Canarium sylvestre* syn.
jawa	Java, island of	1	***kopi jawa***: *Coffea arabica* var.
jepang	Japan (country)	1	***jarak jepang***: castor oil plant, *Ricinus communis*
kuako	Amahai (subdistrict administrative centre)	1	***uri kuako***: *Musa acuminata* var.
makuana	Makuana: village in N. Seram	3	***kasipii makuana***: *Manihot esculenta* var.
malaka	Malacca: port city in Malaysia	1	***kam malaka***: k.o. betel vine, *Piper retrofractum*
malole	Malolo: village in Raja Ampat islands	1	***tapako malole***: *Nicotiana tabacum* var.
manipa	Manipa: island west of Seram	1	***kikuni manipa***: k.o. ginger, *Hornstedtia minor*
nuelo, yainuelo	Yainuelo: promontory and village on S. coast	2	***ai nuelo***: flame of the forest, *Delonix regia*
nuetetue	Nuelitetu: village	1	***mono nuetetue***: k.o. weed, *Syndrella nodiflora*
nusa, nusa inae	island, mother island: Seram	6	***sesene nusa inae***: pineapple *Ananas comosus* var.
pelau	Pelauw: village on Haruku island	3	***uri pelau***: *Musa acuminata x balbisiana* var.
poi, poitai	Irian, Papua, New Guinea	2	***uri poitai***: *Musa* x var.
pokoro, bogor	Bogor: city in West Java	1	***sesene pokoro***: pineapple, *Ananas comusus* var.
punara	Bunara	3	(***ai***) ***punara noene***: *Octomeles sumatrana*
saparuane	Saparua: island SW of Seram	1	***sikeue saparuane***: taro, *Colocasia esculenta* var.
sina	China (country)	1	***napu sina***: *Curcubita moschata*
tamno,(***num***)***tamnone***	Tamilouw	2	***uri tamno***: *Musa acuminata x balbisiana* var.
tombelu(***e***)	village in the Moluccas	1	***sikeue tombelue***: *Colocasia esculenta* var.
upa	Upa: river W. of Rouhua	1	***ana upa***: *Cyperus compressus*
warakai	Waraka: village N. of Elpaputih Bay	1	***uri warakai***: *Musa* x var.
yapono	Ambon	1	***ia yapono***: *Canarium sylvestre* var.
yoko	Yoko: river: W. of Rouhua	1	***ai yoko***: unident. tree, probl. PODOCARPACEAE

Table 2.9 Plant binomials referencing toponyms.

Designation, memory and the visual sense

On 11 April 2015 I was with Heunaka as we collected specimens for between three to four hours along the valleys of the rivers Yoko and Upa, just as we had done for the previous three days, and just as we had done in 2003 and in 1996, when Heunaka had been much younger, and a member of the team I employed to conduct forest-plot surveys. Born in 1950, like most men of his age he is functionally illiterate. I had first become aware of him in my field visits of 1973 and 1975, when we participated together in the making of mineral lime (***nosa***) and the straightening of arrow shafts over an open fire. While not particularly knowledgeable or articulate about Nuaulu symbolic knowledge – concerning which he would always defer to clan elders in positions of ascribed power – Heunaka's memory of plant names and where to find the plants they labelled was always impressive. On the occasion in question, I had drawn up a written list of Nuaulu plant names for which I had no reliable identifications, and which we worked through systematically before setting out for the forest. This list, of between twenty to thirty names, he could memorize without difficulty. Later in the day he would remind me that we still needed to collect such and such, without recourse to the written list that I had previously prepared, or without being prompted by me in any other way. In the following days he would voluntarily bring to me specimens that he had found, of plants that we had failed to secure previously and which I could not even remember being on the list.

I am repeatedly amazed at how many terms non-literate Nuaulu can remember at those times when I have shared my written lists (compiled over months and years) during interviews and focus-group sessions. But as literacy has advanced on Seram since the 1980s, young Nuaulu can now take my lists and simply check them for themselves. The 'descriptive force' of names helps in the learning and remembering process (Berlin 1992:255–9; Hunn 2008:79). Children, especially very young children of two and three, when they are beginning to learn lots of names for things and absorbing them like sponges, have very definite opinions on the 'right' names, which are fixed and emphatic. These are generally the first they learn from their most significant others, and they will often be confused, upset or even cross when another name is suggested for the same entity. Tolerance of alternative pronunciation, names and synonyms comes later, when as their social intelligence develops certainty begins to unwind and doubt becomes a more acceptable part of cognition.

Heunaka is by no means exceptional in his plant-identification skills or in his prodigious memory, but we do need to recognize – like field botanists and plant taxonomists the world over – that there is sometimes disagreement on identification. I would say that this is not simply occasional, but routine and integral to the process. But how do we recognize a mistake, and what kind of mistakes do Nuaulu para-taxonomists make? A Nuaulu may consistently identify the same species using the same term, such that in the context of their own working ontology, similarity is being correctly recognized; but the name used may not be the same as used by others. Often, this only becomes evident in situations where knowledge is exchanged. It is, therefore, possible that the same term might be used for two or more quite different species, but that the social interaction in contexts where the species is discussed is insufficient to resolve the nomenclature and agree on a shared identification with a single term. Such variation in the use of nomenclature depending on social and linguistic context is a phenomenon that has been observed before (e.g. Sillitoe 1995:205), and it is also plausibly a mechanism through which ethnobotanical nomenclature and classification change. We have already noticed the opposite phenomenon, synonymy, where the same plant is described using two or more terms, and it too may be a mechanism through which classifications change.

Nuaulu seldom identify plants on the basis of their flowers or other sexual organs, the underlying principle of the Linnaean scheme for plants. It is rarely the case that a Nuaulu has to identify a plant 'de novo' as it were, devoid of social or ecological context, as a scientist might. A particular context will generally be understood, say a swidden or a generic cultigen – e.g. ***sikeue*** (taro) or ***hueni*** (yams of the species *Dioscorea alata*) – and the critical point of the identification will be to distinguish what kind of ***sikeue***, or what kind of ***hueni***. In this situation it is specific distinctive features that matter. However, in general when identifying plants in contexts where the initial generic identification cannot be taken for granted from an assessment on the basis of an overall gestalt, it will be narrowed down and confirmed by taking into account a range of features: leaf, stem/trunk, fruit, smell, taste, texture, and so on. In the case of some plants, an overall visual assessment is not always possible, for example in close forest, where the form of a single tree is occluded by the presence of others. In this situation, a picture will be built up using a combination of accessible features and then a decision made.

Nuaulu classification of plants does not take place in a cognitive vacuum, within some domain of 'thought' insulated from everyday contexts and the multiple stimuli that might motivate people to classify in different ways. Whatever regularities we might discover are constantly being tested against humdrum experience during production, exchange and consumption, and a variety of – often competing – demands to arrange sense data in different ways. Despite attempts to introduce other sensory stimuli (such as the role of vocalization in animal classification, e.g. Berlin 1992:232–59), the assumption of most researchers of classificatory processes is that these are predominantly motivated by the visual sense. The eye and the gaze, claims Stoller (1989:102), following Foucault, 'have had a lockhold' on interpretations of 'Western thought' more generally, and there in no reason to believe that it should be any different for ethnobiological classification. People in literate cultures see images shaped mainly through physical appearance, and listen to words as if they were seeing them, something which has marginalized the other senses, for example sound (Ellen 2019b). This has meant that within the 'oral-aural' mode, the spoken word has been reified at the expense of the role of other senses but vision and linguistic sounds are integral parts of a more holistic bodily experience in which all the senses are interconnected. Hitherto, we have tended to privilege the lexical over the non-lexical, the vocal over the bodily, and perhaps even the bodily over the ambient. Linguists understandably start with words, and discern other kinds of stimuli like ever-widening concentric spheres of context. An individual Nuaulu is not born into a world of separated, taxononomically determined and visually ordered plant types , but into a phenomenal context of profuse and varied plant growth, one might say of 'vegetativeness', regardless of species, that is experienced multi-sensorially. Heunaka, when he surveys the landscape to find a tree that I am seeking, begins by prehending overarching forms, and only then focuses in on discrete characteristics, to match against a name.

3

From words to categories

Ferns and their allies

Introduction

The purpose of this short chapter is twofold. To begin with, I wish to introduce readers to the conventions and templates I adopt in the systematic and descriptive chapters that follow, which focus on different Nuaulu plant groupings. Secondly, by taking one phylogenetic array not generally considered to be historically subject or susceptible to those complicating processes (social, technical and cognitive) of management that encourage domestication, I seek to show how local categories and names are applied and relate to perceptual differences that Nuaulu recognize as important. Prior attention to this relatively unelaborated 'neutral' assemblage of non-seed producing and non-flowering but nevertheless vascular plants prepares the way for understanding the more complex arrangements and multidimensional strategies that Nuaulu adopt when making sense of the rest of their floral knowledge.

The Berlinian scheme and its modifications

The standard model for understanding ethnobiological classification systems, in its much-reproduced diagramatic form, was devised by Brent Berlin (e.g. Berlin 1992; Berlin, Breedlove and Raven 1974b), and is depicted again here in Figure 3.1. The model has been widely adopted in ethnobotany and cognitive anthropology, and justifiably so, because of its convenience and its bold attempt to provide a simple conceptual basis for how ethnobiological classification might be understood cross-culturally. I will return to some of the general issues that it raises in Chapters 10 and 11, but here I describe its components in outline, and how it might need to be adapted in the Nuaulu case.

The six main ranks as redefined by Hunn (1977:44) can be summarized as follows: (1) UNIQUE BEGINNER is a taxon that includes all other taxa of the domain in question. It is usually applied to folk kingdoms such as 'animal', 'plant' or 'fungi'. In some cases, unique beginners may constitute categories such as 'snake' or 'corn', which based on their psychological and/or linguistic properties might otherwise be considered as LIFE FORMS. (2) LIFE FORMS occupy the level immediately subordinate to the unique beginner, including between 20 and 200 generic taxa, for instance 'bird', 'mammal' and 'tree'. (3) INTERMEDIATE level taxa are categories of greater inclusiveness than folk generics but not as inclusive as LIFE FORM categories, and according to Berlin, Breedlove and Raven (1973:73) are rarely named. (4) GENERIC taxa often take the form of primary lexemes (uninomials), and in many respects constitute a psychologically 'basic' level; while (5) SPECIFIC taxa are the subdivisions of generic taxa, often labelled by secondary lexemes (binomials). Finally, (6) VARIETAL taxa are the subdivisions of specific taxa, often labelled by secondary lexemes (binomials or sometimes trinomials), and generally monothetic, i.e. defined with respect to a single or simple cluster of characteristics, such as colour or size. However, there are many exceptions to the way ranks map on to linguistic features, and I consider that they are best defined through evidence of psychological salience.

It is now almost universally accepted that recognition of biological kinds is essential for survival, in terms of the wider evolutionary history of humankind, and therefore we might expect that how they are cognized is also a critical adaptation (Ellen 2017a). However, there are differences of opinion as to how much of this, if any, is 'hardwired' to give a distinctive shape to the domain of 'living things'. The anthropologist who has done most to disseminate Berlin's framework amongst psychologists, cognitive scientists and philosophers of science is Scott Atran. Atran (1990:5) follows Berlin in conceiving folk classification as essentially 'taxonomic' and 'a rigid hierarchy'. However, there are fundamental issues with the concept and word 'taxonomy' which continue to be unresolved, and which Berlin uses in a particular way. As Friedberg (1979:82) has pointed out, during much of its history taxonomy has simply meant 'an arrangement', but in the present day has come to mean either the logical outcome of a systematic process of class inclusion and division, or refers to any scientific classification of biota, and has often been extended to describe any folk classification of living things (i.e. a 'folk taxonomy'). With Atran (1998) taxonomy becomes a defining feature, specified cognitively, for the domain of living things, with almost memic status.

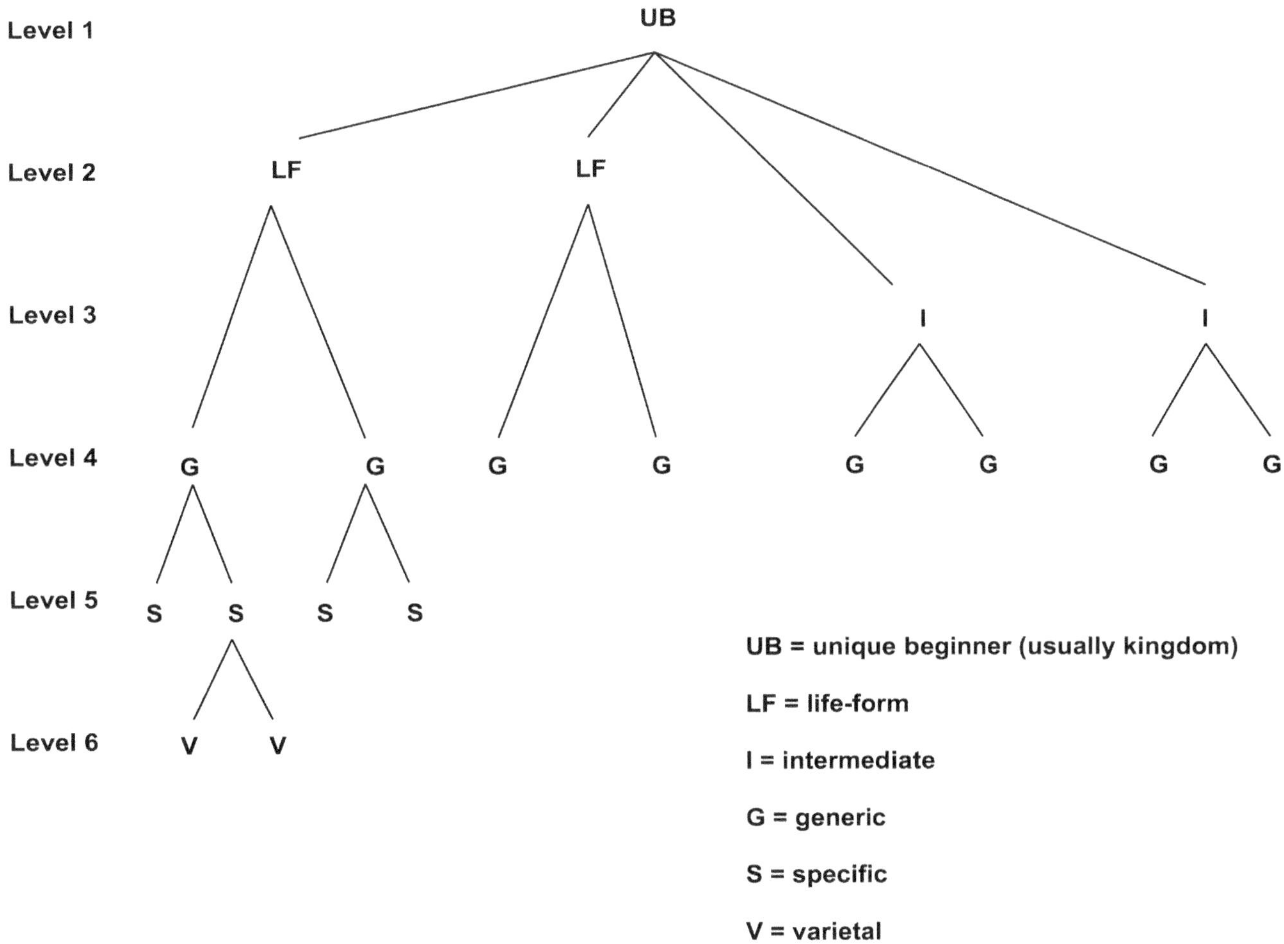

Figure 3.1 Schematic representation of Berlin's six taxonomic ranks as an ideal classificatory system (after Berlin 1992; Berlin, Breedlove and Raven 1974b).

Berlin has somewhat qualified his position on the model in his 1992 book, and Atran too has modified parts of his argument since 1990, particularly in the light of ethnographic work amongst the Itza Maya. I have articulated my objections to aspects of the Berlin and Atran schemes at length in my analysis of Nuaulu animal classification (principally, Ellen 1993), and I do not need to repeat all of these here, or provide references to the supporting literature, accept where I modify a previously held position, or where repetition might be critical to the argument. Although I would not now be as harsh in my judgement of the pioneering formulation of Berlin, especially in empowering several generations of descriptive and theoretical work in ethnobiology, or indeed of the cognitive inferences drawn by Atran in his synthesis, as I was in papers leading up to 1993, my basic methodological and theoretical asssessment remains pretty much as it was then. My earlier relevant ethnography had been predominantly on Nuaulu animal classification, and what has emerged from more recent work on plant classification is evidence for significant differences between the ways plant and animal classification work. Standard theory on 'ethobiological' classificaction, mostly derived through Berlin and Atran, has tended to play down the differences between animals and plants as domains, and emphasize the similarities. Apart from the fact that plants are largely sessile, they do not have obvious breathing organs (breath being often critical to folk definitions of – animal – life). Moreover, sexuality is frequently ambiguous or apparently absent (as in ferns). But most important is utility, which is much more ingrained in how plants are perceived and experienced than it is for most animals. This has consequences for classification. The Nuaulu, for example, do not traditionally domesticate animals other than dogs and occasionally chickens, but they do spend a lot of time manipulating the reproduction of plants. I will be returning to issues of domain specificity in plant classification in Chapters 10 and 11.

For Atran (1990:12, 17), the underlying universals of folk classification are: (a) the concept of species and 'essence'; (b) sequential naming; (c) the regularity of morphological characters; (d) life-form groupings; and

(e) domain-specific cognitive universals. 'Living kinds', he insists, 'are everywhere ranked into transitively structured taxonomies, with no other natural-object domain so structured' (ibid.:47), while folk classification promotes a strong basis for inductive reasoning about biological properties (ibid.:212). He goes beyond Berlin in the claims he makes for domain specificity, and in his identification of the basic units of folk classification, arguing for example, that the species-genus distinction is largely irrelevant to the 'common-sense' view of the world on which he rests his case. He also prefers terminological constructions such as 'generic-specieme' (Atran 1990:6, 27–8; 2007:38), thus combining Berlin's preference for the psychological priority of the GENERIC level with Bulmer's (Bulmer and Tyler 1968) neologism 'specieme' for the basic unit. In this respect I agree with his critique of Berlin (see, e.g. Ellen 1993:67–71), but see no reason to use any term for such categories other than 'basic', since such constructions and indeed any term derived from the established 'levels' of scientific taxonomy, perpetuate the confusion of the folk and scientific schemes, and subtly bias analysis in favour of a view that folk schemes are inevitably taxonomic in the established biological sense. But although Atran claims that the underlying framework of ethnobiological classification is taxonomic, he is critical of anthropological uses of the concept, especially 'where the faulty notion of taxonomy is fused with prototype theory', thus compounding the error (Atran 1990:52). His main point here is that the universals apply to the taxonomic categories not to the taxa themselves (ibid.:56), taxa and categories comprising different logical types (Atran, Medin and Ross 2004:397).

There is one further feature of Atran's polemic that I need to mention at this point, and which has some implications for the conclusions drawn in this book. This is that his basic template for a folk taxonomy of living kinds is a form of 'pre-theoretical' common sense, an intuitive model found everywhere and which in turn is the model on which modern scientific taxonomy is based; in other words, it is a claim that science itself is a kind of 'common sense'. While it is wholly consistent with the position that I argue here that modern science finds its origins in cognate thought patterns and practices observable in the everyday knowledge of pre-literate peoples, Atran's 'common sense', as Hull (1992:55) observes, is not really common sense at all, but rather a logically refined philosophical notion. Moreover, in claiming that there is an intrinsic connection between folk and scientific taxonomies, the question is raised as to what the template might be through which the two could be compared, thus risking some circularity of argument (Lloyd 2007:45; see also, Atran 2007).

So, Berlin's claim that the fundamental building block of all ethnobiological classification is the concept of basic category has been vindicated, but the question remains as to how you would empirically determine them? Berlin's position was that the basic category is usually the folk generic one, but in practice it varies between higher phylogenetic taxa depending on visibility of distinctive features, cultural importance and ecological salience. The variation in whether basic categories are conceptualized at phylum, order, family, genus or species level is much greater in folk classifications of animals (Ellen 1993) than plants, though in Nuaulu classification of plants, basic categories for many cryptogams are designated at the phylum or class level, and even cut across higher level taxa. In vascular plants the correspondence between basic category and GENERIC, in Berlin's sense, is much better.

The empirical method I have selected to determine basic plant categories, in an approximate fashion, is to begin by identifying those that are 'terminal' (see Berlin, Breedlove and Raven 1974b:36), meaning that they are not generally recognized as being further subdivided (n = 870 in the NED). Most terminal categories are labelled, most frequently with a binomial (n = 585), but in a third of instances with a uninomial (n = 321). However, with newly recognized taxa, naming might not be widely shared or evident, and these categories are less likely to enter any of the calculations. Terminal categories among vascular plants can be phylogenetic genera (the fern ***sipoi***: *Lygodium scandens, L. circannatum*), species (the tree cassava ***kasipii sombar***: *Manihot* cf. *glaziovii*) or varieties (bitter cassava ***kasipii paru***: *Manihot esculenta* var.). I have also sought to identify 'primary' categories, that is those that Nuaulu would usually volunteer in freelisting and interviews without further probing, and which were not generally regarded as being a part of any other category other than a life-form or functional category, or rarely an intermediate category (n = 510). Primary categories are almost always labelled with uninomials (n = 402). Where they are labelled with binomials (n = 139), the term is prefixed with ***ai***, that is the life-form term for tree (e.g. what is called in local AM the small 'buas-buas' tree, ***ai suto***: *Premna serratifolia*), or involve a prefix such as ***mono***, indicating grass/weed (e.g. ***mono manahutune***: *Amaranthus viridis*), or are affixed with ***(a)wane***, meaning vine or liana (e.g. ***awane kin-kin***: *Melothria indica*), or where a recent loan-word has been introduced from another language (e.g. AM 'kacang panjang': *Vigna unguiculata*), or where terms are synonyms or lexical variants of a uninomial. In 128 of such cases, the primary category was also terminal. Overall, there were 380 categories that were both primary and terminal, and therefore could only be 'basic'. Atran (1990:53) fundamentally objects to the concept of 'terminal category', as it 'indiscriminately mixes monogeneric life-forms with generic-speciemes', but this is only so if you insist on clinging to the notion of fixed taxonomic ranks and levels. The advantage of 'terminal category' is that the smallest (least inclusive) named classificatory unit can be clearly identified across different languages, whereas there can be

Nuaulu name	Scientific name	Scientific family
soka-sokae marae	*Psilotum nudum*	PSILTOTACEAE
soka-sokae mnatine	*Selaginella* sp.	SELAGINELLACEAE
soka-sokae msinae	*Selaginella* (*caudata* and/or *plana*)	SELAGINELLACEAE
soka-sokae numte (syn. ***mono numte***)	*Lycopodiella cernua,* *Tapeinidium amboynense*	LYCOPODIACAE, DENNSTAEDTIACEAE
soka-sokae onate	*Selaginella willdenowii*	SELAGINELLACEAE

Table 3.1 ***Soka-sokae:*** *Nuaulu named categories for club-mosses (Lycopsida), spike-mosses (Selaginellales), whisk ferns (Psilopsida) and related forms.*

endless arguments about whether what is a folk generic in one language is a folk specific in another.

A related problem for comparative studies is that often it is not the total aggregate number of terms that is significant, but rather the number of terms referring to a particular rank or level, for example the number of life forms or, especially, the basic, folk generic or primary category level. This figure has often been taken as a crucial measure of folk knowledge of plants, and its cross-language stability interpreted as an indication of the optimal knowledge capacity for plants. Berlin (1992:96–101) has estimated that this is generally between 500 and 600. In the Nuaulu case, the figure is around 515, but this is also problematic, as determining what is and is not a basic term is not always self-evident. For example, binomials formed from ***ai*** (tree) and mostly equivalent to scientific species, as in ***ai anawasi*** (*Anisophyllea distichia*) are GENERICS (basic categories) not SPECIFICS in Berlin's sense, and where binomials are used the analyst has to make a judgment as to how to treat them. Similarly, it is not easy to see why cultivars of cassava referred to using binomials should be considered SPECIFICS rather than VARIETALS, and so the distinction seems to fall apart. Indeed, we will return to the special problems of how Nuaulu conceptualize diversity within the same cultivated domesticate towards the end of Chapter 4. In his 1992 book Berlin seems to be acknowledging some of these difficulties. Thus, while we might recognize the practicality of his conceptual apparatus, we must also examine its deficiences, which I will return to in Chapter 10, in relation to theorizing the comparative study of ethnobotanical classification more generally.

Lycophytes

We established in Chapter 2 that pteridophytes and lycophytes are unambiguously included by Nuaulu in the covert category PLANT. In addition, Nuaulu informally group all ferns together with the lycophytes that they encounter. Although all groups of both clades have a widespread altitudinal and geographic distribution, and are often found along waysides, in gardens and cultivated areas, they are particularly abundant on the forest floor and in the margins, and so are best dealt with in this context. Pteridophytes and lycophytes on Seram occur in all major plant communities, and with upwards of 700 species (Kato 1990; Parris 1993). There are 43 Nuaulu names to cover this diversity, which is not surprising, and reflects their relatively low level of use and the fact that most species are typical of habitats (particularly at a higher altitude) with which Nuaulu are less familiar. Club-mosses, spike-mosses and whisk ferns are all placed in a single named category (***soka-sokae***), reflecting closely their phylogenetic relationship (Table 3.1). The strong lexical signature here is not surprising given the morphological similarities and the fact that they are mainly found in the shade of the forest floor, but note that the group includes one fern: *Tapeinidium amboynense.* Interestingly, this is the one term in Table 3.1 indicating a synonym (***mono numte***), flagging the frequency with which this fern is found as a weed in the light open spaces of old swiddens. There are no Nuaulu collections of EQUISETACEAE (horsetails) though Parris (1993:81) reports *Equisetum debile* for Seram.

Ferns

The names that Nuaulu use to describe ferns are listed in Table 3.2. They encounter ferns in three ways: as potential food and medicine, as a source of manufacturing material, and as a characteristic part of the forest floor and wayside flora which, while not regularly a source of human food, is known as food for other animals. Ferns are also seen as a particular feature of the vegetation that colonizes garden sites and early fallow, and through one salient group, the tree-ferns, that serve as topographic and ecological markers.

Ferns can be represented as a semi-covert classificatory complex that will sometimes be referred to as a whole by using the AM terms 'paku' or 'paku-paku'. In terms of names collected, they appear to comprise three main named groupings formed as binomials from ***pupu***, ***mukune*** and ***naene***, plus about twelve basic categories that are optionally without prefixes, giving thirty terminal categories overall. ***Pupu*** is the term that comes closest to incorporating all ferns, distinguishing them by their physical form, but it is not spontaneously used for all types. ***Ahane*** are epiphytic ferns growing on many trees, but most obviously on coconut palms, subdivided into ***aha onie*** and ***aha totu onate***. ***Ahane*** are sometimes described as ***sinsinte*** (a term

Figure 3.2 ***Ai mukune*** *(Cyathea sp.) growing in open stream-side regrowth along upper Yoko valley, NW Rouhua, south Seram; March.*

Figure 3.3 ***Muku sonone*** *(Angiopteris sp.): close-up of a young unfurling leaf crozier, Sune Ukune valley, E. Rouhua, south Seram; February.*

that usually applies to *Codiaeum variegatum*) on account of their delicate decorative appearance, but ferns do not appear to be used for decorative or ritual purposes as such. The homonymous term ***ahane*** is, as we have seen in Chapter 2, also applied to certain lichens (Figure 2.2). ***Oune***, though not prefixed with ***aha***, is widely regarded as a kind of ***ahane***, on account of its close physical resemblance: they are in the same genus (*Microsorum*). The term ***mukune*** labels tree-ferns, their woodiness being recognized in the optional term ***ai mukune*** (tree-fern). Three types are recognized: ***muku putie***, ***muku metene*** and ***muku msinae***. ***Muku putie*** is the type category, referring to various species of *Cyathea* (Figure 3.2). Tree-ferns occur throughout the forest, but especially along streams at lower altitudes, as a feature of other riparian habitats, and in hill forest, for example on the overland path between Bunara and Somau, to the immediate south of the settlement of Nuanea. The tree-ferns of the mossy forests and those occurring in groves above the tree-line in the central mountain range are rarely visited by Nuaulu. They sometimes occur in patches, often with trunks of two metres or more in height. ***Muku metene*** is the most common type, referencing *Cyathea* (probably *amboinensis*), but the term may be extended to other frequently encountered species in the genus (see e.g. Parris 1993:82) found at low elevations and in swamp forest. However, *Angiopteris evecta* (Figure 3.3), while not in the CYATHECEAE, is a huge long-lived evergreen fern with a swollen base and stout stems, and is recognized by some Nuaulu through its unique system of explosively dispersing spores. Its distinctiveness from real tree-ferns is signalled in the use of the term ***sonone***, and its resemblance by inclusion in ***muku metene***. Similarly, ***muku msinae*** (*Pleocnemia*) is not a true tree-fern, but a large terrestrial genus that has prominent basal pinnules and can grow to 2.5 metres. ***Naene*** refers to edible ferns, which are of three types: ***naene tinse***, ***naene putie***, and ***naene totu onate***. However, ferns other than those named ***naene*** may also be eaten (e.g. ***utatone***). Finally, we should note that the large woody cycad *Cycas circinalis revolute* (***bunen***), which can reach a height of 6–7 metres, while not yielding any obvious nomenclatural clues, is seen by some Nuaulu as sharing its gross morphology with tree-ferns in the ***mukune*** group, and so is best mentioned here.

Of those ferns flagged as edible by using the name ***naene***, ***naene tinse*** (*Stenochlaena palustris*) is amongst the most frequently harvested for food, but amounts are

Nuaulu name	Scientific name	Scientific family
ahane	*Microsorum* (syn. *Phymatodes*) *scolopendria*	Polypodiaceae
aha onie	Unident.	Polypodiaceae
aha totu onate	Unident.	Polypodiaceae
oune, ahane	*Microsorum commutatum*	Polypodiaceae
kinekane	*Phymatodes membranifolium*	Polypodiaceae
isane	*Lygodium flexuosum*	Lygodiaceae
kaie	Unident. epiphytic fern	
kau-kau	*Pteris* sp.	Pteridaceae
kopane	*Pteris* sp.	Pteridaceae
mukune	*Pteris semipinnata*	Pteridaceae
(***ai***) ***mukune***[1]	*Pleocnemia leuzeana*, *Cyathea* sp.	Dryopteridaceae, Cyathaceae
muku metene, sonone	*Cyathea* sp. poss. incl. *C. amboinensis, C. caudex* and *C. moluccana; Angiopteris evecta*	Cyathaceae, Marattiaceae
muku msinae	*Pleocnemia leuzeona*	Dryopterdaceae
muku putie[1]	*Cyathea* sp.	Cyathaceae
naene, nae(***ne***) ***tinse***	*Stenochlaena palustris*	Blechnaceae
naene (***totu***) ***onate***	*Acrostichum aureum, Helminthostachys zeylanica*	Ophioglossaceae
naene putie	*Drynaria querifolia, Tectaria repanda*	Polypodiaceae, Dryopteridaceae
nite unue	*Asplenium nidus* and other species of *Asplenium*	Aspleniaceae Clade eupolypoda -II
one[2]	*Pneumatopteris auctipinna, Diplazium* sp. probl. *esculeutum*	Thelypteridaceae, Woodsiaceae
pupu	*Sphaerostephanos unitus* var *mucronatus*	Thelypteridaceae
pupu hanaie	*Sphaerostephanos unitus* var. *mucronatus, Chingia ferox*	Thelypteridaceae, Thelypteridaceae
pupu pina, (***pupu***) ***kaia***	*Nephrolepis biserrata, Nephrolepis hirsutala*	Oleadraceae
pupu onate	*Pteris semipinnata*	Polypodiaceae
sapana	*Pteris vittata*	Pteridaceae
sapuane	*Colyris macrophylla*	Polypodiaceae
senene	Unidentified epiphytic fern	
sina-sina wane	*Pityrogramma calomelanos*	Pteridaceae
sipoi[3]	*Lygodium circinnatum*	Lygodiaceae
utatone	*Helminthostachys zeylanica*	Ophioglossaceae

Table 3.2 Nuaulu named categories for ferns. Notes: 1) tree-ferns; 2) branching ferns; 3) climbing ferns.

Figure 3.4 **Nite unue** *(Asplenium nidas):(a, left) growing on coconut palms in coastal plantation land near Rouhua, south Seram; the bamboo spears are a result of attempts to kill phalangers, which commonly use the fern as a nesting place, May; (b, right) growing on a* **tahenae** *tree (cempedak: Artocarpus integer), Sune Ukune valley, E. of Rouhua, South Seram, February.*

always limited. In a four-month survey of two households in 1970–1, only 25 grams of fern by harvested weight (all ***naene tinse***) was logged, 0.05 per cent of all cooked weight of vegetable matter (Ellen 1978a:73). Note, however, that ***naene tinse*** is ***monne*** (prohibited) for one clan, Sounaue-ainakahata. Of the various ferns harvested as food and used as a green vegetable other than ***naene***, the delicate fronds of *Microsorum scolopendria* and *M. commutatum* (***oune***, ***ahane***) are amongst the most favoured, but these are also ***monne*** for clan Neipani-tomoien, as is ***one*** (*Diplazium*). With ferns generally, it is the young tender leaves burgeoning in curled-up spirals (the croziers or fiddle-heads: Figure 3.3) that are most preferred.

Of those ferns used medicinally, ***aha totu onate*** (an unidentified polypod with a broad but thin leaf) is used in treating bad backs. ***Pupu kaia***, which includes both *Nephrolepis biserrata* and *Nephrolepis hirsutala*, are light-demanding ferns and weeds of near-garden land applied topically for wounds, and also used in childbirth. ***Pupu hanaie*** (*Sphaerostephanos unitus* var *mucronatus)* is used as a poultice, the fine hairs being scraped from the base of young stipes and placed on wounds. The leaf is also used during childbirth. *Angiopteris* is used as a decoction to wash newborn children.

Of the ferns used for binding, the climbers are the most prominent: ***sipoi*** (AM 'paku jari merah': *Lygodium circinnatum, L. scandens*), and ***isane*** (*Lygodium flexuosum*), a climbing fern of the forest margins. ***Sipoi*** is used for axes and arrows, and also for school hats. ***Isane*** is used for arrows and bush-knife handles. In addition, ***kopane*** (*Pteris* sp.: AM 'karancun') may sometimes be used. *Microsorum* (syn. *Phymatodes) scolopendria* (***ahane***) grows in clumps of 5–6 fronds, for example on fallen coconut palms, and is seen as a kind of ***sinsinte***, in reference to its ornamental qualities. The wood of the large thorny tree-fern ***muku putie*** (*Cyathea* sp.) is used for uprights in housebuilding, and is considered stronger than ***nahane*** (*Xanthostemon verus*), one of the more important timber trees. It is also used in scare charms (***wate***), while the exudate found inside the stem is used by children for making sticky rubber balls.

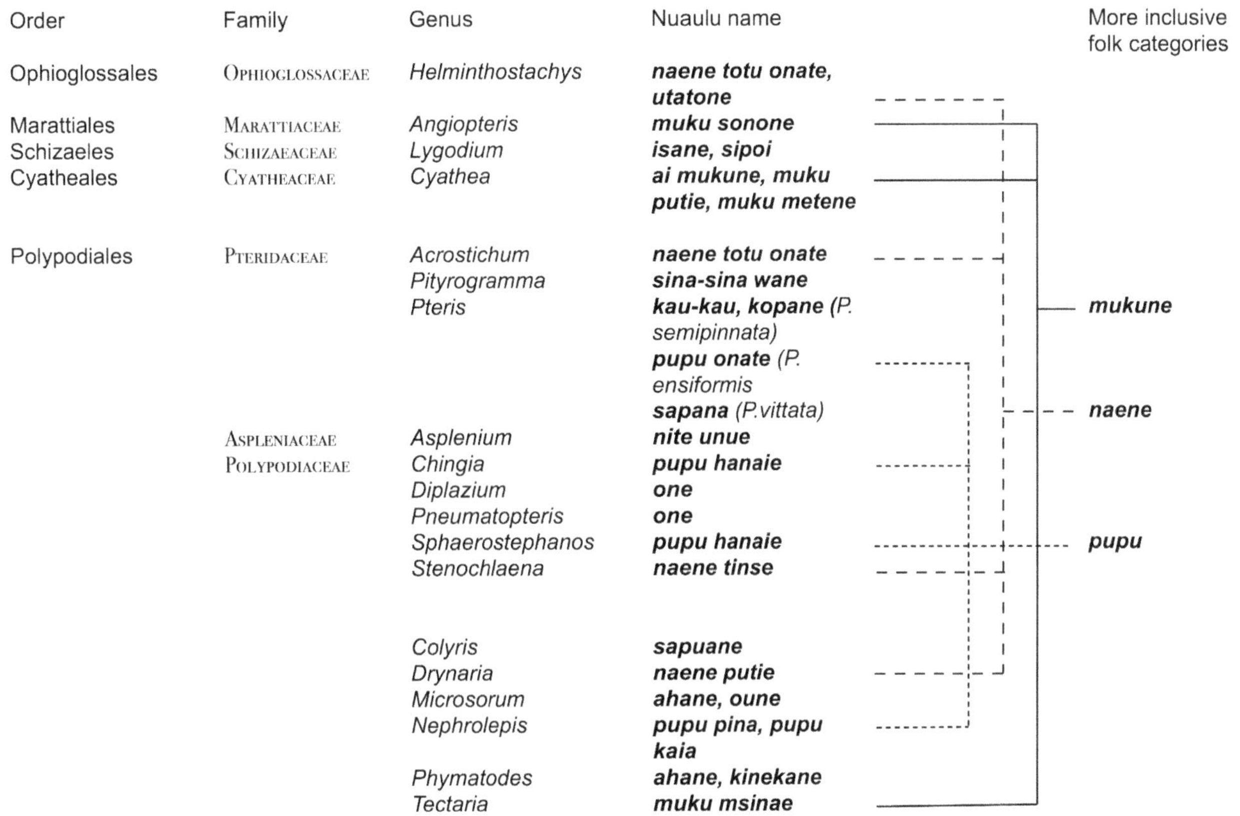

Figure 3.5 Overall pattern of correspondence between phylogenetic and folk grouping for ferns (class Polypodiopsida).

Other ferns are known for their ecological significance. Thus, ***sapuane*** (*Colyris macrophylla*) is commonly seen growing on moss-covered rocks near streams, and is a sign of clear fresh potable water. ***Sina-sina wane*** (*Pityrogramma calomelanos*) is a non-arborescent fern, with a slender purple stem, often eaten by cockatoos (*Cacatua moluccensis*) and purple-naped lories (*Lorius domicella*), but also occasionally used for weaving mats. The epiphytic ***nite unue*** (*Asplenium nidus* and other species of *Asplenium*), the bird's-nest fern, is a feature of more mature closed habitats admitting less light, and is common on coconut palms, where it serves as a nest for *Phalangers*, the nocturnal marsupial cuscus (Figure 3.4a). It is considered harmful to the trees it grows on, as can be seen in Figure 3.4b, where the fern is overwhelming its host, but is encouraged because of its role in cuscus hunting. Among that group of ferns closely associated with swiddens, garden edges and early fallow, and often thought of as weeds, are ***sapana*** (*Pteris* sp. nr. *vittata*), ***oune/ahune*** (*Microsorum commutatum*), a weed of near garden land, ***kinekane*** (*Phymatodes membranifolium*), ***kau-kau*** (*Pteris* sp.) and ***pupu onate*** (*Pteris ensiformis*).

Discussion

In English folk botanical classification, ferns are generally reckoned to be a fairly distinctive group, though other pteridophytes are not so clearly defined as a group, either separately or in relation to ferns. Rumphius, in his *Ambonese Herbal* (2011, vol. 5:120–62), places all ferns together in Chapters 46–58 of Book 10. From his exposition, it is clear also that most ferns in late seventeenth-century Ambon were grouped together using the AM terms 'paku' and 'hahuru', the only exceptions being species of *Adiantum*, *Ophioglossum*, the bird's-nest fern genus *Asplenium*, and some climbing ferns. Other pteridophytes are grouped by Rumphius with mosses and lichens, following on his treatment of ferns (ibid.:163–74). His treatment recognizes their affinities as non-floral seed producers, but is not suggestive of any particular pattern otherwise in terms of Ambonese folk classification.

The overall pattern of correspondence between Nuaulu folk groupings and the phylogenetic scheme for ferns is shown in Figure 3.5. When displayed in this way we can see that the terms ***mukune***, ***pupu*** and ***naene*** link species that are distributed across quite a wide phylogenetic range, a linkage that appears to be motivated by ***mukune*** referring to ferns with woody stems, ***naene*** to edible ferns, and ***pupu*** to some ferns with medical applications, even though such uses are not exclusive to ferns so described. Other pteridophytes do not appear to be attributed with fern-like affinities and seem to work as 'unaffililiated' groups in Berlin's sense. Thus, even with ferns – plants that might be presumed to yield a fairly straightforward folk classification – there is no easy correspondence with phylogenetic categories. Folk groupings cannot be readily inferred from binomials, and the relationship between functional and morphological categories cannot be inferred on any straightforward basis, use being a frequent factor motivating linkage.

4
Villages, swiddens and fallow

Domesticates and the spectrum of plant-management strategies

This chapter deals with Nuaulu knowledge of the plants that are characteristic of those parts of their environment that are obviously and actively managed: growing in the village area, adjacent the village in what we might describe as home gardens, in new and in old swiddens. The chapter also deals with some of the plants found in successional habitats resulting from swiddening, managed fallow and groves, each of which tend to merge over time into lightly managed and unmanaged forest, discussed in Chapter 5. Because it focuses on swiddens, the chapter also examines the category ***monote***, variously 'weed' or 'grass'. And because the chapter deals with grass, we extend it to other grassland habitats also.

Most of the plants used by Nuaulu from these biotopes are domesticates, in that they represent populations of plants that have been modified genetically in ways that improve their usefulness, or species of normally unmanaged plants that are introduced into managed settings in ways that fall short of domestication as it is normally understood. In thinking about the domesticate and analytically complementary concepts, I follow the path-breaking work of David Harris (e.g. 1966, 1973, 1977, 1989, 1996). In doing so, it needs to be noted that in large parts of the tropics, certainly within island South-East Asia and the Pacific, cultivation and wild harvesting are better envisaged as both combined and intertwined, and as also involving other strategies that an over-simplistic dualism disguises. Yen (1989), for example, advocates a 'domestication of environment' model in which the widest range of plant-food-procurement strategies must be considered together for particular locations. It is now no longer tenable to assume that changes taking place in forests after the arrival of the first human populations can automatically be assumed to be the outcome of 'natural' processes (e.g. Flenley 1979). The distinctions between 'wild' and 'domesticated', and between 'wild' and 'cultivated', as applied to some plants, is now seen as a problem, as are schemes of evolutionary stages in which domestication is inevitably linked to the final stage of the emergence of field agriculture (Harris 1996:441). A more nuanced understanding of what we mean by 'cultivation', 'domestication' and 'management' has led to recognition that plant-food production might take place without domestication in the conventional sense (Harris 1989; Rindos 1984), and there has been some lively discussion relating to the idea of the tropical forest as a productive environment, and the extent to which it can support human populations through foraging alone (Headland and Bailey 1991), and concerning the modelling of the subsistence transition from forest foraging to gardening in order to explain the origins of agriculture in the region (Groube 1989:301). Berlin, Breedlove and Raven (1974b:96) in their study of Tenejapa Tzeltal plant classification distinguish cultivated and protected plants and useful wild plants, as a means of testing hypotheses about the relationship between, for example, species polytypy, polytypy in local-language plant names and cultural significance. There is much evidence for a strong correspondence between species polytypy and folk nomenclatural polytypy provided in this chapter.

There is no simple dichotomy drawn by Nuaulu between cultivated and non-cultivated plants, of the kind suggested by others as intrinsic to certain folk classifications (e.g. Balée 1994:figure 7.1, for the Amazonian Ka'apor), and in this respect the situation is similar to that described by Sillitoe (1980:135) for the Wola of the New Guinea highlands. However, in translating Nuaulu ***atuhu*** by Indonesisan 'tanaman' there remains some residual semantic ambiguity between the senses of 'growing' and 'planting', though for Nuaulu all plants can be moved around and manipulated. Place is more important than any abstract notion of functionality. Figure 4.1 provides a simple distribution graphic of all terminal categories grouped etically according to whether they are: (a) domesticates, (b) managed to varying degrees without clearly being domesticates, or (c) 'wild', meaning that they might be said to be extracted without such interventions having a significant impact on their biology. While we can define quite tightly what is meant by a domesticate, the other definitions are problematic and much looser. 'Wild', in particular, comes with all kinds of cultural associations that can be misleading in strictly ecological terms. In addition, we need to distinguish between a plant that appears in, or is introduced to, a garden or swidden, and managed in quite intricate ways, from a true domesticate which has become genetically differentiated over the long term. We can do this by calling such plants 'cultivates' (rather than 'cultivars' or 'cultigens', which have well-established and separate meanings). Obviously, domesticates and cultivates overlap to a considerable extent, but non-domesticates are often introduced into swiddens and gardens, and domesticates become non-cultivates by being absorbed into long-term fallow, and eventually reproducing in unmanaged contexts.

Nuaulu too have difficulty with such distinctions. In fact, they try to avoid them, and there is no specific folk lexicon that reflects these distinctions. They certainly do not have the categories of domesticated plant or of wild plant. They have a concept of planting (***atanane***, (*i*)***rahu***, to plant), of things planted (***rahue***) and of growing plants (***atuhu***); and concepts relating to how plants come to flower, or how leaves develop from a bud (***ruhu***, ***erehui***, ***eretai***), of attending to all things relating to the cultivation cycle (***amrahu***), of a person who tends to growing things (***mamrahue***); and a range of terms to describe the different ways in which plants are cared for, both in cultivated plots and in areas that are not obviously otherwise cultivated. For example, ***ahunata*** means to clear undergrowth, equally in gardens, along pathways or in the forest, there being no term for 'weeding' as such. Nuaulu do, however, distinguish far or deep forest (***apane***) from long fallow (***ahue***), which does embed a conceptual difference of a kind, as ***ahue*** is the final stage of a sequence by which swiddens (***nisi***) become increasingly 'forest-like', while ***apane*** involves no assumption of prior management. We can see from Figure 4.1 that the number of plants managed without being obviously domesticates is relatively small, but it does conceal a great deal of variation in terms of degree of the management impact. Thus, yams of the same species can be actively and deliberately planted in swiddens from seed tubers, but also grow outside gardens, unplanted, usually along pathways, where they are tended by cutting back surrounding vegetation and inserting protective markers. Pandans rarely appear within gardens, but are routinely managed through selective cutting back, as are bushes of different kinds of *Codiaeum variegatum* (***sinsinte***) used ritually and as boundary markers in cultivated areas, and various kinds of fruit tree (e.g. water apples or 'jambu'). Gingers are rarely grown in gardens (except at higher altitudes) because there is too much sunlight and not enough moisture, and are preferably harvested from areas along trails through groves and between gardens. Many fruit trees are planted in old gardens as part of a managed succession but are not harvestable until the land has moved into long fallow, or effectively restored as secondary forest. Thus, the cultivation of plants is one of gradation, both in

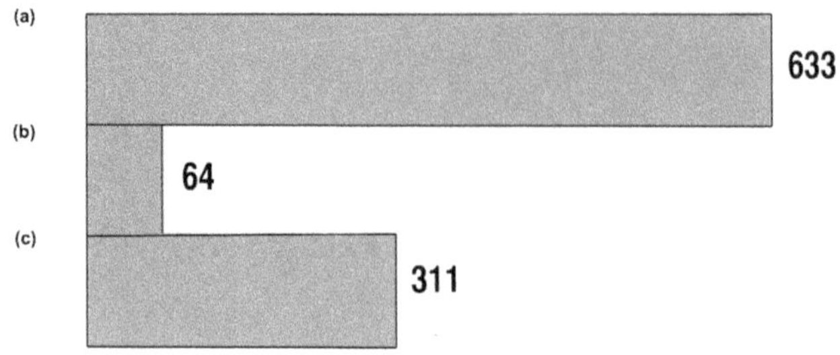

Figure 4.1 Nuaulu labels for terminal plant categories grouped by degree of human management: (a) domesticates, (b) semi-managed, and (c) un-managed.

terms of its intensity, and also spatially, depending on the species, location and life stage. Nuaulu do not fence their swiddens (which distinguishes them from other coastal peoples living in contiguous areas), and so garden edges and margins are often difficult to locate; but these latter nevertheless host a range of harvestable plants that are given a degree of protection. These include medicinal plants and herbs used in cooking.

While the concepts of domesticate and domestication, cultivate and cultivation, are underpinned by necessarily strong analytical distinctions in science, Nuaulu make no clear separation between either domesticate and non-domesticate, or cultivate and non-cultivate. All plants can be managed, moved around, and selected, through deliberate positive actions, or more passively and inadvertently. The only distinctions that come close are in terms of where plants are found. There are adjectival qualifiers for a few plants that imply some kind of wild/domesticated distinction listed in Table 4.1. Of these, some are obviously contrasting pairs (e.g. ***kam niane*** and ***kam wesie***), some are the marked term of a default name for the more common type (e.g. ***amnae/amnae wesie***), and some are binomials for taxa where there are many other contrasting types (e.g. ***kanai apane***, or ***awane wasi***).

The swidden as a site for the acquisition of plant knowledge

Since 1970 I have collected inventories for different Nuaulu swiddens, in various locations, at different stages in their development, and belonging to different individuals. Some of these data have already been summarized (Ellen 1978a). We can take one, not particularly typical, example. Napuae's first year swidden in 2003, when I examined it in August of that year, had been planted from about the previous December, at the end of the wet season. Located high on the north-western slope of the Upa valley at Isonawe, the young swidden contained ***sina manusia*** (*Urticastrum decumanum*) and ***soi mutune*** (*Zingiber papuanum*), a nettle and ginger respectively, both used medicinally; ***kam tunne*** (*Piper betle*) for betel chewing (trained up trees

Nuaulu name	Scientific name	Meaning of adjectival qualifier
ahutau wesie	*Calophyllum* sp.	forest
amnae wesie	*Ficus* sp.	forest
awane apane	k.o. vine	deep forest
awane wasi	*Pericampylus glaucus*	unit of owned land
(*awane*) *munu apane*	*Derris trifoliata* var.	deep forest
(*awane*) *munu wesie*	*Derris trifoliata* var.	forest
hatane apane	*Metroxylon sagu* var	deep forest
hono-hono wesie	*Anischotolype hispida*	forest
kaea wesie	*Terminalia catappa* var.	forest
kam apane	*Piper forstenii*	deep forest
kam niane	*Piper betle* var.	village
kanai apane	*Areca* probl. *glandiformis*	deep forest
kasauni apane	*Casearia glabra*	deep forest
kunie apane	*Curcuma aurantiaca*	deep forest
kupanahane wesie	probl. *Decaspermum fruticosum*	forest
naha niane	*Arenga pinnata* var.	village
niune apane	*Miscanthus* sp.	deep forest
popote wesie	*Pternandra cordata*	forest
sune niane	*Artocarpus altilis*	village
tom-tom wesie	*Phyllanthus borneensis*	forest
tonne wesie	*Solanum torvum*	forest
utatam numa	*Gnetum gnemon* var.	house
utatam wesie	*Gnetum gnemon* var.	forest
wana apane	k.o. bamboo	deep forest

Table 4.1 The distinction between cultivated and wild reflected in plant names.

left standing after clearance), young clove and coconut seedlings, papaya, tomato, chilli, taro and cassava. The main components of any swidden in the first two years of its life are roots and tubers. There is no overarching named folk category for these, neither are they only found in swiddens, but this is their primary association, and so in the next section I discuss taro, yams, sweet potato and cassava. In Ambonese Malay, all starchy tubers and roots are labelled 'ubi' and the way Nuaulu translate their terms is part of the evidence that these roots and tubers are part of a wider covert category. Moreover, over the last 100 years new plant types have been introduced in contexts where Ambonese Malay is the medium of communication. The significance of the history of starchy tubers and roots is discussed further in Chapter 9.

Starchy roots and tubers

Starchy roots and tubers are by far the most important group of cultivated plants grown in a swidden, and the first to be planted after clearing and burning. First amongst them, both in cultural salience and planting order, are taros and yams. These were certainly the most important tuberous plants grown in upland Seramese swiddens until the arrival of Europeans, and still have ritual priority. It is these species that are the focus of first-planting and first-harvest rituals. They are dealt with separately here.

Sikeui (*Colocasia esculenta*; taro)

Colocasia esculenta (syn. *C. antiquorom*, in Ellen 1978a) is nowadays regarded as a polymorphic species containing many phenotypic varieties. At least ten folk varieties are recognized by Nuaulu, distinguished mainly in terms of colour, size and texture of corm (Table 4.1). Most of these are grown in swiddens, though with some in home gardens adjacent to houses and the village area. However, it is worth noting that *Colocasia esculenta* grows and reproduces without human management outside of swiddens, along paths and stream sides in the forest, where they are described as ***pun*** or ***punetene*** (AM 'keladi hutan'). These forms have itchy stems and are not regarded as edible. As Matthews (2014:115) reports, wild *Colocasia esculenta* is rarely mentioned in ethnographic reports. A second wild taro, ***pina totue*** (lit. 'woman's ear'), is not supported by a voucher specimen in my database, but may be a species of *Alocasia*, with a harvest season in March–April.

Nuaulu name	Meaning	Parenchyma colour	Notes
sikeue hatusoke	kind of stone	red	large corm
sikeue hau rihue	aromatic smell	red	
sikeue kunie	(like) turmeric	yellow	
sikeue masikunne	yellow	deep yellow	
sikeue msinae	red	red	fine hairs on stem cause itching
sikeue nakatua	cockatoo, *Cacatua moluccensis*; reference to colour	white	similar to ***s. tekene*** in terms of corm size and quality
sikeue nikate	patterned	variegated leaf	
sikeue saparuane	Saparua, island name; referring to alleged place of origin		
sikeue tekene	snake	red	
sikeue tombelue		reddish	

Table 4.2 Folk varieties of ***sikeue*** *(Colocasia esculenta).*

Yams (*Dioscorea*)

Here I use the term 'yam' to refer to all kinds of *Dioscorea*. Of those known on Seram, and to the Nuaulu, the most familiar are *D. alata* and *D. esculenta*, but also *D. hispida* and *D. pentaphylla*. There is no single term for *Dioscorea*. Instead three species are labelled using three distinctive uninomials: ***hueni*** (*D. alata*), ***siahue*** (*D. esculenta*) and ***loloeno*** (*D. pentaphylla*). ***Hueni*** is divided into at least ten folk varieties (Table 4.3) and ***siahue*** into at least eight (Table 4.4), while ***loloeno*** is not further subdivided. Two further uninomials (***kawasine*** and ***akae***) are subdivided into two named categories each, in which their partial covert similarities are reflected in the overlapping application of unmarked ***akae*** and marked ***kawasine putie*** (lit. white ***kawasine***) to the same species: *Dioscorea hispida* (Table 4.5). The relationship between different Nuaulu categories for yam and their scientific identification is shown in Figure 4.2.

At this point it is instructive to see what Rumphius has to say about yams, given the apparent complexity of Nuaulu classification. Rumphius describes *Dioscorea* in Chapters 7–16 of Book 9 (*Ambonese Herbal* 2011, vol, 4:335–67). He groups them into 'ubi' and 'kombili', following then current Ambonese Malay practice. 'Ubi' includes different varieties of *D. alata* and *D. bulbifera*, and species that Rumphius regards as 'half-wild' (*D. pentaphylla*) and 'truly wild' (*D. hispida*), often described as 'ubi utan'. By contrast, 'kombili' is reserved for *D. esculenta*. 'Kombili' or 'kumbili' is current in modern Ambonese Malay and used by Nuaulu to describe *D. esculenta* (***siahue***), *D. pentaphylla* (***akae***) and, sometimes, *D. alata* (***hueni***). Rumphius notes that in seventeenth-century local Ambonese languages, though not in Ambonese Malay, *D. esculenta* was also called 'siahu' or similar, in other words a cognate of modern Nuaulu ***siahue***.

New World starchy roots

Nuaulu do not separate their root crops in terms of Old World or New World origin, but in considering their practices it makes sense to make that distinction. I have already described how taros and yams alone are the focus of certain planting and harvesting rituals, and the importance of both is reflected in the diversity of folk varieties. Other (introduced) roots and tubers are regarded as being culturally secondary, even when they have become equal to or, in the case of cassava, more important in terms of food intake (Table 4.6).

Tania (***yohoru***: *Xanthosoma sagittifolium*) are a prominent feature of swiddens because of their size, but are planted less by Nuaulu than by coastal groups such as Sepanese, and then generally in flat coastal soil with a higher moisture content. Sweet potato (***kasitena***: *Ipomea batatas*) is planted soon after yams and taro, but is usually harvested earlier. Cassava (***kasipii***: *Manihot esculenta*) is by far the most important root in this group, reflected in both number of folk varieties recognized, the space devoting to growing it and its dietary role. However, compared with many other Indonesian rural populations, the significance of all three cultigens is suppressed, partly by the dietary preference for sago as the main source of starch. Nuaulu cassava-naming practices and management have been discussed at length in a series of papers (Ellen and Soselisa 2012a, 2012b; Ellen, Soselisa and Wulandari 2012). In terms of folk classification, it is significant to note here two additional plants that are classified as ***kasitena*** and ***kasipii***, respectively, though neither have starchy roots: ***kasitena nuae*** (*Ipomoea pes-caprae*) and ***kasipii sombar*** (*Manihot* cf. *glaziovii*). In each case the genus is identical, and leaf and stem habit and flower override the absence of a starchy root in terms of classificatory logic.

Nuaulu name	Meaning	Characteristics and uses
hueni ai arine	buttress roots of tree	large irregular flat tubers
hueni (huae) hatasia	***hatasia*** = Patasiwa	'nine group' fruit, reference to symbolic territorial grouping of this name; probl. referring to the lobes of the tuber
hueni marane	cuscus (Phalangeridae)	these marsupials are important symbolically, and as a source of meat
hueni msinae	red	purple parenchyma
hueni nakatua	cockatoo	very large tuber
hueni nante	sky	very large hairy tuber with red/blackish parenchyma
hueni putie	white	white parenchyma
hueni sune minete	reference to young breadfruit leaf	as ***sune totue***, but with larger tuber
hueni sune totue	breadfruit leaf	reference to appearance of periderm
hueni tumata	human, person	syn. ***hueni maneia***; epidermis is said to resemble human skin in texture and appearance, smooth with no hairs ('naked'); two kinds recognized, white and red

*Table 4.3 Folk varieties of **hueni** (Dioscorea alata).*

Nuaulu name	Meaning	Characteristics and notes
siahue kawei	hole in a tree	
siahue loloeno	referencing similarity to *Dioscorea pentaphylla*	large single tuber
siahue masikunne	yellow	reference to colour of parenchyma
siahue msinae	red	red periderm
siahue nusa ina	'mother island', Seram	white parenchyma, many large 'offspring' tubers
siahue putie	white	white periderm
siahue sapatu	AM 'shoe'	describing shape of tuber
siahue siaie	figurine	reference to shape of tuber

*Table 4.4 Folk varieties of **siahue** (Dioscorea esculenta).*

Nuaulu name	Meaning	Scientific name	Characteristics and notes
akae		*Dioscorea pentaphylla* var.	forest yam, sometimes cultivated, but must be harvested in wet season as weather damages crop; AM 'kumbili, ubi pasir, ubi maraia'
aka hunuhue, hunu hunue	hairy	*Dioscorea pentaphylla* var.	tubers hairy with pale pink parenchyma
kawasine masikunne	yellow	*Dioscorea hispida* var.	yellow tuber with large leaf; used in sorcery; AM 'ubi racong'
kawasine putie	white	*Dioscorea hispida* var.	white parenchyma with large leaf; AM 'ubi racong', k.o. forest yam
loloeno		*Dioscorea pentaphylla*	small hairless tubers with white parenchyma; AM ' betah, akar umbi'

Table 4.5 Other folk varieties of Dioscorea.

Nuaulu name	Scientific term	Scientific family	N folk varieties
kasipii	*Manihot esculenta*	EUPHORBIACEAE	17
kasitena	*Ipomoea batatas*	CONVOLVULACEAE	4
kentang	*Solanum tuberosum*	SOLANACEAE	not grown
yohoru	*Xanthosoma sagittifolium*	ARACEAE	2

Table 4.6 Sweet potato, tania and cassava.

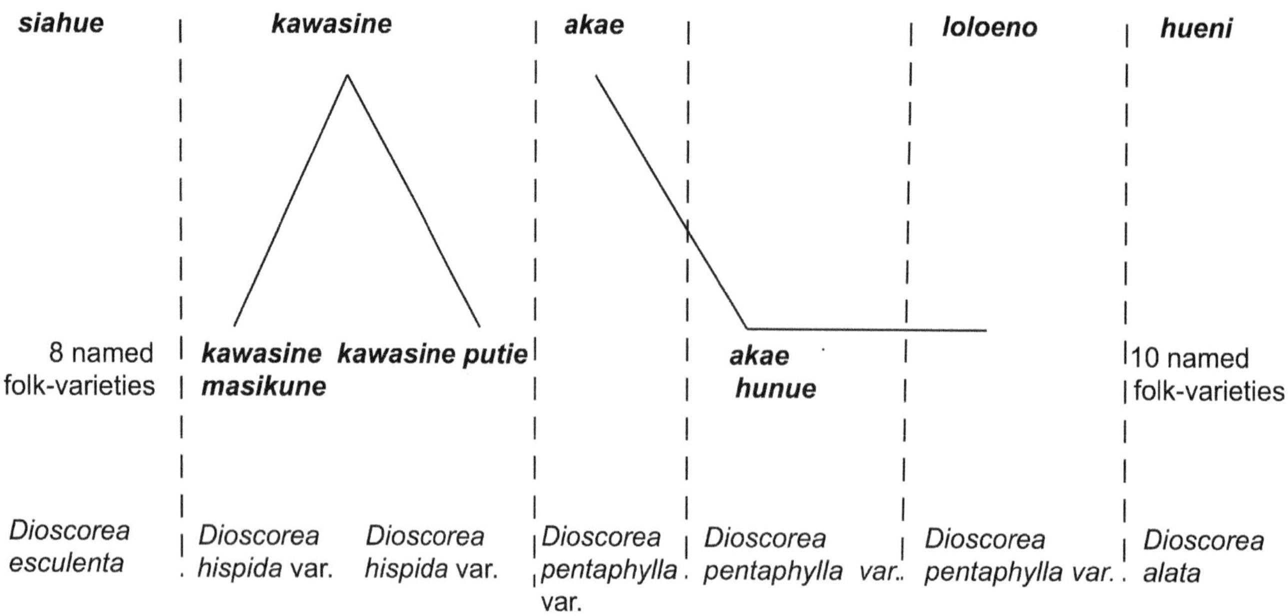

Figure 4.2 Starchy tubers as a folk grouping: endemic yams.

Bananas and plantains

After starchy roots and tubers, the next most important group of plants grown in swiddens are bananas and plantains (***uri***), which are also starchy, and also the most common food plant growing in village areas. Bananas not only provide fruit as food, but the ***uri usue*** (the 'heart' or male flower at the tip of the peduncle, the fruit-bearing stalk: see Figure 2.4b) is also edible, though seldom eaten nowadays. The leaves and stalks are widely used in food preparation, as containers, wrappers, platters and floor coverings when feasting. There are about thirty-six folk varieties, the names of which are listed in Table 4.7 (also Figure 4.3), along with identifications and descriptive characteristics where available. One of these is ***uri poi***, *Musa acuminata* subsp. *banksii*, the only record of this taxon west of New Guinea. Another is ***uri marinanate***, *Musa* x *troglodytarum* L. (Callimusa section), sometimes described as *Musa fehi*, and known as 'pisang tonka langit' in Ambonese Malay. The rest are hybrids, either genomic configurations within *M. acuminata*, or hybrids of *M. acuminata* and *M. balbisiana*, now described as *M.* x *paradisiaca*. Most are described using binomials of ***uri***, but some are known (usually optionally) by a uninomial (e.g. ***hanoate***, ***kainonone***, ***moane***), though ***moe-moe*** never receives a prefix in ordinary speech. This lexical practice, found with some other fruits such as *Citrus* (e.g. ***moni***, ***koromase***), may have something to do with their physical distinctiveness, but is as likely to relate to contextual factors in conversational speech. Despite the large number of cultivars, there are no obvious local groupings of types of banana between the primary and terminal level. However, ***uri pelau*** comes in two forms (***uri pelau hua metene*** and ***uri pelau hua putie***), while the popular fast-growing diploid AA hybrids seem to be all variants of what in Ambonese Malay is described as 'pisang empatpuluh hari' (forty-day banana): ***moe-moe***, ***uri inana hutu ete***, ***uri hutu ate*** and ***uri empatpuluh hari***.

Nuaulu name	Phylogenetic taxa	Description	Meaning of term and notes
(uri) hanoate	*Musa* x var.		
(uri) moane	*Musa acuminata* var. (AAA, Gros Michel)	very large eating var, l. 19–21 cm, circum 13 cm	
moe-moe	*Musa acuminata* (AA, Sucrier subgroup) var		AM 'pisang empatpuluh hari'
uri abu-abu	*Musa* x *paradisiaca* var. (ABB)	thick with red blotches, can be baked or boiled	
uri empatpuluh hari	*Musa acuminata* var. (diploid AA)	may be cut after 40 days	AM 'pisang empatpuluh hari'; seed introduced via ABRI
uri hunan itane	*Musa* x var		***hunan itane***: 'full moon'
uri hunananane	*Musa acuminata* (AA, Sucrier subgroup) var.		***hunahane***: gold; AM 'pisang mas'
uri hunanitane	*Musa* x var.		***hunan***: moon, ***itane***: shine
uri hutuate	*Musa acuminata* var. (diploid AA)		***hutu***: nest; AM 'pisang empatpuluh hari'
uri ibreda	*Musa* x var.		***ibreda***: hybrid; obtained from agricultural research station at Makariki, S. Seram
uri kainonone	*Musa* x var.		AM 'pisang suangi', allusion to sorcery
uri kapananoi	*Musa* x var.		
uri kasupa	*Musa* x *paradisiaca* var. (AAB) or *M. acuminata* var. (AAA)	red ocherous, shape resembles aubergine	reference to *Bixa orellana*, source of a red dye
uri kokune	*Musa* x var	small and fat, like a chubby ***uri warata***; l. 7–8 cm	reference to *Callicarpa longifolia* and *C. candicans*: small tree that is the source of a fish poison
uri kuako	*Musa acuminata* (AA, Sucrier subgroup)		Nuaulu toponym for Amahai
uri lewaka	*Musa* x *paradisiaca* (ABB)		'Lewataka' is a Bandanese clan, now found in east Seram; AM 'pisang Geser'
uri makuana	*Musa* x *paradisiaca* var. (triploid ABB)	fruit large and plump: l. 20 cm, circum 18 cm; white	Makuana (toponym): village in N. Seram; syn. ***uri hua puti(e)***; ***hua putie***: white fruit; AM 'pisang putih'
uri marinanante	*Musa* x *troglodytarum* (TT)	large; red with black blotches	***nante***: sky, ***arina***: 'look above', meaning the clumps of these bananas grow up as opposed to growing down; AM 'pisang tonka langit'
uri nuetetue	*Musa* x var.		***nuatetue***: toponym (Nuelitetue), name of village
uri numanaita	*Musa acuminata* var. (Diploid AA)	l 14 cm	***Numanaita***: name of clan; AM 'pisang ter(a)tau masa'
uri onana hutuete	*Musa acuminata* var. (diploid AA)	orange flesh usually remains on fruit; fruits approx. 40 days after planting, thin skin; l. 11 cm	***hutu***: nest; AM 'pisang empat puluh hari'
uri pau	*Musa* x *paradisiaca* var. (tetraploid ABBB)	plump squarish fruit, pale yellow, l. 11–13 cm	***pau***: sago cooked in wok

*Table 4.7 Nuaulu banana (**uri**) folk varieties, ordered according to best guess of phylogenetic relationship. Where there is sufficient evidence to be confident about an identification this is provided, otherwise cultivars are listed as Musa x var. Sources: Edison et al. 1996; Hapsari et al. 2017; Hermanto et al. 2013; Kennedy 2008; Valmayor et al. 2000. Continued overleaf.*

Nuaulu name	Phylogenetic taxa	Description	Meaning of term and notes
uri pelau	*Musa* x *paradisiaca* (triploid AAB, Silk subgroup) var.	cream fleshy, thick-skinned plantain for cooking, l. 23–4 cm, circum 9 cm	
uri pelau hua metene	*Musa* x *paradisiaca* (triploid AAB, Silk subgroup) var.	black when ripe, smaller than ***pelau hua putie***	***hua metene***: black fruit
uri pelau hua putie	*Musa* x *paradisiaca* var. (triploid AAB, Silk subgroup) var.	white flesh	
uri pesa, geser	*Musa* x var.	small, orange flesh, l. 10 cm; similar to ***uri tamno***	Geser (toponym): small island off east Seram
uri pinate	*Musa acuminata* var. (AA, Sucrier subgroup)	med. size var., red-potted surface	***pinate***: bitter
uri poi	*Musa acuminata* subsp. *banksii*, or *M. lolodensis*	inedible	AM 'pisang hutan'; water extracted from stem; leaf used to wrap stillborn or aborted embryo
uri poitai	*Musa* x var.		***poitai***: Papua, Irian, New Guinea
uri punau	*Musa* x var.		***monne*** (totemic) for Sounaue-ainakahata and Matoke-pina; believed that all ***uri punau*** derived from clan Matoke, reference to myth where 'child' of a Matoke male and Sounaue female was an ***uri punau*** planted by Sounaue woman
uri suaie	*Musa* x var.		
uri susu	*Musa* x *paradisiaca* var. (AAB, Silk subgroup); or non-Cavendish triploid AAA		***susu***: breast, milk; AM, 'pisang susu'
uri tamno	*Musa* x *paradisiaca* var. (AAB) Silk subgroup		***Tamno***: Tamilouw, toponym; boiled bananas (***uri mati***) are usually young ***uri tamno***
uri tonamune	*Musa* x var.		
uri tope	*Musa acuminata* var. (diploid AA)	large fruit, when ripe gold and brown	***tope***: sago (*Metroxylon sagu*) leafstalk
uri wanate	*Musa* x var.		***wanate***: k.o. bamboo; AM 'pisang bambu'
uri warakai	*Musa* x var.	medium size	Waraka (toponym): village on Elpaputih Bay
uri warata	*Musa* x var.	Small fruit, l. 9–10 cm; circum 8–9 cm	***warata***: Dutch; syn. ***uri ikine***, 'small'

*Table 4.7 continued Nuaulu banana (**uri**) folk varieties, ordered according to best guess of phylogenetic relationship. Where there is sufficient evidence to be confident about an identification this is provided, otherwise cultivars are listed as Musa x var. Sources: Edison et al. 1996; Hapsari et al. 2017; Hermanto et al. 2013; Kennedy 2008; Valmayor et al. 2000.*

Figure 4.3a Folk varieties of banana: ***uri hanoate***, Moni, south Seram, April 2015.

Figure 4.3b Folk varieties of banana: ***uri kasupa***, between river Upa and Rouhua, south Seram, March 1996.

Grains

Grains are not recognized as a separate unambiguously labelled category by Nuaulu, though there is evidence that they are treated as such covertly (Table 4.8). Of all grains grown in swiddens, maize (*Zea mays*) is of the most significance. Depending on the term used, it is either labelled as an independent cultigen (***yakoni*** < Ind. 'jagung') or one with specific affinities to other grains (***ana manipa***) (Figure 4.4). *Coix lacrama-jobi* may occasionally be present, but the seeds are used only for decorative purposes. I have never seen rice growing in Nuaulu swiddens, although it is of some symbolic importance, and purchased rice has been of increasing dietary significance throughout my periods in the field. Similarly, both millet (*Setaria italica*) and sorghum (*Sorghum bicolor*) are present in the area, though I have only seen them in Sepa gardens since 1970, and even here they are marginal crops. Rice, millet and maize are grouped together through alternative naming practices (***ana***: rice, ***ana hotone***: millet, ***ana manipa***: maize), suggesting ***ana*** (rice) as a type-reference category (cf. Taylor 1990:50).

We can reasonably assume that the Seram samples of *Setaria italica*, like those reported for North Halmahera (Sakamoto, Kimata and Kawase 1980), are in the southern group, and contrast with strains found in Taiwan northwards, for example in being non-glutinous. The work of these same researchers suggests that the variation of *Sorghulm bicolor* in north Halmahera is more complex, with both glutinous and non-glutinous strains, and with both recent imports and more ancient strains. *Coix lacryma-jobi* may also occur in glutinous form, as on Halmahera. It occurs on the margins of Nuaulu swiddens, in secondary regrowth and along path-sides.

Leafy vegetables

Leafy vegetables (Table 4.9) do not form a distinct category, but the leaves and fleshy stems of many herbaceous plants, ferns, shrubs and trees are harvested for food. Nuaulu would describe all of these in terms of their usefulness using AM 'sayur', which can approximately be translated by Nuaulu ***utane***. However, 'sayur' also includes various fruits

Nuaulu name	Scientific name	Notes
ana	*Oryza sativa*	purchased rice from non-local sources
ana msinae	*Oryza sativa*	red landraces
ana nusa inae	*Oryza sativa*	rice landraces grown on Seram
(ana) hotone	*Setaria italica*	all local types of millet
(ana) manipa, yakoni	*Zea mays*	maize: three named folk varieties
kaihoke	*Coix lachryma-jobi*	Job's tears
salea	*Sorghum bicolor*	sorghum

Table 4.8 Nuaulu folk terminology for grains.

Nuaulu name	Scientific name	Notes
ai polo	*Antiaris toxicaria*	not planted
cintar manis	*Gynura procumbens*	not planted
kamamuri	*Costos speciosus*	
karatu	*Sauropus androgynous*	
kelor	*Moringa oleifera*	occasionally planted
kokohahau	*Dianella javanica*	
kol	*Brassica oleracea*	known, rarely planted
maka sainane	*Jacaranda filicifolia*	not planted
manahu marae	*Amaranthus tricolor, A. blitum*	planted in swiddens
manahu msinae	Amarathus tricolor, Celosia argentea	planted in swiddens
manahu putie	*Amaranthus spinosus, Cyathula argentica*	planted in swiddens
manahu wan anoi	*Celosia argentea, Polygonum*	planted in swiddens
mono manahute	*Amaranthus viridis*	planted in swiddens
mono senane	*Portulaca oleracea*	not planted; leaves eaten
saumahu	*Abroma mollis, A. fastuosa*	planted in swiddens
sesawi	*Brassica juncea* and other BRASSICACEAE	known, rarely planted
utatam	*Gnetum gnemon*	fruit and leaves eaten; several varieties recognized
uta tone	*Helmintostachys zeylanica*	non-cultivated fern of long fallow and swidden margins

Table 4.9 Nuaulu folk terminology for leafy vegetables.

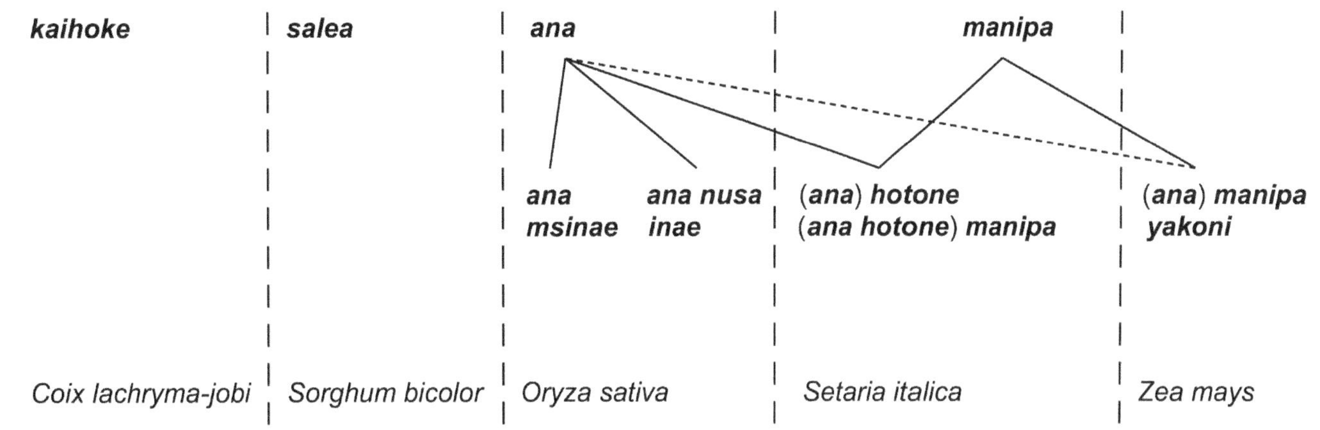

Figure 4.4 Nuaulu folk classification of grains.

cooked as vegetables. Most 'sayur' is harvested from planted crops, but some comes from non-cultivated plants, often species self-set in swiddens of all phases, or growing along swidden margins. Many leafy vegetables are harvested from crops primarily grown for other reasons, the most obvious example being cassava, the leaves of which are a ubiquitous feature of most meal times; but sweet potato leaves are also consumed. The young 'fiddle-heads' of ferns are also significant, especially those of ***naene tinse*** (*Stenoschlaena palustris*) and ***oune*** (*Microsorum commutatum*), both of which are common on swidden margins and as early regrowth (see Chapter 3).

Of those plants grown only as leafy vegetables, amaranths (***manahute***) are the most important, and raise some interesting classificatory issues. ***Manahute*** refers to edible plants of the family AMARANTHACEAE, but may also sometimes include *Stachytarpheta indica*, while ***manahu wan anoi*** refers to aquatic forms of *Polygonum*, but may also include *Celosia argentea*. Only ***mono manahute*** refers to a single species, *Amaranthus viridus*, an edible but virulent weed of older first-year swiddens. Other ***manahute*** are grouped according to colour: white (***putie***): *Amaranthus spinosus, Cyathula argentica*; red (***msinae***): *Celosia argentea, A. tricolor*; and blue-green (***marae***): *A. tricolor, A. blitum*. This suggests strong interference of culinary-use factors when grouping some phylogenetically closely related genera.

Pulses

Pulses (Table 4.10) are best treated as a covert category approximately equivalent to AM 'kacang'. The term ***kihue*** covers a number of species widely grown in island South-East Asia, but excludes folk varieties of ground-nut (***kaisan***), which are accorded a separate primary term, and ***kacang panjang***, which as its Indonesian name suggests, is a relatively recent introduction. ***Kacang panjang*** is now widely grown for sale, for example in Masohi market.

Nuaulu name	Scientific name
kacang panjang	*Vigna unguiculata* grp. *sesquipedalis*
kaisan msinae	*Arachis hypogaea*
kaisan putie	*Arachnis hypogea* var.
kihue leka	*Vigna radiata* var.
kihue manora	*Vigna radiata* var.
kihue msinae	*Vigna radiata* var.
kihue putie	*Cassia fistula*

Table 4.10 Pulses.

Nuaulu name	Scientific name	N named folk varieties
hisae	*Cymbopogon rectus*	1
kanapua	*Gossypium acuminatume*	2
karatupa	*Capsicum annuum*	10
pahane msinae	*Allium cepa* cv. Group *aggretatum*	1
pahane putie	*Allium sativum*	1
papae	*Carica papaya*	2
sesene	*Ananas comosus*	7
sukue	*Kaempferia galanga*	1
tamati	*Lycopersicum esculentum*	2
tamkari waene	*Cyperus elatus*	1
tapako	*Nicotiana tabacum*	4
tohu	*Saccharum officinarum*	9
tonne	*Solanum melongena*	6
topikane	*Saccharum spontaneum* var. *edulis*	1

Table 4.11 Plants other than cucurbits deliberately grown in swiddens and in the village area, but not listed in Tables 3.2 to 3.10.

Nuaulu name	Scientific name	English name	AM or Indonesian name	OW/ NW	Notes
awane kin-kin	*Zehneria japonica*			NW	no known use
awane munie	*Cucurbita maxima*	squash		NW	gourd used as container
napu onate	*Cucurbita moschata*	squash, pumpkin	'labu besar'	NW	eaten and used as container
napu potone	*Lagenaria siceraria*	bottle gourd		OW	gourd used as container
napu sina	*Cucurbita moschata*	squash, pumpkin	'labu sina'	NW	eaten and used as container
napu tekene	*Trichosanthes cucumerina*	snake gourd	'sayor patola'	OW	eaten
papino syn. ***temon***	*Cucumis sativus*	cucumber	'ketimun,' 'papinyo'	OW	seen as related to (***tumno***) ***pateka***
rorong	*Benincasa hispida*		'kondor'	OW	seen as related to (***tumno***) ***pateka***
tumno nawe	*Cucurbita pepo*	squash pumpkin	'patega'	NW	seen as related to ***papino***, eaten
(***tumno***) ***pateka***	*Citrullus lanatus*	watermelon	'semangka'	OW	seen as related to ***papino***, long and round landraces in Sepa gardens

Table 4.12 Nuaulu folk classification of cucurbits (Cucurbitaceae).

Other swidden crops

Other plants deliberately grown in swiddens, or in the village area in spaces adjacent to houses, are listed in Table 4.11, and include shallots, garlic, chillies, pineapple, ginger, tobacco, sugar cane, tomato, aubergine, AM 'kusu-kusu sarai,' papaya and cucurbits. Of these, the cucurbits are of particular interest from a classificatory point of view given their patterns of cultivar diversity and hybridization (Table 4.12, Figure 4.5). As with those terminologies for curcubits found in Euro-American languages, continuous variation in fruit form of *Cucurbita* is reflected in its inclusion in three separate categories: ***awane munie*** (the ***awane*** prefix emphasizing the vine rather than the fruit), ***tumno*** and ***nawe***. Watermelons (a separate species) are nevertheless grouped with *Cucurbita pepo* as ***tumno***, while ***napu*** also incoporates two distinct species of New World origin: *Trichosanthes cucumerina* and *Lagenaria siceraria.*

Old swiddens, managed fallow and groves

Even before the first-year annuals in a swidden have been harvested, longer-term crops will begin to be planted (banana, papaya, mango, coconut), and after the annuals have been harvested some of these will be ready for harvesting as well. Increasingly, also, cassava is replanted at this stage, replacing other starchy tubers such as taro and yam. Much of this activity has already been described in some detail (Ellen 1978a:161–89); suffice to say it is a gradual process such that a first-year garden (***nisi honue***) is gradually transformed after the first ten months into an old garden (***nisi monae***). During the second year, more – and often a greater variety of – species will be planted and generally ***nisi monae*** remain very mixed, and include also the encroachment of unplanted species of secondary growth. However, the ***nisi monae*** stage is also an opportunity for specialization, and increasingly areas will be devoted to tree crops that yield a cash income: coconut (thus, ***nisi nione***), clove (***nisi kupanahane***), chocolate, nutmeg, coffee and so on. Such areas effectively release land from the swidden cycle into permanent groves.

By the end of the second year a garden will already be described as ***nisi ahue*** (what I am calling here long-term fallow). Where the successional growth is actively managed it will continue to be described as ***nisi ahue*** for many years, but some of this land will inevitably be subject to less management as labour-demands shift to yet more new swiddens. In the typical ***nisi ahue*** the species pattern is generally very mixed and diverse, but with a much higher proportion and density of harvestable species than in other kinds of forest. In some areas of ***nisi ahue*** there is a degree of specialization usually planned at an earlier stage. It may, for example, display small clumps or groves of the betel palm *Areca catechu* (plot 4, Appendix 4) or cempedak *Artocarpus integer* (plot 6, Appendix 4), and with the pressures to produce marketable produce, increasingly these areas of ***nisi ahue*** become permanent groves. Some of the technical features of this process of cultivation and transformation have already been published, but what we need to emphasize here is the context in which particular kinds of cultivated plants are experienced and managed, the proportions in which they are planted, the extent of human-plant interaction, and therefore the background factors that have a bearing on their perception and classification.

Village spaces, path-sides and grassy biotopes

All of the spaces and secondary biotopes described so far involve a degree of weeding and cutting back of vegetation in order to maintain them, and this unwanted invasive growth is often generically described as ***monote***. The category ***monote*** is examined in greater detail in Chapter 7, along with other terms defined more by functionality than morphology. It is used to label many plants that Nuaulu experience as 'weeds,' that is as plants that hinder the development, management or harvesting of swidden and plantation crops. Most of these are grasses, so ***monote*** might be seen as an incipient 'grass' category. Certainly, it is

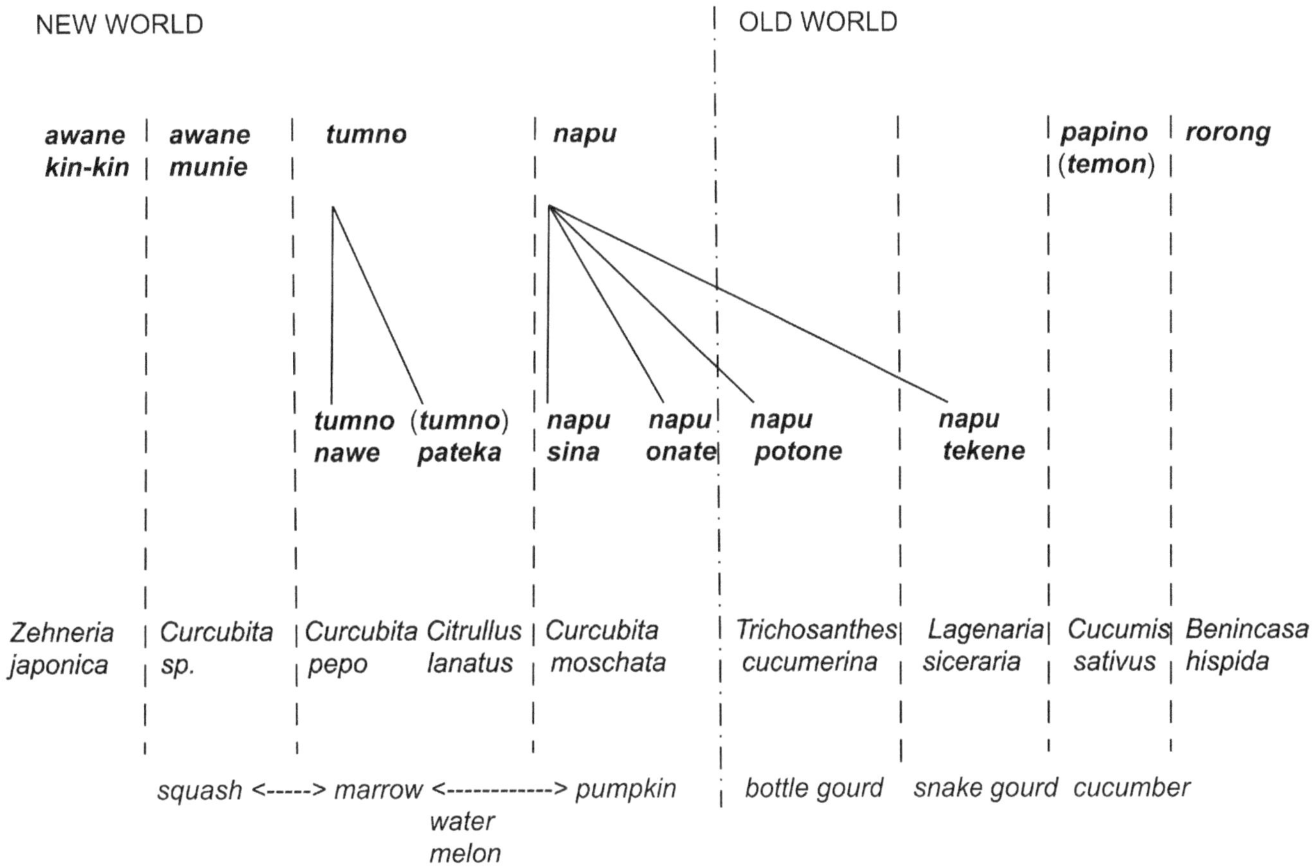

Figure 4.5 Nuaulu folk classification of Cucurbitaceae.

often translated as such by Nuaulu themselves, though the Indonesian gloss 'rumput' too has the extended meaning of weed. Nuaulu will spend time in young swiddens in particular, weeding between growing crops. They will pay special attention to cutting back grass from around young fruit trees, or even around semi-wild yams growing along forest pathways. These activities define ***monote*** in their experience, and even where plant types are not consistently regarded as ***monote*** or clearly fall within other more widely recognized groupings, they will be described as such if they hinder the growth of more useful plants. However, weeding is minimal, and is not the systematic activity found in the neighbouring gardens of Sepa or of the various incoming settler groups (such as Butonese), where it is more time-consuming, and where the order and tidiness of gardens is marked by the use of perimeter fencing. Nuaulu swiddens, by contrast, have unruly edges and no fences. Those patches of grassland that occur within the forest landscape, in some areas to the extent that they might be described as savanna, are not generally described by Nuaulu using the term ***monote.*** Often, though not exclusively, as a result of over-intensification of swiddening and forest extraction, such areas are dominated by grasses Nuaulu call ***oni*** (*Imperata exaltata* and *I. cylindrica*).

How Nuaulu experience the world of cultivated plants

As we shall see in Chapter 5, Nuaulu experience forest as infinite variety. As can be seen from the plot surveys in Appendix 4, different species are distributed widely within any one surveyed plot, though occasionally there are patches of single species, and patchiness more generally leads to an overall uneven distribution of species. In parts of the forest that have previously been cultivated, the natural pattern has been disturbed, sometimes over many centuries; but a consequence of this is even more single-species patchiness, either of trees and other plants that at some point in the past have been deliberately planted, or of those species that emerge as a consequence of ceased cultivation, such as ***oni*** grassland and bamboo. Swamp land, with its domination by sago, is another example of how long-term mono-distribution has been maintained through a combination of deliberate management and inadvertent growth and reproduction patterns, and this we will discuss further in Chapter 6 in the context of anthropic palm habitats.

Although these semi-managed zones are in many ways exceptional, we can see a contrast between how Nuaulu experience individual kinds of plant in the forest, and in swiddens and other cultivated areas, where we find blocks of a single type, distributed geographically, as between a coconut grove, a clove grove, or a swidden dominated by cassava. Over time, a first-year swidden will move from domination by yams and taro, to cassava. As the early starch staples are established, so other vegetables and fruit and tubers will be planted, increasing the diversity of the plot. Within the range of habitats that Nuaulu interact with and experience on a daily basis, the swidden is the one where detailed observation of species is at its most intense, where the entire life cycles of particular kinds can be watched and examined as they unfold over the course of a year or more, where seed is removed, saved and selected, and where the process of separating plants with interesting and useful characteristics is at its most visible. And because Nuaulu must seek to make sense of these differences, it is here that they might truly be said to 'make' classifications.

Conceptualizing diversity below the species level

One prominent feature of those plants occurring in swiddens is the degree of diversity recognized and fostered below the species level. It is common enough to find farmer recognition of biological diversity within crops without the variants being systematically named, but as soon as such variants are transmitted between farmers they generally acquire names. A common proxy for the measurement of diversity within a crop, therefore, becomes the number of distinct names attached to variants below the species level, for example, the 89 names for kinds of rice reported for the Baduy of West Java (Iskandar and Ellen 1999), or the 43 names reported for Debut cassava in the Kei islands (Ellen and Soselisa 2012a:24). However, there is debate in the literature as to how these should be conceptualized and described. Taxonomists use the term 'variety' to describe a taxon below the species or sub-species level that can be distinguished using general morphological indicators and (increasingly) DNA bar-codes. However, agricultural and horticultural scientists have long used the term 'cultivar' to refer to variants of planted crops, usually emphasizing those properties of practical significance (e.g. early or late maturing) or factors important in consumption (e,g. glutinous or non-glutinous). This means that the concept of cultivar does not map comfortably onto the taxonomic notion of 'variety'. Agriculturalists looking at the traditional practices of farmers also speak of 'landraces', which are simply the types of a crop identified by farmers themselves without necessarily having been subjected to the scrutiny of agricultural scientists or other professionals, and which therefore sometimes include distinctive phenotypic or clonal variants important locally, but not always possessing separate genetic profiles.

To avoid constant interchange of terms, I here adopt the term 'folk variety' throughout, except where it is necessary to distinguish the technical sense of this term from other terms. 'Cultivar' (Bailey 1923; Gibson 2009) is usually defined as a subspecific variant of a cultigen (itself generally defined as a cultivated species). However, cultigens are typified by a high degree of hybridization, which makes their representation in taxonomies awkward. Moreover, some folk varieties may cut across several conventional scientific species. We can see this in the Nuaulu case, with respect to bananas and cucurbits.

In the lowland tropics the largest numbers of locally recognized folk varieties have been reported for rice (e.g. Conklin 1954 [Hanunóo: 90], 1988 [Ifugao: 85–7]; Iskandar and Ellen 1999 [Baduy: 89]), though there are also large numbers reported for cassava (Heckler and Zent 2008). In calculations used for comparative purposes it is necessary to avoid confusion with synonyms; dialectal variants for the same type of germplasm; or variants at a regional rather than local level, or for an entire ethno-linguistic group rather than for a local socially interacting population exchanging plant material. Where there are large numbers of folk varieties distinguished, the question arises as to internal classificatory organization. The Berlin scheme would suggest a downward extension of the two-dimensional taxonomic logic of contrast and division, and this may work in some cases. Conklin (1988), for example, has suggested that Ifugao divide rice on the basis of several further subspecific levels of contrasting categories. However, the continuous and overlapping distribution of qualities often makes this difficult. In sexually reproducing grains such as rice, it is often more accurate to represent the groupings of cultivars that farmers distinguish as non-taxonomic cross-cutting categories, based on multiple criteria, with no fixed two-dimensional hierarchy (Iskandar and Ellen 1999). Clonally propagated crops, such as cassava and sago (e.g. Ellen 2006a), pose special problems for folk classification. Local populations often separate cassava folk varieties into 'bitter' and 'non-bitter', as in the Kei island distinction between 'enbal' and 'kasbi', though even in such cases the fine gradation in hydrogen-cyanide toxicity across different named categories, and variation in toxicity according to microecology, makes even such a distinction problematic (Ellen, Soselisa and Wulandari 2012). It is also important to differentiate all locally known named folk varieties and those actually grown in particular locations at any one time. Those Nuaulu crops exhibiting the greatest diversity below species level are listed in Table 4.13, with banana far exceeding all others. What is otherwise noteworthy in this list, are the low levels of diversity compared with grain crops, particularly rice in South-East Asia; the low levels of diversity overall for cultigens compared with some other well-recorded populations, which may stem from the broad-spectrum character of Nuaulu subsistence and heavy reliance on the sago palm; and the large proportion of folk varieties maintained not through seed but through vegetative propagation only.

Species	English name	N folk varieties
Areca catechu	betel palm	5
Capsicum annuum	chilli pepper	10
Cocos nucifera	coconut palm	10
Colocasia esculenta	taro	9
Dioscorea alata	greater yam	11
Dioscorea esculenta	lesser yam	8
Manihot esculenta	cassava	11
Metroxylon sagu	sago palm	11
Musa x *paradisiaca*	banana and plantain	37

Table 4.13 Numbers of lexically recognized folk varieties for most diverse Nuaulu cultivated plants (modified from Ellen 2006a).

5

Trees, lianas and forest

A semi-managed landscape

Chapter 4 dealt with the plants of cultivated spaces, and most of the species implicated were those subject to intensive management. The majority of these would be regarded as domesticates by plant biologists. However, the concept of 'domesticate' in itself makes little sense to Nuaulu. We have seen from Chapter 4 how the composition of swiddens changes after the first year of intensive cultivation of annuals, through a second year in which are planted different longer-term crops that determine the future of the plot. Continued interventions result in agroforestry groves of various kinds (*Areca, Cocos, Durio*), but the default position is the re-establishment of semi-managed long-term resources against the backdrop of gradual colonization by secondary regrowth. This long fallow is what Nuaulu call ***nisi ahue***, as domesticates and cultivates get incorporated into vegetation that will ultimately have the appearance of mature forest, ***wesie***. The result is what we might call a 'semi-managed' agroforestry landscape. Plots 1, 4 and 6 in Appendix 4 provide some empirical exemplification of the diversity in composition of ***nisi ahue***. Some ***nisi ahue*** have a high preponderance of species planted for their usefulness, and in some of these there are clear dominants. Thus, Plot 4 at Ratipenisa, between Rouhua and the river Upa, comprises 9 species divided between 26 trees of over 10 dbh, of which 17 are 'cempedak' *Artocarpus integer*; while Plot 6 at Iana Onate Matai comprises 19 species divided between 54 trees, of which 14 were *Areca catechu* planted for its psychoactive nut, though the next most frequent species was the self-seeding *Ardisia humilis* (5), to which Nuaulu attribute no specific uses. By comparison, Plot 1 had a less skewed species distribution (35 trees from 13 species) with no clear dominants, of which 4 trees were *Mallotus* sp., 3 *Neonauclea lanceolata* and 5 *Syzygium boerlaegi* or *Syzygium* sp. *Syzgium* contains species such as water apples, 'jambutan' and 'jambu hutan' with edible fruits, but none of these nor any of the other species would ordinarily be planted. Similar managed fallow maintained by Alune speakers living in the Lumoli area of west Seram (Suharno 1997; Suharno and Friedberg 1996)[1] contains taxa (families, genera and species) neither reported for the Nuaulu plots nor in the NED (e.g. the families CUNONIACEAE, ICACINACEAE and STAPHYLEACEAE, genera *Buchanania amboinensis, Kopsia arborea*, and *Melanopsis multiglandulosa*). Moreover, the Lumoli study does not report many families found in the Nuaulu plots (e.g. CANNABACEAE, ELASTOMACEAE, PRIMULACEAE, and – surprizingly given its commercial importance – DIPTEROCARPACEAE), and many genera and species of families that are reported for Nuaulu (e.g. *Callicarpa longifolia* [LAMIACEAE], *Neonauclea lanceolata* [RUBIACEAE] and *Syzygium boerlagei* [RUBIACEAE]). There are similar patterns of presence and omission if we look at other studies of managed forest (e.g. Edwards *et al.* 1993; Purwanto and Cosiaux 2013; Ranlund 2011). Although we might expect a degree of incompleteness in these studies, the data do support an earlier observation of the degree of ('patchy') variation in forest composition within quite small areas, as well as across wide geographic ranges, for both secondary and mature forest on Seram (Ellen 2007).

The difficulty of drawing a dividing line between the domesticated and non-domesticated, and between the cultivated and the non-cultivated, is something much emphasized in recent work on landscape and historical ecology, including on Seram (Suharno 1997), but its recognition is hardly new. Rumphius had experienced considerable difficulty in separating cultivated, domesticated and wild species. Fortunately, he usually describes plants in enough detail for us to obtain a fairly clear idea of the extent of human management involved. Consider, for example, book six (Rumphius 2011, vol. 2:179–463), entitled 'Concerning shrubs, domesticall and wild'. These ambiguities created by human practices with regard to trees and forest, are all well reflected in Nuaulu knowledge organization and practice today.

1 The equivalent biotope amongst Alune is called 'kwesie' (Suharno and Friedberg 1996). As we have seen, the cognate Nuaulu term ***wesie*** means forest in general, implying lack of recent human intervention.

The present chapter focuses on trees, knowledge of forest in general terms, and on vines and rattans, but not on the flora of the undergrowth, river edges and rock faces. Pteridophytes have already been covered in Chapter 3, while herbaceous angiosperm vegetation not anticipated in Chapter 4 is discussed in Chapter 7. Neither does it deal with three groups of plants commonly part of forest associations (pandans, palms and bamboo), which are treated separately in Chapter 6. These latter are phyically distinctive groups of plants for Nuaulu with special uses, and it is convenient to treat them separately. Of course, they too are an integral part of forest ecology, and Nuaulu also understand them in forest contexts. Rattans too might have been treated together with these special groups in Chapter 6, given their taxonomic affinities with palms. However, here I have instead decided to follow the logic of ecology, and of their functional affinities with lianas, and treat rattans with other climbing plants as part of the forest canopy.

Forest and long fallow

Nuaulu apply the term ***wesie*** to forest in general, and in many symbolic practices a strong distinction is drawn between ***wesie*** and non-***wesie***. Indeed, I have argued in previous publications that in its widest – most abstract – sense, ***wesie*** becomes the nearest thing Nuaulu have to a concept of nature as a spatial other (Ellen 1996b, 1996c, 2007). But in practice there is evident recognition that the distinctions are never clear-cut. The boundary is drawn differently, depending on whether the contrast is with swiddens, groves, the village or the house. Moreover, forest is internally diverse in its composition and structure. Consisting predominantly of trees, it also contains other plant-life forms, mainly vines (***wane***) but also bamboo, rattans, pandans, palms, ferns and herbs. Until recent years the Indonesian state has adopted a de facto flexible and pragmatic approach to Nuaulu access to forest. Prior to 1998 access was officially governed by a combination of the Basic Agrarian Law and specifically forest legislation, but in practice access was permissive, sustained by locally accepted customary law. Pressure on land increased with logging and transmigration during the 1980s, which resulted in some conflict. The reforms after 1998 gave more formal recognition to Nuaulu land rights, though with increased opportunities for disputes over resources due to mounting pressures, mainly from transmigrants. The Indonesian legislation of 2016 formally recognizing customary forest has yet to impact fully on the situation in Seram. For Nuaulu themselves, in areas where there is no recent history of groves or swiddens, access to mature forest for individuals of all clans is relatively free, though non-Nuaulu wishing to access forest are expected to first seek Nuaulu permission. In the past failure to do so was less consequential than it has become with a growing population, land pressure and rising levels of extraction. Also, where Nuaulu extract – say timber – from non-clan land, they need permission from the relevant clan. Apart from this, a forest user must ensure that the relevant ancestral clan spirits are placated, by making offerings. Most near forest will have been used for swiddens in the past, and be comprised of new plots, mature gardens and fallow forest. Fallow, as we have seen, may contain fruit trees or other trees planted for cash income or subsistence. Such land is divided into ***wasi*** associated with individuals and households, and individual gardens will be cut within a ***wasi***. The unit of ***wasi*** comes closest to what Sasaoka and Laumonier (2012), writing of the Manusela area, call 'forest lots', though Nuaulu do not socially partition unused forest within the clan. Indeed, the notion of exclusive bounded units of forest does not fit the concept of serial rights entailed by Nuaulu practices, in which clans, households, and individuals can simultaneously claim rights in a piece of land. Moreover, swiddening complicates the determination of where long fallow ends and mature forest begins. For example, old settlement sites (e.g. Figure 5.1) – some long abandoned – are still important fruit and sago reserves.

Figure 5.1 Freshwater lake close to Wara-wara, an old Sopanani clan-settlement site, February 1996.

Plot	1	2	3	4	5	6	7	8	9	10	11
Number of trees (N)	35	30	37	26	28	54	124	116	79	35	68
N with Nuaulu name	34[1]	28	37	26	28	54	124[1]	115	74	35	68[2]
Percentage N identified with agreed Nuaulu name	97	93	100	100	100	100	100	99	94	100	100
N Identified to family level	30	26	37	25	16	50	108	100	53	26	68
N identified to generic level	23	22–5	33–5	23	11–16	50	90–107	91	46–8	26	66–8
N identified to species level	10–12	18–20	26–9	23	4	43–4	76–81	72	31–2	13–17	58–60

Table 5.1 Summary of selected 1996 plot data: levels of identification. For profiles of plots see Appendix 4.
Key: N = number of standing trees at 10 cm dbh or above. Where a range of numbers is provided (e.g. 11–16) the first number indicates definite identifications and the second possibles and probables based on voucher data.
1. one doubtful name in each case.

Surveys of eleven plots were undertaken in 1996 in order to capture some sense of the diversity of forest with which Nuaulu interact, see whether there were any significant discontinuities between biotopes, and measure Nuaulu use of forest, the extent to which it is managed and how such discontinuities are mapped onto Nuaulu language and categories. These surveys were conducted in a variety of habitats, including along edges of cultivated areas, depleted mixed forest, 20-year regrowth, 30-year regrowth, old settlement sites, areas of ritually protected forest, plots adjacent to logged forest, on the periphery of transmigration areas, in relatively undisturbed lowland forest and in undisturbed higher-altitude forest (Table 5.1; Appendix 4). In selecting these areas, I also sought to capture a representative variety of habitats and salient species associations reflecting the terminology Nuaulu use to describe forest variation. The results of this work have already been published (Ellen 2007, 2010), and include a list of plots, and maps of plots 7 and 11, respectively, the old village site of Amatene north-west of the Upa estuary, and the Peinisa clan ***sin wesie*** (ritually protected area) at Mon Sanae, together with a discussion of their significance (Ellen 2010). Appendix 4 provides updated details of all eleven plot surveys of different kinds of forest, mainly semi-managed, that are regularly utilized by Nuaulu. Here I will simply summarize the findings and report some additional points that have not previously been elaborated. Not included here is swamp forest dominated by *Metroxylon sagu*, a specialized and anthropic Nuaulu ethnoecological category, more appropriately described alongside sago and other palms in Chapter 6.

The major finding deriving from this methodology was with regard to Nuaulu tree-identification skills, and the completeness of Nuaulu terminological data. Of all the trees over 10 cm dbh in all the plots, there is a mean level of identification of 97.5 per cent, with 7 of the 11 plots yielding 100 per cent agreed identifications (between the usually three male Nuaulu surveyers). As might be expected, identifications in the more remote plots proved more difficult, including the most distant plot 9 in a sub-montane area of Mawoti, which yielded a 94 per cent identification rate. Some species were unfamiliar to Nuaulu assistants in this plot, but a more practical problem proved to be tall trees with little leaf visible below canopy level.

The earlier papers mentioned above (Ellen 2007, 2010) provide a general account of the ecology, structure and composition of lowland forest as Nuaulu experience it, in relation to the recent work of Edwards *et al.* (1990, 1993), which contains useful diagramatic profiles of different kinds of forest (see also Suharno 1997). These papers examined the problems of typologizing forest, given its extremely variable and patchy composition, and noted that Nuaulu approaches to understanding forest diversity anticipated issues raised in recent scientific tropical-forest ecology. Nuaulu utilize a relatively simple lexical framework to indicate basic structural differences, but embed within this a more detailed and nuanced knowledge along a number of classificatory axes. These latter include historical and mythic narratives, different land forms and water features, a 'toponymic grid', land tenure, disturbance histories relating to patterns of extraction (including grassland), and, especially, cycles of swiddening activity. On this basis, it is clear that Nuaulu cognition of forest diversity cannot be reduced to classificatory schemes reminiscent of those that work for folk systematics. Rather, Nuaulu categorization of vegetal biotopes is flexible, depending on place and purpose, and is perhaps better described as 'a dense network of social facts realized in biological form' with 'patches of significance' (Damon 2017:139). Another way of looking at forest is in terms of how Nuaulu move through it (Figure 5.2), the same device employed by Rumphius to organize his systematic account of Ambonese flora. Rumphius's descriptions are by way of a 'journey' (Beekman 2011:98), beginning with the southern beaches of Ambon Bay on the Hitu peninsula, and moving through different habitats: the groves surrounding villages, the fallow along the cultivated edges and then deep forest, across the mountains and down to the coral-fringed coast of north Hitu. It is not always an empirically accurate tracing of a trajectory, and diversions sometimes have to be made to group annual crops of the gardens and useful

1. 400 m asl: Amatene old village site: jambu, durian, bamboo, coconut, sago
2. 350 m asl: open bamboo scrub
3. C 300 asl: *Canarium* belt (with some *Calamus* spp.)
4. 250 m asl: depleted forest with indications of selected extraction for timber
5. 200 m asl: long fallow old gardens with some 'alang-alang' (***oni***: *Imperata*), current garden land
6. 150 m asl: long fallow (with *Pandanus* sp.)
7. 100 m asl: bamboo long fallow — Yana Ikine river
8. 50 m asl: belt of ***tepine*** (*Calyptrocalyx spicatus, Pinanga rumphiana*) and rattan
9. 0 m asl: current garden land — Bunara
10. 0 m asl: mixed secondary forest
11. 0 m asl: old garden land
12. 0 m asl: current garden land — Sepa
13. 0 m asl: coastal groveland: *Cocos nucifera.*
14. 0 m asl: village area: some *C.indicum* — Rouhua

Figure 5.2 A transect through different biotopes encountered on the descent from Amatene to Rouhua, 1996.

herbs, while some of the plants he describes were not to be found along the notional route, or indeed on Ambon at all; but as a means of making sense of disparate data, in the absense of the conventions of modern phylogeny, it was as good as any. We shall return in Chapter 10 to a consideration of the key role played by place in Nuaulu organization of their plant knowledge.

The life form 'tree' and the category *ai*

All large woody plants that are not pandans, palms or bamboo, are generally described by Nuaulu as ***ai***, and 'tree' in this sense is clearly a cognitive prototype. Nuaulu recognize and have discrete names for 322 plant taxa (31 per cent) that we would describe as 'tree'. Of course, part of the problem is knowing at what point a woody shrub becomes a tree, or a non-branching but woody and erect palm likewise. Frake (1962) famously defined trees for the Subanun of Mindanao in terms of their woodiness and erectness, and for many languages these seem to be the minimal conditions for defining a plant as a tree. It also seems to be the case that a branching habit and perhaps other features are necessary in some cultural populations to define trees. Certainly, in the Nuaulu case palms and pandans are not 'trees', and almost as if to underline this point ***ai*** is used as an adjectival qualifier in ***hata ai*** to designate one folk variety of sago palm (*Metroxylon sagu*) that is most like a tree in not having thorns on its trunk. Of the named plants that Nuaulu describe as trees, only 63 (including optional synonyms) are labelled with a binomial the first element of which is ***ai***, e.g. ***ai suto*** (*Premna serratifolia*), though in a few other cases ***ai*** may appear at another position in the name indicating that the plant is classified as a tree (***nana ai***: the hardwood *Heritiera littoralis*), rather than simply resembling or connoting a tree in one of its features (as in the yam ***hueni ai arine***: ***hueni*** – *Dioscorea alata*, yam + 'tree buttress root', referring to shape of tuber). In only four cases of the ***ai*** + qualifier form was the ***ai*** element recorded as being optional (e.g. ***supa*** = ***ai supa***: *Ficus tinctoria*), though it is likely that many more uninomial tree terms can be optionally converted into binomials. In some cases, the binomial is mandatory, the term making no sense as a plant name otherwise (e.g. ***ai msinae***, lit. 'red tree': *Syzygium* sp.; ***ai weti***, tree + 'sago pounder': *Phaleria perrottetiana*). Some binomials have synonyms that are uninomials, which allows us to confirm their status as core ***ai*** categories (e.g. ***hana*** = ***ai unate***: *Bruguiera gymnorrhiza;* ***tahena ukuna*** = ***ai osi***: *Aglaia parviflora*).

A few Nuaulu plant categories included in the life form 'tree' are at the same time prominent members of other intermediate plant categories that are defined expressly by criteria that characterize them as being 'other than trees'. The most obvious examples are certain kinds of tree-fern (e.g. ***ai mukune***: *Cyathea* and a few species of other fern genera), but something similar can be seen in the way

kupanaia (the frangipani *Plumeria rubra*) or ***kupaleno*** (*Cananga odorata*, the decorative and fragrant blooms of which are sometimes worn in the hair of both men and women) are clearly trees in terms of gross morphological structure, but are labelled – and indeed treated – as a 'flower', indicated by the prefix ***kupa-***.

In most human environments trees share with 'herbs' the distinction of being the largest group of plants recognized, and we know from the work of Berlin (1972) and Brown (1977, 1984) that if life forms are recognized lexically in a particular system, then 'trees' are more likely to be recognized than any other. Moreover, 'trees' hold a special symbolic role in the way people model social relations and configure the world (Rival 1998). Yet trees in most environments are internally very diverse as a group; poorly differentiated using intermediate categories between life form and basic level; and, in particular, show very little evidence of 'family-like' groupings comparable, if not equivalent, to scientific botanical families. Examples of the relative absence of mid-range groupings to organize the diversity of trees are found among those terms used by Nuaulu to describe *Macaranga* species. *Macaranga* is a large genus of small trees within the family EUPHORBIACEAE, with at least eight species known from Seram (Whitmore, Tantra and Sutsina 1989). They are fast-growing recolonizers, often found in swidden fallow, where Nuaulu recognize them by the hollow stems and twigs of their saplings, and because many species host ants, and many others exude a reddish gum. Other species display cauliflory, meaning that flowers and fruits occur on the trunk. Many have large leaves that are commonly used as wrappers for bushmeat. Although Nuaulu distinguish five folk categories corresponding to six scientific species – ***hunua*** (*Macraranga* sp. nr. *tanarius*), ***hunori*** or ***nori-nori*** (*M. trichocarpa* and *M. involucrata*), ***ini*** (*M.* sp.), ***nihuhune*** (*M. mappa*), ***ranoi*** (*M. hispida*) and ***kupatonae*** (*M.* sp.) – what is noticeable is that these are all uninomials, seldom prefixed with ***ai*** in default speech. In other words, each of them seems to be accorded a degree of perceptual salience beyond just being ***ai***, but any commonalities are unmarked linguistically. However, although some similarities between the species are recognized by Nuaulu (such as their soft wood, which means that they are not preferred for building), and the diversity of *Macaranga* species is acknowledged, their lack of prominence as useful species other than as ecological indicators means that none appear to form pairs or to be linked in other ways that might suggest some mid-level grouping.

In what follows I explore some cases where there is middle-range differentiation in the tree-knowledge corpus, and ask how it might be that Nuaulu organize systematic knowledge of tree diversity, when, for example, the models of 'cognitive economy' that we associate with Elcanor Rosch (e.g. 1977) and George Miller (1956) predict that remembering large numbers of words requires intermediate categories, which themselves come with size limitations. There is clearly evidence of list-like behaviour, of covert groupings that are not expressed lexically, but what appears to give ad hoc and situational shape to the large number of trees are a number of overlapping and unformalized criteria: where trees are found (location), the relations into which they enter with other plants (ecology), and the uses to which they are put by humans. Use (special purpose) considerations are therefore central to cognition. Some trees are grouped with many herbs on the basis of their medicinal significance, and it is this that gives them cognitive salience rather than their specific morphological attributes. Examples include, ***rene*** (*Erythrina variegata*), which has spectacular red flowers and is used in the treatment of diarrhea; and ***ai turi*** (*Cassia alata*), the leaves of which are boiled and ground as an application for eczema. Similarly, there are those trees providing, say, extracts which can be used for stunning freshwater fish (such as ***asaherane***: *Myristica* sp., probl. *fatua*); or which are salient by virtue of being the distinctive food of animals that are hunted by Nuaulu, and provide clues during the chase (such as ***anapakone***: *Leea* sp., whose fruits are a sought-after delicacy for cassowaries and whose low-hanging leaves are forage for deer). Then there are species linked through their ecological similarity, such as those quick-growing persistent species (e.g. ***hunua***: *Macaranga* sp. nr. *tanarius* and ***kupatonae***: *Macaranga* sp.) which colonize abandoned swiddens and through their large leaves exclude light, and therefore also other species that potentially occupy the same niche. Similar applications for the manufacturing of materials may mark out other species, such as **saumahu** (*Abroma mollis* and *A. fastuosa*), both of which are deliberately planted in gardens, with their leaf fibre being made into thread used, for example, when wrappng a corpse.

The totemic veneration of some trees is shared in common by different clans, and this gives them salience, especially for those clans that respect them – e.g. ***ai suto*** (Peinisa), ***aniate*** (Numanaita) and ***nihasa*** (Sounaue-ainakahata) – but they may also become a generic category for all Nuaulu. Other ritually important species may be similarly marked, such as ***masahatune*** (*Mackinlaya celebica*) rubbed on the bodies of neophytes during both male and female puberty ceremonies; or ***kanapua*** (*Ceiba pentandra*), the kapok tree which features in a narrative concerning a baby hidden within it to avoid being killed. Other tree categories are connected through being paired. Such pairs are often quite different taxonomically, such as ***hukila*** (a species of *Eugenia*) and ***kawasa*** (*Archidendron clypearia*), both used for shields and for ***nene*** (the blocks of wood bound to the head of male ritual specialists and incorporated into the red head-cloth to form a 'horn'). Pairings such as these can connect together to form a chain or series of species – often through polythetic criteria – giving order to classificatory space, but without forming stable formal categories. In addition, the 'kind of' relationship may be employed in a flexible way to express closeness between otherwise contrasting categories. Thus ***ananaine*** (*Litsea resinosa*) is said to be 'a kind of ***kawasa***' (*Archidendron clypearia*), and ***kawasa*** 'a kind of ***ananaine***', though the first is Fabaceae and the second Lauraceae. Indeed, for some individuals ***kawasa pina*** ('female' ***kawasa***) is a synonym for ***ananaine***. In this particular instance, the relative cultural salience of the name ***kawasa*** over ***ananaine***, through its association with sacred shields (Ellen 2019d), most likely makes it the preferred and unmarked superordinate. Many 'wild' forest tree species are defined by their resemblance to prominent cultivated species: thus we have wild ***popote wesie*** (*Pternandra cordata, P. coerulescens*) labelling species of the family Melastomataceae, but associated with the larger group of other domesticated 'jambu' (water apples) described by Nuaulu as ***asapopote***, and grouping taxonomically related species of *Syzygium* within the family Myrtaceae; or wild clove, as opposed to cultivated clove. This process of knowledge organization is enhanced by the overlaps between many of these middle-range functional groups. Thus, mango trees are located both in village spaces and in second-year gardens, durian trees both in deliberately planted groves and in deep forest.

I suggest that what allows Nuaulu to organize, store and retrieve named categories outside these groupings and the complexes I discuss below, are chains of linkages to similar species, locational proximity and ecological interaction, rather than taxonomy based on morphological criteria in the strict sense. Damon (2017:116) in his work on the knowledge and vernacular classification of trees on Muyuw, one of the islands of the Kula Ring in the New Guinea Massim, notes how local people seem to recognize 'groups' of trees. These 'groups' are often 'covert' and in some respects similar to intermediate taxa. He suggests that Muyuw tree names, when part of an overarching 'kay/ke' class conform to the 'received understanding of folk biological systems' (in other words, what I term the Berlin-Atran model), 'yet it can be equally validly argued that these forms represent behaviourally distinct patches of significance, not all of which carry names, and that their place in a system of ranked relations is not very important'.

In what follows I identify a number of the more salient intermediate folk groupings of kinds of tree within the life form ***ai***: *Calophyllum, Canarium* and *Ficus*. These groupings – which I here describe as 'complexes' – mostly conjoin scientific taxa belonging to a single genus, but some conjoin taxa at a more inclusive (supra-generic) level to form quasi-family categories. In some cases, it is clear that the motivation for grouping is botanical affinity, but in other cases the functional or utilitarian salience of the grouping clearly motivates its existence as a recognized (sometimes covert) category. This kind of grouping has been recognized widely for other folk classifications of plants (e.g. Atran 1983; Kocher-Schmid 1991:267), though there is debate as to whether some are truly 'taxonomic' at all when they rely on non-morphological criteria, such as ritual uses. This argument, however, is spurious if the objective of the analysis is to unlock a unified system by which a particular people make classificatory sense of plant diversity, and if we seek to make comparative sense of different systems cross-culturally. In their ordinary lives, when it comes to making sense of the world around them, Nuaulu, no less than other people, do not bother too much with the distinction between taxonomic and non-taxonomic, in the sense that a logician or plant taxanomist would understand. People do not consciously separate out 'fragments' with indeterminate boundaries from other categories, but rather treat these complexes like any other category they use to group plants.

Calophyllum complex

A particular consideration in the cognition of tree type categories is sexual difference. The use of male and female adjectival plant names for a wide number of species has often been noted (see also Chapter 2:37–9 here), both to demonstrate the way in which the male/female distinction is used, for example, to indicate the larger and smaller of two related taxa; and also to demonstrate that people are well aware of the existence of dioeceous or unisexual species. What causes some confusion is when these two combine, especially where clear taxonomic identifications are not available. Further difficulties arise when the dioeceous distinction in a species is recognized, but another species in the same genus is given a separate local name.

Compare the situation among Muyuw in the Massim area of New Guinea (Damon 2017:181) with that of the Nuaulu. Nuaulu use the term ***ahutaune*** to refer to any tree obviously belonging to the genus *Calophyllum*, but within this wider grouping is a core of terminal categories that are always labelled ***ahutaune*** (Table 5.2) and a number of *Calophyllum* species often referred to using distinctive uninomials. Thus, core ***ahutaune*** is divided into five categories: ***ahutaune pina*** (female), ***ahutaune hanaie*** (male), ***ahutaune metene*** (black), ***ahutaune nuae*** (sea) and ***ahutaune numte*** (far forest/mountain). Nuaulu distinguish these through a combination of morphological, ecological and functional features, and we should notice in particular the male/female distinction. Muyuw too recognize female *C. inophyllum* (Damon 2017:236–7), and although there is evidence of male and female in some species, the scientific knowledge of sex forms in the genus *Calophyllum* remains generally poor (Stevens 1974:351). Moreover, ***ahutaune nuae***, the common shore type is more usually referred to as ***hitaio***, and although the resin is used in lamps, it is mainly noted for its twisted wood unsuitable for most construction work. In contrast, 'true (***titue***)' ***ahutaune*** are forest trees (with a smooth yellow

	hitaio (***ahutaune nuae***)	***ahutaune hanaie***	***ahutaune pina***	***ahutaune metene*** (***ahutaune totu onate***)	***ahutaune numte*** (***ahutau mkauke***)	***ahutaune wesie***	***akenue***
Species	*C. inophyllum*	*Calophyllum* sp.	*Calophyllum* sp.	*C. soulattri*	*C. articulatum*	*Calophyllum* sp.	*Calophylum* poss. *sil* and/ or *vitex*
Habitat	sea-shore	near forest	near forest	plot 10	deep uncut forest	far forest	
Bark	rough		yellow	black			rough and reddish
Fruit	large	no fruit	small hard fruit		fruits hollowed out to make whistles		small light-green fruit regarded as inferior
Wood	yellow					yellow	
Tree size							small (similar to ***ahutaune***)
Leaf		long	round, medium size	large	small, short, fine		
Uses	resin	suitable timber	suitable timber		preferred timber		

Table 5.2 How Nuaulu make sense of ***ahutaune*** *(Calophyllum) diversity.*

bark), and may also be contrasted with ***akenue*** (*Calophyllum* sp., possibly *C. sil* and/or *C. vexans*), with rough red bark. Thus, ***ahutaune*** is used in an inclusive sense and slightly differently in the contrast set: ***akenue*** (far forest): ***ahutaune*** (forest): ***hitaio*** (shore). However, when we come to match the use of these names against scientific identifications, we find that while *Calophyllum* of the shoreline (***hitaio***) is always *C. inophyllum*, 'true' ***ahutaune*** of the forest may also sometimes be *C. inophyllum* (both male and female), but also other species such as *C. soulattri* and *C. articulatum*, though ***akenue*** is seldom *C. inophyllum*. As Damon (2017:191) has pointed out, *C. inophyllum* is notorious amongst plant taxonomists for its intra-specific variation.

Canarium complex

As with *Calophyllum*, Nuaulu distinguish different types of *Canarium* that they see as having a family resemblance (Table 5.3, Figure 5.3), and which they describe as ***iane*** in its most extensive sense (***iane***$_1$). However, the term ***iane*** is polysemous, and also contrasts with ***kamine***, giving ***iane***$_2$, which includes *C. asperum*, *C. hirsutum* and *C. decumanum*. This latter is the 'kenari Seran' (or 'damar putih') with a larger edible fruit than all other terminal categories of ***iane***. Its timber is used in housebuilding, but has to be carried long distances, a factor inhibiting systematic harvesting of nuts. The resin is used in lamps, though it is said to yield less light than other types of *Canarium*. In a third, less extensive sense, ***iane***$_3$ contrasts with ***ananate*** (*C. sylvestre*), which has a smaller nut than ***iane***$_3$. For this reason its nuts are seldom harvested, though it is a source of an edible grub and of resin. ***Ananate*** is thus both recognized as a 'kind of' ***iane*** and treated as a separate kind depending on the context of elicitation. Certainly, in our plot surveys in 1996 the tree was always identified as ***ananate*** rather than ***iane***, and the uninomial was used as the default name. By contrast, ***iane***$_3$ comprises two kinds based on the shape and size of the fruit: ***mkauke*** (a small fruit containing three locules) and ***hanate*** (a large fruit usually containing one large locule). Confusingly – but representing a classic ethnobotanical problem – there are other names used synonymously for these two folk categories, by those for whom the main name is prohibited: ***yapono***, ***haruku*** and ***pukune*** for ***mkauke***, and ***nawe*** and ***kahiaue*** for ***hanate***. This suggests more culturally recognized diversity than there actually is. The third segregate of ***iane***$_3$ is ***iane hanaie,*** which refers to *C. indicum* trees bearing male flowers only. Nuaulu recognize that some trees are hermaphrodite (dioecious) and some monoecious, and in using the term ***hanaie*** are simply noting that such trees flower but do not fruit. Figure 5.4 shows cross-sections of the kernel for each of these folk categories, from ***ananate*** (smallest), through ***mkauke*** and ***hanate***, to ***kamine*** (largest), indicating number and size of locules, and mapped on to scientific species. The most significant feature is that both ***ananate*** and ***iane mkauke*** are variants of *C. sylvestre*, while ***mkauke*** and ***hanate*** are

Nuaulu name	Meaning	Synonym	Meaning	Scientific name	Characteristics and notes
ananate				*C. sylvestre*	non-edible: small
iane$_3$				*C. indicum*	
ia hanaie	'male'			*C. indicum*	male: no fruit
ia hanate	'west wind'	***ia nawe***	'long'	*C. indicum*	edible: large
		kahiaue		*C. indicum*	
ia kauke, mkauke	k.o. cricket	***ia pukune***	'short'	*C. sylvestre*	edible: small
		yapono	toponym	*C.* probl. *sylvestre*	
		ia haruku	toponym	*C.* probl. *sylvestre*	
kamine				*C. hirsutum*	edible: hairy
				C.decumanum	edible
				C. asperum	

Table 5.3 Nuaulu named categories for types of ***iane****.*

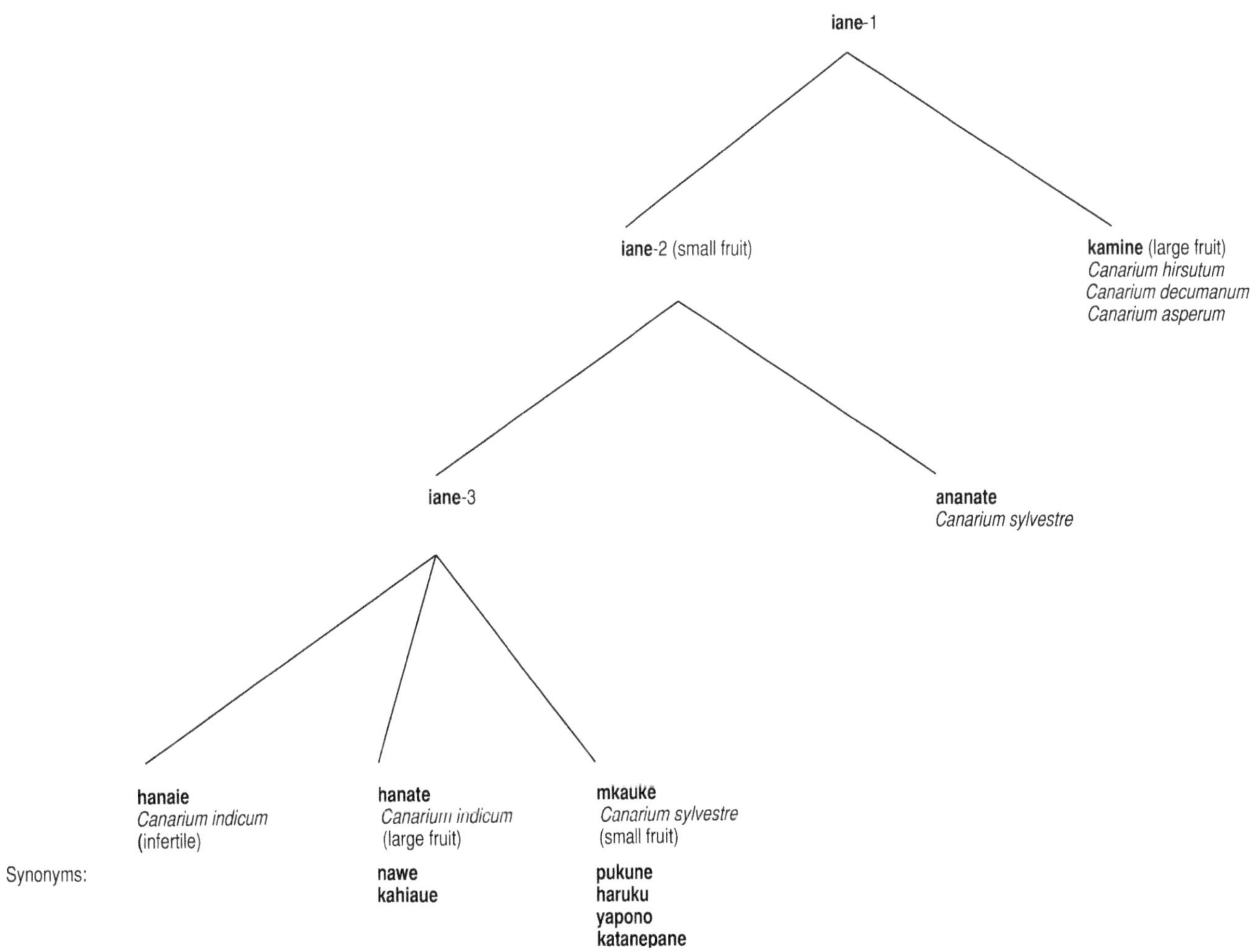

Figure 5.3 Nuaulu folk classification for Canarium species.

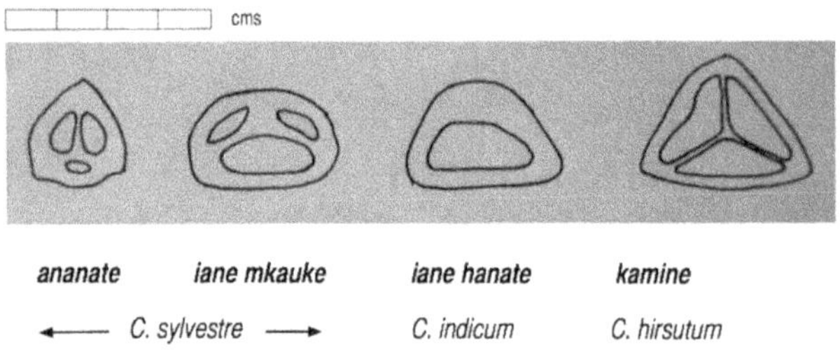

Figure 5.4 Cross-section of the mature fruits of selected species of Canarium showing placement of edible seeds in cavities (locules) with corresponding Nuaulu names.

de Wit (1959)	Rumphian name
Canarium acutifolium	Canarium nigrum
Canarium album	Canarium sinense
Canarium asperum	Dammara nigra legitima
Canarium balsamsiferum	Canarium odiferum leve
Canarium decumanum	Canarium decumanum
Canarium hirsutum	Canarium odiferum hirsutum
Canarium indicum	Canarium vulgare/zephyrinum
Canarium oleosum	Nanarium minimum s. oleosum
Canarium pimela	Canarium sinense Tsjacana
Canarium sylvestre	Canarium sylvestre alterum
Canarium vulgare	Canarium vulgare
Canarium zeylanicum	Arbor zeylanica

Table 5.4 Rumphian and post-Linnaean species of Canarium compared.

different scientific species, but placed together in the folk classification.

There are clues as to how Nuaulu see the relations between folk categories for taxa, in the way Rumphius handled his confrontation with *Canarium* diversity in seventeenth-century Ambon, and in the way present-day Nuaulu use Ambonese Malay or Indonesian terms to describe their own flora. Rumphius (2011, vol. 2:215–51) was the first naturalist to describe *Canarium* and reports thirteen species, eleven of which he calls 'Canarium' (one of which is synonymously 'Dammara', one exclusively 'Dammara', and one 'Nanarium'). De Wit (1959:421–44), following Merrill (1917) and Heyne (1913–17), translates the Rumphian names into twelve post-Linnaean species (Table 5.4).

In contemporary Ambonese Malay the term 'kenari' may refer to *Canarium* other than *C. indicum*, or indeed to non-*Canarium* species. Thus, ***ananate*** is described as a 'kenari', but also distinguished from *C. indicum* by contextually using the name 'nanari', which Rumphius (2011, vol. 2:220, 228, 245) applies to various types of *Canarium* in different places, including the Bandanese landrace. Indeed, in the Moluccas more generally the term 'kenari' is applied to many species of *Canarium*, not just *C. indicum*. Interestingly, the Malay term 'kedongdong' in its area of origin (presumably Riau and peninsula Malaysia) encompasses various species of *Canarium*, but not *C. indicum*, which was introduced during the early modern period, and is the only species described as 'kenari'. In the Moluccas, an area where initially Malay was simply the lingua franca, 'kedongdong' was applied to edible fruits superficially resembling *Canarium*, but quite different phylogenetically. For example, 'kedongdong hutan' (which Nuaulu call ***nihasa***) is *Spondius pinnata* (ANACARDIACEAE).

Plot number	Location	Ecology	Plot size (m^2)	Total trees in plot	Number (percentage) of *Canarium*	Nuaulu *Canarium* name
Plot 1	Nahati	coralline limestone not known to have been previously cultivated	20	34	0 (0)	
Plot 2	Mon	depleted forest on sandstone	20	30	0 (0)	
Plot 3	Nahane Ukune	25 m from logging road	20	37	4 (10)	***ananate***
Plot 4	Yoko	old fallow last cleared 1976	20	26	1 (4)	***iane***$_3$
Plot 5	Benteng	old settlement site, 300 m asl	20	28	0 (0)	
Plot 6	Yana Onate	long-term fallow	20	54	1 (2)	***ananate***
Plot 7	Mon Sanae	***sin wesie***	30	124	7 (5.5)	***ananate***
Plot 8	Sokonana	riparian, partially depleted by cutting	20	116	3 (2.5)	***ananate, iane***$_3$
Plot 9	Rohnesi	1000 m asl, nr Mawoti	25	79	0 (0)	
Plot 10	Amatene	old village site	30	68	1 (1.5)	***kamine***

Table 5.5 Frequency and distribution of Canarium in 1996 plot surveys (see Appendix 4).

Nuaulu name	Scientific name	Notes
amnae (***wesie***)	*Ficus* sp.	
kaenuri, kaenuni	*Ficus polyantha*	
kaiohite	*Ficus botryocarpa* subsp. *hirtella*, *F. callosa*, *F. glandulifera*, *F. copiosa*	
kako atae	*Ficus lepicarpa*	
nana ai	*Ficus botryocarpa* subsp. *hirtella*	1996 plot 10: edible fruit, can withstand sea water, does not crack or split, timber hard and strong, but small
nunu(***e***) ***ina aie***	*Ficus* sp.	'mother of all figs'
nunu(***e***) ***potone***	*Ficus* sp.	
nunu(***e***) ***putie***	*Ficus* (*benjamina?*)	features in cassowary myth
nunu inane	*Ficus* sp.	cassowary food; far forest
nunu tinse	*Ficus* sp.	***tinse*** = k.o. cordage 1996, plot 7
nunu mkauke	*Ficus* sp. probl. *pungens*	small leaf
nusu	*Ficus adenosperma*	milky latex; fruit eaten by bats
pate	*Ficus* sp.	
(***ai***) ***supa***	*Ficus tinctoria*	white bark-cloth: second day of male puberty ceremony
tahola	*Ficus* cf. *septica*	leaves used as medicine for eczema, mixed with white ash
tatane	*Ficus* sp.	white bark-cloth
waronone (***rapa***)	*Ficus pubinervis*	red bark-cloth for elders
waronone kahuite	*Ficus* sp., *F. nervosa* subsp. *pubinervis*	
waruae	*Ficus variegata*	red bark-cloth: first day of male puberty ceremony

Table 5.6 Nuaulu folk classification and nomenclature for figs (Ficus).

Ficus complex

Banyans or strangling figs are a physically distinctive group of plants by virtue of their habit, but also occupy a prominent cultural position in many societies of South-East Asia. For Nuaulu, they often indicate the sites of old settlements, some individual species have medicinal uses, others feature in myth, but by far the most important service they perform is in providing bark-cloth for male ritual attire. Nuaulu terms for *Ficus* species are listed in Table 5.6.

However, whereas with *Calophyllum* and *Canarium* there is a good match between scientific genus and application of a Nuaulu name (***ahutaune*** and ***iane*** respectively, even if sometimes optional), the nearest comparable Nuaulu term for *Ficus*, ***nunue,*** only applies to six of the nineteen species identified. Some *Ficus* not designated ***nunue*** are nevertheless described using the AM term 'waringen', and can therefore be said to be linguistically linked. The four kinds of *Ficus* providing bark-cloth for male-puberty and elder installation ceremonies – ***supa***, ***tatane***, ***warue*** and ***waronone*** (***rapa***) – are not described as ***nunue***, but are associated within the wider grouping by virtue of this important function. ***Nunue*** is a symbolically important category frequently occuring in myth, the bark reputed to have a characteristically bitter taste which makes it appropriate for use in an anti-malarial concoction. It includes both wild trees and semi-domesticates, but the defining member of the group is a cultivated or semi-domesticated species. Looking at useful biota more generally, it is not unusual that the default unmarked defining term should be a domesticate (as in, dog:wild dog; durian:wild durian; clove:wild clove).

Rumphius (2011, vol. 3:3–46) distinguishes about twelve kinds of *Ficus*. Although he avoids any explicit sub-grouping, it is noticeable that while all are regarded as figs, within this group there are those that he habitually calls 'waringen' following current AM practice, trees which are also those described in local Ambonese languages as 'nuno' (i.e. cognate with Nuaulu ***nunue***). The genus is very variable in habit, some being more like climbers, with nearly fifty species in the Moluccas, mostly pollinated by fig wasps. All figs placed in Corner's (1959) subsection *Urostigma* are called 'waringen' in AM and by Nuaulu ***nunue***. However, some trees not habitually treated as ***nunue*** are also called 'waringen' in AM, such as ***ai supa***. ***Tatane*** and ***nunu tinse*** are also regarded by Nuaulu as 'waringen'. Rumphius records the term 'supa' for a type of *Ficus* in seventeenth-century Ambon, and notes that Hituese called the same form 'nuno'. Nuaulu ***waruae*** (*F. variegata*) is described in AM as 'gondal' and not 'waringen'. Nuaulu recognize the affinity of all *Ficus* in terms of the mode of pollination through fig wasps, and the form of the fruit.

Trees with edible fruits

Trees with edible fruits do not constitute a grouping, for the purposes of plant or food classification, that Nuaulu readily volunteer or which might be convincingly inferred to exist. ***Hua*** refers to fruit in the broadest sense of the fruiting body of any plant, but there is a family resemblance in that

Nuaulu name	Scientific name	N named folk varieties	English name and names in other languages
advokat	*Persea americana*	0	avocado
ai pinamo	*Annona reticulata*	2	custard apple
apalam	*Mangifera indica*	5	mango
asamjawa	*Tamarindus indica*	0	tamarind
asapopote	*Syzygium aquem*	2	water apple
kalapasa	*Crescentia cujete*	0	calabash
kanoke	*Syzygium malaccense* var.	0	Malay apple
kuine	*Mangifera odorata*	0	k.o. mango, kuwini
mangosta	*Garcinia mangostana*	0	mangosteen
moni	*Citrus* x sp.	0	k.o. citrus
nasate	*Lansium domesticum*	0	AM 'langsat'
nusi huni	*Citrus* x *aurantium*	0	bitter orange, AM 'limon suangi'
(***nusi***) ***kapane***	*Citrus medica*	0	citron, Ind. 'jeruk sukade'
(***nusi***) ***koramase, nusi masine***	*Citrus* x *sinensis*	0	sweet orange; Ind. 'jeruk manis'
(***nusi***) ***mananai***	*Citrus* poss. *nobilis* or x *reticulata* x *sinensis*	0	AM 'limon matahari, limon papaya'
nusi moune	*Citrus* x *limon*	0	lemon, AM 'limon cina'
(***nusi***) ***peka-peka***	*Citrus maxima*	0	pomelo, AM 'limon besar'
nusi ramatae	*Citrus hystrix*	0	Mauritius papeda, AM 'limon purut'
nusi sina	*Citrus* x *aurantifolia*	0	lime, AM 'limon nipis'
nusi une	*Citrus reticulata*	0	mandarin
nutune	*Syzygium malaccense* var.	0	Malay apple
papae	*Carica papaya*	2	papaya
rambutan	*Nephelium lappaceum*	0	rambutan
sisoe		0	unident. fruit
tahenae	*Artocarpus integer*	3	chempedak
taulela	*Averrhoa carambola*	0	star fruit; AM 'belimbing manis'
taulela maninue	*Averrhoa bilimbi*	0	AM 'belimbing asam', 'belimbing hutan'
tom-tom	*Flacourtia rukam*	0	rukam
tunene hua mainae	*Durio zibethinus*	0	k.o. durian
tunene ikine	*Durio zibethinus*	0	k.o. durian
tunene warata	*Annona muricata*	0	soursop

Table 5.7 Categories for trees yielding edible fruits.

fruits listed in Table 5.7 are eaten generally as snacks or on special occasions. A few have a role as food additives when cooking. Some are used as 'sayur', AM for 'green vegetables, leafy greens'. Edible fruits are not prominently plants of the swidden, except for bananas (which have been described above) and papaya. Given the maturation times of fruit trees, even when planted in second-year swiddens, they are not fruiting until these swiddens are well into fallow (***nisi ahue***). They are therefore widely distributed across ***nisi ahue*** and different kinds of secondary and mature forest, in groves and in the village area. For a variety of reasons wrapped up in its role as a dominant multipurpose species, and in terms of physical and symbolic significance, coconut is rarely grouped with sweet recreational fruits.

Of the categories for fruit trees listed in Table 5.7, there are three points to be made that have a bearing on folk classification more generally. The first is that ***popote*** (*Syzgium aquem*) is recognized as resembling ***asapopote*** (*Syzgium jambolanum*) – as reflected in the name; but since the latter does not bear fruits that are generally regarded as edible it is not listed here. The second category of interest is ***tunene***, joining together fruits which are superficially similar but from two different families: BOMBACEAE and ANNONACEAE. ***Tunene*** is polysemic, referring in its most inclusive sense to all durians and soursops, and in its exclusive sense to durians alone. Clearly, durian is the reference type here, the soursop being referred to as 'Dutch/foreign durian' (***tunene warata***), referencing its smaller size, and possibly a more recent introduction. Finally, all citrus fruits are treated as being strongly closely related. This is reflected in the use of the term ***nusi*** to refer to most of them, but in four cases the prefix is optional in binomials, as in (***nusi***) ***kapane*** and (***nusi***) ***moune***, and in one (***moni***) is in my experience never used at all. This pattern reflects the very distinctive outward appearance and internal anatomy of the fruits, despite similarities in the features of the tree as a whole. *Citrus* species in the central Moluccas are extensively hybridized, and distinguishing individual species can be hazardous (Raharjo *et al.* 2018). This is partly reflected in the way the same AM name may be used for types of *Citrus* that Nuaulu distinguish lexically.

The category *awane*

The labelled category ***awane*** potentially encompasses all vines, lianas and trailers, ranging from the large woody vines of the deep forest – such as ***punara wane*** (*Piper corylistachyon*) and ***sapate wane*** (*Tetracera scandens*) – to the delicate herbaceous vines of coastal scrub and gardens. However, the same term is used for string, cord or rope (as in Malay 'tali'). There is a separate term for very fine string or thread – ***pola***, as might be made from twisted pineapple, ***sesene*** (*Ananas comosus*), fibre and used, for example, to operate a mouth-harp (Ellen 2019b). However, in a general covert sense, all natural fibres that tie, bind or connect (cord, rope, cable, string, twine, thread) are grouped together semantically. It is, therefore, difficult to disentangle the morphological category from the utilitarian category, and impossible to establish which meaning is historically prior – the plant life form or the cultural use. Both must have reinforced each other over time (Figure 5.5).

There are forty-four Nuaulu plant names described by local people as ***awane***. Of these twenty-two contain the element ***wane*** or ***awane***, either as the first element of a binomial (e.g. ***awane kin-kin***: *Melothria indica*) to indicate that the plant is 'a kind of ***awane***', or as the second qualifying element (e.g. ***hunane wane***: *Merremia peltata*), a type of ***hunane*** that happens to have parts that can be described and used as cordage, as in the manufacture of a sago-cask carrying device (***tonar***, see BM 1972 As.1.94). In four cases, the (***a***)***wane*** element is optional, e.g. ***makanitu*** (***wane***), *Flagellaria indica*, and in a few the position of the (***a***)***wane*** element is variable (e.g. ***papate wane*** = ***awane papate***: *Petraeovitex* sp., used to make the tree-climbing device ***sakaenate***, see BM 1972 As.1.19). These instances suggest that the classificatory status of the plant can be seen either way. Names which include the term ***wane*** cover such diverse forms as vines in the family ARACEAE (e.g ***ken awane***: *Rhaphidophora pinnata*), climbing ferns (***sina-sina wane***: *Pityrogramma*), brambles (***awane unie***: *Rubus fraxinifolius*), filamentous creepers, pepper vines such as ***punara wane*** (*Piper corylistachum*), and woody lianas including ***sapate wane*** (*Tetracera scandens*), which can be up to 18 cm thick and will discharge large quantities of water when cut in the forest, a property that enhances its reputation as a source of liquid refreshment during long forest treks.

In the remaining twenty-two cases the name itself does not contain the element ***wane*** in any form, for example ***nonanae*** (*Donax canaeformis*), the bark of which is used to make small baskets (***pakune***). However, other plants with a similar habit and with uses for binding, notably ***meute***

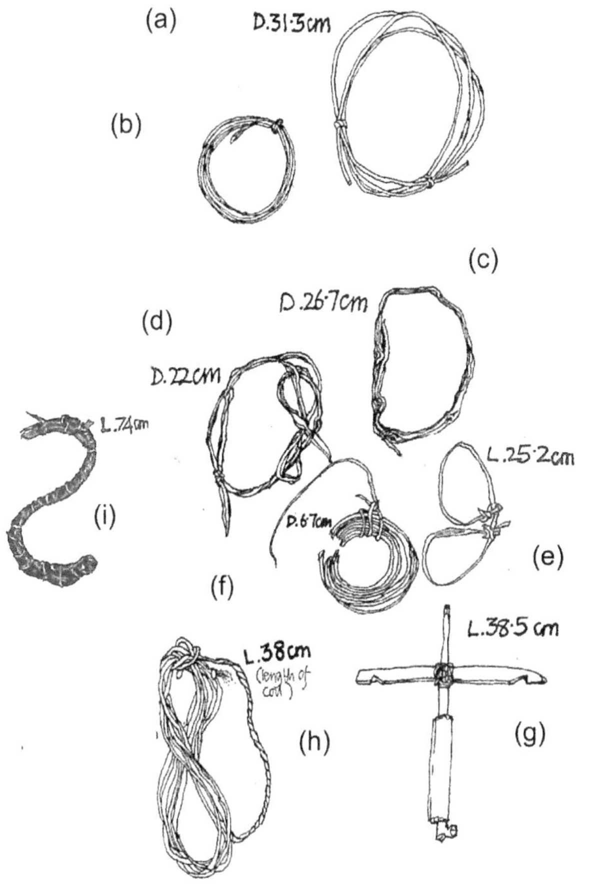

Figure 5.5 Examples of cordage, illustrating relationship between concepts of ***awane*** *(vine, liana),* ***meute*** *(rattan) and* ***naha*** *(Arenga pinnata and its fibre) as plant categories and as homonymous types of processed fibre used for binding. Original image (courtesy of the Trustees of the British Museum) taken of material culture specimens deposited in 1971 (As1972.01, Register 20): (a)* ***meute hehue****: thin coiled rattan used for securing roof thatch to battens and rafters in house construction (As1.1972.15); (b)* ***meute onate****: strong thick rattan, split, for securing main house timbers (As1.1972.16); (c)* ***ken awane*** *(Tetrastigma lanceolarium): tough liana used in house construction (As1.1972.18); (d)* ***sakaenate****: device for climbing coconut palms and similar (As1.1972.19); (e)* ***kahau****: suspension string used in house for storage (As1.1972.20); (f)* ***isane*** *(Lygodium flexuosum): climbing fern used for binding generally, but when coiled in way pictured described as* ***pepe*** *(As1.1972.21); (g)* ***maue nahue****: device for making rope from Arenga pinnata fibre (****nahue****; AM 'gamutu'), consisting of wooden pin loosely encased by bamboo cylinder with notched vegetable-fibre cross-bar secured by rattan (AS1.1972.22); (h)* ***nahue****: course black fibre from around the base of the petioles of the* ***naha*** *palm (Arenga pinnata), used in manufacture of firelighters etc. (As1.1972); (i)* ***sakate****: length of fibres of* ***naha*** *doubled and bound with thin rattan, used as firelighter (As1.1972.30).*

(rattan), which we will consider in more detail in the next section, are not considered ***awane*** at all. Like ***awane***, the term ***meute*** labels both a kind of plant and a material used for binding and other purposes. In a slightly different sense, the course fibre found around the petiole bases of ***naha***, the aren palm, *Arenga pinnata*, used widely for making rope and string, are in utilitarian terms a source of ***awane*** – perhaps the most important source – but are never placed in the plant category of that name. Yams too are not ***awane***, although they may occasionally be described as 'vine-like', either because of their overall habit or because the function of their tubers as food semantically trumps all else.

The category *meute* (rattans)

Rattans are grouped by Nuaulu as a separate and quite distinct kind of named plant, distinguished from standing palms (Table 5.8), but also from other climbing forms. In other words, they are never subsumed under ***awane***, in the sense of vines and lianas in general. Rumphius (2011 [1741–50]), in Volume 3 of the *Ambonese Herbal* calls them 'palm-reeds', a term that emphasizes their interstitial character. Although the folk concept of a rattan is clear enough, it is a problem in scientific palm taxonomy because, despite its apparent distinct morphology as a climbing palm, its other features make it impossible to yield a single separate taxonomic category of rattans that is phylogenetically plausible (Dransfield *et al.* 2008). Rumphius was content to follow his local interlocutors here, and it is true that the Ambonese Malay concept of 'rotan' is transparent, but in the *Ambonese herbal* Rumphius portrays it as a kind of hybrid group related to both palms and rushes. At one point he also describes them as 'degenerate' palms. Interestingly, Nuaulu follow wider Austonesian practice in clearly marking an unambiguous category here. This is partly a consequence of emphasis on the material features of rattan that make it a particularly important binding material, but also because local ecology and palm biogeography do not present any inconvenient intermediate palms to prevent a clear distinction being made. The classification, after all, is based on local knowledge, not on the knowledge we now have as scientists based on the accumulation of global information.

Different kinds of rattan are typical of different kinds of forest. Though not actively managed, by leaving rattan and its host trees intact when clearing forest, Nuaulu impact on its distribution. Nine distinct rattans are named locally, and are mainly distinguished for the qualities of the cane that they yield for various manufacturing and construction activities, though some have parts considered edible. The main use of rattans is, indeed, as ties, cordage and binding in housebuilding, but also for a wide range of equipment and ritual objects, and especially in basket-making. Rattan continues to be mandatory in the construction of sacred clan houses and in the making of baskets used in ritual, despite access to modern equipment and new kinds of containers. Rattan is also collected for exchange, though more so between 1970 and 1990 than subsequently. The tissue surrounding the apical bud (the heart or cabbage) is sometimes eaten, together with the young shoots, but not usually the fruits.

As it happens, there are also approximately nine scientific species confidently known from Seram, though no doubt there are more to be discovered. However, how Nuaulu categories and scientific species map on to each other is not entirely clear. Nuaulu swidden-management practices encourage the species they generally use from near forest, and discriminate against those species they tend not to use. All are said to be readily accessible to Nuaulu in Rouhua (that is, easily reached, cut and stripped within a day's trek), though one (***meu peune***), valued for its edible young shoots, is said only to be found in distant forest north of the Ruatan river. Whatever the case, in terms of Nuaulu conceptualization, strong shared uses for, and

Nuaulu name	Scientific name	Notes
meu hahu kikae	*Calamus*	housebuilding
meu hehue	*Calamus*	thin cane for tying thatch and housebuilding
meu munate	*Calamus*	thick cane; edible shoots termed ***meute wasauru***
meu nunte	*Calamus*	small thin canes
meu onate	*Daemonorops*, probl. *calapparius*, *D. robusta*, poss. *D. beguinii*	medium diameter rattan for securing house timbers, and as lasso for catching pythons
meu patona	*Calamus*,[1] *Daemonorops*	
meu peune	*Calamus*	edible young shoots, distant forest only
meu wa ukune	*Calamus*	basketry, housebuilding
meu werane	*Calamus*	

*Table 5.8 Nuaulu rattans (**meute**).*

Note: 1, The genus Calamus on Seram may include one or more of the following: C. albus Pers. (C. rudentus Roxb.) – a 'rotan putih,' known from Ambon; C. cawa Bl. – Moluccas gen.; C. equestris Willd.; C. robinsonianus Becc. – Seram, Ambon; C. rumphii Blume – Moluccas gen.; C. zollingeri Becc. – AM 'rotan air,' 25–60 mm cane, once exported from Seram; or other species. Source: Kew and Bogor voucher determinations: Dransfield and Manokaran 1994; Mogea 1991; WCMC 1994.

ecological experience of, a relatively small number of rattan species (compared, say, with Borneo), reinforce boundaries of a category that for Nuaulu is physically distinct, despite the views of scientific palm specialists (Dransfield and Manokaran 1994; Dransfield *et al.* 2008). Taxonomists place rattans within the subfamily Calamoideae, along with other standing palms, most notably *Metroxylon sagu* (sago), which for Nuaulu is a singular salient cultural keystone species. Although its trade continues to decline along with rattan use in building, the case of rattan exemplifies a principle I have described elsewhere (Ellen 2009a) as cross-domain resilience: if a resource is used across a range of divergent purposes (say, construction, food and ritual), it is less likely to be vulnerable to knowledge loss than when a resource supports a single purpose.

On not seeing the wood for the trees

To summarize, what might be seen to be problematic when we examine the cognition of tree diversity, though not a major problem for the much smaller vine life form, is that while there are 423 kinds of 'tree' (***ai***), there are very few intermediate complexes that can be reliably established.[2] This is surprizing given theories of cognitive economy and 'chunking', including those subscribed to by Miller (1956), Rosch (1977) and Bloch (1991), which would seem to predict that in order to memorize and retreive plant names and information associated with them, we should expect shared systematic grouping strategies that organize the continuity of tree diversity into manageable portions. Yet, I have identified here only three plausible complexes based on botanical affinity, and a few other tree and vine complexes reinforced by – if not based on – use factors. There remain a large number of tree species that do not appear to be grouped with any others or subdivided further. This is partly evident in the prominence of uninomials rather than binomials, but also in the absence of informant data on how they group these trees. This is not because I neglected to interrogate people on the matter, but because people just seemed resistant to the idea of systematic organization of this kind. It appears that Nuaulu just 'know them'. In the case of the identification skills of bird-watchers, we have learned that the way birds are easily placed in family-type groupings is a powerful deductive aid for identifying individual species (see, e.g. Ellen 1993:72–3; Hunn 1977:46). This process of categorization is based on the ability to discern relatively discrete cognitive prototypes, mainly based on overall visual features. This does not appear to be the case for many trees, and certainly not in a forest context where the continuous character of the canopy and the close proximity of trees of different kinds occludes the visibility of individual trees. Moreover, while tree-species diversity considered at a global level may yield distinctive family-like groups for professional botanists (though not generally families composed of trees only), there are often insufficient examples of members of one family in the local ecologies with which Nuaulu are familiar to make this plausible. If a botanical family is well represented, then the features that taxonomists use to group them are those that Nuaulu regard as unimportant, for example classic Linnaean characteristics relating to the morphology of sexual organs. If we separate all named categories Nuaulu consider to be trees (423), of these 143 are indicated as uninomials and terminal categories referring to species that are plausible non-domesticates. The linguistic evidence would suggest that they are not further subdivided so as to be part of an obvious group. To these can be added non-domesticated-tree categories labelled with binomials of the form ***ai*** + (e.g. ***ai osi***), where ***ai*** is the life form prefix (59). This gives approximately 202 unaffiliated tree categories. Table 5.9 lists examples of these unaffiliated singular categories labelled as uninomials, and Table 5.10 lists examples of unaffiliated binomials with the prefix ***ai***.

So, we need to ask, given the absence of intermediate larger categories into which to group these singular species, how they are memorized, and how information about them is articulated in appropriate contexts. In some cases, we might say that certain categories mark the existence of complexes, but because of their geographic remoteness or lack of direct interest for Nuaulu, these are not linguistically elaborated. Thus ***hana***, although only supported by voucher evidence applying to *Brugeria gymnorrhiza*, does seem to be used for all mangrove species. Mangroves, as species of the brackish estuaries, are seldom encountered by Nuaulu, and although the wood has many virtues, it is rarely used by them. We can also see that in some cases unaffiliated terminal categories seem to occur in pairs (e.g. ***ai yarine*** and ***anametene***, or ***hukila*** and ***kawasa***). Such pairs are not necessarily closely related botanically, but a particular physical feature, ecological association or common use may motivate the classificatory linkage. Another feature of tree naming, is the pairing of types of tree recognized as domesticates or semi-domesticates with types of tree in more remote forest which superficially resemble them, but which are not closely related phylogenetically, such as ***tom-tom*** (Salicaceae: *Flacourtia rukam, F. inermis*) and the marked ***tom-tom wesie*** (Phyllanthaceae: *Phyllanthus borneensis, Glochodion* sp.). Of course, all unaffiliated categories must ultimately be subject to some form of comparison with other species for them to exist at all as categories.

2 Of the 358 terminal plant categories judged to be 'trees' and for the 62 descriptions of different kinds of terminal category specifying ***ai***, we have no vouchers for 16 and 19 per cent respectively. This must reflect to some extent the difficulty in locating trees and obtaining specimens, though in some cases these were easily recognizable species (often bearing edible fruits) that did not require voucher specimens for identification.

Nuaulu name	Scientific name	Scientific family	Notes
ahauni	*Antidesma stipulare*	PHYLLANTHACEAE	used for making house hooks
aikau	*Rhodamnia cinera*	MYRTACEAE	
akioae	*Premna serratifoia*	LAMIACEAE	
anae ate	*Xylopia malayana*	ANNONACEAE	no use recorded
anametene	*Diospyros rostrata*	EBENACEAE	paired with ***ai yarine***; 'parang' handles
ananione	*Horsfeldia bivalvis*	MYRISTICACEAE	
anapakane	*Leea indica*	VITACEAE	eaten by cassowary and deer
aniono	unident.	EUPHORBIACEAE	yellow dye/paint
apunue	*Aglaia sapindina*	MELIACEAE	associated with cuscus (Phalangeridae)
asaherane	*Virola surinamense*	MYRISTICACEAE	extensively used for stunning fish
asuhunne	*Dysoxylum caulostachyum*	MELIACEAE	
atupikae	*Syzygium aquem*	MYRTACEAE	sea-shore; used for wood
	Phaleria perrottetiana	THYMELAEACEAE	
aunasate	*Prunus javanica*	ROSACEAE	planted in gardens
	Cryptocarya	LAURACEAE	
hana	*Brugueria gynmnorrhiza*	RHIZOPHORACEAE	good firewood
***hana*(*panne*)**	*Neolamarckia cadamba*	RUBIACEAE	
hukila	*Eugenia* sp.	MYRTACEAE	paired with ***kawasa***; used as ***nene*** for head-cloths
kasupa	*Bixa orellana*	BIXACEAE	
kinae	*Kleinhovia hospita*	MALVACEAE	
kupanaia	*Plumeria rubra*	APOCYNACEAE	Engl., 'frangipani'
panne	*Nauclea* sp.	RUBIACEAE	
rene	*Erythrina variegata*	FABACEAE	tree of the sea-shore with spectacular red flowers
riane	*Elaeocarpus augustifolius*	ELAEOCARPACEAE	
saumahu	*Abroma augusta*	MALVACEAE	leaf fibre used as thread
tonate	*Intsia bijuga*	FABACEAE	AM 'kayu besi'

Table 5.9 Non-domesticated trees: selected unaffiliated categories labelled with uninomials.

Nuaulu name	Scientific name	Scientific family	Uses
ai msinae	*Syzygium rumphii*	MYRTACEAE	used in male initiation ceremonies
ai suto	*Premna serratifolia*	LAMIACEAE	taboo for Peinisa
ai turi	*Senna alata*	FABACEAE	treatment for eczema and ringworm
au oine	*Endospermum moluccanum*	EUPHORBIACEAE	house construction

*Table 5.10 Non-domesticated trees: selected unaffiliated categories labelled with binomials prefixed with **ai**.*

6

Pandans, palms and bamboos

Introduction

This chapter looks at three large groups of plants that are both physically salient in the landscape of Seram and of high cultural importance: pandans, bamboos and palms. All three happen to be monocotyledons, which may not be entirely coincidental, as in terms of angiosperm phylogeny, monocots tend to form a set of higher taxa that are each decidedly distinctive from a visual point of view, compared with the more continuous distribution of the growth forms of dicotyledons. This reflects a greater evolutionary distance between the different groups of monocots (Chase 2004). Arguably, the reflection of evolutionary distance in physical appearance reinforces their cognitive salience as separate groups for Nuaulu. All three groups form intermediate ethnobiological categories, if not LIFE FORMS by Berlin's definition, but all are covert in the sense that in each case there is no overarching term that comfortably labels the category at the most inclusive level.

While the material distinctiveness of pandans, palms and bamboos might be thought sufficient to warrent their perceptual salience as separate groups, in each case the distinctions are bolstered further by their cultural functionalities. Individual plants and selected species in each group are managed. But Pandans are the least managed and are seldom planted, growing mainly along the coastal strip and paths, being essentially a plant of the margins. The palms can be divided into two groups: the larger number of species, which are mainly solitary inhabitants of secondary forest and are seldom managed or used; and a smaller number of economically significant species that are intensively managed and grow at much higher population levels (coconut palms in large groves and as single trees around the village, sago palms mainly in swamp land, but also individually in gardens and sometimes on the village edge, and *Arenga* and *Areca* palms mainly planted in clumps in old gardens and managed fallow).

Pandans

Pandans (screw pines) are a clear example of where shared cultural recognition of a physically distinctive group is clear from the lexical evidence, even though there is no single term for the entire grouping. Two terms describe most pandans: ***koae*** and ***saune***, though ***koae*** is the most prominent, with ***saune*** sometimes being referred to as a type of ***koae*** (***koae saune***), though never reciprocally. This kind of naming practice is also found with certain bamboos, and – as we shall see – palms, and through a polythetic-chain effect conjoins all distinct names and types into a single group. The word ***koae*** also refers to mats made from pandans, and ***saune*** to the use of the fruits as food and as a dye. So, there is a utilitarian element in both nomenclature and classification, and evidence of the lexical conflation of a term for a plant type and cultural object that we can also see in other areas of Nuaulu ethnobotanical classification. In this connection we need to note that *Pandanus* is, when not sterile, a dioecious genus, with only the female plants bearing fruits that can be used as food and dye-stuff. This is sometimes reflected in the naming.

Pandans are mainly plants of the sandy coasts of Seram, occuring as littoral thickets, on the edge of brackish marshes and mangroves, and along watercourses at low altitude. They are informally managed, and sometimes deliberately moved and propagated vegetatively, especially in more interior areas; only *Pandanus amaryllifolius* does not really grow wild at all, and it rarely flowers. Some species are shrubby, others are coarsely branched trees with huge prop roots. They tend to be quick growing. The young leaves of some species are eaten, like palm cabbage (de Guzman and Siemonsma 1999), and used for food flavouring.

The terms and identifications for Nuaulu pandans and their perceived allies are set out in Table 6.1, with an indicative general folk classification in Figure 6.1. Non-utilitarian characteristics used by Nuaulu when distinguishing pandans include leaf breadth, presence or absence of thorns on leaves, certain features of the fruit, and the general architecture of the plant reflected in branching and prop-root arrangements. ***Koae*** is clearly divided further into four types: ***kokoio*** (Figure 6.2), ***koae putie*** and ***koae metene*** (the male and female respectively of the broad-leafed *Pandanus amaryllifolius*), and ***koae kinoe*** (*Pandanus* sp.), ***kinoe*** being a kind of mat. In addition, ***koae*** optionally includes ***saune*** (as noted in the preceeding

Nuaulu name	Scientific name	AM and Indonesian names	Characteristics	Uses
kinawane	*Agave cantala*	'daun tikar'		
koae	*Pandanus amaryllifolius*	'pandan wangi', 'kokoia'	very tall, taller than coconut palm	produces a red dye, fragrant leaves used as flavouring
koae koio, kokoio, kokoio nuae	*Pandanus* probl. *polycephalous*	'kokoia', 'kakowa', 'daun kikir'	fruits round, branched, narrow leaf; shrubby, much in forest fallow	mat weaving
koae kinoe	*Pandanus krauelianus*	'kakoia'		mat weaving
koae metene	*Pandanus amaryllifolius* ♀	'sanoma', 'daun tikar'	black fruits; said to resemble ***kokoio***	
koae putie	*Pandanus amaryllifolius* ♂		does not fruit; found at Tihun	best for mat weaving, used in male puberty ceremonies
paone	*Pandanus krauelianus*	'daun tikar'	inland, mature regrowth; said to resemble ***kakoio***	mat weaving
sanoma	*Pandanus*		very wide leaf	
(***koae***) ***saune***	*Pandanus conoideus*	'pandan Seran', 'saun', 'pandan hutan'	broad leaves, oily pericarp, crooked trunk	fruit eaten,
saune hahu	*Pandanus tectorius*		narrow leaves, planted in forest fallow	edible pericarp, used for sauce
saune hehue	*Pandanus conoideus*		finer fruit than ***hahu,*** planted	

Table 6.1 Nuaulu pandans and their allies.

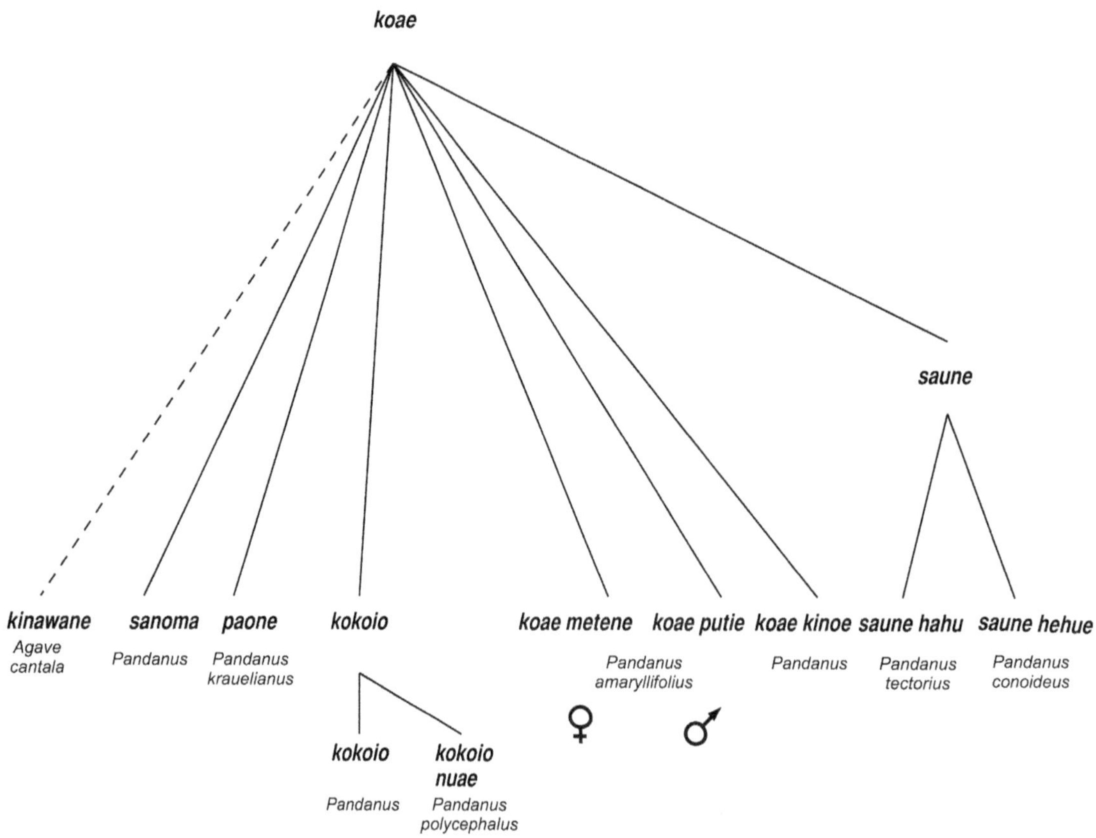

Figure 6.1 The inferred folk classification for pandans and their allies.

Figure 6.2 **Kokoio** *(Pandanus probl. polycephalous): (a) a branching form growing on old garden land behind Rouhua, August 2003; (b) close-up of* **kokoio** *fruit in the mid-Upa valley, near Rouhua, July 1996.*

Figure 6.3 (a) **Paone** *(Pandanus krauelianus) growing in secondary forest near gardens at Iso Nawe, upland and north of River Upa, August 2003; (b) A massive* **koae putie** *(Pandanus amaryllifolius), between river Upa and Rouhua, indicating size of prop roots, March 1996.*

paragraph), while the wide-leafed ***sanoma*** (*Pandanus* sp.) and the slender non-branching ***paone*** (*Pandanus krauelianus*: Figure 6.3a) are considered ***koae*** by virtue of their uses, but do not carry the prefix. ***Saune*** is divided into ***hahu*** and ***hehue***, the first *Pandanus conoideus* having an oily pericarp and the second *Pandanus tectorius* having a pericarp consisting largely of carbohydrate (Verheij and Coronel 1992:240–343). The leaves of various species of pandan are dried, decorated and plaited into objects. ***Koae*** is preferred for hats, mats (***kinoe*** mats, ordinary household mats, bedding rolls, mats for wrapping corpses, baskets, ***tasi matahenne*** – the special betel-requisites pouches used in male puberty ceremonies – and ***tine nene siae*** charms), though certain species are thought superior to others.

One plant category that does not match a pandan phylogenetically is nevertheless linked to this grouping by Nuaulu. This is ***kinawane*** (*Agave cantala*), the leaves of which are used like those of pandans, to make mats and bedding rolls. *Agave* is in the Asparagaceae and both *Agave* and the Pandanaceae, which have a palm-like habit and simple linear leaves, are placed with palms in the order Arecales. On the other hand, *Freycinetia* probl. *ceramensis* (***sina-sina wane***), a climbing pandan, is not considered by Nuaulu as part of this grouping at all and is placed with vines and lianas.

All pandans are glossed by Nuaulu with various Ambonese Malay and Indonesian dialect terms (e.g. 'kakaoio', 'pandan', 'kakowa', 'daun tikar', 'kikir', 'saun'), which reinforce the inference that Nuaulu treat pandans as a clear separate grouping. Rumphius (2011, vol. 3:433–63) too follows AM usage and groups all pandans (for him 'pandang') together in Book 6, and Chapters 80–9 of the *Ambonese Herbal*. He follows the true pandans with 'rope-like pandangs' (e.g. *Freycinetia* spp.) in a departure from contemporary Nuaulu classificatory practice, but (echoing Nuaulu treatment of *Agave cantala*) includes the non-pandan *Scirpiodendron ghaeria*.

Palms

Rumphius (2011, vol. 1) lists numerous palm species for Ambon and Seram, which he presents together at the beginning of Book 1 Chapters 1–20, and in which he also includes *Cycas*. His species were reduced to thirty by Merrill (1917) and thereafter to about twenty by van Slooten (1959:333–6) and to twenty-two by de Wit (1959), after eliminating *Cycas*, synonyms and some cultivars of the larger useful species. However, it is clear that the list can be further shortened by absorbing some subspecific taxa into existing species, and we can also note that some were uncommon even in the seventeenth century, and on the edge of their range. Dransfield *et al.* (2008) report nineteen non-climbing palm genera (i.e. not including rattans) for the Moluccas in general: *Areca, Arenga, Borassus, Calyptrocalyx, Caryota, Drymophloeus, Heterospathe, Hydriastele, Licuala, Livistona, Metroxylon, Nypa, Oncosperma, Orania, Pholidocarpus, Pigafetta, Pinanga, Ptychosperma*, and *Rhopaloblaste*. Today some of these species are only occasionally seen on Seram. In addition to those described and named for the Nuaulu below, other palm species in other sources not reported for Nuaulu include two mentioned by Florey and Wolf (1998) for Alune-speaking areas of west Seram: *Licuala ramsayi* ('libune') and *Borassus flabellifer* ('baune').

Nuaulu recognize and name sixteen scientific genera and seventeen species of palm, and organize their knowledge of these using fifteen uninomials and between thirty-eight and forty binomials, fourteen of which describe folk-varieties of the sago palm, *Metroxylon sagu* (Table 6.2). Nuaulu are well aware of the similarities between what we call standing palms in terms of their overarching architecture and reproductive pattern; and together these might be said to form a weak covert category – a grouping that can be inferred to exist in local thought and practice but which is not named. The recognition of standing palms as a group is reflected in a common series of partonyms that are used distinctively when compared to other plants (Figure 6.4). Rattans, as we have seen from the discussion in Chapter 5, are separated from other palms, and are by contrast explicitly named as a group (***meute***), which includes the genera *Calamus* and *Daemonorops*. This is a classification based on growth form, which contemporary scientific taxonomists and palm biologists recognize as a folk category, useful when engaging with the wider public but not significant for formal scientific classification, and with low phylogenetic meaning (see e.g. Dransfield and Manokaran 1994). *Nypa*, which has a more-or-less prostrate stem and some rattan-like characteristics, as Nuaulu admit, is otherwise grouped with standing palms. Though ***ai samine*** was offered as a synonym of ***mapua*** (*Hydriastele costata*), Nuaulu do not consider palms to be trees (***ai***), which are woody and branching, in the normal sense; but they are disproportionately important symbolically given their high utilitarian value (Ellen 1998a:64). One feature that in part explains the ambiguity of palms as a group, and their weakly covert character, is that it contains three taxa which individually have a high profile in Nuaulu subsistence, economy and ritual life – *Metroxylon sagu, Cocos nucifera* and *Areca catechu* – and this is reflected in the internal diversity of each genus and a lexical polytypy marked by numerous binomials.

Some palm species are grouped together by Nuaulu on account of their edible starch, namely *Metroxylon sagu, Nypa fruticans* and *Caryota rumphiana*. The edible starchy parts of this latter palm, which Nuaulu call ***panuke***, occur only in the mature-growth form, indicated by a swelling at the base of the stem. A perceived resemblance between this palm and *Oncosperma filamentosa* is reflected in a pairing of the two kinds, and in the synonym ***katue panuke***.

Nuaulu name	Scientific name	Notes
hatane	*Metroxylon sagu*	incl. 14 folk varieties
huaena	*Rhopaloblaste ceramica*	AM 'pinang hutan'
inene	*Hydriastele microcarpa*	
kanai	*Areca catechu*	incl. 8 folk varieties
kanai apane	*Areca glandiformis*	
katue	*Oncosperma filamenta*	AM 'nibung'
katue panuke	*Caryota rumphiana*	immature form
korane	*Orania sylvicala*	
makoi	*Pigafetta filaris*	used by Christian non-Nuaulu for making palm beer (AM 'sageru')
mapua syn. ***ai samine***	*Hydriastele costata*	AM 'pinang hutan'
naha	*Arenga pinnata* var.	AM 'mayang', 'aren'
naha niane	*Arenga pinnata* var.	
nasinana	*Saribus rotundifolius*	AM 'daun seribu'
nione	*Cocos nucifera*	incl. 12 folk varieties
panuke	*Caryota rumphiana*	mature form
piasoi	*Drymophloeus oliviformis*	
salak	*Salacca zalacca* var. *amboinensis*	
tepine	*Calyptrocalyx spicatus, Pinanga rumphiana*	
tepine huaena	*Nypa fruticans*	

Table 6.2 Nuaulu categories for species of palm (excluding rattans).

However, *M. sagu* is a multi-purpose mega-species that is so significant for Nuaulu that their identity as a cultural group is strongly connected with its continued use and management (Ellen 2004b, 2006a). A second group of palms (Ellen 1991a) is identified by Nuaulu because its fruits can be chewed with other ingredients to produce or simulate mild psychoactive effects, the sharing of which are important in many ritual and other social contexts: *Areca catechu*, *Areca glandiformis*, *Calyptrocalyx spicatus*, *Drymophloeus oliviformis*, *Hydriastele microcarpa* and *Pinanga rumphiana*. Separated from the starch- and betel-nut-producing palms is coconut (*Cocos nucifera*), which is consumed with virtually every meal, and which is also of major commercial importance. I have elicited twelve coconut folk-varieties named by Nuaulu, including ***nione nene sanae***, the branching coconut, which Rumphius (2011, vol. 1:219) had earlier noted and regarded as a 'miscreation'.

Amongst the lesser palms, ***korane*** (*Orania macrocladus*) is planted and used for bows. Pith (sago) is extracted from older palms of *Oncosperma filamentosa* (***katue***), while edible 'cabbage' is harvested from the younger palm, and the trunk wood of both provides material for door bolts. ***Nasinana*** (*Saribus rotundifolius*) wood (seen as related to ***panuke*** and ***kanai***) is used for bows. The nut of ***tepine*** is similar to ***kanai***, and can be employed as a betel-nut substitute. ***Panuke*** is sometimes described as a kind of ***tepine*** (*Calyptrocalyx spicatus*, *Pinanga rumphiana*). The use of binomial synonyms for certain palms otherwise known by uninomials betrays Nuaulu recognition of resemblance. This can be seen in Figure 6.4, where ***tepine*** is used as an overarching term to group ***huaena*** (*Rhopalablaste*) through ***tepine huaena*** (*Nypa fruticans*), ***panuke*** through ***tepine panuke*** (*Caryota rumphiana*) and ***tepine*** itself (*Calyptrocalyx*, *Pinanga*). *Caryota rumphiana* is, as we have seen, lexically distinguished into mature (***panuke***) and immature (***katue***) forms, as well as being called ***tepine panuke*** to show its affinities with other palms called ***tepine***. Finally, ***katue*** is used separately to refer to *Oncosperma filamentosa*, to distinguish it from the ***katue panuke*** phase of *Caryota rumphina*. Thus, through overlapping and synonymous terms that emphasize different characteristic of the palms, they form links in a chain from which we can infer a covert subgrouping. As some of the functional groups of palm species overlap, we have further evidence for the existence of the larger covert grouping of PALMS, following the principle of analytic induction (Ellen 1998a). Moreover, Nuaulu terms for palm morphology vary slightly, but in ways that point to a subgrouping within the larger covert category. For example, there are three different terms used for inflorescence, that used for: a) coconut palm and betel palm; b) sugar and 'nibung' palms; and c) sago palm.

Nuaulu, therefore, seem to recognize a covert grouping of PALMS polythetically by virtue of differential synonymy (Figure 6.5), with characteristics shared between groups of species constituting a network of resemblances that are sometimes morphological, sometimes ecological and sometimes utilitarian. Interestingly, this is just how Rumphius understood their connections. Through his nomenclature, Rumphius recognizes palms as a group

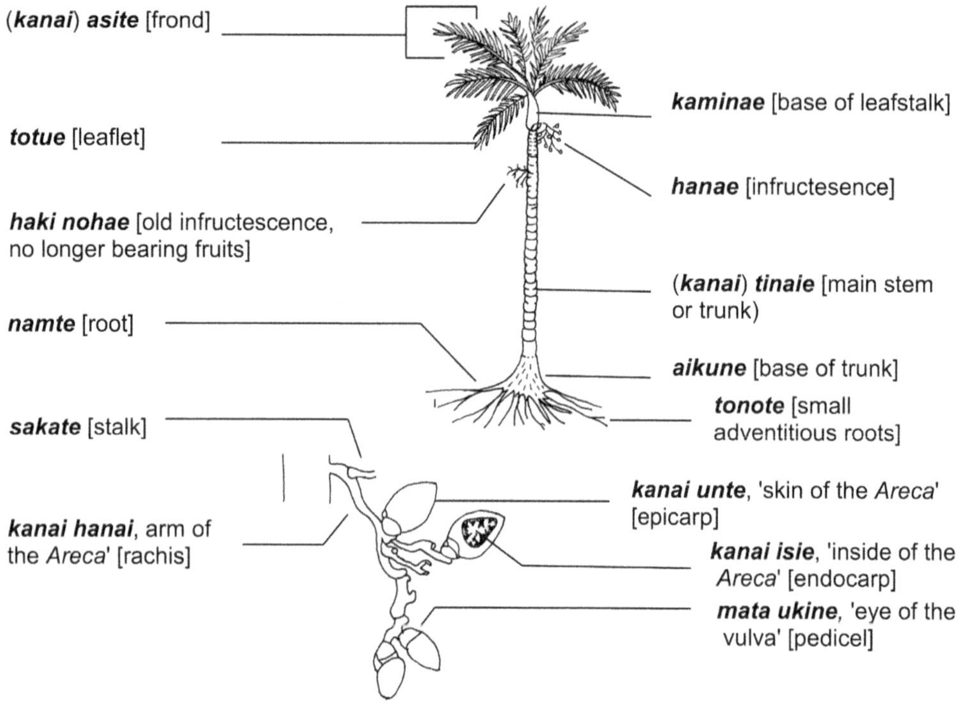

Figure 6.4 Some special features of palms: (a) partonyms as applied to entire Areca catechu plant, and (b) terminology for parts of the Areca infructescence.

based on morphology, but places most of them at the beginning of Book 1, 'Containing all sorts of trees that bear edible fruit and are husbanded by people.' *Cocos nucifera* is placed at the beginning of the series and is clearly regarded by Rumphius as the most salient and defining of palms. I could find no evidence that Nuaulu envisage the coconut in this way, and although physically salient and dominant in the coastal landscape, and with many uses, it does not have the cultural centrality of the sago palm.

Of course, for Rumphius the category PALM was not at all covert, and covered all those trees that in Latin Europeans of the time called 'palma', including species that were common in the Indies but which even today are unfamiliar to Nuaulu. The coconut was pre-eminent in his experience and provided – as it were – the 'type' for the botanical family, a template for the understanding of other palms, and represented what today we might call a kind of cultural keystone species. As we have seen, Nuaulu would certainly argue that for them the *primus-inter-pares* amongst palms is sago rather than coconut, and from a historical perspective coconut has only begun to challenge the prominence of sago in their cultural experience since the movement of clan hamlets from the mountain to the coast, and the growth in the importance of copra as a source of cash income (Broersma 1934), two developments that occurred almost simultaneously during the last two decades of the nineteenth century. The large-scale cultivation of the coconut can be dated only to the beginning of the commercial copra trade in the Moluccas, towards the end of the nineteenth century and peaking between 1920 and 1930. Nuaulu only grew coconut for subsistence before moving to the coast.

After *Cocos* (3 chapters) Rumphius has one chapter on *Cocos* 'miscreations', then (2) 'pinang' (*Areca*); (3) 'wild pinang', followed by a subgroup of 'vinous palms', important for the production of alcohol; (4) *Borassus*; (5) 'wild *Borassus*'; (6) 'sagueer' (*Arenga*); (7) 'wild sagueer' (*Caryota rumphiana*); (8) 'small wild sagueer' (*Drymophloeus oliviformis*); and (9) *Nypa*. Interestingly, while Rumphius treats *Nypa* as connected to other 'vinous palms' because of its drinkable sap, he recognizes its ecological and morphological similarities with sago (*Metroxylon sagu*), providing in his understanding a kind of 'conceptual bridge' between the 'vinous palms' and the starchy palms, pre-eminently *Metroxylon sagu*. Rumphius devotes two long chapters to sago (10), follows this (11) with a chapter on the 'thready sagu' (*Pigafetta filiaris*), which he sees as resembling in some respects rattan. He concludes the account of palms with a chapter (12) on *Cycas*. Like Nuaulu, Rumphius saw the similarities between palms and cycads (*Cycas*: Nua. ***punen***) in a way that is common among many non-botanists today.

Anthropic palm-dominated habitats

As we have noted, most wild palms occur as solitary trees, or as small clumps, in forest and occasionally in open grassland. In forest, given their physical appearance, palms tend to stand out against a backdrop of mainly dicotyledonous trees. However, collectivities of palms can be transformed into two distinctive anthropogenic

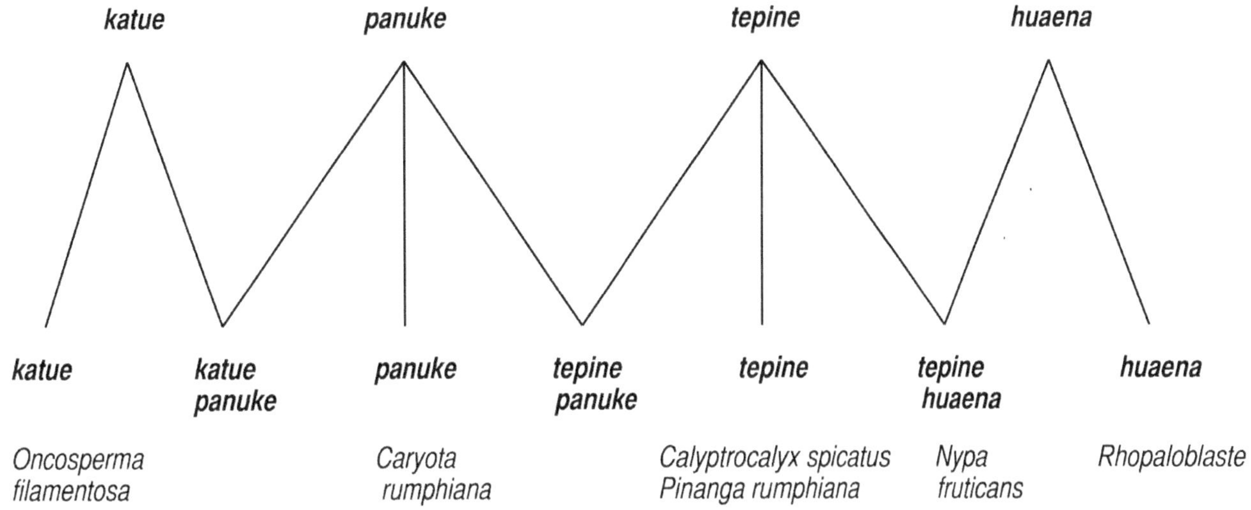

Figure 6.5 Polythetic subgrouping of minor palm categories inferred from synonymy.

habitats in which they are the dominants: in swamps and groves. These managed alterations in their pattern of distribution influence the way in which they are perceived, and inevitably in both cases the changes relate to the ways in which the relevant palm species are utilized.

Nuaulu call swamp ***kopute*** (***kopu*** = putrid, rotting), and apart from the brackish watery zones of estuary and littoral that merge with mangrove, where *Nypa* is often a characteristic feature, swamp land constitutes a special kind of forest dominated by *Metroxylon sagu*, and central to Nuaulu ecological experience. In fact, apart from those plants that are symbiotic or parasitical with the sago palm, there is little additional vegetation to be found in these areas, which are permanently or semi-permanently under water. In south Seram, these swamp-forest areas are largely located along the banks of the rivers around the confluence of the Nua and Ruatan, and at the mouth of the Yala, which empties directly into the Banda Sea. As *Metroxylon sagu* may have been introduced into Seram from New Guinea early in the Holocene (Ellen 2004c), and as the historic activity of human management of these areas is well documented, it is impossible to say what kind of habitat might have preceeded them. In large part, they can now be described as actively managed, though in some larger areas the sago is still considerably in excess of what local people need, and reproduces in part naturally, both sexually and vegetatively. Not only is *M. sagu* the dominant in such areas, a keystone species in terms of its pivotal role in the movement of energy and materials within the swamp ecosystem, but it also constitutes a cultural (or ethnobiological) keystone species in terms of its contribution to Nuaulu diet as the most significant starch staple, and of its role as a ubiquitous construction material. I have documented all of this in a series of separate publications (Ellen 2004b, 2006a), and the details do not need not be repeated here. What I do need to elaborate, however, are the implications of the cultural importance of the sago palm for Nuaulu understanding of their plant world more generally, and of ethnobotanical classification in particular.

First of all, the conceptual salience of sago is reflected in the internal diversity that is labelled (fourteen terminal categories, compared with twelve for coconut and eight for *Areca catechu*). Secondly, the physical prominence of sago in daily subsistence privileges it as an interpretative frame of reference and template for other plants, especially palms. Thus, Nuaulu are more likely to have learned about plant morphology, growth and ecology from encounters with sago than with any other plant, together with the lexical partonymy that accompanies it. No wonder that it correspondingly features in core origin myths, in ritual practice and general cultural imagery. The consequences of this for plant classification are that its sheer overarching presence sets up a contrast between sago and all other palms, making it the conceptual lens through which other palms are understood – particular species being perceived as being more (or less) like sago. A more nuanced view provided by some Nuaulu is to imagine a triangle of salient palm types, with *Metroxylon, Cocos* and *Areca* at the three corners (Figure 6.6), *Metroxylon* contributing the key food, coconut the key liquid and source of cash income, and *Areca* the source of psychoactive stimulation.

If swamp forest is heavily biased in terms of sago management, then the palm associated in a comparable way with groves and plantations is coconut. The main tree crops grown in groves are coconut (***nione***) and clove (***kupanahane***), and to a lesser extent nutmeg (***pala***), coffee (***kopi***) and cacao (***sakalate***). These are all planted primarily as cash-crops, though coconut has a significant role in subsistence, for food and other useful materials. During the first half of the twentieth century, Hevea rubber (***kataa***:

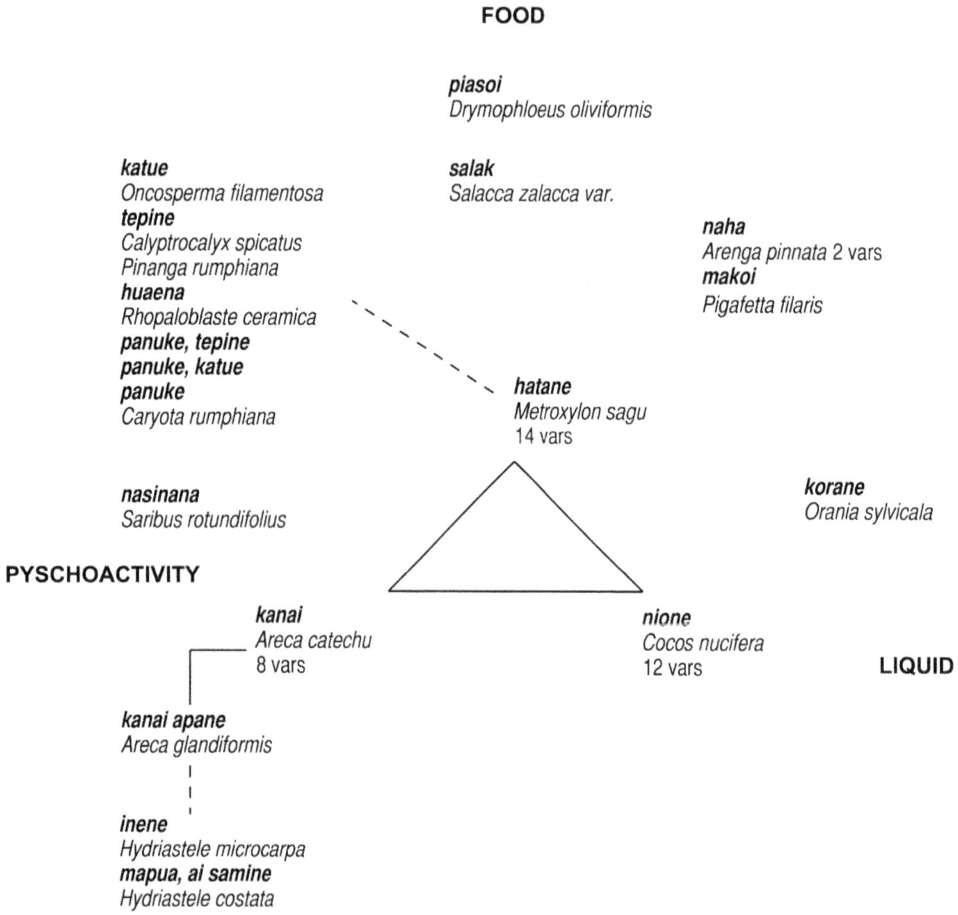

Figure 6.6 Representation of linkages between Nuaulu palm categories based on cultural function.

Hevea brasielensis) plantations were established along Elpaputih Bay, but there is no evidence of it ever being important in the Nuaulu area. In the contemporary period durian (***tunene***: *Durio zebethinus*) has become an important cash-crop for Nuaulu. Though not grown on anything like a plantation scale, it is abundant in clumps and single trees in the managed fallow, agroforests and secondary forests of the upper Upa valley. More recently still, Nuaulu living in the Nuanea area have experimented with *Citrus*. Apart from the focal crops, groves are characterized by grass and *Selaginella* undergrowth, occasional other useful trees, such as 'jambu' and 'kenari' and the odd pandan, plus the important ornamentals, ***sinsinte*** (*Codiaeum variegatum*) and ***sone*** *(Cordyline fruticosa)*. Both these latter are managed and used for ritual purposes, with the first also serving as a boundary plant.

Bamboos

Like palms, the different kinds of bamboo recognized by Nuaulu do not have a collective designation (Table 6.3), but bamboos and some other plants that resemble them physically – and which share similar uses – are treated as a covert grouping. As with pandans, this is partly reflected in agreement on the appropriate terms in other languages with which to gloss them. Thus, in Indonesian and AM both 'bulu' and 'bambu' are used to describe bamboos as a whole, though in spoken Ambonese Malay 'bambu' is preferentially used for certain species rather than others. Rumphius (2011, vol. 3:179–216) reports the Ambonese term 'utte' being used for most of the bamboos he describes, and interestingly uses the term 'tree reeds' or 'reed trees' to describe bamboos and functionally related reeds more generally.

Nuaulu see strong use patterns in this group of plants, and alternative terms for the same plant (e.g. ***ona/wana ona*** and ***tomo seane/wana tomo seane***) link different groups to suggest a more inclusive covert category (Figure 6.7). Of all Nuaulu categories for bamboo taxa, ***wanate*** is the most salient and to some extent serves as a type category to which other bamboo species may be referenced, as in the response 'X is like ***wanate***'. Taylor (1990:50) reports a similar usage in Tobelo bamboo classification, except that the reference category here refers to *Schizostachyum* rather than *Dendrocalamus/Gigantochlea*. Moreover, like palms, distinctive partonyms are used for bamboos and their graminous allies, and general plant terms are applied in a distinctive way (Figure 6.8). I think there is good evidence that in the wider schemes through which Nuaulu perceive

Nuaulu name	Scientific name	Notes
niune	incl. *Miscanthus floridulus*	
niune apane,	*Sorghum propinquum*,	
niwapane	*Thysanolaena maxima*	
niu ikine	*Miscanthus floridulus*	
niu tunne	*Saccharum spontaneum*	AM 'loleba'; thin-walled stem, best for arrows; also for basketry, mats and binding
suenie	*Schizostachyum*	small rangey bamboo, with a rough surface
tapine	*Schizostachyum brachycladum*	AM 'buluh sero'; in N. Seram said to be large type of ***tapine***.
tenne	*Gigantochlea apus*	AM 'buluh teni'; largest, rough; Alune: *Dendrocalamus asper*, 'tenine'
tomone	*Schizostachyum lima*	very smooth, short node
tomo kainun	*Bambusa bambos* var. *spinosa*	AM 'bambu duri'
tomo onate (***kehane***)	*Schizostachyum lima*	
tomo seane	*Schizostachyum lima*	AM 'buluh toi'; rough surface
wanate, ***wana apane***	*Dendrocalamus*	found in long fallow and mature forest
wana hatu	*Dendrocalamus*	***wanate*** are all smooth; all leaves similar
wana ona, ***ona***	*Dendrocalamus asper*	AM 'bamboo kecil', 'bambu betung'; a giant bamboo naturalized on Seram and sometimes planted; its thick walls make it ideal for cooking and building; the shoots are rarely eaten
wana tunne	*Gigantochlea atter*	AM 'buluh jawa'; large var. with internode l. 46–51 cm., circum 26–30 cm.; small var. internode l. 42–6 cm., circum 10–14 cm.
(***wana***) ***tomo seane***	*Bambusa vulgaris* var *striata*	AM 'domar', 'buluh suangi', 'bambu nitu'; used for string and rope

Table 6.3 Nuaulu bamboos and their allies.

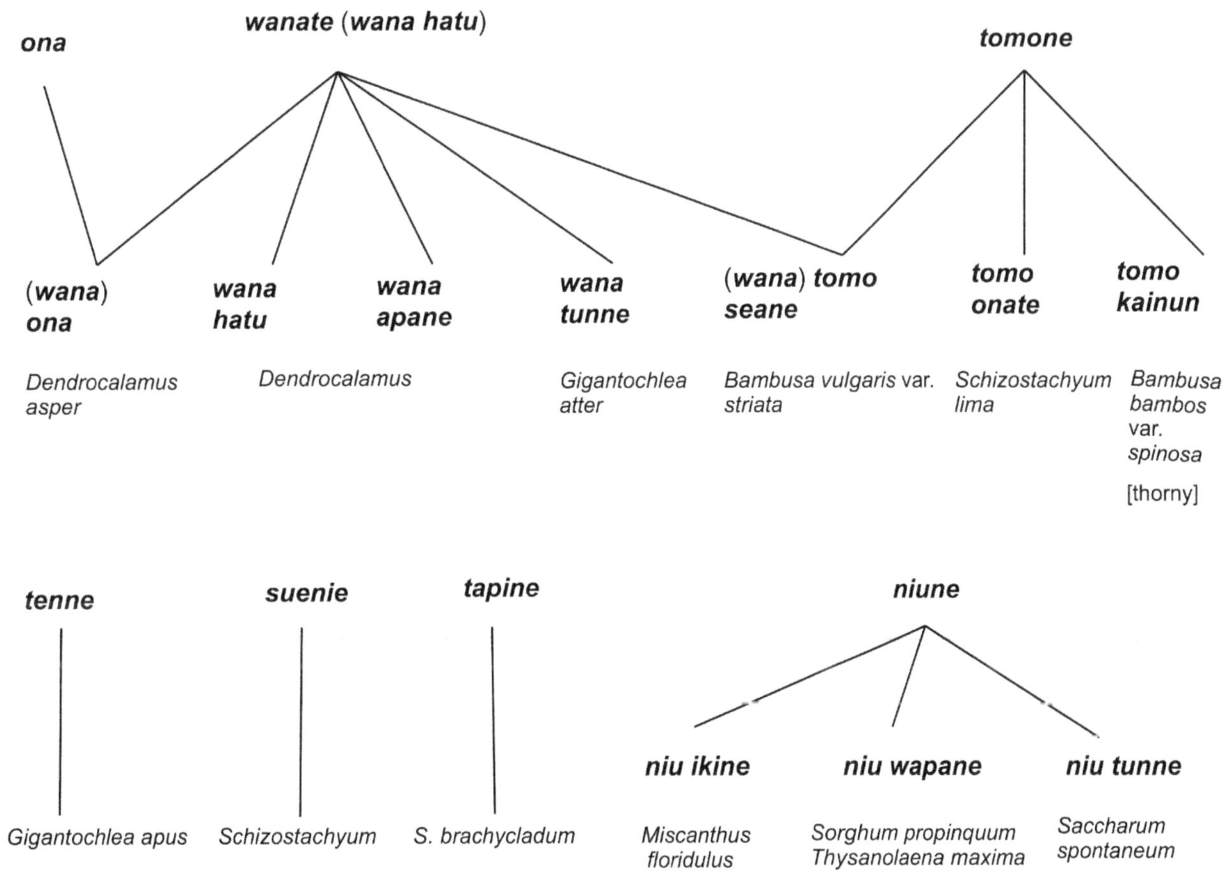

Figure 6.7 Nuaulu folk classification of bamboos and their allies: tree-reeds.

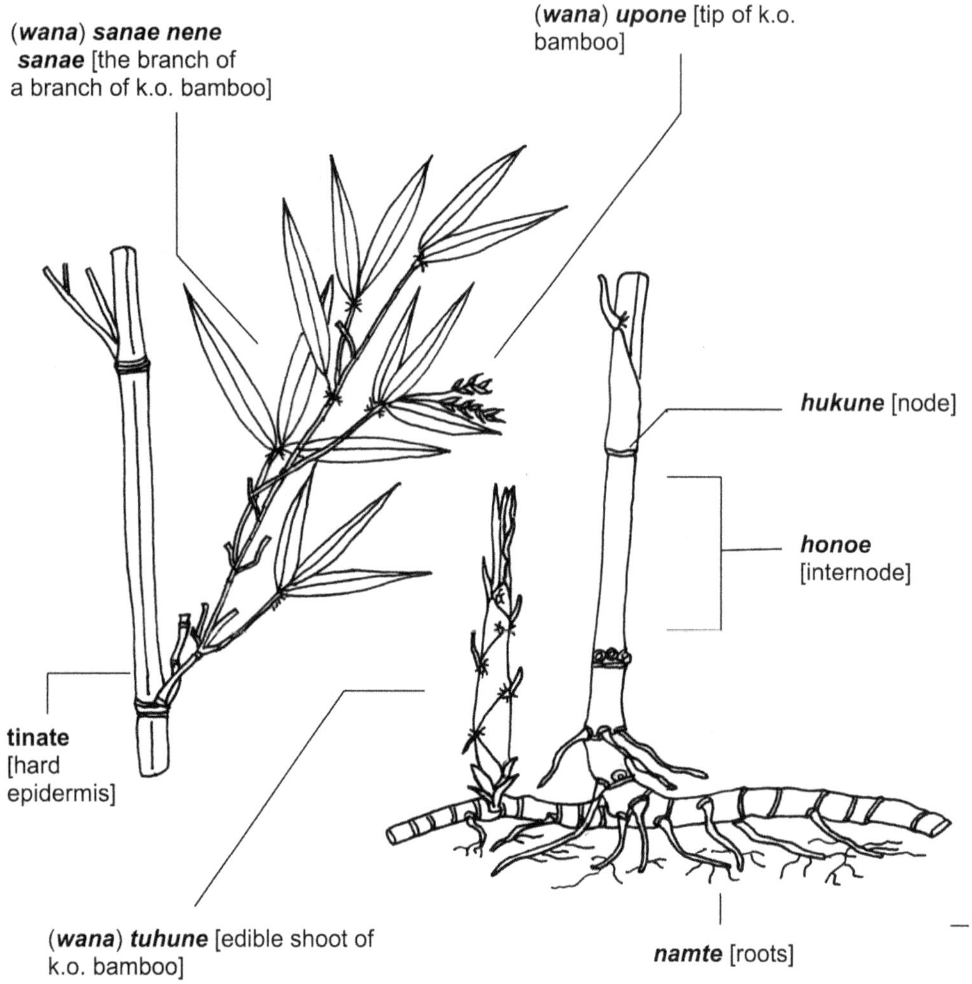

Figure 6.8 Some special features of Nuaulu bamboo partonymy. Reproduced with modifications, courtesy Royal Botanic Gardens Kew.

the plant world, bamboos sit between trees (***ai***) on the one hand, and grass (***monote***) on the other, as connected through reeds (***niune***).

Nuaulu folk classification of bamboo and related reeds comprises three main groupings – ***wanate***, ***tomone*** and ***niune*** – plus five additional named types: ***ona***, ***tenne***, ***suenie*** and ***tapine***. The distinctive features most frequently mentioned by Nuaulu in separating different bamboos were maximum growth height and diameter, length of internode, thickness of internode wall, roughness and smoothness of the epidermis, and in a few cases, the presence of thorns. ***Wanate*** and ***tomone*** account for most bamboos used in construction, and these are both subdivided further. In addition to construction, bamboos are a source of twine (***aitone***) made from immature culms, are used in basket weaving, and as food. The most important bamboo in terms of food is ***wanate***, the shoots of which, known in AM as 'rebong', are called ***wana tuhune***. Large shoots are also sliced and dried to make a preservative. The shoots of other bamboos (e.g. *Schizostachyum*) are seldom eaten, being generally too small and bitter.

Wanate is the most frequently encountered bamboo term, often used interchangeably with its most important subtype, ***wana hatu*** (Figure 6.9). This latter has a hard epidermis, is important in basket weaving, for twine, for arrow points, as a construction material (the only bamboo allowed for sacred houses), and for making the knife used to cut the umbilicus of a newborn child. Most ***wanate*** are species of *Dendrocalamus*, and include ***wana hatu*** (the 'type' folk category), ***wana apane*** (not managed by Nuaulu and regarded as a feature of distant forest) and ***wana ona*** (*Dendrocalamus asper*, a distinctive soft, small and slender species set apart by the alternative use of the uninominal ***ona***). ***Ona*** is used for baskets and containers. ***Wana tunne*** (a long tapering bamboo used to make fishing rods, and resistant to woodworm), confusingly, applies to *Gigantochlea atter*, but is reckoned to provide amongst the largest and best shoots for eating (Figure 6.10). The term ***wana tomo seane*** (alternatively ***tomo seane***) is used for *Bambusa vulgaris* var. *striata* (the yellow culm group) or *Schizostachyum lima*. The young haulms of ***wana tomo seane*** are bright yellow, and are sometimes seen flowering on the village edge.

Figure 6.9 ***Wana hatu*** *(Dendrocalamus) in various contexts: (a) in* ***nisi ahue*** *(long fallow) along the Upa valley, February 1996; (b) on edge of cleared garden in the headwaters of the river Mon, February 1996; and (c) in* ***sin wesie*** *(protected forest) along upper reaches of Mon valley, February 1996.*

Figure 6.10 Freshly harvested shoots of ***wana tunne*** *(Gigantochlea atter) growing in long fallow, mid-Upa valley, July 1996.*

The category ***tomone*** is divided into four: ***tomo seane*** (which we have discussed as part of ***wanate***), ***tomo kainun*** (the small thorny bamboos *Bambusa bambos* var. *spinosa*), ***tomo temine*** (Figure 6.11: a slender rangey bamboo which Nuaulu claim to find only in north Seram), and the large bamboo ***tomo onate*** (syn. ***kehane***), applied to *Schizostachyum lima*. ***Tomone*** has a smooth and glossy surface and short internode. It is used for construction, for example for the sitting-platform of a house, and together with ***tenne*** is considered best for the blade of a sago pounder. There is a prohibition on the use of ***tomone*** by the clan Neipani-tomoien, for whom it is totemic and features in an origin myth.

The remaining categories covering true bamboos (Bambusoidae), those that evidently contrast with ***tomone*** and ***wanate***, are ***tenne*** (*Gigantochlea apus*), ***suenie*** (*Schizostachyum*) and ***tapine*** (*S. brachycladum*). ***Tenne*** (Figure 6.12) is a large bamboo, of about 40–50 cm maximum circumferance, found on the more distant fringes of gardens and groves. It is prohibited for some Nuaulu clans (e.g. Neipani-manunte, in Bunara), but is mentioned in myths and stories as a source material for a tool used when climbing trees. Apart from sago-pounder blades, I have seen ***tenne*** used for ***kawa*** (containers for keeping live birds when in transit), and harvested for its large edible shoots (AM 'rebong'). ***Suenie*** is a small thin bamboo, a little rough to touch, with a brittle stem and splitting easily. It is used for building (for example temporary structures such as ***posune*** menstrual seclusion huts), various tools, and as firewood. I have seen spikes of ***suenie*** stuck in the

Figure 6.11 ***Tomo temine**, a slender bamboo growing on edge of Manusela national park, north Seram, August 2003.*

ground to protect copra from marauding animals, and as a ***kasipate***, a long stick used to knock fruit from trees, in children's blowpipes (***kouomoi***), and placed over developing maize cobs to protect against bird damage. ***Tapine*** is used for arrowheads.

Niune labels a category grouped with and overlapping bamboos. It includes mainly rigid reeds with technological applications for arrow shafts and similar: ***niu ikine*** (*Miscanthus floridulus*) used in basketry (e.g. BM 1972 As 1.319, BM 1972 As 1.532, BM 1972 As 1.60–66), ***niu wapane*** or ***niwapane*** (which includes *Sorghum propinquum* and *Thysanolaena maxima*) and ***niu tunne*** (Figure 6.13), regarded as the best reed for arrow shafts (*Saccharum spontaneum*), and which may be deliberately planted in swiddens for this purpose. The term ***niune*** is used as a synonym for certain kinds of arrow, providing an example of homonymous convergence in plant-category and technological-object names, a phenomenon discussed further in Chapter 7.

Conclusion

Pandans, palms and bamboos are conceptually striking as groups of plants in the context of Nuaulu plant classifiation because of their physical distinctiveness, especially in the way they stand out singly or as multiples against the backdrop of the landscapes in which they feature. While forming covert rather than clearly named categories – PANDAN, PALM and BAMBOO – they are treated as distinctive macro-groupings overall, and none are especially unambiguous conceptually. Each of them is found across

Figure 6.12 *The bamboo **tenne** (Schizostachyum brachycladum): (a) a clump in **nisi ahue** at Sama, north of Rouhua, February 1996; (b) base of clump, showing harvestable shoot (**uhune**) at Sama, February 1996; and (c) base of clump showing distinctive ridged nodes on Yana Ikine tributary of the Upa, March 1996.*

Figure 6.13 The reed ***niune tunne*** *(Saccharum spontaneum): (a) characteristic habit; and (b) close-up of stems indicating utility as arrow shafts; both between Rouhua and Upa river, March 1996.*

a variety of biotopes, but particular species in all three groups are also stereotypically associated with one major biotope: pandans with the coastal strip, palms with groves (*Cocos nucifera*) and swamp land (*Metroxylon sagu*), and bamboo with swidden regrowth and path-sides. However, in terms of the clarity with which classification, naming and identification practices work – compared with pandans and palms – the folk classification of bamboo is more fuzzy, flexible and uncertain. This is mainly related to the physical properties of the plant architecture, stems and leaves (and the infrequency of situations in which seed-bearing organs can be inspected) – all reflecting the practical difficulties of making easy rapid decisions about individual plant types and how to assign them to terminal categories.

I emphasize these points because an underlying but not always explicit assumption of the evolutionist-cum-universalist interpretation of natural-history perception is that plants – and animals – are somehow perceived and conceptualized independently of the context (ecologies) in which they are found. I think that at the very least we need to test this hypothesis. While animals as mobile entities, are – from a perceptual point of view – arguably semi-detached from their immediate environment; this is not true of plants, which are literally rooted in a place and therefore conjoined physically with those other sessile species providing a context in which individual types are prehended. As we shall see in Chapter 10, place is an important consideration in other ways regarding perception of the plant world.

7

Shrubs, herbs and utilitarian groups

Utility and classification

The origins of modern ethnobotany lie in a kind of economic botany that was initially concerned with listing useful plants. Though not without its practical applications, the aridity and limitations of this approach have been much critiqued (e.g. Ellen 1996a), and are one reason why Ford in 1978 could still assert the paucity of ethnobotanical theory (but see Gaoue *et al.* 2017). At the same time, classifications of plant uses have been equally problematic, in part because plants have multiple uses and so the use categories into which specialists place them are often overlapping; and in part because many of the use categories adopted as international standards (e.g. Cook 1995) are highly ethnocentric. For example, the notion of 'social' use of plants (discussed later in this chapter) often fails to reflect local valuations, and generally only weakly reflects local categorization.

The influence of Berlin's approach to ethnobiological classification was to emphasize not so much the cognitive priority of local classifications based on use, but rather general classifications based mainly on visual resemblance – so-called 'natural' classification, or what Atran (1990:212) calls 'common sense' classification. This assertion has meant that special-purpose classifications logically had to exist as a type, in order to accommodate residual groupings that could not be classified as 'general purpose'. While this distinction has led to stimulating debate, and progressed our understanding of how ethnobiological classification works, it is often difficult to sustain in particular cases. In this chapter I examine evidence for Nuaulu groupings of plants based on use.

Nuaulu are fully aware of the difference between a classification based on physical form or habit, and one based on use, but in most cases freely move from one to the other and combine features of each as appropriate to the circumstances. So, while groupings plausibly exist in the sense envisaged by Berlin, I argue here that in many cases the uses to which plants are put is not so much an 'after thought', or even a 'separate thought', but rather an intrinsic part of how plants are conceptualized as 'natural' groups; and that genuinely utilitarian special-purpose groups are routinely incorporated within general classificatory space of the plant world. When special-purpose groupings are not permanent features of the way the domain of plants is conceptualized, they are ad hoc responses to particular situations, and no less important for understanding classificatory practice. The most obvious and widespread instance of this is the way the distinction between cultivated or domesticated plants and non-cultivated and non-domesticated plants not only cuts across the classifications based on morphology, but for all intents and purposes is integral to a unitary cognitive process, as classically reported by Conklin (1954) for the Hanunóo.

Shrubs and herbs

Nuaulu do not have a distinct life-form term for HERB, in contrast to TREE or VINE. One term (***rahue***) can be heard referring to herbs and shrubby plants as a group, but is usually employed in the context of things planted. The Nuaulu do not make a distinction between a shrub and a herb, though there is a cline of resemblance in terms of woodiness between a tree and a herb. Some shrubby trees are clearly ***ai***. However, within the semantic space left over when we remove plants that are clearly subsumed under ***ai*** (tree), ***awane*** (vine) or ***monote*** (weed, grass), certain groups seem to form quasi-categories. Terms that cannot be included under the life forms of tree (***ai***) or vine (***awane***) constitute 568 types, that is over half the total number of Nuaulu plant categories reported. Of these, 23 have names that place them within ***monote***, the only other named category that approaches the definition of LIFE FORM. HERB itself is not recognized as a life form, covert or otherwise. If we subtract from the total number of labelled categories in the NED (1014), those registered as 'bamboo' (17), 'pandans' (13), 'palms' (52), 'ferns' (33), 'trees' (423) and 'vines' (69), for which there is evidence – including linguistic evidence – that Nuaulu recognize them as life-form-like intermediate groupings based mainly on morphology, we are left with 407. Included in this, though the numbers are rather tentative, are those categories which may be considered to fall outside a putative HERB grouping because they are well defined in other ways: starchy roots, bananas, grains, leafy vegetables and pulses, in other words most crops planted in swiddens (***rahue***). This leaves a residue of approximately 275 terms. To describe this diverse group of other vascular plants as HERBS simply on the grounds of progressive elimination and because they cannot be included in any of the foregoing specific groups seems hardly credible. Thus, there are a lot of plants that are not obviously included in life-form or intermediate categories, and which are therefore 'residual' in a technical classificatory sense, and among these are some individually important species in terms of their uses.

I argue here that Nuaulu life-form and intermediate categories are very elastic, with a small hard core of plants that will always be placed in a particular primary category (usually indicated lexically), but also many that will be placed in several categories depending on situation of use or linguistic context. This large residue of plants that do not obviously fall under a clearly defined life form appears to violate Miller's (1956) 'rule of 7' stipulating that the mind will need to interpose more encompassing categories in order to more easily handle sets of subcategories that consist of (roughly) seven or more differentiated entities. In this case it could be argued that the flexible use of multiple overlapping use categories serves to achieve this requirement. The situation with unaffiliated HERBS is therefore very similar to that we found with unaffiliated trees in Chapter 5. There is no Nuaulu term for HERB, and the number of unaffiliated categories that apply to what we would describe as herbs is much smaller than that for trees, where we have the overarching life form term ***ai***. Herb, in this sense, also excludes not only ferns, but also ***soka-sokae*** (club-mosses, spike-mosses, whisk ferns and related forms), mosses and other cryptogams that constitute a considerable proportion of vegetative ground cover in forests and other associated biotopes. Some of these terms, where they have no obvious uses, can be absorbed into ***monote***; others are either domesticates or wild plants that are often protected and sometimes moved into swiddens and door-yard gardens. This leaves less than 275 named categories that do not easily fit into any other grouping, a selection of which are listed in Table 7.1.

There is also an ecological and methodological consideration underlying the small number of unaffiliated herbs recognized by Nuaulu: in tropical-forest environments non-woody herbaceous plants are rare (though they may be underestimated because they are researched less). Shrubby plants of the forest often grow into small trees, so most of the diversity is within the tree-life-form range. With little herbaceous plant cover on rainforest floors, tree seedlings tend to dominate this layer (Lugo and Lowe 1995).

Monote: weeds and grasses

Twenty-three plant terms contain the prefix ***mono***, though other plants may be described as ***monote*** or include the prefix optionally. I have discussed this category in an earlier paper (Ellen 1991b), but Table 7.2 here updates the data already published. As I have previously indicated, to the extent that the term ***monote*** conveys the sense of 'weed', it is a special-purpose or functional category. Although sometimes glossed as 'grass' (AM 'rumput'), it is surprising how few species in the grass family (POACEAE) are actually described as such. Some plants described as ***monote*** will also be systematically included in other overarching or life-form categories, for example ***kupako***, ***ai*** or ***awane***, and include edible leafy greens (*Cyathula prostrata*), useful cooking herbs (*Mentha arvensis*) and medicinals (***mono panu-panu***). Weed/grass categories are reported for other languages in the Moluccas (e.g. Taylor 1990), and it is plausible to interpret it as a proto-life form.

Kupako: plants with salient decorative flowers

As with its semantic equivalents in many other languages, ***kupako*** is used in two senses: to refer to the reproductive organs of flowering plants (though in the Nuaulu case not to all), and to group together particular plant categories that are renowned for their distinctive flowering parts in either functional or decorative terms (including size, colour and fragrance). In this second sense it is increasingly acquiring the sense of English 'flower' (as in 'a bunch of flowers' or a 'flower garden') or 'fleur' in French. However, whereas in some languages the European term has been introduced to accommodate this meaning (e.g. 'fleur' in certain languages of francophone West Africa: Komaromi 2009:31, 43), or a term in another lingua franca such as Malay/Indonesian (as in the use of 'bunga' in Manggarai: Verheijen 1984:9), in the Nuaulu case an existing term has been semantically expanded. In this second sense the other Nuaulu terms for plant organs of reproduction (e.g. ***hunne***, ***noene***) are excluded. This dual meaning and divergence has been explored by Goody (1993:1–27), and with it the historical particularity of the idea that some plants should be distinguished from others on the basis of aesthetics rather than on the basis of morphology, ecology or other functional characters.

The Nuaulu category ***kupako*** includes many plants that have ***kupa***- as a prefix (Table 7.3). These include both endemic species, such as the epiphytic aroid ***kupawane*** (*Rhapidophora pinnata*), and introduced species, such as the sunflower ***kupa rani*** (*Helianthus annus*), but also, conspicuously, *Syzygium aromaticum* (***kupanahane***), the unopened flower of which produces the clove of commerce. This natural product has dominated the seaborne economy of parts of the Moluccas for over a thousand years, although has only been important to Nuaulu for a little over 100 years. However, why the term should apparently be formed by adding the flower prefix to ***nahane***, otherwise used for the ironwood tree *Xanthostemon verus*, is unclear. Apart from its super-salience in commerce, the clove flower

Nuaulu name	Scientific name	Scientific family	Number of terminal categories	Notes
anahuai	*Leucas zeylanica*	APIACEAE	1	
anaupua	*Cyperus compressus*	CYPERACEAE	1	often placed in ***monote***
hahusina	*Laportea interupta*	URTICACEAE	1	
hono-hono	*Pollia secundiflora*	COMMELINACEAE	5	
	Aneilema sp.	COMMELINACEAE		
	Amischotolype hispida	COMMELINACEAE		
	Andrographis paniculata	ACANTHACEAE		
kamruku	*Pseuderanthemum*	ACANTHACEAE	2	often placed in ***monote***
kasipehe	*Canna indica*	CANNACEAE	1	
	Phrynium macrocephalum	MARANTACEAE		
kaute		MALVACEAE	1	
kanna	*Aglaonema simplex*	ARACEAE	1	no known use
kamkari	*Alpinia nutans*	ZINGIBERACEAE	1	no known use
kio nenane	no ID		1	herb for cooking fish, decorative yellow flower
manapesi mau	*Orthosiphon aristatus*	LAMIACEAE	1	medicinal use
mosone	*Justica gendarussa*	ACANTHACEAE	2	ritual and medicinal use
moune	*Peristrophe bivalvis* nr. *Dicliptera* sp. *Peristrophe* sp.	ACANTHACEAE	3	source of red dye (***moune msinae***)
numte	*Hepatica*	RANUNCULACEAE	1	no known use
ohone		ACANTHACEAE	1	
panu-panu	*Cyathula prostrata*	AMARANTHACEAE	2	medicinal and secondary food use, sometimes placed in ***monote***
pasite	*Begonia isoptera*	BEGONIACEAE	1	herb for cooking fish
pitiri malu	*Mimosa pudica*	FABACEAE	1	secondary food
rorie	*Begonia isoptera* (***rorie***) *Begonia* (***rorie msinae***)	BEGONIACEAE	3	
	Cyrtandra sp. nr. *C. decurrens*, *poss. C. pilosa* (***rorie waene***)	GESNERACEAE		medicinal use
tonne hanaie	*Solanum aguivi*	SOLONACEAE	1	
wainite	*Phrynium pubinerve* and other species of *Phrynium*	MARANTACEAE	5	medicinal and ritual uses; large leaf used as wrapper

Table 7.1 Selected unaffiliated herbaceous plant categories.

Nuaulu name	Scientific name	Scientific family
ahane	*Phymatodes scolopendria*	Polypodiaceae
ahone	*Microsorum commutatum*	Polypodiaceae
anahuai	*Leucas zeylanica*	Lamiaceae
anaupua	*Cyperus compresssus*	Cyperaceae
bubu kaiya	*Nephrolepsis hirsutala*	Polypodiaceae
ekene	*Centotheca lappacea*	Poaceae
hahuseite	*Cyperus diffusus, Scleria* sp.	Cyperaceae
hahu sina	*Fleurya interrupta* *Coleusa atropurpurens*	Urticaceae Lamiaceae
hono-hono	*Aneilema* spp.	Commelinaceae
hunua	*Homalanthus populifoliua*	Euphorbiaceae
kamruku	*Pseuderanthemum* sp. *Ophiorrhiza canescens*	Acanthaceae Rubiaceae
kamruku sororione	*Pseuderanthemum* sp.	Acanthaceae
kamruku wane	*Pseuderanthemum* sp.	Acanthaceae
kau-kau	*Pteris* sp.	Polypodiaceae
kauote	*Urena lobata*	Malvaceae
kopane	*Pteris* sp.	Polypodiaceae
kin ekane	*Phymatodes membranifolium*	Polypodiaceae
kupa rosi	*Rubus moluccanus*	Rosaceae
matapete onone	*Asystasia gangetica*	Acanthaceae
mono hau mesie	probl. *Mentha arvensis*	Lamiaceae
mono karanate	*Clidemia hirta*	Melastomataceae
mono mahusine		Asteraceae
mono maianane		Fabaceae
mono makae		k.o. herb
mono manuate	*Amaranthus viridis*	Amaranthaceae
mono mar*(*a*) *osu	*Ageratum conyzoides*	Asteraceae
mono maune		Acanthaceae
mono monatotu	*Cyathula prostrata* *Synedrella nodiflora*	Amaranthaceae Asteraceae
mono nuae	*Desmodium heterocarpon, D. sequax*	Fabaceae
mono nute	*Euphorbia hirta*	Euphorbiaceae
(***mono***) ***panu-panu***	*D. sequax* *Leptaspis urceolata*	Fabaceae Poaceae
mono nione senene		Malvaceae
mono numte (***soka-sokae numte***)	*Lycopodiella cernua* *Tapeinidium amboynense*	Lycopodiaceae Dennstaediaceae Selaginellaceae
mono nuetetue	*Crassocephalum crepidioides* *Ageratum conyzoides* *Synedrella nodiflora*	Asteraceae Asteraceae Asteraceae
mononute	*Euphorbia hirta*	Euphorbiaceae
mono seane	*Eupatorium*	Asteraceae
mono senane	*Portulaca oleracea*	Portulaceae
mono tohu	*Synedrella nodiflora*	Asteraceae
mono totu mkauke	*Ludwigia* sp.	Onagraceae
(***mono***) ***hari moue***		k.o. herb
(***mono***) ***une-une taia***[1]	*Synedrella nodiflora*	Asteraceae
mo wapane	*Pseuderanthemum* sp.	Acanthaceae
nemonone	*Melanthera biflora*	Asteraceae
nesa-nesa hanaie	*Chloris barbatas*	Poaceae
nesa-nesa pina	*Digitaria setigera*	Poaceae
noha	*Melanopsis multiglandulosa*	Euphorbiaceae
ohone		Acanthaceae
oni	*Imperata exaltata*	Poaceae
papate wane	*Petraeovitex* sp.	Verbenaceae
pupu onate	*Pteris semipinnata*	Polypodiaceae
sapana	*Pteris vittata*	Polypodiaceae
sapane	*Trema orientalis*	Urticaceae
soka-sokae marae	*Psilotum nudum*	Psilotaceae
soka-sokae msinae	*Selaginella* sp.	Selaginellaceae
uta numa		Poaceae

Table 7.2 The category ***monote.***

Nuaulu name	Scientific name	Notes
kupahaia	*Plumeria rubra*	syn. ***kupa kupero hanae***; Engl. 'frangipani' tree
kupa rosi	*Rubus moluccanus*	broad-leaf bramble
kupawane	*Rhapidophora pinnata*	epiphytic aroid
kupa rani	*Helianthus annus*	sunflower
kupanahane	*Syzgium aromaticum*	domesticated clove tree
kupanahane wesie	*Decaspermum fruticosum*	forest clove
kupaleno	*Cananga odorata*	perfume tree
kuparuso	*Caesalpinia pulcherrima*	peacock flower
kupatonae	*Macaranga*	AM 'kayu tongkat lanit' (Bolton and Matoke 2005)
kupa unie	*Pilocereus plachyclades*	an introduced cactus

Table 7.3 Examples of plants grouped as ***kupako****.*

is not especially attractive, and so it does not readily conform to the emerging definition of ***kupako*** on aesthetic grounds. However, given that ***kupanahane*** is placed in the category it is not surprizing that we should – by extension – find ***kupanahane wesie*** there as well, the 'forest clove' (*Decaspermum fruticosum*), taxonomically related in MYRTACEAE but known mainly to Nuaulu not for the commercial value of its flowers but for its fruit, a favorite food of the passeriform ***hiru*** bird. Thus, what emerges here is a mode of recruitment to a more inclusive plant category based on semantic 'chaining' or polythesis, whereby a plant may share a feature with another plant in the category, but not necessarily with all, such that there may be no overall unithetic definition.

The term ***kupako*** is also used to refer to other plants that are used or appreciated for the decorative qualities of their flowers, the names of which do not incorporate the prefix ***kupa-***. Some of these are non-native ornamentals, and the term is also applied to plastic flowers, a ubiquitous indicator of aesthetic modernity placed on tables in the reception spaces of officials, including village officials, in many parts of contemporary Indonesia. In either sense, the category cuts across other life forms and intermediate categories, such as TREE, and is therefore not exclusive to HERB. It also includes plants with names derived from Ambonese Malay and other languages (e.g. *Cassia soplera*: ***ai saja***), as well as those recognized but not specifically named in Nuaulu: 'they are only flowers'. For example, *Euphorbia antiquorum* is an unmistakeably salient species planted near to houses in some Nuaulu villages, and first reported for Ambon by Rumphius. It is, however, rarely named, 'just ***kupako***' (***kupako mani***), though I have heard it called ***kupa unie*** along with the true cactus *Pilocereus plachyclades*. Nuaulu put flowers in their hair (e.g. *Cananga odorata*), grow orchids on the sides of their houses (which they also sell), and plant shrubs in the village to achieve decorative ends (e.g. ***ai yainuelo***, *Delonix regia*, the magnificent 'Flame of the forest'). Orchids collected for sale or display are not grouped collectively, other than as Ind. 'bunga anggerek', but are placed in four main categories: ***kopuni*** (*Phreatia secunda*, *Oberania* spp.), ***koriwane*** (*Dendrobium anosmosum*), ***sipane hanaie*** (*Robiquetia gracilistipes*, *Pomatocalpa* sp.), ***sipa nione*** (*Dendrobium padangensis*, *D. crumenatum* and *Phalaenopsis ambilis*); the last two collectively being ***sipane***. The term ***hono-hono*** groups a number of wayside plants with distinctive flowers, all with individual binomials, but otherwise no indicated uses. ***Hono-hono*** includes an orchid and several ACANTHACEAE (e.g. *Andrographis paniculata*), but comprises mainly species of COMMELINACEAE (*Amischotolype hispida*, *Pollia* including *secundiflora*, and *Aneilema*). Significantly, plants selected on account of their visual properties for use in traditional ritual, such as the highly differentiated ***sinsinte*** (*Codiaeum variegatum*) and ***sone*** (*Cordyline fruticosa*), are never considered ***kupako***.

Classifying plants as food

The detailed data presented in Chapters 3 through to 6 show clearly that categories for plants recognized for their food value tend to be more internally lexically differentiated, complying with the Nida-Conklin hypothesis (Conklin 1962:132; Frake 1961:121). The most spectacular examples of these are ***uri*** (banana) with 36 labelled subcategories, ***kasipii*** (cassava) with 17 subcategories, ***hatane*** (sago) with 14 subcategories and ***nione*** (coconut) with 14 subcategories (Table 7.4). In all cases these are likely to be underestimates. However, the numbers do not correlate with cultural significance in any straightforward way, as labelling of subcategories will also reflect the ease with which these cultigens yield new types, ecological factors such as the local advantages of cultivating increased diversity, the role of sexual reproduction versus vegetative propagation, and the facility for social circulation of germplasm. Thus, patterns of use and ecological opportunities as reflected in the lexical elaboration of important categories of cultivated plants influence classifying activity even before we consider categories that might be explicitly linked to types of use.

Nuaulu words for food (***muaine***) intrude to some extent on categories that divide up the domain of plants. There is no single word for meat, though ***minai*** is a portmanteau term

Nuaulu name	Scientific name	English name	N labelled folk varieties
uri	*Musa* hybrids	banana	36
kasipii	*Manihot esculenta*	cassava	17
hatane	*Metroxylon sagu*	sago	14
nione	*Cocos nucifera*	coconut	14
sikeue	*Colocasia esculenta*	taro	10
hueni	*Dioscorea alata*	greater yam	10
siahue	*Dioscorea esculenta*	lesser yam	8

Table 7.4 Lexical differentiation of major food plants ordered by number of labelled folk varieties.

that covers meat, grease, fat and oil; while ***peni*** is a salient category that simultaneously refers to the meat of pig, deer and cassowary, and also serves as an intermediate animal category, but with strong ritual connotations. It provides a good example of the interconnection and overlap between utilitarian, ritual and ethnobiological categories (Ellen 1993:114–18). Vegetables, meaning mainly leafy greens, are ***utane*** (approximately equivalent to Indonesian and AM 'sayuran'), and fruit are ***huae*** (referring both to edible fruit and generically to the mature ovary-encapsulating seeds found in all vascular plants). As in Indonesian 'buah', ***hua*** is also a classifier used for a wide range of objects other than fruit in the strict sense. There is no single term or concept for tubers and roots, comparable to Indonesian or AM 'ubi', and instead the various root and tuber crops are known by their individual primary terms, such as ***hueni***, ***kasitena***, ***kasipii*** and ***sikeue***. 'Ubi' is a strong collective term in Ambonese Malay folk classification, at least since the time of Rumphius, and Nuaulu will use the term in a way that suggests that a covert category exists. A similar position may apply to pulses, collectively 'kacang' in AM but covered by the Nuaulu terms ***kihue*** (*Cassia fistula*, *Vigna radiata*) and ***kaisan*** (*Arachis hypogea*), with the addition of certain borrowed AM terms such as 'kacang panjang' > ***kacang panjang*** (*Vigna unguiculata*) and 'buncis' > ***buncis*** (*Phaseolus vulgaris*). Sago always stands alone, but the starches of other palms that are very occasionally eaten may sometimes be grouped with it as AM 'sagu', but never as ***hatane***. ***Muaine***, meaning solid foods of whatever kind, can be distinguished from liquid foods (***ininiu***). Spices, relishes and other food additives are known as ***haun huae*** (lit. 'smelling fruits'), and are arguably a loose grouping that does not constitute a strong ethnobotanical category. Part of the reason for this may be that so many plants are recognized as food additives, amongst other uses they have, and that as with plants used as fuel and medicines (see below), and as with trees used for timber, an encompassing category based on this feature alone would be clumsy and unhelpful.

The utilitarian/ecological distinction between plants deliberately cultivated and growing say, in swiddens, and those physically similar plants growing spontaneously in the forest, significantly influences perceptual distinctions based on morphology alone. These are usually expressed through pairing or contrast, where the domesticated or cultivated element in a pair or contrast set is the default category. Thus, ***sikeue*** (domesicated taro, *Colocasia esculenta*) may be paired with a wild variety of the same species (***punetene***), or with a completely different aroid genus (***puni***: *Homalomena pendula*). The former has fine hairs on the stem which cause itching, and its tubers are not harvested for food. The latter is employed to protect people from spirits of children who have died in childbirth; sometimes described as 'the cloth of an evil spirit', it can be used to make a scare charm (***wate***) and will be wrapped around a stillborn child or an aborted foetus which is then taken to a burial site by a coastal route rather than by the usual path (Ellen 2012:184–8).

Classification and the healing properties of plants

To heal is ***suisene***. This word also refers specifically to the act of expelling a fine spray from the mouth of a spirit medium over a patient, one derived from a chewed mixture of betel-quid ingredients, coconut oil and the ginger *Kaempferia galanga* (Ellen 2019b). Medicine is described either as ***upati*** (< Ind. 'obat') or ***ai totue***, the latter meaning also (and literally) 'tree leaf'. The first term for medicines is usually associated with purchased drugs or medicaments, and the second with materia medica used in traditional healing. Thus, there is a strong identification between plants and medicine. There is no generic term for healing plants, which are sourced from across a wide range of phylogenetically very different species, and are by no means restricted to particular taxa. Indeed, the concept of medicinal plant is quite flexible and it would be difficult to come up with a fixed or definitive list. As Raj Puri and I have argued elsewhere (Ellen and Puri 2016), plants used medicinally in rural Indonesia are often geographically specific and there is a surprizing lack of overlap, even on the same island, of individual species used. Moreover, plants used medicinally are also often used for a variety of other purposes. All these factors make it difficult to see how healing properties might separately influence classification. Almost all plants are potentially medicinals for Nuaulu, and determining their virtues is in a sense always 'work in progress'. At the last count there were 109 reports of

Nuaulu name	Scientific name	Notes
ai kam, ai tohu	*Piper aduncum*	AM 'sirih hutan'; tree habit, NW origin
kam apane	*Piper forstenii*	
kam hanate	*Piper* probl. *caninum*	
kam hau	*Piper corylistachyon* also probl. *P. amboinense*	
kam inate	*Piper* sp.	
kam malaka	*Piper* probl. *retrofractum*	
kam minne	*Piper caninum, Piper sundaicum*	inedible, causes intoxication
kam niane	*Piper betle*	
kam putie	*Piper* sp.	
kam sisinate	*Piper reinwardtianum*	
kam tunne	*Piper corylistachyon*	
kam(u) wane	*Piper* probl. *sarmentosum*	
kam wanae ukune	*Piper* sp.	
lada	*Piper nigrum*	

*Table 7.5 Relationship between the Nuaulu category **kam** and the scientific genus Piper.*

different plants being used medicinally, but only 28 species reported as being solely medicinals. However, there are certain species and folk categories that repeatedly occur in records of treatment episodes, and the most important of these are plants that are also important in ritual and as food, e.g. ***kanai*** (mainly *Areca catechu*), ***kam*** (mainly *Piper betle*), ***nione*** (*Cocos nucifera*), ***soi*** (various species of *Zingiber*), ***sukue*** (*Kaemferia galanga*) and ***masahahute*** (*Mackinlaya celebica*). Medicinal plants may also be grouped with other species necessary for a particular treatment (***nori-nori***), but are not otherwise associated functionally, other than by distinguishing those that are 'hot' from those that are 'cold', though these characteristics do not in themselves yield stable classificatory groups, much like the role of the male-female distinction described in Chapter 2.

One feature of healing plants considered as a whole is that while there is clearly a process of selection taking place that must impact on their distribution and the accessibility of their useful medicinal features, they do not – unlike most food plants – constitute groups of closely related cultivars in scientific terms, or groups that are perceived as being closely related in a folk classification. Medicinal plants cut across many other groups, especially food plants and – as is widely appreciated – the boundary between food and medicine is often difficult to establish (e.g. Etkin 1994). Two genera that might be said to go against this generalization are *Areca* and *Piper*, species of which provide the two main ingredients of the betel quid. We have already looked at *Areca* in the context of palms (Chapter 6), where it was noted that while the fruits of several related palms (including species of *Areca*) can be combined with various species of *Piper*, it is most usual to use *Areca catechu*, of which eight binomial folk varieties are regularly recognized. The way species of these genera come together as a psychoactive substance with some distinct physical effects is deeply significant in social interaction, in recreational drug use, and in Nuaulu healing (usually in the form of saliva or spittle supplemented with additional substances); and there are consequences stemming from this in the ways plants are classified (Ellen 1991a). The types of *Piper* recognized in Nuaulu classification are listed in Table 7.5.

We might compare this degree of organization and lexical differentiation with another group of plants, gingers of the family ZINGIBERACEAE. These are disproportionately well represented in the Nuaulu ethnopharmacopeia, but weakly subgrouped. Thus, *Alpinia* and *Riedelia* of various species are known by at least five separate uninomials, *Amomum* and some *Hornstedtia* are grouped as ***kikun***, ***kunie*** is mostly *Curcuma*, while *Kaempferia galanga* is unambiguously ***sukue***. The genus *Zingiber* is mostly divided as different kinds of ***soi***, though with some species named using separate uninomials (Appendix 1:187–8). Exceptionally, *Codiaeum variegatum* (***sinsinte***), an important decorative and ritual plant, also used medicinally and for cleansing, is divided into seven lexically differentiated folk varieties, overall unambiguously labelled as a single category, despite the fact that its folk varieties are managed phenotypical clones or life stages of the same species (Ellen in press).

Plants grouped on the basis on their properties as manufacturing materials

Nuaulu strongly conceptualize many plants and plant groupings through their uses. In a few cases the name of a cultural object and a plant are identical, and we might assume the meanings to have merged. We have already noted in Chapter 6 how amongst pandans, ***koae*** refers both to the whole plant (incl. *Pandanus amaryllifolius, P. krauelianus* and related species) and to mats made from these and other pandans, while ***saune*** refers polysemically to a whole plant (*Pandanus conoideus, P. tectorius* and their subtaxa) and to the red sauce made from the flesh of the oily pericarp and seeds. In the same chapter we saw how the term ***niune*** is used both as the name for a certain type of arrow and to label a plant category that includes various species of reed used to make such arrows, including *Sorghum propinquum, Thysanolaena maxima, Miscanthus floridulus* and *Saccharum spontaneum.* Similarly, amongst dicotyledonous trees, ***pasane*** is both the name of a type of tree (*Vitex cofassus*) and the main post of a Nuaulu sacred house: (***hinai***) ***pasane***. Such usages (though reported elsewhere, e.g. Kocher-Schmid 1991:72) can be confusing to the ethnographer, and it has certainly taken me repeated field trips and multiple collections of vouchers to sort it out. Among dye-producing plants, ***kasupa*** is a source of yellow and red colouring matter used for baskets and mats. At the present time the only species described by Nuaulu using this term is *Bixa orellana*, an introduced plant of American origin. However, the AM term 'kasumba' is also applied to some other dye-producing plants, particularly *Carthamus tinctorius* (Rumphius 2011, vol. 2:115 and vol. 4:135; Heyne 1913–17, vol. 4:253). In this case, it is possible that the more recent New World introduction has replaced the earlier western Eurasian species as a source of dye, but the name conveying the same usage has been retained. The elaborate protective charm that is erected at the front of a birthing hut to protect mother and child is known as an ***asinokoe***. It is usually made from ***akunin hanaie*** wood (*Actinodaphne* sp.), but for some clans it is ***akunin pina*** or ***kawasa*** (*Archidendron clypearia*), a timber also used for sacred shields. The term ***asinokoe*** is also used synonymously to refer to the tree species (Ellen 2012:86–7). Sometimes ***akunin*** (or ***kawasa***) might be said to be the 'jenis' (AM) name and ***asinokoe*** the ***monne*** (sacred) name.

Perhaps more easily observed are situations where utility and plant morphology merge lexically at more inclusive degrees of classification and particularly at life-form level. Thus, all three Nuaulu life-form terms clearly implicate use in some way. ***Rahue***, as we have already seen in Chapter 4, refers to something planted, and thereby utilized; and ***awane*** to anything that can potentially be used for binding.

Ai (tree) is perhaps less obvious. Though the basic image prototype of 'tree' may have existed for millions of years, the life-form category and label seem relatively recent (Witkowski, Brown and Chase 1981), its earliest manifestation appearing to have involved functional considerations reflected in tree/wood polysemy. The implications of this claim have been widely discussed (e.g. Atran 1985:306; Ellen 1998a), with Atran (1990:34) asserting that while tree/wood polysemy undeniably exists, 'wood' and 'tree' are conceptually quite different. Thus, Atran (1985:306) asserts that there is an inconsistency between Brown's (1977, 1984) claim for the definition of tree in terms of size, and his functional definition in terms of wood. Atran (1985:307) continues:

> it will not do to argue that referential expansion simply consists in relaxing the functional component of the definition and stressing only the perceptual component ... The perceptual features attaching to the artifact 'timber' and the living kind 'tree' are quite different in nature: those that attach to 'tree' are virtual properties that include phenomenally typical attributes of trees, whereas the only perceptual logically characteristic of 'timber' relate to being wooden and cut.

The Proto-Austronesian word for tree/wood, 'ayu*', has a long and complex history, and by the time it reached its Malay iteration 'pohon' (tree) and 'kayu' (wood) had separated, though the term 'pohon kayu' is sometimes heard for a woody tree. Indeed, some wood terms do appear to have emerged before tree terms (e.g. Blust 1974:5, 11–12). Malay 'pohon' and 'kayu' seem to distinguish the sense of the morphology of the living plant from the utility of the prepared timber; yet many trees renowned for their timber will be known in Indonesian or Malay not as 'pohon X' but as 'kayu X'. There are, however, no separate tree and wood terms in contemporary Nuaulu, where the single term ***ai*** incorporates both senses, such that we might define a tree as something that produces wood. Thus, the less inclusive item – the reference of which appears to have been extended – is the one which acquires its salience from cultural use rather than from intrinsic perceptual qualities: so, wood as a material with many technological applications exemplifying trees in general; or (as in Austronesian bird/chicken polysemy) chickens as domesticates which have co-evolved closely with human populations and which provide food and sacrificial objects exemplifying birds as a whole (Ellen 1993:81). In either case, conceptual difference is expressed adequately despite using a single word.

Another eastern Indonesian language (though this time in the West Papuan phylum) displaying tree/wood polysemy is Tobelo, reported by Taylor (1990:26). Tobelo also displays vine/rope or vine/string polysemy, which we find in the use of the Malay word 'tali' and in Nuaulu, where ***awane*** can refer to any string or 'something you can tie things with', and until the arrival of polyester nylon or other plastic twines, was always of plant origin. However, the term for rattan (***meute***), arguably the most important kind of ***awane*** in the sense of cordage, does not display this kind of polysemy. By comparison, though pandans share a crucial physical resemblance, their grouping is strongly linked semantically through the materiality associated with the use of a particular kind of leaf in mat making. There is no single term for pandans (the genus *Pandanus*), as we noted in Chapter 6, but ***koae***, which also refers to a kind of mat, is a prefix in many names and may also be used to include pandans that have distinctive uninomials, such as ***sanoma*** and ***paone***. Bamboo is a bit more complicated, but seems to exist as a covert grouping through the utility of a

Scientific name	Nuaulu name	Modern AM or Indonesian name	Malay name provided by Rumphius
ARAUCARIACEAE			
Agathis dammara	***kama putie***	'damar putih'	damar putih' or 'damar batu'
Agathis dammara	***kama wae***	'damar putih'	'damar putih' or 'damar batu'
BURSERACEAE			
Canarium acutifolium	***kama metene***	'damar hitam'	'damar hitam'
Canarium asperum	***kama kamine***	'damar hitam'	'damar hitam'
Canarium decumanum	***kama kamine***		
Canarium oleosum		'damar putih'	
Haplobus moluccanus		'damar tai babi'	
DIPTEROCARPACEAE			
Anisoptera thurifera		'damar utan'	
Shorea selanica	***kama onie***	'kayu meranti'	'damar selan'

Table 7.6 Cross-language evidence for the existence of 'resin-producing tree' as an ethnobotanical category (data source for Haplobus and Anisoptera, Whitmore, Tantra and Sutisna 1989; all other data from herbarium voucher labels, field notes and Rumphius 2011).

rigid stem for making things. The concept of BAMBOO therefore implies its uses.

Resin-producing plants

Tree species regularly providing resin or similar exudates for use as illumination (nowadays almost entirely in sacred clan houses), used as a filler or adhesive when making equipment, or as a cash-crop, are collectively called ***kamane*** in Nuaulu. The extraction of *Agathis dammara* and other resins for the market has virtually ceased, though 250 metric tons were reported as having been produced in West Seram during 1971 (Badan Pusat Statistik, Kabupaten Maluku Tengah 1971). Historically, *Agathis dammara* was also planted, and within living memory, residents of Tamilouw planted *Agathis* groves around the headwaters of the river Lata on the easternmost flank of the Nuaulu extractive area.

The term ***kamane*** groups together significant resin-producing trees, cutting across botanical families, and the prefix ***kama*** appears as the first element in a binomial that contrasts with other basic level plant categories. The category ***kamane*** therefore appears not to be serving simply as a special-purpose grouping. Thus, we have ***kama wae*** and ***kama putie*** (different kinds of *Agathis dammara* resin, but with the second being essentially a kind of ***kama wae***), ***kama metene*** (*Agathis* and *Canarium asperum*), ***kama kamine*** (*C. asperum* and *decumanum*), which we have already met in Chapter 5 as part of the intermediate *Canarium* complex, and ***kama onie*** (*Shorea selanica*) alternatively described simply as ***onie***. ***Kama wae*** and ***kama onie*** are prohibited as resin for lamps used in sacred houses, which must at all times be harvested from ***kama kamine***. Other kinds of ***kamane*** can, however, be used to provide illumination when in the forest and garden. By contrast with the preceeding categories, ***ai kamane*** is *Jatropha curcas* and not placed within ***kamane*** at all, though the exudate from the cut stem is given to newborn children orally to encourage the production of mucous, and to adults to expel heat from the bodies of malaria patients.

As with several other cases already discussed, this classificatory arrangement finds support from the descriptions of Rumphius, and also from what we know of contemporary Ambonese folk classification. Rumphius (2011, vol. 2:213–70) places important resin-producing trees with *Canarium* in book 3, chapters 1–10. He notes that the AM label 'damar' is not applied to all *Canarium* species (see p. 80 here), and that 'damar' is used for both particular species of tree and collectively for resin. Thus, Rumphian 'damar hitam' includes *Canarium acutifolium*, 'damar selan' is *Shorea selanica* and 'damar putih' or 'damar batu' is *Agathis dammara*. These distinctions for all intents and purposes survive in modern Ambonese Malay, and we can see the approximate correspondences in Table 7.6, though the inclusion of particular species in the AM categories almost certainly varies from place to place and over time. 'Damar' was already an item of trade in Rumphius's time, but it is plausible to speculate that where resin has become economically important amongst local forest producers, as was once the case on Seram, then there is greater reason to elaborate the category of 'resin-producing tree' and incorporate it as an intermediate in local ethnobotanical classifications (see, e.g., Gianno 1990).

The irrelevance of fuel properties in yielding robust ethnobotanical categories

Overall, the uses Nuaulu make of plants as fuel fail to provide evidence for convincing ethnobotanical categories, even when conceived of in special-purpose terms. On the whole, the uses of plants as firewood by Nuaulu show that although 'use' is often relevant to understanding unified folk systems of plant classification, it is not necessarily systematically encoded.

A few plants are singled out by Nuaulu and named because of their strong association with use as fuel. For example, ***hatae*** (*Allophylus* sp.) is used for burning shells to make lime, important in the practice of chewing betel, while the only reported use of ***asihata*** (an unidentified euphorb) is its ability to be used green as fuel because of its high oil content. However, to list named categories of woods used by Nuaulu as fuel provides a very partial and misleading representation of how Nuaulu think about and select firewood, or of the extent to which they specify individual species; and firewood use in general is a poor basis for constituting an ethnobotanical grouping. Nuaulu call firewood ***menie*** (lit. 'that which is dry'), and we need to understand that ***menie*** is not just sourced from trees (***ai***) but comprises other burnable vegetable material as well, kindling and firelighters. Most species of woody plant provide Nuaulu with potential fuel, and therefore a large number of species could be listed as being used for firewood, and entered in the relevant 'use' field in a database. But whether and when Nuaulu use particular species depends on the options and opportunities available, as much as the particular purpose to which they will be put. If you asked someone to list all woody plants that could potentially be used as fuel, you would likely get a number around 340, that is almost all trees and some bamboos and palms. If you ask what fuel woods had actually been used in the last 12 months, you would likely arrive at a number considerably less than this, and in my intermittent records of species actually observed on particular occasions, the number was 24. Regardless of whether the number of species used as fuel is thereby underestimated or overestimated, any form of enumeration is likely to be distorting and misleading.

Thus, although a list I have provided elsewhere (Ellen 2017c) samples the diversity of species used, there are other materials it does not refer to; nor does it make much sense as a basis for the division of a semantic domain in the form sometimes expected from studies of folk classification, comprising instead a number of convergent and cross-cutting evaluative criteria, for example:

hardness (hard/high density–soft/low density),
burning quality (quick–slow),
drying quality (fast–slow),
heat quality (cold–warm–hot),
harvesting characteristics (ease of cutting, breaking and splitting), and
size (small–large).

Fast-growing trees, for example, provide low-density wood and instant heat when burned, but release a lot of smoke. The advantage of these fuels for Nuaulu is that they are rarely in short supply, being readily to hand. However, for slow-burning hot fires with less smoke for cooking certain foods and keeping fires going overnight, more dense woods are appropriate, derived from slow-growing species. Examples of such are *Casuarina equisetifolia*, which provides an excellent fierce heat and a long-lasting charcoal, and *Erythrina variegata*, which keeps the fire in, as it does not burst into flames, and also burns completely to ashes. *Ficus altissima* and *Avicennia marina* are also regarded as being effective in keeping a fire going; both burn slowly and well in a dry wind. Consequently, it is not that the cultural domain Nuaulu call 'firewood' does not have a distinctive internal organization reflecting a broad and relevant knowledge base; it is just that its complex polythetic character cannot easily be represented as a quasi-taxonomy of 'special-purpose' use categories of the kind that we have become familiar with, say from studies of medicinal plants, dye plants or culinary herbs, and others of the standard codifications of indigenous knowledge.

As a general rule, the species used by Nuaulu as fuel, and the physical form the wood takes, only occasionally depends on the purpose of the fire or on the seeking out of particular species for their qualities. Most woody species used as fuel are not explicitly tagged as such: most are useable in some circumstances, depending on their condition, and Nuaulu apply generic judgments of acceptability based on the condition of the wood (including the diagnostic criteria listed above) and what is readily available. The fuels of preference, therefore, are often the most humble (including bamboo and palm leaves – what Nuaulu call 'rubbish fuel', ***menie ai yehue***). But even where materials are available they have to be harvested, and these costs can be high. Ease of harvesting is therefore a factor in preferring particular species. Most Nuaulu firewood is collected by women and moved around in large back-baskets from near forest and the edges of swidden land. Piles of firewood are a common sight along the edge of paths, where they are protected with prominent scare charms (***wate***). Until the 1980s the only tool used for cutting firewood was the 47 cm or 68 cm 'parang'. Most households would have access to a felling axe, but these would seldom be used when harvesting firewood. However, with the arrival of the petrol-driven chain saw in the 1980s trees could be considered for firewood that would have previously been avoided, for example the coconut palm, sawn into short lengths and then split.

Beyond this, factors determining species are heavily circumstantial. For example, fires lit when travelling to distant sago swamps are made on floating rafts of sago-palm bark, and given that these areas are almost mono-specifically composed of sago palm (*Metroxylon sagu*), most fuel for these fires is derived from parts of the palm itself. However, since sufficient dry tinder may not be reliably available in swamp land, some dry bamboo for this purpose will actually be taken to the location where the palms are to be cut, along with lengths of green bamboo in which to cook. A similar example is the disproportionate use of dry coconut husks and shells as fuel when drying copra (both *Cocos nucifera*).

However, outside of this core of default firewood used in village fires, where species is less important than the application of generic knowledge, and the special circumstances of the kind just illustrated, there are instances of species-specific related knowledge. For example, some species are flagged precisely because they should be avoided. It is important not only to know what good firewood to prefer but also what bad firewood to avoid, either due to poor physical burning and heating qualities (e.g. ***ito***, *Alstonia scholaris*), or because there is a taboo placed upon it (as in the case of ***ai suto***, *Premna serratifolia*, used commonly as fuel but prohibited for the clan Peinisa). On the other hand, there are species that are singled out as excellent fuels for particular purposes, especially as charcoal (***okone***). Different species of woody plant collected as ***menie*** can be transformed into distinctively different types of charcoal. For example, the stony endocarp (shell) of the coconut makes excellent compact charcoal, as does mangrove wood (in this case ***hana:*** *Bruguiera gymnonorrhiza*). Nuaulu have few applications for high-temperature fires, as they do not work metal, but there is a market demand for suitable dense woods that Nuaulu occasionally exploit. However, they do have other specialist applications for fire, for which different fuels are appropriate. Smoking is important for preserving meat, and to a lesser extent for driving away insects and other animals. Coconut husks are excellent for this purpose. Nuaulu also require a special kind of fire (***usa nosa***) for producing the mineral lime (***kotu nosa***) used in betel chewing. This involves burning mollusc shells, coral or soft limestone. The fuel of preference for this is coconut shells (on top), and a pyramid of stacked split bamboo raised off the ground on several lengths of sago leaf stalk to allow for good ventilation. Bamboo fires, though not of this structural complexity, are also used of preference for fixing the vegetable dyes used to prepare *Pandanus* leaf strips when making ritual objects, for hardening arrow points, for roasting plantain or yams, or for heating stones used when cooking ***maea*** (a ritually important comestible made from sago flour and crushed *Canarium* nuts).

Plant classification and ritual

One of the most problematic use categories used in economic botany is that sometimes called 'social', 'symbolic' or 'ritual'. This is not only because most plants are of multiple use in the technical narrow sense, but also because all plants that are in any way used, are used 'socially', and because symbolic and technical use are often simultaneously one and the same thing in the eyes of local people. Ritual influences classification by privileging various species or groupings by virtue of their ceremonial role, or in avoidance practices or beliefs connected with them. This is reflected in many of the plant pairings already noted or alluded to (e.g. the edible aroid ***sikeue*** in contrast to the wild aroid *Homalomena pendula* used as a wrapper for aborted foetuses and stillbirths). In other cases, ritual plants themselves are paired, as with ***akunin*** and ***kawasa*** (see Chapter 7, p. 107). Some non-uses are as significant as uses. Thus, many plant taxa are prohibited for certain groups in certain contexts, and in these situations those on whom the prohibitions are placed have to use other synonyms or circumlocutions. For example, if someone is in a ***saumonne*** (avoidance) relationship with someone called Urimahu, then the person cannot use the term ***uri*** for banana in any context (Bolton 1997; Florey and Bolton 1997). However, there are no intermediate-level groupings of plants motivated by their common function in ritual, or terms that might label such groups, as I found with the prominent term ***peni*** which groups cassowary, deer and pig in Nuaulu animal classification (Ellen 1993:115).

Sinsinte (*Codiaeum variegatum*), referred to above in the context of medicinal uses, is another plant used decoratively and in ritual (Ellen in press), the classificatory constitution and internal diversity of which are strongly associated with its ritual uses. The word ***sinsinte*** in addition may refer to any decoration that is used in the same way as *Codiaeum variegatum* (usually referring to the semantically privileged ***sinsinte masikunne***, with its long trailing variegated leaves), such as decorations made from coconut or ***wainite*** (*Phrynium*) leaves shredded and tucked into armbands, or used to decorate temporary pavilions used in rituals. ***Tapako sinsinte*** and ***kasipii sinsinte*** are folk varieties of tobacco and cassava, respectively, with leaf variegation said to resemble that found in *C. variegatum*, and these too are sometimes co-opted in ritual contexts.

The concept of use

The cases and issues discussed in this chapter are reminiscent of earlier and well-worn utilitarianist versus universalist arguments (Hunn 1982; Randall and Hunn 1984) of the 'chicken or egg' variety, but they also require that we re-examine what we might mean by 'use' in the context of ethnobotany and folk-classification studies more generally. For example, how might we resolve the statement by Berlin (1978:10–11, reiterated by Atran 1990:20), that over one third of Tzeltal and Aguaruna plants 'have no uses', with the Nuaulu claim that all plants have potential uses, and no plants have no uses.

Some ethnobotanists and economic botanists have adopted a simple unproblematized concept of a 'use', but an anthropological treatment requires that we interrogate the concept more radically. One of the difficulties in measuring utility is to know what a 'use' is. In ethnobotany and economic botany it has become standard to employ a tight empirical definition of what might constitute 'a use' (e.g. Cook 1995), with established protocols to assess the number of uses a plant has, such as 'multiple use curves' (Balick 1994:10–11). Such protocols permit enumeration and other kinds of measurement, while meeting the requirements of science and politically driven accountancy. But this is not how ordinary people living in tropical forest, who seldom think in terms of fixed finite species uses, view the utility of their plant world. Rather, all plants have uses, but particular plants can have multiple but often ad hoc and indefinite uses, not all of which may be the best application of their properties. What is useful often depends on what is available. For example, there is a widely held but generally unstated assumption in studies of medicinal plants, that not only is the local knowledge corpus something like a fixed list of species with particular applications, but that somehow many species and the treatments with which they are associated are found over a wide area, such that we might speak of 'regional pharmacopeias'. However, in a recent comparative study of data drawn from eight ethnographic locations in Borneo and Seram (including the Nuaulu), Raj Puri and myself (Ellen and Puri 2016) found – counter-intuitively – that medicinals used for particular treatments were extremely locally specific. While genera used medicinally overlapped widely, there was little overlap in terms of species used. Even within local cultural groups defined by language boundaries there was variation in species harvested and used for particular purposes.

The notion of multiple use is relevant to plant classification in another way. Thus, there are some plants – sago palms are the best example in the Nuaulu corpus – that have so many uses, are so central to cultural practice and identity, that to place them within any wider generic grouping somehow fails to do justice to their overall significance. Where these plant species are pivotal ecologically as well as culturally, they have been described as biocultural or cultural keystone species. (e.g. Ellen 2006a). Thus, the multiple valencies of sago – as food, a useful construction material, a ritual and symbolic resource, and in terms of being pivotal ecologically – place it apart from other palms that may be otherwise perceptually similar. Other palms, though sharing a strong physical resemblance, do not have the overwhelmingly strong cultural associations of sago, and therefore exist only weakly as a covert grouping.

The most important connotative and therefore perceptual quality pertaining to plants as far as humans are concerned is that they are all potentially useful, at least for people who directly interface with their environment as subsistence producers. In other words, they are 'affordances' in the sense developed by James Gibson (1979:127). From an evolutionary point of view, so-called 'natural' classifications as specified by Berlin and developed by Atran, are therefore only a framework convenient for organizing data about plants that are potentially useful. But if utility is not relevant to how classifications evolve, then – in another sense – 'natural' classifications are simply what are left once the need to prioritize everyday usefulness has disappeared. This is so amongst those who have given up a subsistence way of life and use 'natural' schemes from which considerations of utility have been expunged, as happened to scientific plant classification as a result of the Linnaean revolution. The utility here has therefore been transferred from an old utility, with regard to day-to-day subsistence, to a new, utility, with regard to achieving the recent and rather specialized purposes of science. In a way, this is parallel to the argument proposed by Atran and Medin (2010) concerning the historical erosion of detailed knowledge of nature in favour of abstract knowledge at the life-form level, sometimes called the 'nature-deficit' syndrome (Ellen 2009b).

But there is one final issue that needs to be addressed in this chapter on the role of use in everyday classification, which arises from Berlin's insistence (e.g. Berlin, Breedlove and Raven 1966) on the separation of special-purpose and general-purpose classifications, and Atran's (1990:212) claim that anthropologists regularly conflate 'common sense' (propositional understanding) with symbolism. Atran supports his claim by referring to the perfectly sensible distinction made by Friedberg (1979:86–7) between identification and representation, and his point that when we identify or distinguish one plant from any other we necessarily focus on morphology and habit rather than on use. Thus, both betel nut (*Areca*) and betel pepper (*Piper*) are highly important symbolically in both the Nuaulu and Bunaq systems of representation, but in identifying the various forms that are acceptable, in the Nuaulu case, as ***kanai*** and ***kam*** respectively, or in the Bunaq case as 'pu' and 'molo', the only features that are relevant are those directly accessible through physical examination. However, such distinctions and evaluations are made in the context of both fruits being of cultural value, and particular attributes being important because of those values and practices. Identification does not take place in a vacuum, and often in the process of making an identification various cultural representations or classifications will be part of that context, and therefore inform decisions.

8

The correspondence between Nuaulu plant names, categories and scientific taxa

Introduction

An early preoccupation of cognitive and linguistic ethnobiology was with the precise correspondence between local and phylogenetic names and categories. As I have discussed elsewhere (Ellen 2006b:s4–s7, 201613–15), ethnobiology has for most of its existence been a project of translation, first in order to accurately render locally important plants and animals in scientific terms, so as to identify them; but secondly to address fundamental questions about the existence of a universal human capacity to divide up the natural world into units of a comparable type (Atran 1998; Berlin, Breedlove and Raven 1966), and to do so in broadly the same way. Underpinning this have been certain assumptions relating to what Ludwig (2018) has referred to as 'convergence metaphysics', an idea to which I shall return in Chapter 10. Conklin (1954:165) had previously argued for a loose correspondence between Hanunóo categories and scientific taxa, with no exact one-to-one correspondence, but later writers (e.g. Atran 1999; Berlin, Breedlove and Raven 1974a; also Hunn 1999) were to present data suggesting tighter correspondence, at least at the genus-species level and for macroscopic forms. In ethnobotany the methods for addressing this issue ultimately and most persuasively involve careful collection of supporting voucher specimens to enable taxonomic identification of living plants through the triangulation of names and descriptions provided *in situ* by local language speakers. The appendices of this work supply these data ordered in three different ways: Appendix 1, by phylogenetic order; Appendix 2, by local name in alphabetical order; and Appendix 3, by scientific binomial in alphabetical order.

Nuaulu recognize 870 named terminal categories for vascular plants that they apply to approximately 647 scientific species. When looked at in terms of phylogenetic content placed in local categories, it is evident that while in many cases there is a good correspondence between the two, in some instances local names apply to two or more scientific taxa, often different species in the same genus, sometimes to different genera in the same family, and sometimes even to species in different families. In some cases it is clear that the inclusion of (especially) different families in the same category is due to corruption introduced during the course of processing between field site and the receiving herbarium: numbers get detached, specimens get muddled in bundles and inevitable clerical errors are made. Given that there are many examples where Nuaulu do group different genera and even families in the same local category, it is therefore not always easy to distinguish the genuine cases from the consequences of corruption. In these circumstances I have had to make judgments based on all the available evidence: duplicate specimens, notes and different herbarium identifications. Thus, voucher 96-745 was identified by BOGOR as the fig tree *Ficus pungens*. Given the character of duplicate material, it is unlikely that this is an inaccurate identification, but the accompanying label indicates the local name ***meu onate***, a kind of rattan. In this case I reasonably judged the data as almost certainly corrupt, and I re-organized the master lists accordingly. Similarly, 96-923 and 96-924a were identified respectively by both BOGOR and KEW as *Barringtonia*, but since these were part of a set of nine vouchers for one folk category – ***kokune*** – that was otherwise identified consistently as various species of *Callicarpa*, and since *Barringtonia* was otherwise placed mostly (10 specimens) in the category labelled ***hanasane***, I judged that the *Barringtonia* specimens assigned to ***kokune*** had probably been mixed up during the processing of specimens.

There are two important issues that arise from a consideration of data on the correspondence (or lack of correspondence) between Nuaulu plant categories, folk names and phylogenetic taxa. The first is the surprising lack of agreement in some areas among individual Nuaulu on identification, particularly for terminal categories, and especially the application of binomials to different cultivars of the same species. As Sillitoe (1980:138–40) has pointed out in relation to similarly conflicting responses by Wola farmers in the New Guinea highlands, some of this apparent confusion might be accounted for by methodological indeterminacy, or may reflect the fact that some cultivars are much less common than others and therefore provide fewer opportunities for the sharing of knowledge, or may derive from some crop species hybridizing and naturally

Type of correspondence	N
Binomial labelling sex type of the same scientific species	2
Uninomial labelling variety or cultivar of the same scientific species	4
Binomial labelling variety or cultivar of the same scientific species	230
Trinomial labelling scientific variety or cultivar	5
Uninomial labelling one scientific species	306
Binomial labelling one scientific species	313
Trinomial labelling one scientific species	13
Uninomial labelling two or more species of the same scientific genus	38
Binomial labelling two or more species of the same scientific genus	20
Uninomial labelling two or more genera from the same scientific family	26
Binomial labelling two or more genera from the same scientific family	11
Uninomial labelling genera from two or more scientific families	17
Binomial labelling genera from two or more scientific families	14
Uninomial cutting across many scientific families	5
Unknown	8
Total	**1,012**

Table 8.1 Degrees of correspondence between Nuaulu folk nomenclature and scientific taxa: type of correspondence and number of cases.

diversifying more readily than others, presenting difficulties in allocating a particular plant to one category rather than to another. All this echoes the problems discussed at the end of Chapter 4 concerning the local importance of hybrids and clones, and the way phenotypic variation in local cultivated plants exacerbates the difficulties of achieving fixed correspondences between local categories and subspecific and specific taxa.

The second issue concerns the concept of basic and extended range (or category extension), introduced by Berlin, Breedlove and Raven (1974b:56–8), and widely adopted (e.g. Ellen 1993:75–82). This is often regarded as a common means of accommodating biological variants and similar taxa in a broader category defined by a main or focal type. From the Nuaulu data, this seems to work where several different species of the same genus are placed in a single Nuaulu category, such as ***ai yarane***, where *Diospyros maritima* is the most common (focal) species placed in the category, and where *D. elliptifolia* and *D. ebenaster* are less commonly encountered and might be considered to be in the extended range; or more convincingly, the plum ***ai ane***, with *Prunus arborea* as the basic identification and *P. wallaceana* included by extension. In terms of different genera placed in the same folk category, we might use the example of ***hanna*** (***panne***), where *Nauclea* sp. is plausibly the focal taxon and *Neolamarckia cadamba* the extension (both being RUBIACEAE); or the 'wild nutmeg' ***asaherane***, where *Virola surinamensis* (MYRISTICACEAE) is the focal taxon and *Myristica* sp (MYRISTICACEAE) and *Syzygium boerlagei* (MYRTACEAE) are in the extended range, exemplifying cases that include two or more botanical families. But there are other examples where the basic and extended-range model does not work so well. Thus, the Nuaulu term ***kaiohite*** was used without binomial qualification by informants in Rouhua to identify figs of at least four species (*Ficus botryocarpa* subsp. *hirtella*, *F. callosa*, *F. gladulifera* and *P. copiosa*), but it is difficult to see what might be basic and what extended in this category, and my suspicion is that any variation between the species in terms of features was not regarded as diagnostically significant, and that no one species could be held to be the truest of its type. For Sillitoe (1980:146) the notion of basic and extended range is a device to explain the inconsistency of identification rather than always being a deliberate strategy used by informants, but I think that these examples, while illustrating a classifying strategy that is convincingly widespread, and can be clearly demonstrated from our own cultural experience and under laboratory conditions, is less easy to pin down using field data, where different individuals are employing different strategies.

Table 8.1 provides some data on the degree of correspondence between Nuaulu folk categories and scientific taxa. However, we need to be very careful about how such quantitative measures are interpreted, quite apart from any consequences of the kind of data corruption discussed above. For one thing, there is linguistic variation not only between individuals, but also by the same individual on different occasions, as discussed in Chapter 1. People may, for example, use both binomials and uninomials for the same plant, either the superordinate (generic) or subordinate (specific) term being dropped depending on context. In each case where there is an alternative term, I have had to make a judgment based on my experience of the most usual form in everyday speech. There are similar uncertainties generated through the process of scientific identification. Not only might there be errors made by taxonomists or legitimate differences of opinion between taxonomists, but in addition I do not always know whether more voucher specimens linked to a particular folk term might have increased the range of taxonomic content. Thus, where I have only one specimen for a particular folk term, I may assume that there is a good one-to-one correspondence. But, I do not know whether if I were to collect additional specimens the range of taxonomic correspondence might increase. Of course, the obvious way to guard against this problem is to ensure that you collect multiple specimens. In many cases, collecting – say – three in different contexts might be sufficient, but where the range of taxonomic reference is much greater, more specimens would be required. Even where the specimens are collected, they may not be identified, and judgments have to be formed on the basis of less than ideal data. Moreover, in many cases (214, that is 21 per cent) there are no plant identifications for particular labelled categories to species level. Where this is so, I have in Table 8.1 presumed in favour of their being a one-to-one correspondence. However, if we remove the entries in the database for which no species-level identifications are available, then the number of one-to-one correspondences falls from 632 to 418, or from 62 to 41 per cent.

One-to-one species correspondence, under-differentiation and over-differentiation

Once all possibilities for data corruption and synonymy have been accounted for, and erring on the safe side, I think we can say that there was a direct correspondence between confirmed or suspected scientific species and a labelled Nuaulu folk category in 632 cases (62 per cent of all labelled categories). We obtain this number by adding together the frequencies for rows 5, 6 and 7 in Table 8.1. In 18 cases scientific species were under-differentiated, and two or more species were grouped within a single Nuaulu terminal category. For example, the term ***ahau uni*** was applied equally to specimens of both *Antidesma stipulare* and *A. montanum* without there being any indication that difference was systematically recognized in the folk names that were being used. Of these, ten were uninomials (e.g. the climbing fern ***sipoi*** = *Lygodium scandens* and *L. circinnatum*), seven binomials (e.g. ***ai ane*** = *Prunus arborea* and *P. wallaceana*) and one (though probably more) optional, namely (***ai***) ***supa*** = *Ficus tinctoria* and *F. callosa*. As we have seen from the discussion in the introduction to this chapter, one binomial was applied to three species of ebony, *Diospyros elliptifolia, D. maritima* and *D. ebenaster*, and one uninomial (***kaiohite***) to various fig trees without further lexical differentiation (*Ficus botryocarpa* subsp. *hirtella, F. callosa, F. glandulifera* and *F. copiosa*), and another, ***nihasa***, included *Spondias novoguineensis, S. pinnata* and *S. cythera*.

In 230 cases a scientific species was designated by a Nuaulu uninomial that was further differentiated (using binomials); in other words, Nuaulu categories over-differentiated scientific taxa at the specific level, recognizing differences that were not accorded species-level distinction in scientific taxonomy. Of Nuaulu binomials, between 227 and about 245 are applied to subspecific taxa, usually varieties or cultivars. These were mostly for cultigens of commonly planted crops (various yams, mangoes, sago etc.) but also included some species that would not normally be viewed as domesticates, such as the edible seed-bearing and highly anthropic almond *Terminalia catappa*, and *Sterculia foetida* (Table 8.2). In terms of locally recognized subspecific diversity, these ranged from two (e.g. calabash: *Crescentia cujete*) to thirty-four (banana: *Musa acuminata* and hybrids). In four cases, one uninomial and one binomial were reported as distinguishing separate types of a scientific species, with the binomial distinguishing a special type of the default uninomially labelled category, for example the aren palm *Arenga pinnata*.

There are a number of features of the data reported in Table 8.2 that warrant comment. The first is that some very salient cultigens, while mostly subdivided by folk binomials equivalent to cultivars of a single species, can be shown to exhibit anomalous features (marked here with an asterisk*). Thus, the Nuaulu banana label ***uri*** applies mainly to hybrids of *Musa acuminata* and *M. balbisiana*, though also to at least one distinct edible species (*Musa troglodytarum*) and *Musa acuminata* subsp. *banksii*, or *M. lolodensis*. Similarly, ***kasipii*** labels most cultivars of *Manihot esculenta*, but also includes one cultivar of *M.* cf. *carthaginensis* subsp. *glaziovii*, grown as a shade trees and for its leaves. In addition, a subgroup of *M. esculenta* cultivars are sometimes identified using the separate uninomial ***postel***. ***Kanai***, by comparison, mostly labels cultivars of the betel palm *Areca catechu*, but also includes other species of *Areca* with both edible and inedible fruits. Other examples are ***iane***, which refers mainly to cultivars of *Canarium indicum* but also includes other species of *Canarium;* and *Durio zibethinus*, folk varieties of durian that are mainly labelled using the uninomial ***tunene***, except one, labelled ***hutuine***. These anomalies reflect a combination of the role of hybridity in selecting cultivars, and also the practice of highlighting the importance of particular cultivars by using a separate uninomial, sometimes by dropping the classifier and converting the qualifier to a new uninomial (e.g. ***kasipii postel*** > ***postel***), which then becomes the basis for the construction of further binomials for other cultivars.

Another feature of Table 8.2 is the large number of uninomials registering just two binomial segregates. On the one hand, this may point to cases where two distinct types of a cultigen or other plant are being labelled; in other words, the two binomials accurately match two types that would be distinguished in scientific taxonomy. On

Scientific name	Nuaulu uninomial	N terminal categories
Musa hybrids*	***uri***	36
Metroxylon sagu	***hatane***	14
Cocos nucifera	***nione***	11
*Manihot esculenta**	***kasipii***	10
Colocasia esculenta	***sikekue***	10
Dioscorea alata	***hueni***	10
Capsicum anuum	***karatupa***	9
Dioscorea esculenta	***siahue***	9
Saccharum officinarum	***tohu***	8
Ananas comosus	***sesene***	7
Codiaeum variegatum	***sinsinte***	7
*Areca catechu**	***kanai***	7
Solanum melongena	***tonne***	6
Mangifera indica	***apalam***	5
*Manihot esculenta**	***postel***	5
Ipomoea batatas	***kasitena***	4
Nicotiana tabacum	***tapako***	4
Calophyllum inophyllum	***ahutaune***	3
*Canarium indicum**	***iane***	3
Arocarpus indica	***tahenae***	3
Aracaceae sp.	***punetene***	3
Justica gendarussa	***mosone***	3
Syzygium aqueum	***popote***	2
Arachis hypogaea	***kaisan***	2
*Durio zibethinus**	***tunene***	2
Sterculia foetida	***pomane***	2
Zea mays	***yakon***	2
Xanthosoma sagittifolium	***yohoru***	2
Falcataria sp. moluccana	***nisoae***	2
Homalomena cordata	***puni***	2
Aleurites moluccana	***kahea***	2
Syzygium aromaticum	***kapanahane***	2
Archidendron clypearia	***kawasa***	2
Dioscorea hispida	***kawasine***	2
Terminalia catappa	***kaeone***	2
Coffea arabica	***kopi***	2
Hevea brasiliensis	***kataa***	2
Syzygium malaccense	***nutune***	2
Carica papaya	***papae***	2
Artocarpus altilis	***sune***	2
Cyathula prostrata	***panu-panu***	2
Crescentia cujete	***kalapasa***	2
Theobroma cacao	***sakalate***	2
Lycopersicum esculentum	***tamati***	2
Oryza sativa	***ana***	2
Palaquium obovatum	***katou***	1
Arenga pinnata	***naha***	1
Gnetum gnemon	***utatam***	1
Durio zibethinus	***hutiene***	1

Table 8.2 Number of named cultivars for major Nuaulu cultigens in decreasing order of number.

the other hand, we should note that any population of a species of plants is going to exhibit a range of variation, even where this is not formalized into official scientific taxa, and one major strategy used by the human mind to capture this diversity is to distinguish some 'other' variant from the more usual default type. Moreover, a binary contrast of the kind 'main (true) type: other' is so much a feature of human cognition that it may emerge with respect to virtually any plant population. As I have noted above, there are some species for which only one binomial segregate is distinguished. This situation emerges where the default type is not labelled by an adjectival qualifier, and where the marked term is always associated with the 'other' less normal type (***naha***: ***naha niane*** for *Arenga pinnata*, and ***utatam***: ***utatam marane*** for the gnetum *Gnetum gnemon*). Of course, in these cases it also reflects the prevalence of binary thinking. In a few cases, the recognition of a single binomial may reflect loss of one cultivar, or ignorance on the part of an individual informant concerning additional binomials for the same type.

Different genera placed in the same terminal category

Nuaulu plant categories that are not routinely further subdivided by the application of labels, what I have described as 'terminal categories', may be either uninomials or binomials. They may be VARIETAL, SPECIFIC or GENERIC folk taxa in Berlin's sense (see Chapter 3, p. 48), and highlight problems in distinguishing these as distinct 'ranks' or 'levels'. Of all Nuaulu terminal categories applied to plants, seventy-three contain two or more scientific genera, though only four contained more than two. Of these, thirty-seven comprise genera from just one family and thirty-six comprise genera from more than one. Table 8.3 lists those terminal categories containing multiple genera belonging to just one phylogenetic family.

Why should Nuaulu labels for terminal categories sometimes refer to plants from more than one phylogenetic genus in the same family? In some cases it is simply because the features that are diagnostic for the phylogenetic taxonomist are not significant for Nuaulu, who on the whole are more concerned with overall morphology and see in the genera co-labelled a helpful superficial resemblance. Nuaulu may recognize variation between plants named using the same label, but regard any difference as insufficiently significant to warrant a separate name. In Table 8.3 all genera grouped together are from the same phylogenetic family, so at least there is correspondence at this level of taxonomic determination. However, for some of these genera, if we look at the history of scientific nomenclature, it is evident that taxonomists too have changed terminologies in ways that suggest a closer resemblance, and uncertainty as to how this should be reflected in the names. Thus, ***hihane wane***, a liana listed in Table 8.3, indicates a category

Nuaulu terminal category	Scientific genera	Scientific family
ai kihoke	*Breynia, Glochidion*	PHYLLANTHACEAE
ai osi	*Chisocheton, Aglaia*	MELIACEAE
ekene	*Centotheca, Setaria*	POACEAE
hahuseite	*Cyperus, Scleria*	CYPERACEAE
hahusina	*Laportea, Fleurya*	URTICACEAE
hana	*Neolamarckia, Nauclea*	RUBIACEAE
hihane wane	*Bauhinia, Phanera*	FABACEAE
hono-hono hanaie	*Aneilema, Pollia*	COMMELINACEAE
manahute putie	*Cyathula, Amaranthus*	AMARANTHACEAE
mataponane	*Cerbera, Rauwolfia*	APOCYNACEAE
meute patona	*Daemonorops, Calamus*	ARECACEAE
mono nuetetue	*Crassocephalum, Syndrella, Ageratum*	ASTERACEAE
nakuasi	*Alpinia, Riedella*	ZINGIBERACEAE
niwapane	*Sorghum, Thysanoiaena*	POACEAE
panne	*Nauclea, Neolamarckia*	RUBIACEAE
pupu hanaie	*Chingia, Sphaerostephanos*	THELYPTERIDACEAE
sapuane	*Colysis, Drynaria*	POLYPODACEAE
sinatane urone	*Dendrocnide, Procris*	URTICACEAE
sioe	*Triadica, Homalanthus*	EUPHORBIACEAE
sipa nione	*Dendrobium, Grammatophyllum, Oberonia*	ORCHIDACEAE
tepine	*Calyptrocalyx, Pinanga*	ARECACEAE
tom-tom atae	*Glochidion, Phyllanthus*	PHYLANTHACEAE

Table 8.3 Nuaulu terminal plant categories comprising multiple genera from one scientific family.

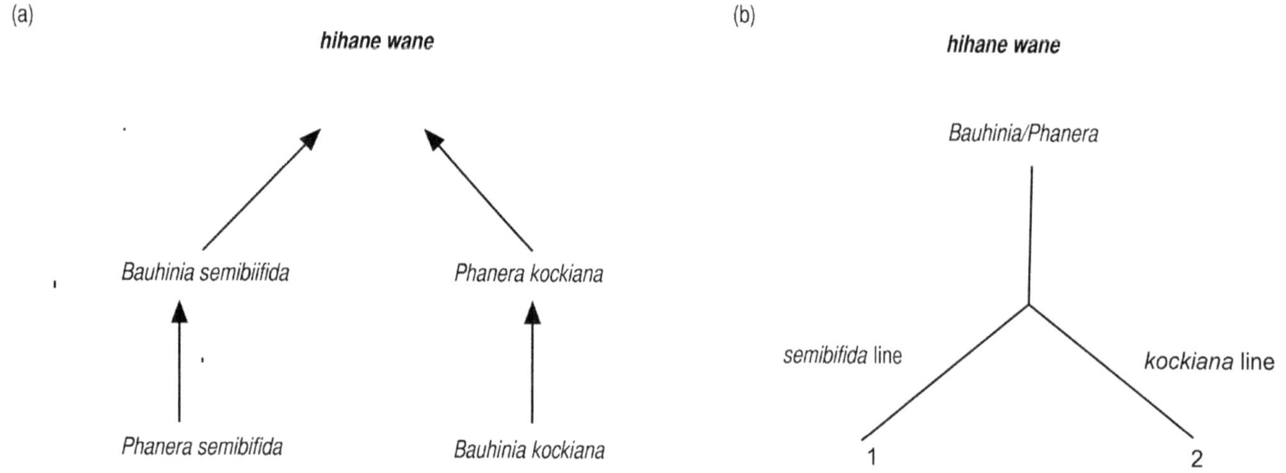

*Figure 8.1 Phylogenetic content of Nuaulu category **hihane wane** expressed in terms of (a) successive name changes for contrasting scientific taxonomic categories, and (b) as converging 'cladistical' lines.*

applied to *Bauhinia semibifida* and *Phanera kockiana* following THE PLANT LIST nomenclature, though the original determinations provided by Kew were *Phanera semibifida* and *Bauhinia kockiana* respectively (Figure 8.1). Such switching of names for closely related species is not uncommon in scientific plant taxonomy, and it suggests that whatever the precise nomenclature adopted, taxonomically the degree of resemblance is very close. Such practice also suggests that for both Nuaulu and biological taxonomists, the cognitive process of establishing classificatory resemblance is essentially cladistic and processual, being a pattern based on a perception of shared 'descent' rather than one of placing types in boxes which automatically can be assumed to share some horizontal alignment or 'level'. We usually associate conventional scientific taxonomy with a model of fixed categories in a spatial universe. After all, pre-Darwinian natural history assumed a fixed creation, and the model had to be altered once evolutionary change was considered the default underlying process. However, cladistic or phenetic classifications assume evolutionary change as standard, and have difficulty in placing the point in time when one species turns into another. Nevertheless, here is evidence suggesting that Nuaulu pragmatically assign plants to some categories, not on the basis of a firm digital logic of characters being either present or absent, but in terms of processual resemblance.

Different phylogenetic families placed in the same Nuaulu terminal category

We can perform a similar analysis to that demonstrated in Table 8.3 for multiple genera of a single phylogenetic family placed in a single Nuaulu terminal category, for genera from two or more families placed in a single Nuaulu terminal category, and which provides examples of a greater degree of under-differentiation. The results are shown in Table 8.4. We can also look at the same problem in a different way by listing Nuaulu uninomials encompassing species from two or more scientific families (Table 8.5).

The first stand out group are ferns, where there is a strong morphological prototype. Thus, tree ferns from three families (CYATHACEAE, DRYOPTERIDACEAE, PTERIDACEAE) are placed in ***ai mukune***, on account of their woody stems (Figure 3.6). Similarly, the terminal fern uninomial ***one*** is applied to representatives of both THELYPTERIDACEAE and WOODSIACEAE. An interesting case is the primary or folk-generic category ***naene***, divided into ***naene totu onate*** comprising species from ADIANTIACEAE and OPHIOGLOSSACEAE, and ***naene putie*** comprising species from POLYPODIACEAE and DRYOPTERIDACEAE. The pairing of these types in Nuaulu classificatory practice indeed reflects phylogenetic similarity, the first having simple or single leaves, and the second pair fronds that are entire, pinnate or variously forked and pinnate. A rather different case is the uninomial ***tunene***, presumably originally used for both cultivated and wild forms of durian (*Durio zibethinus* [MALVACEAE]). With the introduction of soursops and custard apples from the Americas as early as the seventeenth century, we find that many Moluccan languages place them in the 'durian' category because of a superficial resemblance (spines) on the exterior of the fruit Thus, *Annona muricata* [ANONNACEAE] is ***tunene warata*** ('Dutch durian') in Nuaulu, and in the related Saleman language 'duria arata' (Cosieaux 2012:43). Rumphius notes that European authorities, such as Gonzalo Fernandez de Oviedo, also mistook superficial resemblance for genetic affiliation, though Rumphius himself (2011, 1:340, Beekman note 35) did not do so. By contrast, the related custard apple or sweetsop (*Annona reticulata*), which may have been in the Moluccas before the arrival of other *Annona*, is described using a quite separate term, ***ai pina ikine***.

Nuaulu terminal category	Scientific genera	Scientific families
ai mukune (f)	*Cyathea*	CYATHACEAE
	Tectoria	DRYOPTERIDACEAE
	Pteris	PTERIDACEAE
ai pina(mo)	*Annona* (***ikine***)	ANNONACEAE
	Microcos (***onate***)	MALVACEAE
ai turi nuae	*Andrographis*	ACANTHACEAE
	Vitex	VERBENACEAE
ai weti	*Phaleria*	THYMELAEACEAE
	Steganthera	MONIMIACEAE
anino	*Syzygium*	MYRTACEAE
	Walsura	MELIACEAE
asaherane	*Syzygium*	MYRTACEAE
	Virola	MYRISTICACEAE
atiponate	*Pterocarpus*	FABACEAE
	Vitex	LAMIACEAE
atupikae	*Syzygium*	MYRTACEAE
	Phaleria	THYMELAEACEAE
aunasate	*Prunus*	ROSACEAE
	Cryptocarya	LAURACEAE
hunori	*Macaranga*	EUPHORBIACEAE
	Cheilocostus	COSTACEAE
kasipehe	*Canna*	CANNACEAE
	Phrynium	MARANTACEAE
kasitena nuae	*Ipomoea*	CONVOLVULACEAE
	Canavalia	FABACEAE
katinoie numte	*Melastoma*	MELASTOMATACEAE
	Rhodamnia	MYRTACEAE
katou totu ikine	*Palaquium*	SAPOTACEA
	Amomum	ZINGIBERACEAE
kikun marae	*Amomum*	ZINGIBERACEAE
	Sauaiu	ACTINIDACEAE
manahute msinae	*Stachytarpheta*	VERBENACEAE
	Celosia, Amaranthus	AMARANTHACEAE
mono monatohu	*Cyathula*	AMARANTHACEAE
	Syndrella	ASTERACEAE
mono panu-panu	*Desmodium*	FABACEAE
	Leptaspis	POACEAE
naene totu onate	*Helminthostachys*	OPHIOGLOSSACEAE
	Acrostichum	PTERIDACEAE
naene putie	*Drynaria*	POLYPODIACEAE
	Tectoria	DRYOPTERIDACEAE
nana ai	*Ficus*	MORACEAE
	Zehneria	CUCURBITACEAE
one	*Pneumatopteris*	THELYPTERIDACEAE
	Diplazium	WOODSIACEAE
sina-sina wane	*Pityrogramma*	PTERIDACEAE
	Freycinetia	PANDANACEAE
soka soka numte	*Lycopodiela*	LYCOPODIACEAE
	Selaginella	SELAGINACEAE
	Tapeinidium	LYCOPODIACEAE
sonae	*Campnosperma*	ANACARDIACEAE
	Phyllanthus	EUPHORBIACEAE

Table 8.4 Genera from two or more scientific families placed in a single Nuaulu terminal category.

Nuaulu uninomial	Scientific families
ai unate	RHIZOPHORACEAE; also several species of mangrove from other families
hono-hono	COMMELINACEAE, ORCHIDACEAE, ACANTHACEAE
kasipehe	CANNACEAE, MARANTACEAE
mukune	CYATHACEAE, MARATTIACEAE, DRYOPTERIDACEAE, PTERIDACEAE; also refers to other woody ferns
naka-naka	LYTHRACEAE, FABACEAE
pupu	THELYPTERIDACEAE, PTERIDACEAE
soka-sokae	SELLAGINELLACEAE, LYCOPODIACEAE

Table 8.5 Nuaulu uninomials encompassing species from two or more scientific families.

Nuaulu uninomial	Scientific family	Genus
anakoine	PRIMULACEAE	*Tapeinosperma*
anametene	EBENACEAE	*Diospyros*
anine	ERYTHROXACEAE	*Erythroxylum*
asora	DILLENACEAE	*Dillenia*
dalima	PUNICACEAE	*Punica*
hota/okar	GOODENIACEAE	*Scaevola*
iane	BURSERACEAE	*Canarium*
kaeane	COMBRETACEAE	*Terminalia*
kamkaune	LOGANACEAE	*Fagraea*
kanapone	BORAGINACEAE	*Cordia*
kasupa	BIXACEAE	*Bixa*
katinoie	MELASTOMACEAE	*Melastoma*
katoe	ACTINIDACEAE	*Saurauia*
kelor	MORINGACEAE	*Moringa*
koae	PANDANACEAE	*Pandanus*, including pandans labelled with different uninomials but recognized as affiliates (***paone, sanona***)
koitipi	NEPENTHACEAE	*Nepenthes*
neune	CASUARINACEAE	*Casuarina, Gymnostoma*
nunte	LYCOPODIACEAE	*Lycopodiella*
onie	DIPTERACARPACEAE	*Shorea*
papae	CARICACEAE	*Carica*
pitiri	PASSIFLORACEAE	*Passiflora*
punen	CYCADACEAE	*Cycas*
riane	ELAEOCARPACEAE	*Elaeocarpus*
***sapa*(*ne*)**	ULMACEAE	*Trema*
sesene	BROMELIACEAE	*Ananas*
sosone	CANNABACEAE	*Celtis*
taulela	OXALIDACEAE	*Averrhoa*
uri	MUSACEAE	*Musa*, including bananas labelled with different uninomials (***moe-moe***)
utatam	GNETACEAE	*Gnetum*
utatone	OPHIOGLOSSACEAE	*Helmintostachys*

Table 8.6 Nuaulu uninomials encompassing a single genus that is the only local representative of a scientific family.

Correspondence between uninomials, basic categories, and phylogenetic families

The evidence from pre-Linnaean European botany and folk botany generally suggests that recognition of groupings at what we would now call the biological family level is limited but significant, and may be the origin of family groupings in modern botany (Atran 1983, 1990:11). Some of these groupings are explicitly labelled, but others, such as the proto-legume 'bean' in Nuaulu (glossed in Indonesian as 'kacang'), are covert, and while they sometimes correspond to Berlinian INTERMEDIATES, this is not necessarily the case. Looked at more widely, family-level distinctions are more apparent in some groups of plants and other organisms than others. Thus, bird names in general (Alcántara-Salinas, Ellen and Rivera-Hernández 2016; Berlin, Boster and O'Neill 1980; Bulmer 1979; Hage and Miller 1976; Hunn 1977), and in Nuaulu both bird and fish names (Ellen 1993:57–80, 128–43), seem to correspond quite well with phylogenetic family distinctions, where these are often congruent with distinctive outline shapes. A systematic correspondence between terminal categories and even higher taxa, such as orders, can be found almost universally for certain groups of invertebrates. Table 8.6 lists cases where Nuaulu vascular plant uninomials map on to, but are not necessarily the precise equivalent of, scientific families. There are many cases where the 'equivalence' between folk uninomial and scientific family is an artefact of biogeography, in that only one genus of the family is to be found in the flora of Seram that Nuaulu are familiar with (e.g. ***kasupa*** in BIXACEAE, ***naka-naka*** in LYTHRACEAE, or ***koitipi*** in NEPENTHACEAE), yielding what might be called 'pseudo-families'.

In some cases uninomials are not further divided, or only weakly divided, but nevertheless reflect a lot of recognized diversity within a single biological family. This is often apparent with cryptogams, such as ***soka-sokae*** for SELAGINELLACEAE. However, in other cases it can be plausibly argued that the forms are so physically distinct from the rest of the plants known to Nuaulu, and contain so much diversity, that they are better portrayed as family-like categories than genus-like categories. Among flowering plants, pandans are a good example of a folk category coming cognitively close to a scientific family: most pandans are labelled ***koae,*** though a few are given other distinct uninomials but are obviously seen by Nuaulu as perceptually affiliated. These, if you like, are covert categories operating at a family-like level. The same is true of the uninomial ***uri*** and the biological family MUSACEAE.

On the non-correspondence of life forms to phylogenetic taxa

We have noted that seventeen Nuaulu plant uninomials label scientific genera from two or more families and fourteen binomials label genera from two or more families, but this is not evidence for more inclusive classificatory groupings above the family level. The family-like groupings correspond in some respects with Berlin's concept of INTERMEDIATE, but as many studies of folk classification have shown, above the botanical family level there is considerable divergence between scientific taxonomy and folk classification: orders, series, classes and divisions are just not relevant. In Chapter 1 we looked at the Nuaulu concept of plant (an example of Berlin's UNIQUE BEGINNER) and its degree of correspondence with the scientific notion of kingdom. There are, however, a small number of what Berlin has conceptualized as LIFE FORMS in Nuaulu, certainly ***ai*** and ***awane***, plus possibly ***monote***. There are other uninomials labelling categories that cut across many scientific families, such as ***kupako*** and ***kamane***, as discussed in Chapter 7. Of these, ***ai*** and ***awane*** unambiguously correspond to tree and vine respectively, and are clearly LIFE FORMS according to the definition of Berlin (1992), and as adopted by Brown (1984) in his comparative language survey. There is no term that comes really close to HERB in the sense of Berlin, Frake (1962) or Brown. ***Monote*** has the sense of weed or grass, but includes many species that are not taxonomically grasses. It does seem to work at LIFE FORM level, in that it may be contrasted with other life forms in ordinary conversation. ***Kupako*** is a term used for flower in both the sense of the specific organs of reproduction of many angiosperms, and, increasingly, of a group of plants distinguished for their ornamental features. In some folk classifications of francophone West Africa 'fleur' has been introduced to cover this latter sense, as has Indonesian 'bunga' to some extent amongst Nuaulu. Whether or not we can accept it as a LIFE FORM is doubtful. ***Kamane*** is used to describe resin-producing trees, and is thus possibly a special-purpose intermediate grouping but again it is hardly a LIFE FORM.

There is additionally a question as to whether certain groupings labelled by uninomials that we have already looked at in Chapter 6 and 7, and which match phylogenetically defined groupings quite well (such as ***meute*** for rattans and ***koae*** for pandans), might also be seen as 'life form like', in that they will be contrasted with, say, trees and vines. Finally, there is at least one major covert grouping, PALM, that is not named, though there is evidence that Nuaulu recognize a family resemblance with classificatory implications, and which could be said to function as a pseudo life form. Such groups conceptually resemble those life forms that I have previously identified for Nuaulu animal classification, but which do not meet the strict Berlinian definition, such as frogs (Ellen 1993:102). We can agree with Atran (1985:311) that life forms are not phylogenetically 'natural', but then while they are certainly 'anthropocentrically biased', they are neither more artificial nor 'special purpose' than higher order scientific taxa. We will return to the concept of life form in the Chapter 10.

How much of the flora of Seram do Nuaulu know and name?

How can we measure the extent of the flora of Seram that Nuaulu claim to know and to use? One way is to see how many botanical families are represented in the corpus, and the number of Nuaulu labels that apply to each of these families. Tables 8.7a and 8.7b list all botanical families for vascular plants found in the Nuaulu (NED) corpus, the number of Nuaulu labels that subdivide them, and the number of scientific species these represent. In some cases where identifications for a single Nuaulu category suggest inclusion of species from more than one family, these families are also included in the calculation. The Nuaulu labels are those appearing at various degrees of classificatory inclusion; so, uninomials, binomials and – in some cases – trinomials applied to the same taxon are included. Thus, records for ***utatam*** and ***utatam marane*** both indicate GNETACEAE, so in the calculations this counts twice. This is not a perfect measure of cognitive salience, but it is sufficiently so for my purpose here. Figure 8.2 plots number of botanical families against number of Nuaulu names, and it is immediately apparent from this that most families are only represented by one or two Nuaulu names, and that the greatest density of names are concentrated in just four families (ARECACEAE, EUPHORBIACEAE, FABACEAE and POACEAE): 62, 56, 55 and 53 folk taxa respectively. Although these are large families on a global scale anyway, looking at the distribution as a whole, it is remarkable just how asymmetric Nuaulu knowledge is when measured in these terms.

Table 8.7a Scientific families represented in the NED alongside approximate numbers of Nuaulu categories and number of represented scientific species (ordered alphabetically by family).

Scientific family	N Nuaulu named categories	N species to which Nuaulu named categories apply
ACANTHACEAE	21	8
ACTINIDIACEAE	2	2
ALANGIACEAE	1	1
AMARANTHACEAE	9	8
AMARYLLIDACEAE	4	3
ANACARDIACEAE	14	12
ANISOPHYLLEACEAE	1	1
ANNONACEAE	8	6
APIACEAE	6	6
APOCYNACEAE	8	8
ARACEAE	28	9
ARALIACEAE	3	3
ARAUCARIACEAE	4	3
ARECACEAE (PALMAE)	62	27
ASPARAGACEAE	6	3
ASPLENIACEAE	5	6
ASTERACEAE (COMPOSITAE)	12	10
BEGONIACEAE	3	3
BIGNONIACEAE	5	3
BIXACEAE	1	1
BORAGINACEAE	1	1
BRASSICACEAE	2	5
BROMELIACEAE	8	1
BURSERACEAE	12	6
CACTACEAE	1	1
CALOPHYLLACEAE	9	6
CANNABACEAE	1	1
CANNACEAE	1	1
CARICACEAE	2	1
CASUARINACEAE	3	3
CLUSIACEAE	7	6
CONVOLVULACEAE	9	5
COMBRETACEAE	4	2
COMMELINACEAE	5	4
CORNACEAE	1	2
CUCURBITACEAE	14	12
CYATHACEAE	5	4
CYCADACEAE	1	1
CYPERACEAE	3	4
DICHAPETALACEAE	1	1
DILLENIACEAE	2	2
DIOSCOREACEAE	26	4
DIPTEROCARPACEAE	5	1
EBENACEAE	3	5
ELAEOCARPACEAE	2	2
ERYTHROXYLACEAE	1	1
EUPHORBIACEAE	55	33
FABACEAE	55	49

Scientific family	N Nuaulu named categories	N species to which Nuaulu named categories apply
Fagaceae	1	1
Flagellariaceae	1	1
Gentianaceae	1	1
Gesneraceae	1	2
Gnetaceae	3	2
Goodeniaceae	2	2
Heliconiaceae	1	1
Lamiaceae	12	13
Lauraceae	13	12
Lecythidaceae	2	3
Lycopodiaceae	2	2
Lythraceae	3	4
Malvaceae	25	19
Marantaceae	11	6
Marattiaceae	1	1
Melastomataceae	6	9
Meliaceae	7	6
Menispermaceae	2	2
Monimiaceae	1	1
Moraceae	33	33
Moringaceae	1	1
Musaceae	39	21 (excluding hybrids)
Myristaceae	6	7
Myrtaceae	33	24
Nepenthaceae	3	2
Oleaceae	1	1
Onagraceae	1	1
Ophioglossaceae	2	9
Orchidaceae	7	11
Oxalidaceae	3	2
Pandanaceae	11	8
Passifloraceae	1	1
Phyllanthaceae	6	8
Piperaceae	17	15
Pittosporaceae	1	1
Poaceae	53	38
Podocarpaceae	1	1
Polygonaceae	1	1
Polypodiaceae	8	9
Portulaceae	1	1
Primulaceae (incl. Myrsinaceae)	3	3
Psilotaceae	1	1
Pteridaceae	8	7
Ranunculaceae	1	1
Rhamnaceae	1	1
Rhizophoraceae	1	2
Rosaceae	5	5
Rubiaceae	25	20
Rutaceae	16	14
Salicaceae	3	4
Sapindaceae	2	2
Sapotaceae	3	2
Schizaeaceae	2	3
Selaginellaceae	4	2
Simaroubaceae	2	1
Smilacaceae	1	1
Solanaceae	28	7
Tectariaceae	1	1
Tetramelaceae	2	2
Thymelaeaceae	6	4
Ulmaceae	1	1
Urticaceae	8	8
Verbenaceae	8	7
Vitaceae	8	8
Xanthorrhoeaceae	1	1
Zingiberaceae	26	23

Scientific family	N Nuaulu named categories	N species to which named categories apply
Arecaceae	62	27
Euphorbiaceae	56	35
Fabaceae	55	49
Poaceae	53	38
Musaceae	40	5 (excluding hybrids)
Moraceae	33	33
Myrtaceae	33	24
Araceae	28	9
Solanaceae	28	7
Zingiberaceae	26	23
Dioscoreaceae	26	4
Rubiaceae	25	20
Malvaceae	25	19
Acanthaceae	21	8
Piperaceae	17	15
Rutaceae	16	14
Anacardiaceae	14	12
Cucurbitaceae	14	12
Lauraceae	13	12
Lamiaceae	12	13
Asteraceae (Compositae)	12	10
Burseraceae	12	6
Pandanaceae	11	8
Marantaceae	11	6
Convolvulaceae	9	11
Amaranthaceae	9	8
Calophyllaceae	9	6
Polypodiaceae	8	9
Apocynaceae, Urticaceae, Vitaceae	8	8
Pteridaceae, Verbenaceae	8	7
Annonaceae	8	6
Bromeliaceae	8	1
Orchidaceae	7	9
Clusiaceae, Meliaceae	7	6

Scientific family	N Nuaulu named categories	N species to which named categories apply
Melastomataceae	6	9
Phyllanthaceae	6	8
Myristicaceae	6	7
Apiaceae	6	6
Thymelaeaceae	6	4
Asparagaceae	6	3
Aspleniaceae	5	6
Rosaceae	5	5
Commelinaceae, Cyathaceae	5	4
Bignoniaceae	5	3
Dipterocarpaceae	5	1
Amaryllidaceae, Araucariaceae	4	3
Combretaceae, Selaginellaceae	4	2
Ebenaceae	3	5
Cyperaceae, Lythraceae, Salicaceae	3	4
Araliaceae, Begoniaceae, Casuarinaceae, Primulaceae	3	3
Gnetaceae, Nepenthaceae, Oxalidaceae, Sapotaceae	3	2
Brassicaceae	2	5
Lecythidaceae, Schizaeaceae	2	3
Actinidiaceae, Dilleniaceae, Elaeocarpaceae, Goodeniaceae, Lycopodiaceae, Menispermaceae, Sapindaceae, Tetramelaceae	2	2

Table 8.7b Scientific families represented in the NED alongside approximate number of Nuaulu categories and scientific species to which they apply (ordered according to number of Nuaulu names).

Scientific family	N Nuaulu named categories	N species to which named categories apply
Caricaceae, Ophioglossaceae, Simaroubaceae	2	1
Cornaceae, Gesneraceae, Rhizophoraceae	1	2
Alangiaceae, Anisophylleaceae, Bixaceae, Boraginaceae, Cactaceae, Cannabaceae, Cannaceae, Cycadaceae, Dichapetolaceae, Erythroxylaceae Fagaceae, Flagellariaceae, Helioconaceae, Marattiaceae, Monimiaceae, Moringaceae, Oleaceae, Onagraceae, Passifloraceae, Pittosporaceae, Podocarpaceae, Polygonaceae, Portulaceae, Psilotaceae, Ranunculaceae, Rhamnaceae, Smilacaceae, Tectariaceae, Ulmaceae, Xanthorrhicaceae	1	1

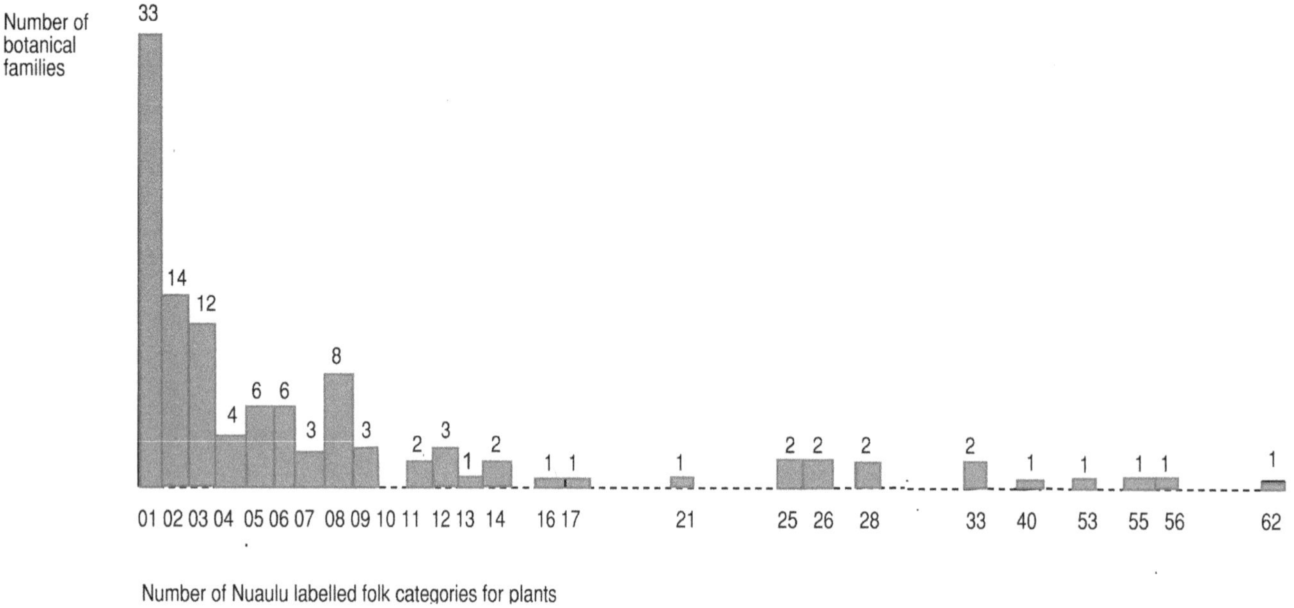

Figure 8.2 Plot of frequencies of Nuaulu labels against number of botanical families. Each vertical bar indicates the number of botanical families represented by the given frequency of Nuaulu names.

The asymmetry between Nuaulu classificatory knowledge of the Seram flora and the way it is rendered phylogenetically partly reflects their physical experience of particular biotopes rather than others. Thus, knowledge of the highest altitude areas (land over 1000 m asl in Figure 1.4a) is restricted, as these are seldom frequented. Note, however, that scientific family revisions can also alter the ratios of correspondence. For example, the moving of many species to PHYLLANTHACEAE from EUPHORBIACEAE has raised the profile of the former family compared with the latter. Under recent revisions, GUTTIFERAE has been completely absorbed into CLUSIACEAE, but CALOPHYLLACEAE (containing the culturally significant *Calophyllum* genus) separated from it. Similarly, much of VERBENACEAE has moved into LAMIACEAE, while placing *Allium* in AMARYLLIDACEAE removes LILIACEAE as a family altogether. These changes can have dramatic effects on graphs of the kind illustrated in Figure 8.2, where we are dealing with small numbers of species and local categories.

It is also true that we have to take into account the fact that Nuaulu are aware of many plants that they themselves do not use, but which are used in other parts of Seram. They will, for example, say, 'We know it, but we do not use it.' And despite the impact of technical taxonomic revisions on the interpretation of the scope and pattern of Nuaulu systematic botanical knowledge, the underlying reason for the asymmetry of their knowledge with what we might discover from an exhaustive scientific survey is prioritization and selection for attention of not only the plants that are ecologically accessible, but also those which are repeatedly and routinely useful. This conclusion might appear to run counter to some of Lévi-Strauss's famous observations concerning the existence of a universal human 'science of the concrete' (which we will return to in Chapter 10), and places utility once again centre stage.

Problematic correspondences

Conklin (1954), in his pioneering study of Hanunóo plant classification on the Philippine island of Mindoro, reports 1,625 terminal categories corresponding to 1,100 scientific species of 650 genera in 150 families, plus 90 folk varieties of rice; Berlin, Breedlove and Raven (1974b:100–7) working amongst the Tzeltal Maya report 720 terminal categories, in 471 folk generics corresponding to nearly 1,700 plant species. Tobelo on Halmahera (Taylor 1990) recognize 746 basic terms and 1,122 terminal categories that they apply to 647 scientific plant taxa, plus 89 where the family is undetermined. In her study of the Bunaq of central Timor, Friedberg (1990:120) reports that 240 basic terms ('termes de base') correspond to 532 scientific species of 399 genera, 25 to multiple species of the same genus, 34 to multiple species in the same family, 43 to species in different families, and 12 to different forms of the same species. By comparison, 870 Nuaulu terminal categories map onto approximately 647 scientific species of 391 genera. Using data from Table 8.1, the comparable Nuaulu numbers to those provided by Friedberg for basic terms are 632 (equivalent to scientific species), 58 (multiple species of the same genus), 37 (multiple species of the same family), 31 (species of different families) and 239 (different forms of the same species). In percentage terms the ratios are close (Figure 8.3), the main difference being that Nuaulu have

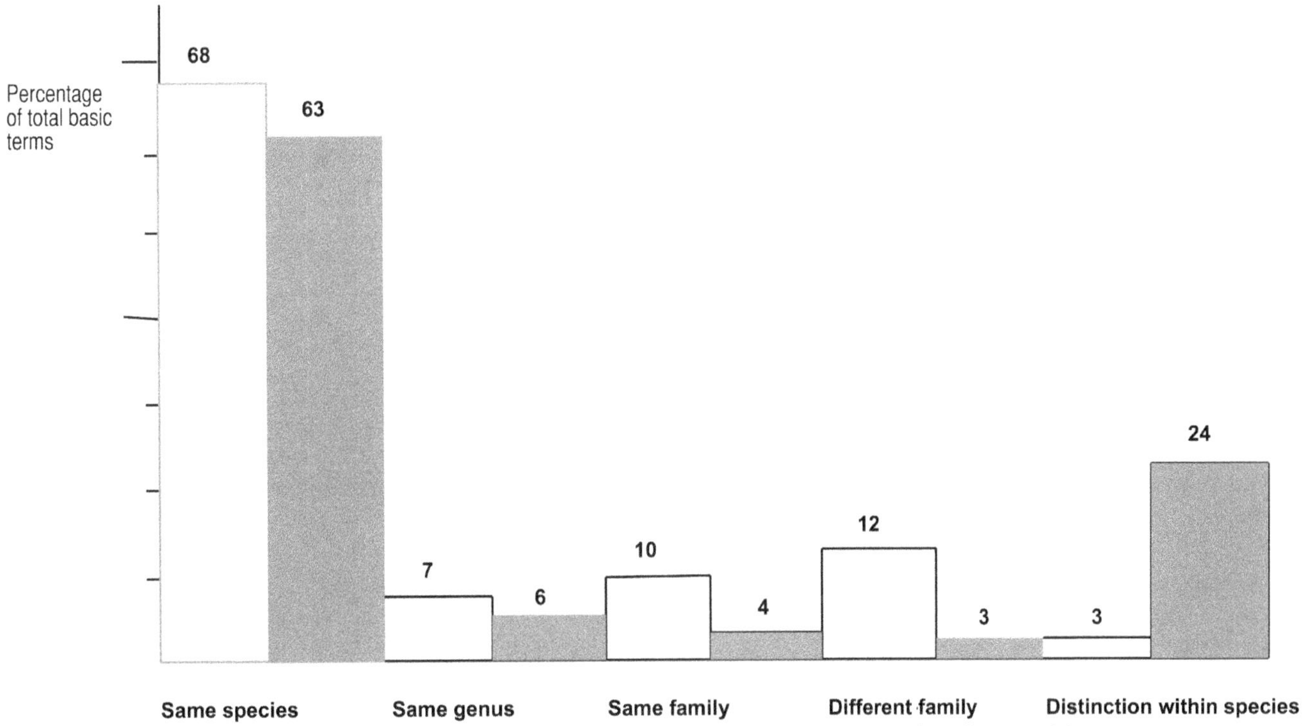

Figure 8.3 *Patterns of correspondence between folk and scientific categories: Nuaulu and Bunaq (Friedberg 1990) basic categories compared. Nuaulu data in shaded column.*

many more categories corresponding to subspecific forms which are not regarded as basic categories in Friedberg's sense. But mapping the phylogenetic grid onto the Nuaulu logic of plant classification as exemplified in naming practices is not simply a question of over- and under-differentiation, or of interposition of categories based on use or morphology that cut right across the logic of the phylogenetic system.

One problem is where a uninomial can label both a folk generic or primary term and a specific or terminal category. For example, ***ahane*** indicates both a generic that includes different kinds of fern, such as ***aha onie*** and ***aha totu onate***, but also a specific type in its own right, in this case *Phymatosorus scolopendria*. The sense in which it is being used here, is to indicate that ***ahane*** is strictly the default kind that Nuaulu experience, whereas those terminal categories that are expressed as binomials are to be distinguished from this default type, though related to it. Earlier in this chapter I described a similar situation in relation to ***utatam*** (*Gnetum gnemon*) and ***naha*** (*Arenga pinnata*).

In the light of the preceding discussion, and certainly in the light of recent work on ontology and convergence metaphysics, it seems to me that it is useful to ask yet again what scientific taxa actually are, and I do so in Chapter 10. In our own interaction with the world we tend to assume that a name will always refer to one particular kind of plant, a model we adopt most emphatically as young children acquiring language. But although it is necessary to make such assumptions in order to navigate our way through the endless variety of plant forms, the correspondences are not always stable. We can see this not only from our own folk classifications, but also from the history of systematic botany as it developed in the West, as plants were moved between taxa and given different names, as more was found out about them and their relations with other plants. The idea of a firm correspondence between a particular physical type, moved from being a cognitive convenience in the pre-Linnaean world to being an *idée fixe* in scientific classification after Linnaeus. In most folk classifications, the identification of plants by name would certainly assume that a fixed number of names are to be applied to a fixed number of items, but pragmatically this is often an approximation based on a necessary working hypothesis. In contemporary scientific taxonomy the certainties underpinning post-Linnaean classification, including the assumption that progressive discovery of increasing numbers of plant species would eventually lead to a definitive set of taxa to accurately describe the world, have been replaced by a much more tentative set of models and procedures that cope better with the complexities of plant genetics and evolution that we now need to take into account.

9
Plant knowledge and time

Introduction

Most studies of ethnobiological classification still adopt a broadly synchronic model, even though there has been an interest in the change and long-term development of classifications and their lexical referents since the early 1970s (Berlin 1972; Brown 1984). However, it is now common to assert almost as a matter of dogma that folk classifications are 'dynamic', and there is an increasing interest in transmission studies and studies of knowledge erosion (Nabhan and St Antoine 1993; Zent 2014), in which the measure of loss is basically one of names. It is therefore surprising to discover that there is relatively little published work on how ethnobiological classification systems respond to the introduction of new percepts of different types (species or varieties usually), or of changes, including loss, in the frequency and salience of names for those percepts, or generally about the non-stable and adaptational aspects of classifying behaviour (but see Wiebusch 2009).

This chapter is about historical ethnobotany: how plant use and knowledge has changed over time, particularly in relation to folk classification. All folk classifications are inevitably unstable, reflecting the responses of people at different times in different places to different problems. However, this is not to say that for various reasons – ecological, cognitive and social – they do not show certain widely repeated regularities. I argue that in looking at changes in representations and the properties of nomenclatural evidence, a multidimensional model of what is meant by a classificatory system provides a better way of describing – and then understanding – the dynamics of the processes of change in cognitive and linguistic terms. The data presented here illustrate patterns of change prompted by biocultural contact stimuli, especially the introduction of species associated with the great Columbian exchange.

There have been several overarching frameworks proposed for examining the shifting relationships between plants and people, none of them in my view mutually exclusive, and all placing the emphasis in a slightly different place. Each offers different insights, and therefore each has something to commend it. All are broadly causal, focusing on modes of adaptation, and all are evolutionary by persuasion. First, we have an approach, historically the earliest, that treats plants more or less as a backdrop to human development, setting constraints and providing opportunities as to how human populations can adapt. In this model plants have priority. Second, there is an approach in which people are seen as actively seeking the affordances of nature and bending them to their advantage. Third, there is an approach in which equal causal weight is given to plant species and human populations, the relationship being seen as one of dynamic but mechanistic co-evolution. Finally, and most recently, the relationship between people and plants has been seen as one of interspecies collaboration, a 'more-than-human' process (e.g. Lewis-Jones 2016) stressing mutual intentionality, where unilinear causation is seen as an oversimplification, even a distortion, of a complex set of events and relations that gives rise to a unique and not always predictable dynamic.

As people and plants interact, so one leaves a near indelible trace on the other. Encounters with plants translate into knowledge, often encoded linguistically, particularly in names, while humans leave micro-ecological and genetic marks on plants. In both cases the marks may be passed on intergenerationally, and in both cases the accumulation of individual instances can result in significant biocultural change that, with human population growth, impacts on the constitution of what we now call the Anthropocene.

Historical ethnobotany: pre-1500 CE

Of the various domesticated and cultivated crops still important today on Seram, some of the most important were available to the pre-Austronesian predecessors of the Nuaulu. Consider, for example, carbohydrate roots and tubers. Taro was probably the earliest of these, particularly *Colocasia esculenta*, which arrived from the west during the Holocene. However, Yen (1995:835) has suggested that there may have been an earlier Pleistocene domestication of *Alocasia macrorrhiza* (L.) Schott. At the species level, the situation with respect to yams is very close to what we know for the island of New Guinea (Powell 1976:119). Coursey (1967:7), following Vavilov (1951), claims that *Dioscorea alata* and *D. esculenta* derive from a centre of endemism in Burma-Assam, while *D. alata* has an Indo-Malayan centre of secondary distribution, where it separated from *D. hispida, D. pentaphylla* and *D. bulbifera*. Linguistically, both taro and yam names include proto-Austronesian forms (Hays 2005:654), and as in New Guinea, we might expect fewer Indonesian or Iberian (Spanish or Portuguese) reflexes for older crops than for those adopted more recently (Hays 2005:633). While edible taro and yam appear to have moved eastwards into Wallacea during early prehistory, *Metroxylon sagu* (the main Nuaulu starch staple) has moved westwards from a centre of dispersal probably located in the Sepik valley of New Guinea, gradually becoming less diverse below the species level as it moved towards the periphery of its western distribution in the coastal islands off west Sumatra and mainland South-East Asia (Ellen 2004c).

One way of inferring changing plant knowledge over time is to compare plant names in different languages. For evidence of the oldest Nuaulu names we might look at proto-Austronesian (PAN) roots. Austronesian languages began to arrive in the Moluccas (and possibly Seram) around 3000–4000 BP, proto-Austronesian dividing into various proto central Malayo-Polynesian (PCMP) languages in the Nusa Tenggara and the central Moluccas area at around this time. However, if we compare plant words reconstructed for PAN and PCMP (Table 9.1), though there is a good measure of correspondence between PAN and PCMP, there appears to be relatively slight correspondence with modern Nuaulu. Many modern Austronesian languages of Nusa Tenggara and the central Moluccas display a high concentration of terms cognate with PAN and PCMP, which suggests slow language change and possibly an ecology that could support many of the plants that Austronesians had domesticated or encountered to the west and north-west. Words for grain crops (including rice and millet) and a taro of the genus *Alocasia* can be relatively safely reconstructed for proto-Austronesian (Donahue and Denham 2010). By contrast, languages with a low concentration of widely distributed PAN and PCMP words indicate rapid language change and possibly reflect new ecologies and encounters with pre-existing speakers of non-Austronesian (presumably Papuan) languages who had already developed terminologies appropriate to the local ecology. This fits in with what we know of the very complex and hybrid character of human DNA and genetics in this area, its mixed socio-cultural composition, and its accompanying transitional biodiversity. Of course, many of the Austronesian terms were for plants that Austronesians would have brought with them, or with which they were familiar from previous areas where they had lived, but some of these crops were either completely dropped or gradually declined in importance (as is the case, for Nuaulu, of rice and millet). Instead, new crops such as sugarcane, banana, *Metroxylon* sago, *Canarium* and *Aleurites* became more important. Only one Nuaulu word in Table 9.1, that for coconut, is obviously cognate with PCMP; and many of the PCMP terms are closer to Malay than to Nuaulu, e.g. *rambia > rumbia, *tales > tales, *pandan > pandan, *bitaquR > bintangor. A further consideration (and possible complication) is that at some point the Nuaulu or their genetic predecessors moved from the coast to the interior of Seram, a different topography and ecology requiring further adjustment in their management of resources, while Austronesian is an essentially maritime culture. For all these reasons, we might be persuaded by Blench (2016:78) of the dangers of being 'seduced by the lure of coherence' and acknowledge that in the absence of evidence for a 'tidy sequence' of Austronesian migration we need to recognize the process as one of 'chaotic expansion', complicated in the case of the Moluccas by ecological disjunction, culture contact and genetic outmating, plus secondary movements between contrasting habitats. As Donahue and Denham (2010) suggest, there is evidence of considerable maritime interraction before the appearance of Austronesian languages, through a mosaic of local networks, and that we probably need a more nuanced model to understand the relationship between farming, material culture and language dispersal in this area.

	PAN	PCMP	Nuaulu
Areca catechu: betelnut	*sawiki	*buaq	***kanai***
Calophyllum inophyllum		*bitaquR	***ahutaune***
Cocos nucifera: coconut		*niuR	***nione***
Colocasia esculenta: taro	*cali	*tales	***sikeue***
Dioscorea alata: greater yam		*qubi	***hueni***
Metroxylon sagu: sago		*rambia	***hatane***
Millet, probl. *Setaria italica*	*beCen, *zawa	*betem, *beten	***hotone***
Oryza sativa: rice plant	*pajay	*pajay	***ana***
Pandanus	*panudaN	*pandan	***koae, saune***

Table 9.1 Comparison of words for selected domesticated and useful plants in proto-Austronesian, proto central Malayo-Polynesian, and Nuaulu (data source: Blust 2018).

For more recent periods we might compare plant names in the different languages of Seram (Figure 2.1). However, this too is a hazardous exercize. On Seram some of the best evidence we have for this is from juxtaposing Nuaulu language data with data collected for Alune by Suharno (1997), and Florey and Wolff (1998; Wolff and Florey 1998). Alune as a language is also well documented through the early work of Niggemeyer (1951–2) and more recently Florey (e.g. 1998). Both Nuaulu and Alune are in the Seram group of east central Maluku languages, but diverged early, Alune being Nunusaku and Nuaulu Sawai-Nuaulu. Within Nunusaku, Alune is classified as Three Rivers Amalumute

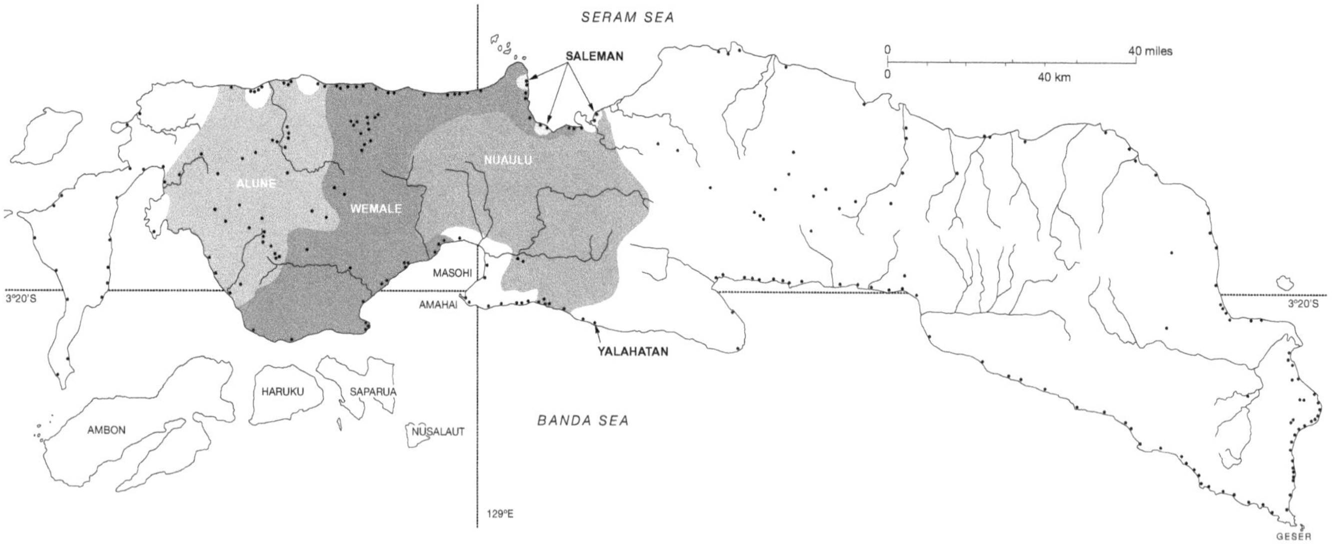

Figure 9.1 Map showing approximate spatial juxtaposition of Alune, Wemale, Nuaulu and other selected languages on Seram, indicating maximum historical spread of Nuaulu and some overlaps (based on Atlas Bahasa Tanah Maluku 1996, and field data on Nuaulu oral history). Note that there were in addition several Wemale-speaking villages in south Seram, east of Elpaputih Bay (such as the now abandoned settlement of Nuelitetu) the result of displacement during the early colonial period.

north-west Seram Ulat Inai. It is close to Wemale (also a Three Rivers language), the core area of which has a boundary with the historic Nuaulu language area, while Wemale speakers also settled in the east Elpaputih Bay area and overlap with Nuaulu as a result of the Dutch wars of the late seventeenth century (Bahasa Tanah Maluku 1996:52). In addition, Yalahatan (with whom Nuaulu have close continuing social exchange relations) and Nuelitetu (Wemale) are Three Rivers languages, so perhaps we should not be surprised to find close lexical similarities between them (Figure 9.1). Consider, for example, the following cases of common salient plants, most of which involve a common shift in which Alune /l/ is replaced by Nuaulu /n/:

'nunue', ***nunue***: referring to many species of *Ficus* in both Alune and Nuaulu

Alune 'nala' > Nuaulu ***nana*** for *Pterocarpus indicus*, the Amboina timber of commerce

Alune 'pasale' > Nuaulu ***pasane***, the timber *Vitex cofassus*

Alune 'lakuase' > Nuaulu ***nakuasi*** for the medicinal and culinary herb *Alpinia galanga*

Alune 'sapale' > Nuaulu ***sapane*** for the common weed of secondary swidden growth *Trema orientalis*, but also for less common plants, for example *Inocarpus fagifer* (Nuaulu: ***kaine***, Alune: 'kaene').

In other cases, we find resemblances between an Alune and Nuaulu term, but applying to different species of a common genus. For example, Alune 'ai mitene' for *Diospyros cauliflora* and Nuaulu ***anametene*** for *Diospyros rostrata*.

In contrast to such striking similarites that seem to present clear evidence of a common origin for a shared term for a plant taxon, the same name in Alune and Nuaulu may sometimes be used for quite different species (e.g. ***pate***: Nuaulu *Ficus* sp, Alune *Albizia saponaria*; or ***ai turi***: Nuaulu *Senna alata*, Alune *Sesbania grandiflora*; ***popote puti***: Nuaulu *Syzygium aqueum*, Alune the reed *Phragmites karka*). Perhaps more surprisingly the names for regionally salient biocultural keystone species such as *Metroxylon sagu* switch in curious ways. Thus, ***hatane*** in Nuaulu becomes Alune 'pia' (cognate with PAN *rambia), while ***mpiae/piae*** in Nuaulu is not a plant term itself but rather raw sago pith. Different again is the shift between Alune 'nikite' and Nuaulu ***mukune*** for tree ferns of the genus *Cyathea*, the first a term composed of phonetically closed sounds and the second of open sounds, suggesting the expression of cultural difference through phonetic reversal, comparable to the semantic reversal found for other words, e.g. sun: Wemale 'tuale', 'tuamane' > Nuaulu ***ranie***; earth: Wemale 'rahie' > Nuaulu ***tuamane*** – which is also a switch in gender symbolism, male to female in the first instance and female to male in the second (Ellen 1983:34; Stresemann 1923:331). Thus, in some cases the differences appear to be linguistically arbitrary, while in others they seem to have been deliberately encouraged so as to mark cultural differences.

Suharno reports 477 Alune names applied to approximately 388 scientific species. Of these latter, only 247 are reported for both Alune and Nuaulu. This may seem very low for two populations with such biocultural and ecological similarities living in close geographical proximity, and is partly explained by errors, incomplete fieldwork and voucher identifications with different taxonomic interpretations (such as Suharno splitting *Metroxylon sagu* into several different species). However, it is also noteable that while there is a high overlap in reported use of the same scientific genera, species often vary. Thus, consider the following:

Alune	Nuaulu
Spondias dulcis	*S. pinnata, S. cytherea, S. novoguineensis*
Polyalthia rumphii	*P. glauca, P. lateriflora*
Cerbera floribunda	*C. manghas*
Aglaonema haenkii	*A. simplex*
Cymbopogon nardus	*C. rectus*

This is consistent with the hypothesis put forward by Ellen and Puri (2016) with respect to medicinal plants: that while we might expect genera to be widely shared culturally, species can be predicted to vary much more in response to local conditions. There may be ecological differences affecting availability; and the process of selection of useful plants often influences local conditions and practices, which then become embedded in a knowledge system.

Another study (Cosiaux 2012; Purwanto and Cosiaux 2013) provides us with some data on Saleman plant terms. Saleman – located on Seram's north coast – is closer linguistically to Nuaulu than to other coastal languages, with strong historical connections (e.g. Ellen 2018a:15–20) and nearby the North Nuaulu language settlements of Opin and Rumah Olat. Both Nuaulu and Saleman are placed in the Sawai-Nuaulu grouping of Seramese languages (Figure 2.1; Atlas Bahasa Tanah Maluku 1996:61), in contrast to Alune, which is located within the Nunusaku grouping. Nuaulu and Saleman share the term ***uri*** for banana and ***tom-tom*** for *Flacourtia rukam*. Although the data do not permit systematic comparison, Saleman and Nuaulu clearly share many plant terms. Thus Saleman 'kayohiti' (*Ficus copiosa*) is close to Nuaulu ***kaiohite*** (various *Ficus*, including *F. botryocarpa, F. callosa* and *F. glandulifera*); Saleman 'ai sinatan' (*Urticastrun decumanum*) is close to the Nuaulu generic term for nettles (***sinatane***), which also includes this common species (***sina manusia***), while Saleman 'papalam' is cognate with Nuaulu ***apalam*** (*Mangifera indica*) and Saleman 'kuin' with Nuaulu ***kuine*** (*Mangifera odorata*). Consider also:

Saleman 'ai pasan' < > Nuaulu ***pasane*** (*Vitex cofassus*)
Saleman 'ai naka' < > Nuaulu ***naka-naka*** (*Lawsonia inermis*)
Saleman 'ai taun' < > Nuaulu ***ahutaune*** (*Calophyllum inophyllum*)
Saleman 'alan' < > Nuaulu ***ana*** (*Oryza sativa*)
Saleman 'poputi' < > Nuaulu ***popote*** (*Syzygium aqueum*)
Saleman 'kuni' < > Nuaulu ***kunini*** (*Curcuma domestica*)

However, while Saleman, Alune and Nuaulu share ***ai turi***, in Saleman (as in Alune) the term refers to *Sesbania grandiflora* rather than *Senna alata*.

The Columbian exchange and its lexical and classificatory consequences

One way of focusing on change in the Nuaulu system of classification is to look at a period of time beginning with major historical consequences for the content of the domain being analysed. For the Nuaulu domain of plants the contact between Europe and the Americas from 1492 onwards resulted in the redistribution to the rest of the world, including eastern Indonesia, of a large number of plant species previously found only in the Americas. The total number of species released in what Alfred Crosby (1972) called the 'Columbian exchange' (though for some species and places it might better be described as the post-Magellenic exchange) has been estimated at many thousands. Of these, in the region of forty-eight species (mainly of the major starch-producing cultigens) have had a subsequently massive global impact on ecological and social history. For obvious reasons, most of the New World species diffusing through the Old World after 1492 were food crops competing with existing indigenous starch crops and grains. Many started out in the diaspora in what Bates (1985) describes as 'secondary' or 'tertiary' plant-resource pools, but others, such as the Irish potato (*Solanum tuberosum*) and cassava (*Manihot esculenta*), have moved in the diaspora to occupy a major position in 'primary' resource pools. There are however, in addition, New World plant species which have remained peripheral in the diaspora, and many (including pan-tropical weeds) which were inadvertent but consequential migrants. A further interesting feature of the Columbian exchange, with significant impacts, including for systems of folk classification, was that most of the major crop species that moved out of the Americas into the diaspora contained little subspecific genetic variation. So, most European varieties of *Solanum tuberosum* are derived from a small number of imported landraces, and the same is true for *Manihot esculenta* in island South-East Asia.

Some of the most important plant species of New World origin currently grown or known by Nuaulu are listed in Table 9.2. However, there is some evidence that plants of American origin may actually have arrived in island South-East Asia before 1492, including sweet potato (e.g. Conklin 1963; Roullier *et al.* 2013), beans and chillis (Wright 2007). Rumphius was well aware of the impact that introduced plants were making on the Moluccas during the second half of the seventeenth century, and notes for example that sweet potato (*Ipomoea batatas*: AM 'patatas' or 'ubi kastela') was of Spanish (Castillian) origin (Rumphius 2011, 4:369), coming from the Americas via Manila. He reports that sweet potato was everywhere,

Scientific name	English name	Estimated date of arrival in Moluccas
Agave cantala	cantala	AH, 1700–1800
Anacardium occidentale	cashew	AH, before 1650
Ananas comosus	pineapple	AH, before 1650
Annona muricata, A. reticulata	soursop, custard apple	AH, 1500–1600
Arachis hypogaea	groundnut	AH, though Rumphius is clear that this was not found in the Moluccas in his time; 1700–1800
Canna indica	Indian shot	AH, 1600–1650
Capsicum annuum	chilli	AH, well-established by time of Rumphius; 1500–1600
Carica papaya	papaya	AH, before 1650
Cucurbita	squash, pumpkin, marrow	AH, 1600–1800, different species arriving at different times
Ipomoea batatas	sweet potato	AH, poss. pre-1500, but later re-introductions from sixteenth century
Lycopersicon esculentum	tomato	AH, 1550–1650
Manihot esculenta	cassava	AH, but mentioned as a NW crop; 1810–1880
Maranta arundinacea	arrowroot	1800–1900
Passiflora	passionfruit	AH, though probably not in the Moluccas until 1800–1900
Persea americana	avocado	1800–1900
Phaseolus	beans	AH, 1550–1650
Psidium guajava	guava	AH, 1550–1650
Solanum tuberosum	Irish potato	1800–1900
Theobroma cacao	cacao	1850–1900
Xanthosoma	cocoyam	1850–1900
Zea mays	maize	AH, 1500–1550

Table 9.2 Selected crops of New World origin grown or known to Nuaulu (following Hill 2004). Estimated dates of arrival from PROSEA, unless otherwise indicated. AH indicates a mention in Rumphius's Ambonese Herbal.

even deep in the mountains (ibid., 370). The evidence points strongly to sweet potato being the first major New World crop to arrive in the Moluccas (Conklin 1963:133; Yen 1974), and we can be certain that it was in Banda by 1633 and widely cultivated by the 1670s (Boomgaard 2003:589). However, Scaglion and Soto (1994:265–6) suggest on the basis of linguistic evidence that the sweet potato could not have moved into New Guinea from Indonesia only.

The chronology of arrival of new crops in the region after 1492 was a consequence of European trans-oceanic trading activity and differed between species. Species that arrived in some areas quite early, did not arrive elsewhere for decades or sometimes centuries. This was the experience of Nuaulu and their predecessors for many of the crops listed in Table 9.2. While maize reached the north Moluccas by 1521–46 (Wigboldus 1979) and tobacco by 1599 (Merrill 1946), despite claims otherwise, and some inferential evidence of attempted earlier introduction, cassava was not established in the Moluccas until the nineteenth century. We tend to know about the main food crops, which came from the Americas, but there were also many other species, including pan-tropical weeds, that were introduced at this time. It is likely that the weedy species would have found their way into Nuaulu villages inadvertently; however, many New World species would have only have had a presence in coastal locations until after the 1880s (e.g. the frangipani *Plumeria rubra*).

Fifty-seven per cent of Nuaulu plants of New World origin have names which indicate that they are not native, though because their introduction spanned a period of nearly 400 years they have not left a clear linguistic or classificatory footprint. Times and cultural routes of introduction are varied, and this has influenced the way they have been incorporated. Of all the records in the Nuaulu Ethnobotanical Database (NED), sixty-four refer to plants that we know to have been restricted to the Americas before 1492 (Table 9.2). While the status of *Ipomaea batatas* (the sweet potato) is disputed, it certainly does not appear to have arrived in eastern Indonesia much before the sixteenth century, though current levels of subspecific variation in the highlands of Papua New Guinea (e.g. sixty-four landraces for the Wola: Sillitoe 1983:29) suggest arrival there much earlier. Of course, plants of New World origin did not immediately enter the Nuaulu folk classificatory orbit, although some did arrive surprisingly early and others surprisingly late. Thus Rumphius, in his *Ambonese Herbal*, writing in the late seventeenth century on the nearby island of Ambon, but reporting on the period 1654–1702, already lists twenty-eight of the NED New World species. From Rumphius's descriptions it is clear that use of chilli, tomato, cashew, peanut, cucurbits and tomato were well established in Ambon by the time he had arrived. However, six cultigens now important had either not arrived in Ambon, or were not salient at the time Rumphius was conducting his investigations. These include *Canna, Hevea* (para rubber), *Solanum tuberosum* (the 'Irish' potato), *Xanthosoma* (an aroid which is currently a very important source of starch), *Theobroma cacao* (chocolate) and *Persea americana* (avocado) (van Slooten 1959:318).

We need to remember that before 1880 Nuaulu were a highly dispersed highland population with little integration into the market economy and less reliance on field cultivation than was later the case. Thus, while reports by Rumphius and others for new species present in Ambon might be held to indicate their presence also in most nearby coastal areas of other islands, including Seram, there must have been a lag in their diffusion into the highlands of the major islands. Moreover, we cannot assume that the order of introduction to Ambon was identical to the order of introduction into the Nuaulu area, nor can we assume that the appearance of a new species meant that it was automatically important. Thus, maize, cassava, sweet potato, peanuts and capsicums did not really become important in Java until after 1800. After 1880, with the relocation of most Nuaulu on the coast, I would expect the rapid assimilation of new species and the growing importance of species already present, as Nuaulu increasingly expanded into swidden cultivation and cash-cropping at the expense of hunting and collecting strategies.

In addition to the New World species, new Old World species were also being added to the Nuaulu resource pool from 1500 onwards. These were not only of European and Mediterranean origin, though mainly so (including, for example, various BRASSICACEAE, coffee and some beans) but included exotics endemic to South-East Asia, such as the rambutan (*Nephelium lappaceum*) and Old World cucurbits from South Asia.

Using this corpus, I now wish to look at some selected groups of species, to examine how they have been incorporated as classificatory knowledge within the Nuaulu ethnobiological system, as a way of exemplifying generally common patterns of lexical and categorical change, but also to note special issues relating to the history and use of the groupings. I shall look at: (1) endemic lexical patterns indicating incorporation of new species and varieties (grain crops and fruits); (2) the many species incorporated with loan-words derived from Ambonese Malay and non-Austronesian languages; (3) the special issues relating to naming and diversification in important polymorphic species (cassava and chilli); (4) starchy tubers as a folk array; and (5) chillies and cucurbits as special groupings.

Endemic lexical patterns indicating incorporation of new species and varieties

The Moluccan islands are on the edge of the grain-growing area of the Old World, and even today the staple food and crops are roots, tubers and palm starch rather than grains. Many grain species have been experimented with, but few have really established themselves. Millet is still important in parts of the south-eastern islands. Of the grains, rice is the most dominant, and has been grown in a few parts of the Moluccas for many hundreds of years. On Seram rice is only known to have been grown on any scale in the Nama language area on Teluti Bay (Valeri 2001:328fn10), and in West Seram (Jensen 1948:34–5, table 10a).[1] However, its social and symbolic profile has been mainly acquired through its role as a trade food. Other grain-crop species are perceived, therefore, in relation to rice, and this is to a certain extent reflected in classificatory practice. Thus, Nuaulu ***ana*** is applied focally to rice (*Oryza sativa*), distinguishing broadly red and endemic landraces, but it is also sometimes used to refer to millet (thus ***ana hotone***, or ***ana hotone manipa***, or otherwise just ***hotone***). It is not surprising, therefore, that NW maize is also sometimes referred to as ***ana manipa***, though more generally nowadays by the term ***yakoni***, from AM 'jagung'. Thus, depending on the term used, maize is classified as an independent cultigen or one with specific affinities to other grains (Figure 4.4).

Other NW species are incorporated more obviously by allusion to endemic plants, by creating binomials where the adjectival qualifier is a generic way of expressing foreign or 'other'. Thus, soursop, *Annona muricata*, becomes the binomial ***tunene warata***, where ***tunene*** is an existing unmarked term for durian, *Durio zibenenthis*, and ***warata*** means Dutch or foreigner. Adjectival qualifiers of apparent specific derivation (as in ***warata***, 'Dutch', or ***cina***, 'China') may become generic ways of indicating alterity.

Species incorporated with loan-words derived from Ambonese Malay and non-Austronesian languages

This is the most recognizable way in which post-Columbian plant additions can be recognized, but it provides a by no means infallible rule. It may work in various ways. The first is where a word or adjectival qualifier is cognate, but phonetic features indicate external origin. Thus, ***tapako malole*** (*Nicotiana tabacum* var.) must indicate external origin as /l/ in Nuaulu would normally be reconstructed as /n/, thus 'malole' > ***manone***. The second is where a uninomial is adopted from another language. This may be in the form of a lexically unaltered term, e.g. ***yakoni*** < AM

1 Since the establishment of settlements of transmigrants in west, central and northern Seram from the early 1980s, rice production has risen dramatically, despite the difficulties of establishing irrigation.

Nuaulu name	Meaning	Synonyms	Meaning	AM names	Notes
kasipii hata metene	refers to black stem, or to Hatumeten (toponym)				possible lexical conflation of two terms
kasipii kanapua	refers to *Gossypium acuminatum* (k.o. cotton plant)	***kasipii tiga bulan***	three month (Ind.)	'kasbi tiga bulan'	refers to quick maturation period
kasipii kasitena	refers to resemblance of roots and root system to sweet potato *Ipomaea batatas*			'kasbi patatas'	k.o. ***kasipii putie***, from Wai Pia
kasipii masikunne	yellow	***kasipii mantega***	margarine	'kasbi kuning'	refers to colour of root parenchyma; eaten only as detoxified flour.
kasipii mata rase					folk variety no longer cultivated in Rouhua, but found along stream edges
kasipii msinae	red	***kasipii postel***			introduced through Ruatan transmigration area by Bandanese or from Pelauw (Haruku)
kasipii paru	from AM 'paru', grater	***kasipii honue***	new	'kasbi paru'	eaten only as detoxified flour; introduced by Butonese; possible semantic conflation between AM 'baru' (new) and 'paru' (grater)
kasipii puna	refers to variegated leaves, as in ***puni*** or ***pun*** (the aroid *Homalomena pendula* used to protect people from the spirits of children who have died in childbirth)				
kasipii putie	white			'kasbih putih'	refers to colour of root parenchyma
kasipii sinsinte	refers to variegated leaves of *Codiaeum variegatum*	***kasipii Makuana***	refers to village of Makuana in north Seram		resemblance to *Codiaeum* permits their use as arm decorations in ritual
kasipii Super Mie	refers to instant noodles with the brand name 'Super Mie', which take just a few minutes to boil				from Baubau (Buton)
kasipii unuhutu	refers to *Clerodendron* (incl. *blumeanum*)				used medicinally; wood sacred and prohibited for clan Soumori; k.o. ***kasipii putie***
kasipii wanate	refers to resemblance of leaf to k.o. bamboo, probl. *Gigantochloa atter*				
kasipii wasaena	***wasa***, house rafter				

Table 9.3 Nuaulu terms for folk varieties of Manihot esculenta, Rouhua 2009. (Continued over.)

Nuaulu name	Meaning	Synonyms	Meaning	AM names	Notes
(***kasipii***) ***posteli***	possible elision of 'bastel' (early twentieth-century variety developed at Buitenzorg), from 'bestelling' or from 'post' or 'posteren', meaning guard post				introduced by Butonese, but name of Dutch origin
posteli putie	white				
posteli tone	k.o. garden fence				
posteli upone	plant tip				

Table 9.3 (continued) Nuaulu terms for folk varieties of Manihot esculenta, Rouhua 2009.

'jagung' (*Zea mays*), ***katar*** < AM 'katar' (*Hevea brasiliensis*, para rubber), ***biji keluar*** < AM 'biji keluar', meaning 'seeds come out' or 'testicles' (*Anacardium occidentale*; cashew); Nuaulu ***kalabasa*** < Port. 'kalabasa'; Nuaulu ***kasipii*** < AM 'kasbi' (*Manihot esculenta*, cassava), Nuaulu ***sakalate*** < chocolate (*Theobroma cacao*); Nuaulu ***tamati*** < AM 'tomati' (*Lycopersicon esculentum*). Sometimes this may be more complicated, thus Nuaulu ***yohoru*** is from AM 'keladi johor' (*Xanthosoma*), meaning a taro from Johore (Malay peninsula). Alternatively, a new percept can be incorporated into an existing basic, intermediate or life-form category. Thus, various gourds, ***awane munie*** (*Cucurbita* spp.) are incorporated, by habit rather than function or fruit morphology, with other vines (Figure 4.5), along with apparent endemic cucurbits such as ***awane kin-kin*** (*Melothria indica*). Overall, of twenty-eight Nuaulu plants of New World origin for which there are good data, sixteen have names that indicate foreign origin in these ways.

Special issues related to naming and diversification in important polymorphic species

With the opening up of new and faster sea connections (albeit in many cases indirect) between the Americas and island South-East Asia, a few NW species very quickly obtained a disproportionate salience in their new ecological and cultural setting. This is reflected in diversification below the species level. In both the case of *Capsicum annuum* (chilli, introduced early and mentioned by Rumphius) and *Manihot esculenta* (cassava, inroduced during the nineteenth century), the original diasporic gene pool must have been quite restricted compared with their regions of origin – Central and tropical South America, respectively. Thus, Nuaulu have 18 named landraces of *Manihot esculenta*, compared with estimates of between 31 and 140 for bitter cassava, and between 100 and 117 names for sweet landraces, recorded in different studies from lowland South America (e.g. Elias, Rival and McKey 2000). From Table 9.3 we can see that Nuaulu do not divide cassava folk varieties according to bitter or sweet, neither do they subdivide them on the basis of other processing or eating qualities. Nevertheless, the adjectival qualifiers each indicate some salient feature that makes them instantly recognizable. Nuaulu cassava diversity is low compared with that found in Kei islands populations (Ellen and Soselisa 2012a:24), where thirty-three named cultivars in a single population is common. This is in part due to landscape aridity and more restricted farming conditions in parts of Kei, stimulating selection for cassava diversity. On the other hand, Nuaulu have ten named kinds of chilli, while the Tzeltal of the Mexican Chiapas, surprisingly, only have four or five names (Berlin, Breedlove and Raven 1974b:482). General studies of *Capsicum* for north and central American populations would suggest greater local diversity (Long-Solis 1986; Votava, Baral and Bosland 2005), but as both the cassava and chilli examples indicate, an initially restricted gene pool for a cultigen in a diaspora can result in secondary diversification on a considerable scale, quite quickly.

Starchy tubers as a folk grouping

Starchy roots and tubers (which of course include cassava) are important mainly because of their role in food security, though less it must be said for the Nuaulu than for some other eastern Indonesian populations living in less forgiving environments. Prior to the arrival of NW species, Nuaulu depended on two main genera: *Dioscorea* and *Colocasia*, and upwards of seven species of the same. These map on to folk terms in an interesting way. There is no single term for *Dioscorea*. Instead, one species has a distinctive uninomial that is not further subdivided: ***loloeno*** (*Dioscorea pentaphylla*); two further uninomials (***kawasine*** and ***akae***) are subdivided into two named categories, where their partial covert similarities are reflected in the overlapping application of unmarked ***akae*** and marked ***kawasine putie*** (lit. 'white' ***kawasine***) to the same species. *Dioscorea hispida* (Table 4.5). By comparison, *D. esculenta* is divided into eight folk varieties (Table 4.4), and *D. alata* into ten folk varieties (Table 4.3). *Colocasia esculenta* (taro) is divided into eight folk varieties (Table 4.2) and *Xanthosoma sagittifolium* into two. These numbers are less than we find for many New Guinea peoples, who

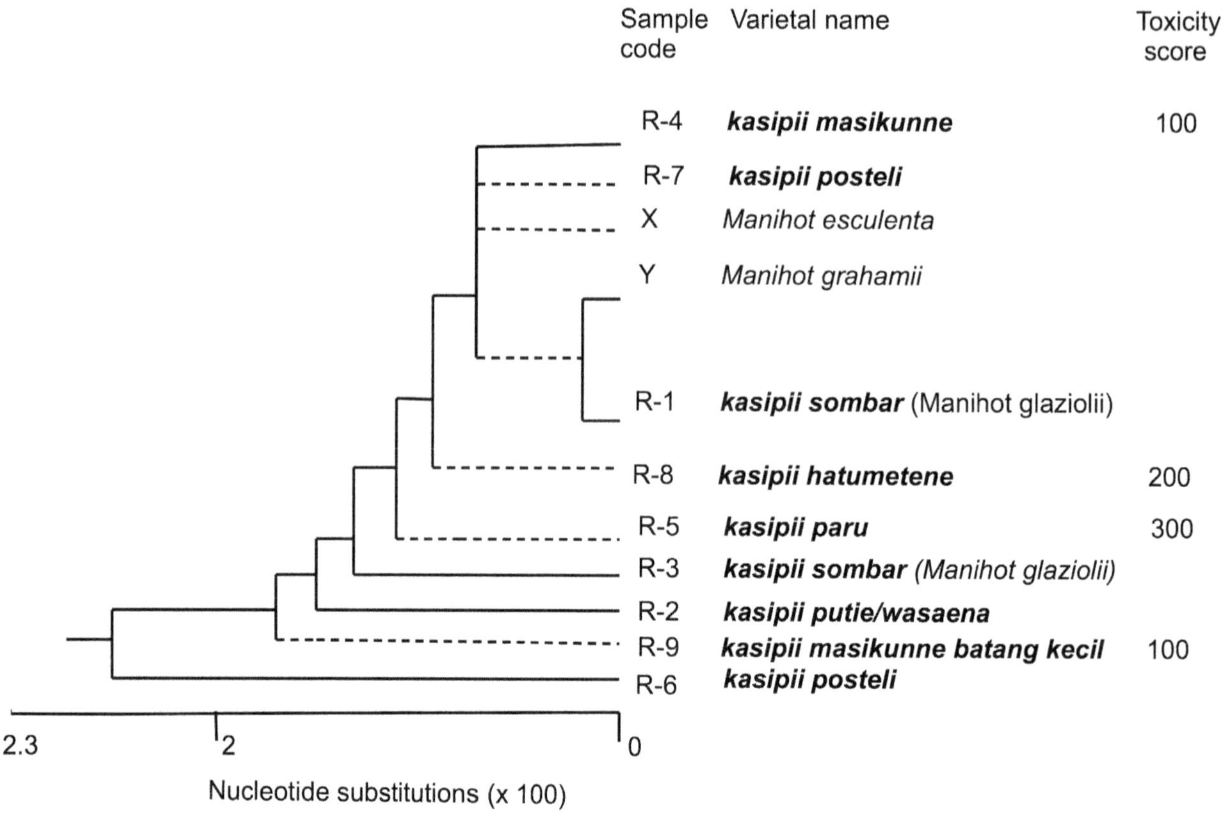

Figure 9.2 Phylogenetic tree based on DNA collected from leaf samples of Manihot esculenta from different folk varieties in Rouhua. The tree was determined using the Clustal method, based on alignment of 1181-bp 18S rRNA. Field measurements for folk-variety toxicity (ppm HCN) appear on the right-hand side. The original laboratory analysis was conducted by Asri Peni Wulandari.

were introduced to American species much later, and we might well assume that there was considerable erosion of Nuaulu cultivar diversity following the arrival of New World tubers. The starchy tubers introduced were *Manihot esculenta, Ipomoea batatas* and *Xanthosoma sagittifolium*, each labelled with terms which unambiguously map on to a single phylogenetic polymorphic species, and which clearly indicate something of their external origins: ***kasipii*** (< 'kasbi'), ***kasitena*** (< 'kasitela' < 'Castiliano') and ***yohoru*** (< Johore), respectively. Of these, *Manihot esculenta* now represents the root crop with the highest level of productivity in swiddens, and all species tend to be more productive than the indigenous crops.

As farmers interact with cassava – uprooting, varying, cooking, eating, planting and moving it, replanting stem cuttings or neglecting it – so they leave a trace of their activity in the DNA profile. Assiduous and frequent interference with cassava plants broadens the base of the DNA dendrogram as farmers make small changes in their management strategies for closely related clones (as we found for Kei islanders in Debut), while the less assiduous and infrequent interactions of Nuaulu farmers leave a DNA profile that has more depth, indicating that DNA diversifies longer without being especially marked by humans, a reflection of Nuaulu being less dependent on cassava as a source of carbohydrate (Figure 9.2; Ellen, Soselisa and Wulandari 2012).

Chillies and cucurbits as special groupings

Capsicum annuum is currently an important culinary ingredient throughout Indonesia, and must have arrived in the Nuaulu area fairly early. It is reported by Rumphius, but not by Galvão (1971), so presumably arrived in the Moluccas between 1544 and 1660. Nuaulu consume what they call ***karatupa*** (a uninomial) as a side dish at most meals, use it medicinally and incorporate it into ritual foods. As ritual practice is generally more resistant to change than mundane practice, this indicates relative long-term availability, as does: (a) the term itself (for which there are no obvious cognates in non-central Maluku languages: Collins 1980, 1984); (b) the local polymorphism of the species (eight folk varieties being recognized); and (c) the fact that the names of terminal categories show no particular evidence of outside derivation (Table 9.4), except for the term ***warata***, which we have noted is a common adjectival qualifier indicating external origin.

The biocultural history of cucurbits is more complex than that for chillies, as there are both Old World

Nuaulu name	Synonym	Meaning	Notes
karatupa mkauke		grasshopper	
karatupa mnaha etute		mousetail	
karatupa hua metene		black fruit	
karatupa konopu		button	var. *cordiforme*
karatupa manesane	***k.*** (***hua***) ***marae***	***marae*** = blue-green	
karatupa onate		large	
karatupa pokotuene		full	
karatupa utane		vegetable	Grossum group: bell pepper
karatupa walon			
karatupa warata	***k. putie***	***warata*** = Dutch, foreign; ***putie*** = white	

*Table 9.4 Nuaulu names and folk classification for Capsicum annuum (**karatupa**).*

and diasporic New World genera and species (Figure 4.4, Table 4.12). However, the New World species are exclusively of the genus *Cucurbita* that hybridize easily, active management and selection being reflected in the larger numbers of cultivars distinguished locally, where species boundaries between what in English we call squash, pumpkin and marrow are no longer easy to discern (Figure 4.4). By contrast, there are no *Cucurbita* among Old World species, although significant similarities between some Old World and New World species are reflected in Nuaulu folk-classificatory practice. Thus, the uninomials ***napu*** and ***tumno*** include both Old World and New World species, ***tumno/pateko*** synonymy suggesting in the case of *Citrullus lanatus* (the watermelon) that recent growth in the significance of New World *Cucurbita pepo* is beginning to eclipse the Old World species.

Species and varieties introduced from the late colonial period

During the late colonial period Dutch government policy extended admistration into the remoter parts of the Moluccas, and provided the conditions for agricultural development. This involved plantation agriculture (of para rubber *Hevea brasiliensis*, coconut, cacao and copal; see, e.g. Tichelman 1918, 1918–19), while at the same time the agricultural extension service introduced new crops to local farmers, and developed varieties of existing crops that appeared to show more productivity, for example cassava. In addition, the internal migratory movement of peoples eastwards from Sulawesi, and the greater movement of individuals generally made possible through the Pax Neerlandica and improved transport infrastructure, allowed more and faster movement of plant germplasm than had hitherto been possible.

Although we can draw reasonable inferences as to how Nuaulu responded to new plants entering their world in earlier times, it is only between 1970 and 2015 (a period of forty-five years) that I have been able to observe such changes directly. In general terms, there has been remarkable stability in the significance of different components of the Nuaulu plant world, both in terms of the vegetal environment and the uses to which species are put. The main starch crops remain the same, and in much the same order of significance, though cassava has progressively become more important in terms of the proportion of land it occupies and its role in the diet. Increasing consumption of purchased rice – in 1970 coming mainly from Thailand, but after that increasingly from Indonesia when the REPELITA five-year plans stimulated national production (Fox 1991) – had only a superficial impact on local consumption patterns, being mainly for feasting and hosting visitors. The main shifts have been in the area of cash-crops, as the national and international markets have altered or fluctuated. In particular, the demand for (and price of) clove declined between 1970 and 2009, leading to new experiments with cacao, coffee, vanilla and even ginseng (this latter having been acquired from Filipino loggers in the 1980s). Clove was also affected by decreasing rainfall during the El Niño year of 1997, and cacao production declined in the first decade of the twenty-first century due to *Phytopthora* pod rot. Such factors have increased the significance of some local folk categories and in some cases introduced entirely new named categories. In addition, new cultivars of existing types have been introduced, such as various types of cassava (***kasipii paru***, [***kasipii***] ***posteli*** and ***kasipii super mie***) and hybrid varieties of banana (***uri ibreda***). The bitter cassava varieties accompanied Butonese migrants from south-east Sulawesi who have a food culture focused on processing bitter cultivars into a kind of polenta (AM 'suami'), and the bananas derived from a local agricultural research facility.

The main plants reported as commodities by the local government in 2014 were *Allium cepa* (***pahane msinae***: shallot), various amaranths (***manahute***: Ind. 'bayam'), *Brassica juncea* (***sesawi***: 'petsai'), *Brassica oleracea* (***kol***: Ind. 'kubi'), *Vigna unguiculata* (***kacang panjang*** < Ind. 'kacang panjang'), *Capsicum annuum* (***karatupa***), *Lycopersicum* *esculentum* (***tamati***), *Solanum melongena* (***tonne***), *Phaseolus vulgaris* (***buncis*** < Ind. 'buncis'), *Cucumis sativa* (***papino***), *Ipomoea aquatica* (***tanko***), *Sechium edule* (Ind. 'labu siam'), *Persea americana* (***advokat*** < Ind. 'advokat'), *Mangifera foetida, M. indica, M. odorata* (***aune, apalam, kuine***), *Nephelium lappaceum* (***rambutan***), ***duku, nasate*** (*Lansium parasiticum*: Ind. 'langsat'), *Citrus* spp. (***nusi***), *Durio zibethinus* (***hutiene, tunene***: durian), *Syzygium aqueum* (***popote***: Ind. 'jambu'), ***rorie waene*** (*Cyrtandra decurrens*: AM 'daun wawo'), *Carica papaya* (***papae***: papaya), *Musa* hybrids (***uri***: banana), ***sesene*** (*Ananas comosus*: pineapple), *Salacca zalacca* (***salak*** < Ind. 'salak'), and *Annona muricata* (***tunene warata***: soursop). Nuaulu have experimented with most of these at one time or another as cash-crops, though the main fruits entering the market in 2015 were langsat, durian and banana (Badan Pusat Statistik 2014:48–51). Of the plantation crops, coconut, clove, nutmeg, cacao and coffee remained important in 2015, though with cacao having eclipsed clove in terms of total hectatres, and coffee in significant decline (Badan Pusat Statistik 2014:52).

The movement of plants and patterns of change in the content of the Nuaulu ethnoflora

Plant germplasm and knowledge of plants are constantly on the move, and part of the evidence for this is found by comparing inventories of plants and plant terms from different places. The comparative study of plant terminologies for Seram is still in its infancy, but we can get some idea of the similarities and differences by examining those two cultural groups for which we have best data: Alune and Nuaulu. In the case of Alune speakers we have two studies: that of Suharno (1997), and that of Florey and Wolff (Florey and Wolff 1998 Wolff and Florey 1998), the first conducted with the Lumoli population on the north-east interior of Piru Bay, and the second conducted in the upland village of Lohiasapalewa and in the coastal settlement of Lohiatala on the southern flank of the mountains of West Seram, where these latter meet the Ambon Strait.

There are close linguistic similarities, where terms correlate well with botanical identifications, for recent introductions, and the common borrowing of the same loan-word (often for fruit and vegetable crops). We have already examined some of these earlier in this chapter. In these cases, the immediate donor language is usually Ambonese Malay or Indonesian (e.g. Alune 'kwini' > Nuaulu ***kuine*** = *Mangifera odorata*; Alune 'ai gandaria' > Nuaulu ***gandaria*** = *Bouea macrophylla*; Alune 'ai manggise' > Nuaulu ***mankositaa*** = *Garcina mangostana*). Consider also the medicinal herb *Orthosiphon aristatus*, 'kumis kucing urei' in Alune and in Nuaulu ***manapesi mau***, an exact translation from the Malay 'kumis kucing' (cat's whiskers). In some cases, the common root is not so obvious, as in: Alune 'atnana(le) katela' > ***kasitena*** = *Ipomoea batatas*, and Alune 'atnane(le) kahbi' > Nuaulu ***kasipii*** = *Manihot esculenta*. In words for plants that have a common origin of greater antiquity, the clue is sometimes in the phonemic shift from /l/ to /n/, as in Alune 'ai palara' > Nuaulu ***ai punara*** = *Octomeles sumatrana*; Alune 'ai iale' > Nuaulu ***iane*** = *Canarium indicum*; Alune 'ai tulene' > Nuaulu ***tunene*** = *Durio zibethinus*; Alune 'kamu' > Nuaulu ***kam*** = *Piper betle* and related species. Moreover, ***nunue*** in both languages refers to various species of *Ficus*; ***saune*** in Alune refers to *Pandanus conoideus* while in Nuaulu it refers to *P. connoideus*, and also to other pandan species such as *P. tectorius*. But similar terms are not always evidence of the same plant taxon: thus 'soi' in Alune refers to betel palms such as *Areca catechu* and *Actinorhytis calapparia*, while in Nuaulu ***soi*** refers to certain gingers of the genus *Zingiber*; while ***ai turi*** in both languages refers to *Sesbania grandiflora* in Alune and *Senna alata* in Nuaulu, and ***ai pate*** to *Albizia saponaria* in Alune and *Ficus* sp. in Nuaulu. For common cultivated plants, where you might expect a close cognate, terms can be completely different, as in Alune 'tema' and Nuaulu ***uri*** for *Musa*. The lesson is obvious: cross-language inference as a means of reliably identifying plants is hazardous.

We can see from the preceeding examples how change in classification behaviour derived from external stimuli is motivated by the arrival of a new percept, an icon of such a percept, or a new word. Change usually occurs simultaneously in naming behaviour (nomenclature) and in classifying behaviour (cognition), and there are just a small number of general rules that seem to underly naming innovation and classificatory change in ethnobotanical knowledge systems.

From variation to change

The principles of language change as they apply to categorization are pretty well understood. They mostly start from the observation of variation in speech practices, and more specifically of how people apply words to categories. We note that some practices are becoming more common, and some less common, depending on the selective impact of different ecological and socio-cultural factors. In other words, the process is akin to genetic change, even allowing for the analogue of genetic 'drift', whereby the ratios between different speech practices alter as a response to demographic shift within otherwise isolated populations.

Marking behaviour and the binomial principle

Examples discussed earlier in this chapter demonstrate how marking behaviour works with respect to introduced species, and the lexical evidence shows us something of the pattern of introduction of new cultigens of New World origin. I think we can also see that while the evidence in some South-East Asian languages, particularly the various dialects of Malay that served as vehicles for the introduction of many of these new species, provide us with classic examples of the principle of category extension (e.g. absorption of *Xanthosoma sagittifolium* into AM 'keladi', previously exclusive to *Colocasia*; and *Manihot esculenta* into 'ubi'), this is not necessarily reflected in Nuaulu. This is not because Nuaulu regard these species as unimportant economically, as some have now eclipsed Old World species in their economic and dietary significance. And as most Nuaulu are now bilingual in Ambonese Malay, the practice is in resistance to what we might expect to be the flow of linguistic influence. Rather, as Nuaulu are still predominantly animist and conduct planting and harvest rituals that focus exclusively on pre-Columbian genera such as *Dioscorea* and *Colocasia*, classificatory practice seems to have something to do with the ritually privileged status of these species.

It is also apparent that the criteria for grouping tuberous crops into more inclusive named categories are limited and exclusively morphological. For the Old World starchy tubers (Figure 4.2) we only have the curious linkage of ***kawasine*** (*Dioscorea hispida*) and ***akae*** (*D. bulbifera*, and possibly its association with monotypical ***loloeno***, *D. pentaphylla*), which appears to relate to the common characteristics of special toxicity and ambiguous wild/cultivated status. By contrast, ***siahue*** (*D. esculenta*) and ***hueni*** (*D. alata*) are long established polymorphic cultigens with large numbers of cultivars, and with low levels of toxicity apparently selected for. In New World *Manihot esculenta*, where it is so important to distinguish bitter from sweet types, there is no overarching grouping that reflects this. This, I suggest, is because toxicity is not always a particularly reliable diagnostic feature in the species, because particularly toxic cultivars are relatively recent arrivals, because local ecology does not especially favour more toxic cultivars, and because it is far more efficient to simply encode this information in binomials at the terminal level.

In a multilingual environment, the ability of most people in a speech community to have access to classifiers in their non-birth languages means that they can group ethnobiological data in ways that may not be reflected in the patterns of linguistic encoding in their birth language. In other words, this knowledge and the lexical practices through which it is expressed become part of the classificatory system of the Nuaulu, even before those practices are formally absorbed into the language. Thus, while Nuaulu have no overarching term for *Dioscorea*, the fact that AM 'kumbılı' refers to all cultivars of the two most widely cultivated types (*D. alata* and *D. esculenta*) is relevant; as is the use of AM 'keladi' to group both *Colocasia* and *Xanthosoma*, and indeed other aroids of both Old World and New World origin.

Another form of flexibility in classifying behaviour is in relation to general-purpose (so-called 'natural') and special-purpose (functional) polysemy. This form of conflation is common in the use of some Malay words, for example 'ubi' to refer both to any edible tuber or root (a functional distinction) and also to *Dioscorea* in particular (a perceptual or morphological distinction); or in the case of 'kayu', which simultaneously refers to 'wood' and 'tree' (see also discussion on p. 107). Both practices suggest a long-embedded usage, but also possibly the gradual adoption of the functional term as a general-purpose name, in the second case replacing old Austronesian cognates for 'tree' (***ai***). In Nuaulu, there is an emergent functional/morphological polysemy in the category labelled ***monote***. This is simultaneously a morphological grouping (something like 'grass'), but at the same time a category that has increasingly acquired the sense of 'weed' (Ellen 1991b). It is possible that over time incorporation of pan-tropical invasive weeds, though not particularly of New World origin, is redefining this category in functional terms. Interestingly, in Nuaulu *Dioscorea* species starch organ tuberosity (a functional characteristic) is always privileged over their habit characteristic as vines. This is something which, as we have seen, is repeated with non-edible gourds (which are vines) and other edible cucurbits (which tend not to be so classified).

Shifting plant significance and selection of type species as descriptive templates

Nuaulu perception and interaction with their plant world is in flux, as certain species and parts of the vegetal landscape come to prominence and other parts decline. This flux results in a shift in how people experience plants, and what species they use to interpret other species. Thus, while coconut has always been a significant food and material resource, its development as a plantation crop from the early twentieth century gave it a visual dominance in the cultivated landscape and day-to-day experience of ordinary Nuaulu that made it an understandable template in referring to other species. Similarly, the rise in the importance of cassava in subsistence at the same time, led to its prominence as a template for thinking about other root and tuber crops. We can see how this notion of botanical ideal type works by noting how Rumphius begins his *Ambonese Herbal* (2011, vol. 1:189–99) with a chapter on coconut, precisely because of the impact it had on him as the most salient feature in the Ambonese vegetal landscape, noting that in this landscape it replaced the role of the date palm in the imaginative landscape of the European or north African. Its precedence in the narrative shows how it serves as a model or 'type' in describing other plants (and of course, in particular what we call palms). The notion of 'type' or 'type specimen' as the reference template with which all others of a similar kind are compared is well established in herbarium-based botanical taxonomy

(Bridson and Forman 1998:87, 135–6), and we will discuss this idea further in Chapter 10.

Linguistic flux and cognitive economy

What allows us to measure the characteristics and rate of change in plant knowledge is, curiously, a paradox: although language is in a constant state of flux, classifying behaviour is phonologically, semantically and cognitively conservative. Change generally only becomes permanent and shared when it becomes indispensible, when reversible contextual shifts become non-reversible (historical) changes. A principle of lexemic and cognitive economy is in operation, which some have assumed to be an evolved characteristic, or at least the by-product of an evolved characteristic (Ellen 2006c:13; Rosch 1977). Thus, if a new percept (that is a perceptual image of a concrete object) can be incorporated into an existing category, it will be – the principle of category extension (Barrau 1979) – and if a new percept is placed in an existing category, the difference between endemic and new forms will most likely be expressed in a binomial. For example, the tomato *Lycopersicon esculentum* becomes 'pomme d'amour' on its arrival in sixteenth-century France, that is 'a kind of apple'. In other words, the introduced term is marked. As the new percept becomes increasingly salient, the label for the more inclusive category may disappear.

Category extension

The principle of cognitive or lexical economy predicts change by category extension, and the existence of new categories may be recognized by marking behaviour, new names and loan-words (Witkowski and Brown 1983), a new uninomial loan-word or innovative local word for a new percept, or a new adjectival qualifier. These latter may include a word such as 'new' in the local language (e.g. ***kasipii honue***, *Manihot esculenta* var.), or a loan-word or adaptation of a loan-word indicating external derivation, as in a common synonym for ***kasipii honue***, ***kasipii paru*** (Nua: cassava + Ind. 'new' or 'grater'), or ***uri ibreda*** (Nua: banana + Ind. 'hybrid'). Similarly, some categories may disappear due to local plant extinction, while others may reappear because of the ecological movements of plants or the introduction of new species, the movement of human populations into areas with new floras, or changes in the cultural significance of particular species. Then there is linguistic and cultural contact and its transfer of knowledge of species not physically present (known, say, through their icons or reputation), new names and categories, following habitual multilingualism and the inability to maintain complete lexical separation; introduction of words and categories due to Dutch and Indonesian cultural domination, for example, through words for the plant kingdom itself (Ind. 'tanam-tanaman', 'tumbuh-tumbuhan'), words for life forms (though absence of a life form in one language may be an obstacle to acceptance, Brown and Chase 1981) and changing degrees of literacy. Finally, there may be changes in the material conditions of life: for example, as people move towards farming they may have less need to classify minor wild resources; and as people move away from direct contact and reliance on the natural world so they tend to rely increasingly on categories at life-form level (Atran and Medin 2010). In other words, classificatory decay seems to operate from the bottom up, as observed early in debates about change in folk classification (Berlin 1972:83).

Exotic crops that are known about but which fail to establish themselves may be referred to by Nuaulu using residual categories, often using unmodified loan-words. We can take as an example the various BRASSICACEAE. *Brassica juncea* is reported by Rumphius (2011, vol. 4:235) as present in Ambon in the second half of the seventeenth century, though it is considered foreign (i.e., of likely Chinese origin), while *Nasturtium indicum* was known to him as 'sesawi Ambon', the native mustard. *B. oleracea* was not known, and Rumphius does not report the garden nasturtium *Tropaeolum majus*. Various *Brassica* are known to Nuaulu today, and there have been some experiments growing them, but they are not widely encountered and neither is the internal diversity of the various evolutionary groups registered. Consequently, two terms of AM derivation, ***kol*** and ***sesawi***, tend to be used rather flexibly to refer to cultivars and hybrids of B. *juncea and B. oleracea*. ***Sesawi*** also includes various mustards such as *B. napus*, *Rorippa indica* and *Tropaeolum majus*.

Change in representational form, social storage and the technologies of communication

In 2010 Henrich, Heine and Norenzayan published a paper which seemed to confirm what some anthropologists had thought they had known for a long time, namely that the assumptions made in a lot of experimental work in psychology were based on subjects sharing the same underlying habits of mind, deriving from being Western, educated, industrialized, rich and democratic (i.e., WEIRD). WEIRD populations have been vastly over-represented in psychological research, while findings from psychological research utilizing primarily WEIRD populations are often labelled as 'universal' theories and inaccurately applied to other cultures. In order to obtain field data that are amenable to cross-cultural comparison, it has sometimes been the case that informants are 'trained' in WEIRD habits of mind in order to perform in tests which may on may not require literacy, and which can give rise to confounding or distorting results (Lloyd 2007:46). Some of the underlying assumptions at work here also relate to problems of understanding that arise with any cross-cultural comparison (see also Atran and Medin 2010). What is helpful about ethnobiological examples is that they allow us to stand back and reflect on (often arbitrary) 'common sense' assumptions and to work things through using recursiveness. As Nuaulu gradually became familiar with the organization of knowledge employed by colonial administrators from the late nineteenth century onwards, and thereafter to some extent with the conventions of science embedded in administrative practice (such as working with map-makers conducting the 1917 topographic survey – Ellen 2007), so some of them acquired habits of mind that became recognizably WEIRD.

Systems change not only through the addition and removal of content, but through the influence of new styles of representation and new technologies of communication, which alter the way we think about classifications. This has been an interest of mine since I first appropriated the work of Jack Goody (e.g. 1977) to this end (Ellen 1979c). In societies where literacy and graphic representation are culturally dominant, the privileged classifications tend to be those serving a particular utilitarian purpose (such as a medical system) or which underpin cosmology (as in holy books, religious manuals and liturgy). Thus, Glover (2018) has approached classification as a kind of narrative, showing how the Tibetan texts that she has analysed reflect changing influences, arguably moving from a classification based on ideas of nature and essence, as related to efficacy, to classifications based on physical characteristics and morphology, arguing that the more recent texts betray the influence of Western Linnaean systems. Similarly, Tsing (2005) has remarked upon the popularity of 'the list' among her informants in Borneo, and Nuaulu too are now eager to offer lists of plants (see Chapter 2, p. 46).

In the Nuaulu case, the history of literacy has been short (effectively no more that forty years), and its consequences have been less to alter the way in which plant classifications are thought about, than to marginalize Nuaulu language and traditional folk classification altogether. The language of the written word, the internet and social media is Indonesian, despite attempts by SIL and other NGOs to produce examples of books about plants and their cultivation in the Nuaulu tongue (Bolton n.d.; Matoke and Matoke 2004; Oh Swee Cheng 2005); and even these, though good technical translations of Nuaulu language, are somehow underpinned by a set of globalized assumptions of plant knowledge that incompletely preserve a rather different way in which Nuaulu think about plants. Nevertheless, the prehistory of literacy (meaning in this case the long coexistence between a dominant literary culture elaborated during the colonial period through adminstrative practices, schooling and the Christian church) has resulted in the seepage into the oral mode of conventions typical of the written (or read) word. The idea of 'the list' is a good example of this, as is the very idea that there might be a single classificatory framework to which all should subscribe, and a consciousness of something called 'plant classification' stripped of its multifarious contextual underpinnings. There has been, therefore, a gradual movement from pragmatic relatively unreflexive utilitarianism to more abstract reflexive general-purpose ideas as to how plants might be grouped.

Early European visitors to the Moluccas introduced the idea of lists of useful plants. Rumphius arrives with the technologies of the written word that reduce oral knowledge to lists (see also Tsing 2005:155–70), and which can then reorganize the content of those lists in novel ways (e.g. Goody 1977), opening up new forms of intellectual scrutiny. However, this same transformation radically separates the names of individual plants and their uses from their context as part of a local physical landscape, ecology and cultural world, in a set of representations that are not fixed or tightly bounded, but rather fluid and complex. Thus, Rumphius' description (2011, vol. 3:86–7) of *Albizia falcataria* is beautiful, but the simple reference to its use in making shields does little justice to, say, the Nuaulu perception of 'the shield tree', which is intricately bound up in the way this species is sought out for making sacred shields (***aniaue monne***) that are then invested with symbolic qualities and used in ritual contexts (Ellen 1990, 2018c). Thus, different kinds of representation significantly change the cultural construction of Nuaulu plant knowledge. As Latour (1993) might put it, whereas modern botanical ontologies separate things, nature and society through the activity of science, other ontologies bring them together. The development of European botanical knowledge depended on local field collaborations that were subsequently erased in histories of science and the creation of manuals of technical practice, while the multidimensional complexities of human-plant interaction were simplified in the pursuit of global knowledge (Ellen and Harris 2000). As Tsing (2005:94) says, 'the system takes precedence over the plant' in a number of respects: in the ways in which local people are increasingly motivated to document their patrimony, to produce 'indigenous knowledge' as part of identity politics and claims on resources, both in the abstract and as 'lists', and to use it for political ends (Miller *et al.* 2016). It has also become a legally and morally contested domain. For these reasons we should always regret the loss of such knowledge, attempt to preserve it in some form before it disappears, but not – I think – to compartmentalize it. By 2015 literacy was beginning to make considerable inroads into Nuaulu life, with all positions that required competence in dealing with the outside world held by literate and sometimes well-schooled individuals, although it cannot be said that its accompanying features had much influenced how Nuaulu thought about plants and their relations. For the ways in which scientific literacy might influence how folk classifications are conceptualized, we can do not better than look at the tranformations in the European study of natural history represented in the writings of Rumphius and which continued apace thereafter, a matter that I address further towards the end of the next chapter.

The degree to which social storage of cultural information, in this instance the extent to which plant knowledge is preserved through distribution in the minds and practices of a large number of individuals, impacts upon how we model knowledge transmission more generally. Nuaulu have historically, like other preliterate societies, often taken advantage of various forms of social storage, such as knowledge devolved in others (particularly specialists), in objects and in art. But it is writing that has had the most significant influence on the human ability to store knowledge outside individual human minds, followed by various other forms of mechanical recording of sound and image. Most folk knowledge of the natural world is transmitted through the oral mode, except when it is transformed through scientific and medical practices involving texts. But the introduction of writing can have major (if ambiguous) consequences for how knowledge systems are organized, as well as for forms of authority and legitimation (Goody 1977; Ong 1982). Technically, writing knowledge down tends to freeze it, which preserves a (usually partial) corpus at a particular point in time. But writing down knowledge also entails simplification at the point of instantiation, when it often has to be selective. Paradoxically, thereafter it permits continual expansion that can lead to complexification within the text. Writing in two dimensions permits the making of new connections between elements, the use of devices such as tables, lists and diagrams, with consequences for understanding causality; but it also denies other connections found in the oral mode, due to loss of flexibility. And writing influences the way we use oral knowledge, giving rise to the social supremacy of what Maurice Bloch (1991) calls the 'linear-sentential' mode: dividing knowledge up into sentences and representing it in a text-determined order. These effects are even found in societies with minimal literacy. Because literacy itself is a skill available to a small fraction of the population in many societies, and competence is highly variable, it provides a new source of control and legitimacy. As long as knowledge remains orally articulated, or even devolved in non-linguistically coded tacit experience and in practical interactive demonstrations, of which language may be the lesser part, it often poses obstacles to effective reproduction through the literate mode, inviting serious oversimplification, straining the limits of ordinary language as a medium of transmission, and hindering the development of specialized forms of language use (such as mathematical notation and theological reasoning).

One conceptual step up from lexical knowledge is textual knowledge, not simply words organized and reorganized (for example as lists), but where the source words are already organized into sequences of utterance that betray features of the organization of knowledge. To some extent we acquire knowledge by mastering scripts that are in part encoded in language, as when Frake (1980) famously invites us 'how to ask for a drink' in Subanun, a language of Mindanao. The oral texts that compress emergent consensus and rules may be narratives such as myths, though they may also take on a more permanent written form. Each of these forms can be transmitted through mimicry, copying or imprinting, but also through language-mediated instruction (telling) and text-mediated learning, through institutionalized teaching and learning (pedagogy, apprenticeship), and through combinations such as informal learning in institutional contexts (e.g. in labs or offices), or the formal learning of religious liturgies. But transmission by example, which all of these exemplify, involves not simply copying but the inculcation of a set of principles and practices that can be used to instantiate behaviour that is logically equivalent in a given context, not materially equivalent behaviour as copying implies. Formulaic sequences of lexical expressions are a proper subset of these, a subcategory rather than another category. So mimicry, copying and imprinting have a role, dealing with the material manifestations or data that individuals require in order to induce knowledge, but which are not copies of the knowledge itself, simply exemplars of what the knowledge once acquired should be capable of producing. Moreover, if a text is written, the knowledge it contains may bypass the oral mode and move between individuals and social groups as an object, emerging in quite different contexts with different assumptions of interpretation.

Temporality, serendipity and the reproduction of knowledge

The corpus of data assembled in the Nuaulu Ethnobotanical Database was obtained through fieldwork conducted between 1970 and 2015, during a number of relatively short fieldwork episodes, some of which were more productive in terms of ethnobotanical documentation than others. Although already discussed in some detail in Chapter 1, I reiterate this information here because it might seem to contradict the emphasis I have been placing on variability and change. Indeed, in the various tables, charts and calculations, I have assumed that it was methodologically acceptable to treat the period 1970–2015 as one in which knowledge remained more or less static. Of course this is not empirically altogether true, and in collating data I have adopted a device that we might call the 'ethnobotanical present', along the lines of the 'ethnographic present': that is an arbitrary time period when processes of change are for some purposes strategically ignored in order to describe a bioculture as if it were a stable system.

Any kind of fieldwork, and any kind of data collection, is inevitably time-phase dependent, but if synchronic description is to be possible, and if generalizations and comparisons are to be made, then time has to be divided into chunks to enable them to take place (Figure 9.3a). Discussions of the classic issues involved, what they mean

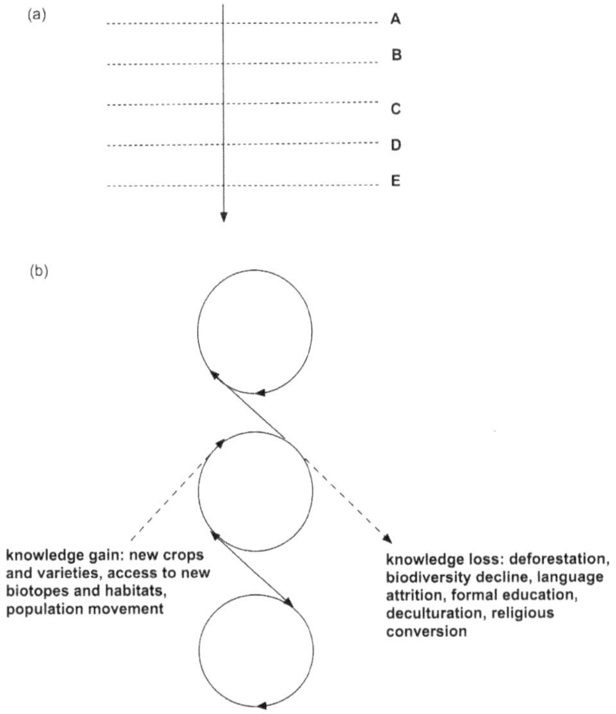

Figure 9.3 Modelling change in systems of ethnobotanical knowledge: (a) the arbitrariness of time-phase boundaries; (b) knowledge modelled as a relatively stable system of reproduction, gain and loss.

for conceptual units such as 'species' and 'cultures', and how they play out in different disciplinatry contexts, have been well covered in the literature, for example, in ethnography (Hastrup 1990; Pina-Cabral 2000), archaeology (Clarke 1968) and palaeontology (Cain 1971:106–20), but not I think in ethnobotany. The distortion introduced by collapsing data drawn from different time frames will depend on rates of change and variability within the time phase selected. Although the boundaries of time-phase datasets are necessarily mostly arbitrary, some may be determined in part by significant events influencing variability and change. Moreover, across any two generations it is typical for most knowledge to be reproduced, as younger people are actively instructed or learn for themselves. Indeed, most change is marginal other than at exceptional moments, either as knowledge is lost (following deforestation, ecological simplification, language erosion, deculturation – say through religious conversion – and education); or gained (as new crops and varieties appear, with access to new biotopes and habitats, and through population movement) (Figure 9.3b). For pragmatic reasons we insert boundaries around sets of data varying through time in order to make generalization and comparison possible at all. If we are dealing with long time periods viewed – as it were – from afar, then the boundaries can be clearer, but the shorter the time period the more arbitrary the boundary.

So, the tabulated results and calculations presented here are determined by conjoining data gathered over a forty-five-year period from all informants consulted during this time frame. Thus, the notion of 'omniscient speaker-hearer' (Berlin, Breedlove and Raven 1974b:58–9) and its critique appear to be relevant not only when considering spatio-social variation at a given time and place, but across time as well. We must accept variation between informants at any given sampling moment, and changes over time as some shared knowledge is lost or gained, but if we have no antecedent cohort with which to compare, assessing rates of change is not easy.

10

Rethinking ethnobotanical classification

The final two chapters of this book look at the implications of my work on Nuaulu ethnobotanical classification for theory, methodology and, especially, existing models of the way plant classification works. I ask how we can make sense of plant-classification systems that we know – from the earliest observations of Conklin (1954:164) – to be 'in a constant state of flux'. If such systems are intrinsically dynamic, how might we achieve some degree of understanding of the general principles that underly ethnobotanical classification more widely, engage in comparison, and defend the kind of radical simplification that schemes such as that proposed by Berlin rely upon, but which have proven in practice to be problematic?

Ethnobotanical classification as a system

We need to begin by scrutinizing the forms of representation we use to collect and analyse data (whether lists, card indexes, pile sorts, picture-recognition exercizes, hierarchies, tree-diagrams, network graphs or Venn diagrams) and the ways in which our informants, research subjects or local co-researchers choose, or are encouraged, to present data to the ethnographer (individually, in groups, in different contexts of collecting and doing, as texts, graphics and so on). Early models of ethnobiological classification (Ellen 2003a:52–4) were heavily constrained by adherence to linguistically defined entities and a language-based interpretation of how classification might operate, even if formal recognition was given to the separation of category and label. Such a model has been described as the 'linear-sentential' model of culture (e.g. Bloch 1991), being based on the same kinds of rules as organize a text in Western literate societies. With a shift away from a lexicographic model of meaning which presumes the listing of serial distinctive features, and a new emphasis on core-periphery models and cognitive prototypes (where meaning is more graded and less contrastive), greater recognition has been given to how we might classify and engage with objective differences in the natural world without necessarily using language as an intermediary (e.g. Hunn 1976). It became possible to talk about the 'basic' and 'extended' ranges of categories (Berlin, Breedlove and Raven 1974b), something could be 'an oak' or a perceptually related form 'like an oak'. Berlin's approach was rooted in the linguistic model, but the influence of theorists such as Rosch (e.g. 1977) and concurrent developments in cognitive anthropology took it in an increasingly cognitivist direction. In other words, there has been a growth in psychological approaches at the expense of linguistic approaches. I think it would currently be accepted that we in fact need elements of both approaches to make sense of the classification process, and in relation to a wider socio-cultural context.

Geoffrey Lloyd (2007:41) warns us against 'expectations of neatness and simplicity' which 'blind us to the actual multidimensionality of ... phenomena'. In order to avoid the temptation of being drawn towards aesthetically and methodologically pleasing – but ultimately misleading – monothetic assumptions and representations, I suggest we need to return to the notion of 'system'. Synchronic folk-classification studies of animals and plants have thus far barely got to grips with this idea. The term 'classification system' is employed often enough, but the notion of 'system' implied is most often understood as a formal collection of entities with a common ontological property or properties (Rappaport 1971:59), rather than an 'interlocking complex of processes characterized by many reciprocal cause-effect pathways' (Watt 1966:2). This second definition is that most consistent with contemporary systems theory and systems-based approaches, and it is the definition most relevant to understanding the dynamic character of classifying behaviour. Most reports of classification utilize two-dimensional schemes of representation, of which taxonomies, pictured as tree-diagrams, are the most common and iconic (Figure 3.1; Ellen 1993:219). One reason for this is that taxonomies in folk-classification studies are derived from their use in scientific biology as developed since the time of Linnaeus. Many have claimed that taxonomic organization with its underlying logic of contrast and inclusion is a feature of all folk classifications of biota. This is partly because of the frequent occurrence of binary opposition and transitive hierarchies. But contrast as a general principle can sometimes be an illusion. The consensus emerging out of various critiques of the Berlinian ranked model, and classic definitions of what taxonomy entails cognitively and semantically (Kay 1971), is summed up neatly by Hunn (2008:94; cf. Hunn 1977, 1982): 'a formal taxonomic structure does not adequately capture the psychological reality of folk biological classification'. Indeed, Friedberg (1968, 1979:85) prefers to speak of 'complex webs of resemblance' with particular forms being assigned to different higher order categories on different occasions. At the same time, we might note that Conklin (1954) in his pioneering work did not use the word 'taxa' at all for folk categories of biota, implicitly acknowledging that we could not predetermine that categories applied to plant (or animal) types were necessarily taxa in the scientifically accepted sense. We shall see why this should be so in the following pages.

Now, while I have problems with the received Berlinian model, I do not reject as psychological phenomena conceptual ranks or the process of ranking per se (contra Holman 2005:79), but rather the arrangement of taxonomic ranks proposed by Berlin. Thus, if I ask Nuaulu informants to list certain trees described as ***nunue*** (a group of fig species) in terms of their familiarity, they will 'rank' them and we can infer from the results of elicitation exercizes of the 'kind-of' variety, for example those using pile sorts, that certain categories are a more inclusive – that is are ranked 'higher' than others, and moreover that transitivity is self-evidently important (Ellen 1993:88). All this is evidence for ranking as a cognitive process using linguistic cues, but not necessarily of a fixed system of ranks where all categories in the same rank are also at the same 'level'. We might also ask, with Holman (2005:74,78), whether taxonomies are 'flat' or 'hierarchical', since the notion of taxonomic rank must be distinct from hierarchical level if taxa at the same rank are supposed to be comparable even if at different levels.

Irrespective of any overall structure we might find in the arrangement of Nuaulu plant categories, there are various cognitive strategies that operate to organize the immensity of information involved and to pattern data in terms of similarity and difference. Pairing is one such strategy, which often works through the device of male-female contrast (Chapter 2), but need not always work this way. Thus, ***hukila*** (*Eugenia* sp.) and ***kawasa*** (*Archidendron clypearia*) are paired on account of both species being used in the making of shields. Often, the familiar default type is paired with a less common type (or types) of the same genus, on the self-other analogy. These may either be expressed linguistically, by marking the non-default type, e.g. ***katinoie*** (*Melastoma imbricatum*): ***katinoie wananoi*** (*M. malabathicum*), or through naming two species of a common genus, e.g. ***anametene*** (*Diospyros rostrata*) and ***ai yarane*** (*D. elliptifolia*), or by pairing two genera with a perceived overarching resemblance, such as the palms ***inene*** (*Hydriastele microcarpa*) and ***panuke*** (*Caryota rumphiana*). Sometimes the 'type' taxon – generally drawn from a terminal category – is designated by a word meaning 'real', 'genuine', 'true', 'authentic', 'proper', 'usual' or 'of the place' (see Chapter 2, on ***tunne***). In early European scientific classification this phenomenon was also reflected in use of the adjectival qualifier 'common' in English, or *vulgaris* in botanical Latin. Thus, Linnaeus used *vulgaris* for the most frequently occuring species in a genus that he describes, subsequent types being given names that contrast with *vulgaris* (Stearn 1973). For Nuaulu one named category in a set is considered as the focal and most dominant reference member, similar to the use of 'type' (and 'typification') in scientific taxonomic practice (Beentje 2010:125; Bridson and Forman 1998:87; cf. Berlin, Breedlove and Raven 1974b:42–3; Conklin 1954:131; Kocher-Schmid 1991:72). Of course, this usage need not only occur when plant categories are paired, but also where the 'authentic' form is compared with an array of three or more other similar terminal forms, and what is 'common' or 'true' in terms of nomenclature may vary as the uses and availability of particular related taxa alter (e.g. Nabhan and Rhea 1987).

An abiding preoccupation with dualist thinking in interpretations of symbolic classification in eastern Indonesia has involved much reflection on what 'binary contrast' actually means. Some studies have shown that sometimes what might look like contrast in one elicitory context, in another is better described as 'encompassment', the bringing of two concepts together as a pair: at one level the pair are a unity, at another in contrast. The notion of encompassment in symbolic classification has been discussed, for example, by Fox (1989), Forth (2001:7–9) and others, following Dumont (1986:228), but it seems to me that it also regularly crops up in Nuaulu plant classification. Even though anthropologists may find it convenient to make clear analytic distinctions between symbolic classificaton and mundane ethnobiological classification, in the minds of people who use and share classificatory knowledge on a minute-to-minute, hour-by-hour or day-to-day basis, it can hardly be credible that classificatory strategies in the symbolic realm are hermetically sealed from those operating in other domains of knowledge. Let us take a concrete example. Dumont suggested that encompassment implied inequality and asymmetry, but the evidence suggests that this is certainly not always the case, and Forth (2001:7) wonders whether it can ever articulate a single value opposition. We can see this in the symbolic pairing of the 'nine' and 'five', respectively AM 'siwa' and 'lima' (Nuaulu ***siwa/nima***), symbolic groupings found widely among central Moluccan peoples, and discussed at length by Valeri (1989:158). In this system a conjoined 'siwalima' encompasses the separate entities of 'siwa' and 'lima', and emphasizes a bringing together rather than an opposition. Thus, the way in which part and whole are identified is in terms otherwise construed as contraries. Moreover, Valeri shows how in certain contexts 'sima' groups see themselves

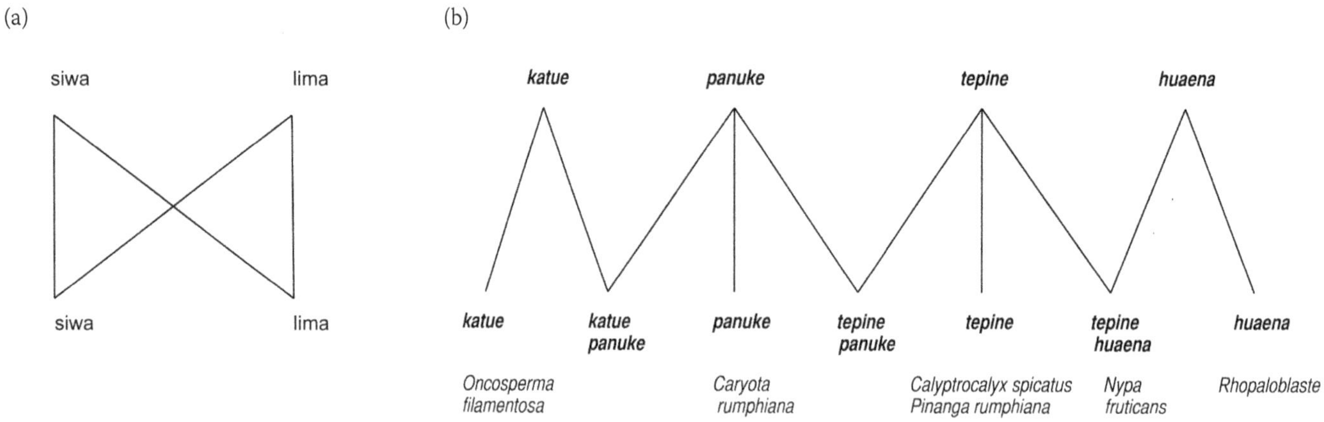

Figure 10.1 Encompassment providing for alternative hierarchical inclusions in: (a) ***siwa-lima*** *symbolic classification; and (b) Nuaulu palm classification.*

superior and encompassing both 'siwa' and 'lima', while in others 'lima' is superior and encompassing both 'lima' and 'siwa'. Figure 10.1a shows these alternative schemes as a simple line diagram, implying two opposite hierarchical evaluations, and it is quite clear that the pattern resembles what we routinely find in Nuaulu plant classification, and which I have also drawn attention to in Nuaulu animal classification (Ellen 1993:81–2). Thus, in the case of palms depicted in Figure 6.5 and repeated here as Figure 10.1b, depending on context of elicitation, ***panuke*** (*Caryota rumphiana*) can be expressed as a kind of ***tepine***, ***tepine*** (*Calyptrocalyx* and *Pinanga*) can be expressed as a kind of ***panuke***, while ***katue*** (*Oncosperma filamenta*) and ***huaena*** (*Nypa fruticans*) can be expressed as kinds of ***panuke*** and ***tepine*** respectively.

Related to encompassment in this sense is the repeated observation that the same terms may apply at several different – sometimes cross-cutting – 'levels' (e.g. Taylor 1990:41). The concept of 'level' itself is drawn from an assumption of hierarchy as being intrinsic to how classification works; but rather than some observations suggesting a problem with the taxonomic model, it is sometimes implied that they instead reflect a weakness in the methods of elicitation adopted. But we can see from certain examples in this monograph that this is not always a credible position to defend, for example in the way in which Nuaulu use the term ***iane*** for several different degrees of inclusiveness for species and subspecific types of *Canarium*, or in the way ***ahutaune*** can be used in a number of different senses to express resemblance and contrast between otherwise separate kinds of *Calophyllum* (Chapter 5.3). Here again we can juxtapose instances of symbolic and ethnobotanical classification to illustrate just how pervasive such forms of naming and classifying can be: in this case Keo classification of symbolic villages in Flores (Figure 10.2a) and Nuaulu bamboo classification (Figure 10.2b, part of Figure 6.7), in which ***ona*** (*Dendrocalamus asper*) and ***tomo seane*** (*Bambusa vulgaris* and *Schizostachyum lima*) are optionally treated as kinds of ***wanate***.

If it is to be adaptive, classification must comprise flexible heterarchies; that is, complex systems in which the elements have 'the potential of being unranked (relative to other elements) or ranked in a number of ways, depending on requirements' (Crumley 1979:144). In such a conception 'hierarchy is an analytic strategy imposed on the heterarchical face of nature' (Crumley 1994:13). Such an approach is consistent with talk of 'classification space' (Hunn 1976:513), and the recognition that the brain can group and link sense data in more than two dimensions. Rather than different schemes treating the same objects being analysed independently, or some schemes being seen as 'secondary' and cutting across some culturally prioritized scheme, it is far less tendentious if we work with a model that acknowledges the intrinsic systematicity of the classifying process; in other words, one that takes it for granted that humans operating in socio-cultural contexts have available to them many classifying resources organized in terms of a set of alternative possibilities, or options that are constantly juxtaposed and reconnected in different ways. Thus, if I encounter a particular natural object (percept) in one context (linguistic, cognitive, physical or social), I may be predisposed to classify it in one way; if I encounter it in a different context, I may be predisposed to classify it in another way; but even in the same context, I may classify and reclassify the same percept in different ways depending on how I think about the situation and my purposes with respect to the percept. Thus, Nuaulu may classify the cassowary (*Casuarius casuarius*) as a bird (***manue***), large terrestrial animal, forest animal (as opposed to an animal of sea or village), food, or member of the category ***peni***, a ritual grouping that also includes deer and

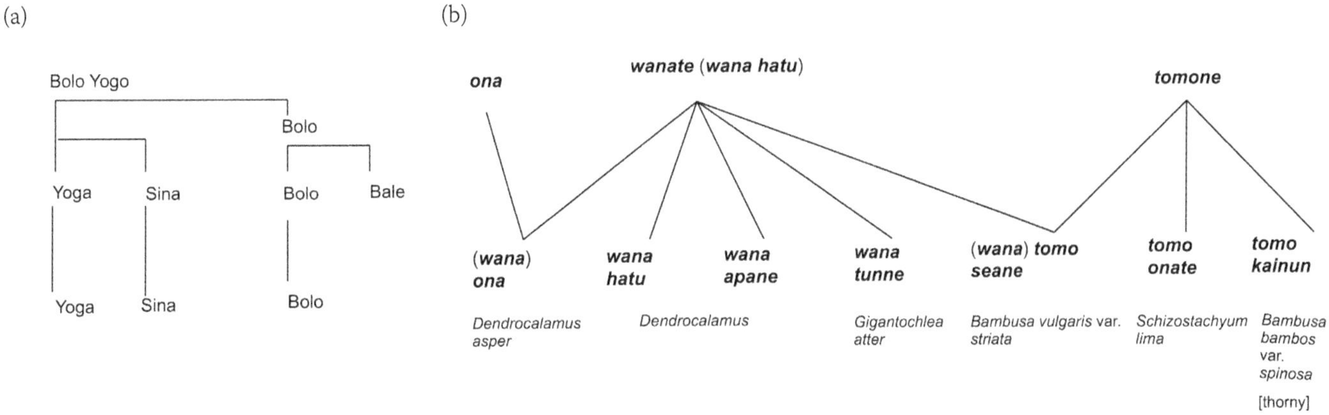

Figure 10.2 Comparison of instances of symbolic and ethnobotanical classification where the same name appears at multiple levels of inclusion: (a) the case of Keo classification of symbolic villages in Flores (Forth 2001: fig. 7) and (b) Nuaulu bamboo classification.

wild pig, depending on context. The same applies to natural objects assigned to the domain of plants.

Moreover, problems arise when the processes of classifying (the cultural and cognitive mechanisms by which the assignation of objects, concepts and relations to categories is achieved) are conflated with 'classifications' (the linguistic, mental and other cultural representations which result). If we prioritize the end result over the process, there is a risk of reifying schemes as permanent cultural artefacts or mentally stored old knowledge, when they are more often properly understood as the spontaneous and often transient end-products of underlying processes in an individual classifying act. I have called this 'the classificatory fallacy' (Ellen 2006c:27, 31). To extend the distinction, and make it more productive, it is useful to see it as an instance of the model of agency and structure (structuration) that we owe in its sociological form to Anthony Giddens (e.g. 1986). Thus, the relationship between classifying as a cognitive and cultural process, and 'a classification' as a representation is recursive and dialectical: you cannot have one without the other (Figure 10.3). The classifying process is also situated within, and assumes some context from, previous classifications, while itself modifying the context for the next classifying act. As this largely operates within the constraints of human culture and memory, it is clearly a matter of degree, depending on the knowledgeability of the classifier, the variability of the contexts, and the entities being classified. In a similar vein, Ingold has distinguished classificatory knowledge (something fixed and mimetic) from what he calls 'storied' knowledge (with echoes of the

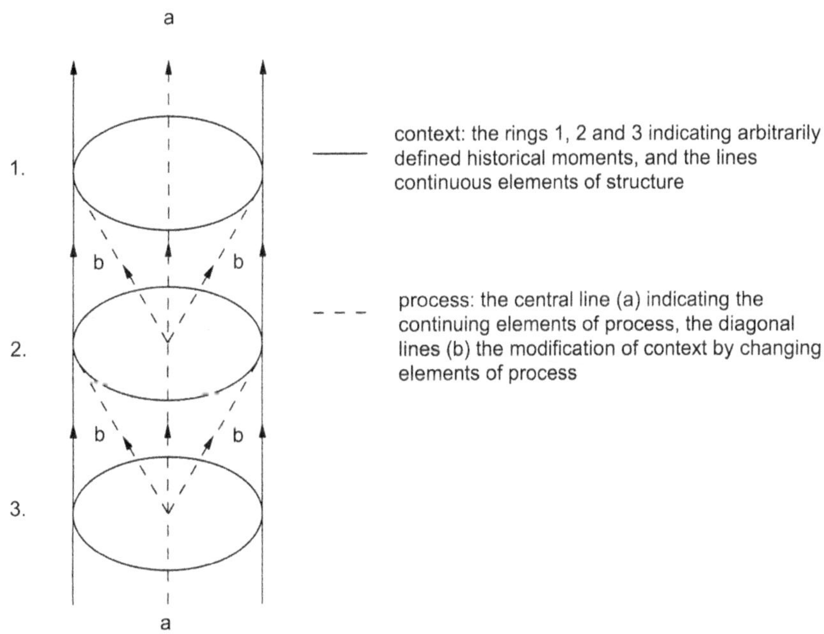

Figure 10.3 The interelationship between agency and structure over time.

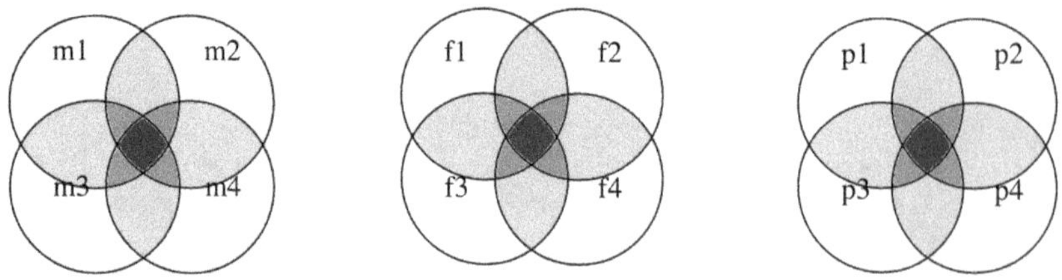

Figure 10.4 Model of overlapping domain membership for a single multipurpose culturally salient species of plant defined in terms of: (a) morphology, (b) function, and (c) technical processes related to use of the species (after Ellen 2009a).

notion of episodic memory). Knowledge, he argues, is not a complex 'structure', but rather an 'ever-emergent product of a complex process' (Ingold 2011:159). We acquire knowledge on a kind of journey or path through space and time, through a network of connections that resembles more Darwin's 'entangled bank' (1928:462) than the formal hierarchy of a taxonomy. As a process, for Nuaulu, classifying plants is not about seeking some perfect fixed definitive scheme, but involves constant switching between different mimetic schemes and context-rich episodic memories.

Thus, classifying is more than an act or succession of acts, it is a contextually bound process (cf. also Casagrande 2017), the sum total of empirical processes determined by interaction between knowledge, context, purpose and cognitive architecture, which gives rise to particular classificatory outcomes, in scientific taxonomy no less than folk classification. This is what I have described in previous publications as 'prehension':

> (a) those processes which through various cultural and other constraints give rise to particular classifications, designations and representations; (b) which integrate the context of the classifying activity, including the interactive interplay with others in that context (including the elicitory techniques of the ethnographic interview) with past experience; (c) which emerge from a sequence of individual 'acts' of perception, but cannot be confined to them; (d) is a model that can cope explicitly with imperfect processing and communication of information, and with the capacity of humans to adapt and modify their store of information and the interrelations between that information; and (e) suggests that there can be no adequate model of classification that attempts to separate the structure of classification from its context and content.
>
> (Ellen 2006c:28)

In applying the notion of systematicity to the folk classification of plants, we can appreciate the process as one contingent upon knowledge in multiple overlapping domains or modules of knowledge, and seeing it as such helps us obtain a more realistic picture of how knowledge transmission works, both in terms of cognition and enskillment (the passing on of performative knowledge). Figure 10.4 illustrates an abstract model of overlapping domain membership for a single multipurpose culturally salient species of plant defined in terms of: (a) morphology, (b) function, and (c) technical processes related to use of the species. An example from my own work appropriate to this kind of modelling is the sago palm, *Metroxylon sagu* (Ellen 2006a). Overlapping spaces indicate areas where knowledge is common to two or more domains. In times of stability this overlap reinforces knowledge transmission, but where domains collapse due to 'the extinction of experience' (Nabhan and St Antoine 1993) the area of overlap indicates a zone of weakness. In this model the domains are diagrammed as if they were perfectly symmetrical in their extent of overlap, though in real-world cases this will vary, and the greater the overlap the greater resilience in the system when it comes to knowledge transmission. There is no implied congruence between the domains diagrammed in (m), (f) and (p). Indeed, processes separated in (p) may be common to making or using elements defined as members of a single morphological domain, or processes identified as being of a single domain may underlie several domains for the item defined morphologically or functionally. No new percept is ever introduced into a local population as a member of only one domain. It is always a member of at least two domains. So, when maize (*Zea mays*) was introduced into eastern Indonesia during the late seventeenth century it was simultaneously introduced conceptually as plant, food and food plant, and possibly also as a grain food plant.

Part of the process of establishing domain specificity (e.g. Hirschfeld and Gelman 1994) involves the pragmatic denial of the extent of overlap between domains. Thus, to take another ethnobotanical example, a particular species of fig (*Ficus*) might simultaneously be a member of numerous domains and subdomains: a plant, food, medicine, constructional material, element in a landscape, totem, and so on. Studies of knowledge transmission and erosion have tended to focus on a transmitted or eroded element as a member of just one domain. Thus, we consider the transmission of plant knowledge, ethnomedical knowledge, food knowledge, and so on. What such an approach ignores is the relevance to transmission of the simultaneous occurrence of that element in several domains. Thus, erosion of knowledge in one domain – in one dimension of the total aggregate cultural profile – may accelerate erosion in another domain of knowledge of which that plant is a member; or alternatively, that maintenance of knowledge of the plant in the context of one domain will enable retention of knowledge in another. The more complex the cultural domain, the more this kind of overlap is likely to be significant.

Some empirical evidence for the existence of such a dynamic multidimensional classification system lies in the results we get when we invite informants to perform identification tasks or open pile sorts; for if we use picture or word cards of animals or plants, then research subjects will utilize a variety of criteria that seem to suggest the conflation of different cross-cutting polythetic sets, but which we cannot necessarily assume to be such. The lexical sets established can be interpreted as an artefact of performing free listing exercizes for a particular domain. Formal kinds of consensus analysis do not prejudge the criteria people use in allocating categories or percepts to particular groupings, and often tend to merge classifications based on disparate criteria. Unidimensional tree-diagrams are representations of just one aspect of that system. Thus, the Nuaulu ethnobiological classification system must be understood as all relevant cultural information and all possible ways of classifying plants and animals available to those we label as Nuaulu, even though there is variation in what elements of this entirety are shared between particular individuals and subgroups.

Place and the localization of knowledge

Plants in every culture are grouped in a variety of different cross-cutting ways depending on context. In the absence of a set of conventions that allow abstract grouping of plants of the kind possible in the virtual spaces of the written page, an important factor in Nuaulu imaginings of their plant world is their positioning in physical space. Place (and therefore also habitat: e.g. Sillitoe 1995:211) connects different kinds of plants and enables groupings that are much more meaningful for most ordinary Nuaulu than abstract perceptual features. Using only the 'kind of' semantic relation, Nuaulu seemingly acquire plant names and link them to actual organisms in particular places, and this tends to be the basis for their organization of a great deal of plant knowledge that may be subsequently rearranged in other ways. For example, the well-reported findings on the 'precocious' learning of plant names by children in subsistence-reliant populations strongly suggests that this takes place in the place-based context of swidden, field or home-garden activity (Dougherty 1979; Hunn 2002; Stross 1973), in other words in locations where useful diversity is concentrated, and that the erosion of knowledge in developing populations is strongly linked to decontextualization. Whether a plant is in a swidden or in distant forest has consequences in ways that determining whether it is 'tree' or a 'vine' do not. PLANT as a cognitive or semantic domain is a very abstract concept, but individual plants are fundamentally associated with place. Flora organizes place, and place organizes flora. In terms of vascular plants, spores, pollen and seeds may have remarkable properties when it comes to disperal. Similarly, systems of vegetative reproduction may exhibit prodigous expansion through space; with the assistance of various vectors, including humans, individual taxa have moved backwoods and forwards around the globe, diversifying and hybridizing. Yet, vascular plants, in contrast to animals and humans, are basically sessile, literally 'rooted' in a particular piece of land, and this is mostly how we experience them, and questions about the places in which a plant is found determine how it is identified. Place is a major variable in naming contingencies, and while people widely accept names for plants found in other places, outside their normal extractive area, they will contest names occuring in places they know well, and where therefore there is an expectation of variation (Damon 2017:108–9). During my fieldwork, Nuaulu informants on the whole took the view that they could not expect to 'know' the plants from other places, that is outside normal Nuaulu spatial experience, for example in the Alune or Manusela language areas.

Plant diversification also happens in particular places, and it is the constraints and opportunities of location that encourage the evolution of new species. The same applies to human knowledge of plants. While knowledge of plants, like plants themselves, can move around, we experience plant diversity and manage them and are educated by them in particular places. In developmental terms, the earliest encounters between children and plants are, as growing organisms, in and around the village; or as modified as foods, in domestic settings; or as material for making things that are situated in a particular place. But places are more than GPS coordinates. We have seen in Chapter 5 how the composition of forest varies between different extractive environments that have been studied – Nuaulu, the Alune village of Lumoli, and Saleman – and how patchiness is a characteristic of lowland rainforest (managed or unmanaged) everywhere on Seram. This localization of objective ecological variation must have implications for what knowledge people can acquire from it. The way that we experience plants, therefore, may be associated with different kinds of recognized habitat and different kinds of humanly managed biotope.

To appreciate how swidden crops can vary from place to place, we can take the example of leafy greens mostly belonging to the family AMARANTHACEAE. Among both Nuaulu and Alune speakers all amaranths are placed within categories labelled ***manahute*** and 'atnanele', respectively, but we can see from Table 10.1 that the genera and species named, and presumably utilized, vary. At the generic level inhabitants of the Nuaulu village of Rouhua and the

Scientific name	Rouhua-Nuaulu name	Lumoli-Alune name
Amaranthus spinosus	***manahute putie*** syn. ***manahu wan anoi, manahute waene anoi***	
Amaranthus tricolor	***manahute marae***	1. 'atnane ndabu oti putile' 2. 'atnane ndabu oti lalakwe' 3. 'atnane ndabu makalikate'
Amaranthus viridis	***mono manuate, mono manahutane, mono manhutanane, mono mana hutanane***	
Amaranthus sp.		1. 'atnane ndabu salimala lalakwe' 2. 'atnane ndabuu salimala putile'
Celosia argentea	***manahu(te) msinae, manahu wan anoi***	'atnane ndabu kwele lalei'
Celosia cristata	***manahu wan anoi***	'atnane ndabu manutulai'
Celosia sp.	***manahu wan anoi***	
Cyathula argentica	***manahute putie*** syn. ***manahu wan anoi, manahute waene anoi***	
Cyathula prostrata	***panu-panu*** 1. ***panu-panu totu ikine*** 2. ***panu-panu totu onate***	
Deeringia amaranthoides		'atnane ndabu ai lalakwe' 'atnane ndabu ai putile' 'atnane ndabu ai palar`a'
Polygonum sp.	***manahu wan anoi***	

Table 10.1 Comparison of AMARANTHACEAE *taxa recognized by Rouhua-Nuaulu and Lumoli-Alune (NED, and Suharno 1997).*

Alune village of Lumoli both recognize different types of *Amaranthus* and *Celosia*, but in addition, Rouhua recognizes *Cyathula* and *Polygonum* (POLYGONACEAE) and Lumoli *Deeringia*. Rouhua report nine species overall and Lumoli five, though Lumoli label more varietal differences within three species. At the level of folk variety we might assume some variation and clear correspondences in popular varieties, though with the present data set we are unable to confirm these. No doubt, this range of amaranths is found throughout Seram, but with some specific local differences in combinations, and probably including other taxa not represented here.

To take a different example, for Nuaulu ***hueni*** (taro) is quintessentially understood in the context of first-year swiddens, and when the same species or a variant of the species (*Colocasia esculenta*) are found in stream-sides along paths, they are known by a different name (***pina totua***); while aroids in the same context, but never grown in swiddens (***punetene***), are nevertheless seen as a kind of taro. In a different way, and as described in Chapter 7, plants used medicinally by Nuaulu, while often similar to those used elsewhere on Seram at the genus level, and in their general bioactivity, are often different species, say the medicinals *Erythrina variegata* and *Senna alata* in Nuaulu but *E. subumbrans* and *S. occidentalis* in Manusela, *Curcuma aurantica* in Nuaulu but *Curcuma longa* in Alune. And the locality-specific character of species becomes even more marked when we compare those used medicinally on different, geographically distant, islands (Ellen and Puri 2016:appendix). Such observations not only vindicate the truism of the ethnographic specificity of local knowledge, but demonstrate how this is a variation not simply in the distribution of knowledge of the plant or of the local environment, but also of the actual phylogenetic context. We cannot understand folk plant classification in the abstract, only in the contexts in which it is produced.

Life forms and major plant groupings

Having established (in Chapter 2) the provisional parameters and qualities of the Nuaulu domain of plants as a generality, we need to outline the ways in which it is then divided up at its most inclusive extent – at what some would call (invoking comparison with social hierarchy and two-dimensional literary representation) 'the highest level'. Despite Lévi-Strauss's (1966) claim that humans have an intrinsic capacity and penchant for classifying natural diversity – 'a science of the concrete' – and despite the plausibility of the existence of some underlying urge to make sense of nature, this simple assumption cannot always explain the variation and commonalities in what is classified, the way objects and types are culturally selected, how they are organized, and what is remembered and transmitted to others. The mind mainly recalls what it repeatedly encounters in particular instantiations, otherwise knowledge is in danger of being lost. There must be some means of sifting incoming information, and through this to evaluate its utility. What is crucial for human adaptation through accumulated knowledge of plants, is the sharing between the minds of others and the external storage in artefacts, which massively increases capacity. But how the necessary shared organization comes about is not always clear, as it is neither intrinsically symmetric nor logical. Not all entities in a plant-classification system have conceptual or perceptual equivalence, and not all plants have the same salience in people's lives, or in connections within an ecological system. For example, as we saw in the discussion of Nuaulu palms, some species are of such overwhelming significance that they redefine people's relations with plants in general, as well as the interpretation of their natural and social worlds, so-called cultural-keystone species (see e.g. Ellen, 2006a; also Garibaldi and Turner 2004). There is therefore an uneveness in the attention given to different species or groups of species, with areas in which knowledge is dense and highly connected, but also with empty zones of classificatory space. Let us first return to the highly theorized (e.g. Atran 1985) concept of life form.

The Nuaulu concept of life form is not a taxonomic category in the strict sense at all, but rather a grouping of types of plant displaying a similar habit or growth form at the time of an elicitation event. We know from the work of Berlin (1972) and Brown (1977, 1984) that, overall, most LIFE FORM concepts are likely to be the most recent of the 'ranks' to have evolved, though a few – such as TREE – may be semantically primitive and ancient in terms of the evolution of human cognition. But in terms of application to identifiable species located within particular cultural traditions and ecologies, there can be much flexibility. Thus, Nuaulu ***nunue*** (a kind of *Ficus*) is generally thought of as a tree (***ai***), but begins life very much as a vine (***wane***), a 'strangling fig'. The seed of a fig is usually established by a bird or a bat dropping it in a host tree, it then pushes out a root down the trunk of that tree, or an aerial root down from a branch. The fig continues to extend its roots downwards, and leafy branches upwards and outwards, surrounding the trunk, eventually killing the host that has sustained it hitherto. The various figs known to Nuaulu take on the habit of climbers, creepers, epiphytes or ordinary trees. Similarly, ***kam*** and species of the genus *Piper* (pepper) are generally reckoned as vines, the prototype being the common domesticated betel vine, *Piper betle*, but the family Piperaceae contains other species that are clearly trees, such as ***ai kam*** (*Piper aduncan*). Sago palms (***hatane***: *Metroxylon sagu*) may begin life as suckers developing into prostrate horizontal stems, only latterly growing vertically. Only one sago folk variety is always described as a tree (***hata ai***), and then only using ***ai*** as an adjectival qualifier to indicate its lack of spines. Sago palms in general are never trees. Indeed, many taxonomic families contain species that assume various habits, and therefore in no sense can the folk-classificatory level of LIFE FORM easily map on to a common phylogenetic content. So, why do we treat this as a rank in a unified system of ethnobotanical classification?

Despite a series of critiques (e.g. Atran 1990:38), Berlin (1992:15, 24) provides us with the most influential definition of LIFE FORM used in studies of ethnobiological classification. As categories incorporated at the most inclusive level within the domain of 'plant', Nuaulu language contains and uses the widely reported terms of 'tree', 'vine' and 'herb': ***ai***, ***awane*** and ***rahue***. Nuaulu seems therefore to conform to what Brown (1984:149, 163) describes as a 'Stage 4' language, referring to the order in which it is proposed terms are added to languages over time. But as Berlin himself points out, we should not expect the conceptualizations of these terms to be standardized across languages, or for these core terms to reflect the only categories that work as LIFE FORMS. Thus, Nuaulu ***rahue*** (***rahu*** = 'to plant', ***amrahu*** = 'to work the soil', ***mamrahue*** = 'farmer') refers to things 'planted' and managed. ***Rahue*** therefore seems to cover only herbs that are planted, and is otherwise used informally as a residual category to refer to plants that are not semantically captured by the sense of ***ai*** or ***wane***. Brown might therefore envisage it as a kind of proto-'herb'. To a certain extent the idea of 'herb' is covered by another term, ***monote*** (grass/weed; locally glossed in Ambonese Malay as 'rumputan'), while the neologism 'grerb', introduced by Brown precisely because of the difficulties incurred by using 'herb' to cover the required meaning, might be a better description here.

But even if we can agree on the three most salient life-form terms, there are doubts about how some other major Nuaulu plant categories might be included within them. Thus, we have seen (Chapters 5 and 6) that ***ai*** (***ukune***) in the strict sense excludes both palms and bamboo, and we have also seen (Chapter 7) that ***monote***, though grouping some grasses and weeds, does not encompass all the other plants that Berlin envisages by the LIFE-FORM concept 'herb', or which are covered by the functional category ***rahue***. In addition (Chapter 5), the term ***meute*** (rattan) fits uncomfortably within ***awane*** (vine), although we might expect Berlin to place both in his 'vine' LIFE FORM. Such terms label the most inclusive plant categories in the sense that they contrast with other categories implicated as 'life forms'. Berlin and others have got round this by using the concept of 'non-affiliated generic'. But if we include these less extensive 'non-affiliated generics' as LIFE FORMS then Brown's stages begin to look problematic, even though we might still agree that core life forms reflect something about the order in which folk biological categories of great semantic breadth are added to languages. In this respect, the picture is much as I described it for Nuaulu animal categories (Ellen 1993b:116–19).

Often, life-form terms are incorporated into binomial nomenclature, as in ***ai kam*** (betel pepper tree: *Piper aduncum*) or ***awane unie*** (thorny vine: *Rubus fraxinifolius*), which shows them to be understood as functioning fully as folk-botanical terminology, but working to label basic categories referring usually to plants at the level of scientific genus or species. Of course, such plants labelled using binomials do not constitute an exhaustive enumeration of those that may be considered perfectly good members of the life form they refer to. And in some cases, life forms might even be said to exist conceptually, but not (in the ordinary way) linguistically, as covert categories, such as BAMBOO or PALM. Neither BAMBOO nor PALM are necessarily kinds of TREE or HERB, but in the Nuaulu case their existence can be inferred from informant narratives, from pile sorting and from interview elicitation using formal question frames which permit inferential logic. Thus, if the bamboo ***ona*** is described by an informant as a kind of ***wanate***, and the palms ***katue*** and ***tepine*** are described as kinds of ***panuke***, and vice versa, then we might reasonably conclude that each should be grouped together within a covert category BAMBOO or PALM, respectively, and if this is done repeatedly for other folk terms for bamboos and palms, then we might reasonably describe them as life forms, even if there is no consistent name. Taylor (1990:42–51) has paid a good deal of attention to the methods for securely establishing the existence of covert categories.

What Berlin and others call LIFE FORMS, therefore, seem to exist on a continuum of size and inner complexity, some conforming well to his definition and others – while covering large areas of classificatory space – would not normally be considered life forms at all. I have tried to avoid the problems posed by tight rank definitions by using the term 'primary' to refer to any category that, while referring to a plant, is not subsumed in any other more encompassing named category (cf. Ellen 1993:91). Others have used different ruses, which seem to suggest that how such categories are identified is really rather ambiguous, such as Kocher-Schmid's (1991:260–2, 266) 'structural categories' – presumably those that define much of the overall shape of the classificatory content of PLANT, of which she identifies eight for Nokopo on the north-eastern Papua New Guinea mainland. Of these, 'trees' absorb 138 of the 409 generics (34 per cent) for vascular plants, 'grass and herbs' absorb 133, while 40 folk generics are unaffiliated. Kocher-Schmid's life forms include 'staple', 'vegetables' and 'epiphytes'. Representing such diverse, overlapping and flexible primary categories, which contrast imperfectly, as part of a taxonomic ranking can be misleading, but representing these categories as a Venn diagram often works much better (e.g. Hunn 2008:83, table 4.2; Kocher-Schmid 1991). For Hanunóo on Mindoro in the Philippines, life-form distinctions between lignaceous, herbaceous and vine-like plants are more commonly made for non-domesticates rather than domesticates (Conklin 1954:95), 'trees' absorbing 38 per cent of plants, 'herbs' 35 per cent and 'vines' 24 per cent, with only 3 per cent of unaffiliated generics not absorbed into life forms. Taylor (1990:84) notes that for Tobelo on Halmahera 'trees' absorb 435 of the 1,122 terminal categories (39 per cent). This compares with 75 per cent of the 471 GENERICS Tzeltal are reported to include in four LIFE FORMS, 178 of which are included in 'trees' (40 per cent), and where most are monotypic terminal taxa (Berlin, Breedlove and Raven 1974b:30–1). According to my calculations, the Nuaulu ratio of terms absorbed into the TREE life form rises to 42 per cent (425 out of a total of 1,013 terminal categories).

In other cases we might presume that it is difficult to make such calculations in a reliable way at all. Thus, Friedberg (1979:86) argues that there are no LIFE FORMS in Berlin's sense amongst the Bunaq, while Sillitoe (1980:135) says that there is no evidence at all for mid-level groupings (such as pandans) amongst the Wola, and that what Berlin might have called LIFE FORMS are perhaps instead better understood as something like a 'primary category' as used by Bulmer (e.g. Bulmer and Tyler 1968). Sillitoe (1995:204) reports thirteen categories of this kind, but includes 'cultivated plants', 'mosses', 'fungi' and 'lichens', only seven of which are explicitly named; but if all these are considered 'plants' by Wola then their construction of this folk kingdom is rather different from Nuaulu PLANT. Obviously, the broader the definition of PLANT, the more likely it is to admit more life forms.

Some limitations of the 'new normal'

As a general rule in cognitive ethnobiology, it is accepted that while we need to be careful when inferring categorical status from nomenclature, names do provide some general guidelines. However, some cases are always particularly problematic. The Nuaulu phytonyms ***punara noene*** and ***punara wane*** are bi-segmental and share the common element ***punara*** (referring to the Nuaulu village of Bunara), but these are working as GENERICS in Berlin's sense rather than as SPECIFICS, especially as the first is the pepper vine *Piper corylistachyon* and the second the tree *Octomeles sumatrana*. What complicates matters further is that a third term, ***ai punara***, a clear binomial, indicates a species of the same genus as ***punara noene***, namely *O. moluccana*. Where a binomial encodes the life form prefix ***ai*** or ***awane***, it is usually the case that we can infer a GENERIC in Berlin's sense. But this is not always the case. Consider for example the binomial (e.g. ***mono manahutane***). Whether this is referring to a GENERIC or SPECIFIC in Berlin's sense is often unclear. Consider also ***awane papate*** (*Petraevitex multiflora*), optionally called ***papate wane***. Where there are alternative and optional terminological possibilities for the same plant (as there are in around fifty-nine cases), some alternatives may suggest a GENERIC and some a SPECIFIC. To a less inclusive degree, ***popote*** (quite distinct from ***papate***) groups together different folk varieties of *Syzygium aqueum*). However, ***popote wesie*** (forest ***popote***) is the term for *Pternandra cordata*. Is ***popote wesie***, therefore, working here as a GENERIC or as a SPECIFIC? Similarly, is ***wana onar*** when applied to the bamboo *Dendrocalamus asper* a SPECIFIC or a GENERIC? And is ***uri poi*** (*Musa* poss. *acuminata* subsp. *banksii*, or *M.*

lolodensis) being treated as a 'real' banana (a kind of ***uri***), or is it rather understood as a separate GENERIC given this name to indicate a superficial resemblance only?

Quite a few folk varieties of important food plants are known by a uninomial, even though most are referred to using a binomial. This happens, presumably, because the said folk variety has eclipsed many of the others in its importance; for example, the coconut folk variety ***wekete*** might be seen as having the hypothetical history: ***nione wekete*** > (***nione***) ***wekete*** > ***wekete***. A more complex case is presented by ***posteli***, a term which groups a number of named folk varieties of cassava (*Manihot esculenta*) that were introduced during the late colonial peried. These are often simply referred to as ***posteli putie***, ***posteli metene*** and so on, but sometimes the entire group will be referred to as ***kasipii posteli***, ***kasipii*** being the usual term for all cassava. I have not come across any trinomial constructions involving these terms, as syllogistic logic and hierarchical reasoning might seem to admit. In such a case should we treat ***posteli*** as a GENERIC or a SPECIFIC, and the binomial forms as SPECIFICS or VARIETALS? When does a SPECIFIC become a VARIETAL? And what happens in the case of species that extensively hybridize, such as cucurbits? Faced with these sorts of problems, the logical certainties of Berlin's model and his terminology begin to unravel.

Despite these difficulties, the Berlinian scheme has powerfully demonstrated a number of things: the priority of basic categories within classification systems, that basic categories match mostly scientific genera reflecting discontinuities in the environment, and that there is a good measure of correspondence between cultural populations in how this is achieved through cognitive prototypes. This view quickly fed through into cognitive psychology and linguistics, where it has been influential. We can see this, for example, in the work of Lakoff (1987). The paradox has been, however, that while basic categories are not conceptual primitives – indeed they show a great deal of internal complexity – they are cognitively simple to process, while categories more or less inclusive than the basic level can deviate from the taxonomic discontinuities described by biologists in quite significant ways.

How, therefore, can we measure classificatory knowledge? Something like the Berlinian scheme provides assumptions that make cross-language and cross-cultural comparison possible. Certainly, we need stable units of comparison to assess with confidence the similarities and differences between populations in terms of their classificatory knowledge of plants. However, quantifying data on plant categorization is no easy matter. In Chapter 2 I provided some data on the quantification of lexical features of words used to describe plant types by the Nuaulu. This is difficult enough, given variation within a language community and the flexibility with which people use plant terms. Plant words are our main route into understanding plant categorization, and even within a single study there are problems inferring one from the other, as we have seen. When it comes to cross-language comparison, the problems are much greater. Surprizingly, we still have very few major studies of the plant classifications for particular peoples that are sufficiently comprehensive and robust, or which are presented in a way that makes comparison between them straightforward. The studies we do have are of variable quality and make different working assumptions about how we might best conceptualize classificatory space. These problems can even be found in those studies that follow Berlin's guidelines and use his terminology, in part because use of the same terminology sometimes masks different working assumptions. To get some sense of these issues and what is possible comparatively speaking, Table 10.2 tries to organize some numerical data drawn from six studies of populations speaking different languages, each arranged for eight analytic category types. These types are: all named categories, life forms, terminal categories, primary categories, basic categories, GENERICS and SPECIFICS. Note that it is not possible in every case to provide numbers for these analytic category types, either because authors dispute their validity or usefulness, or because it is not possible to derive quantitative data for other reasons. A column for VARIETAL has been omitted altogether because of the potential confusion between Berlin's restricted definition and a more generally recognized concept of folk variety as referring to cultivars of domesticated species. Thus, for the Nuaulu data there would be just 16 instances of VARIETAL sensu Berlin, but in excess of 236 instances of categories referring to folk varieties of domesticated species.

To begin with, different writers define the limits of the botanical domain in different ways, sometimes justifying this with reference to local definitions of what constitutes the PLANT folk kingdom, and sometimes preferring to work with etic distinctions. The problem with the latter is that in the context of contemporary scientific taxonomy, a 'kingdom' is no longer held to be reliably monophyletic (e.g. Roger and Simpson 2009). For this reason, I have in this study deliberately confined myself to vascular plants, which – as Nuaulu PLANT coincides with the phylogenetic grouping of tracheophytes (those land plants with lignified tissue for conducting water and minerals) – allows for a reasonably secure boundary. However, Berlin, Breedlove and Raven (1974b) include bryophytes (mosses, liverworts, lichens) as PLANTS, while Atran (1999) deliberately excludes them following Itzaj Maya practice, while others still (Christensen 2002; Conklin 1954) include fungi and even algae. Quite apart from anything else, and as we noted in the previous section, where you draw the boundary of PLANT determines the number of life forms that might be recognized.

Population	Location	ALL	LF	TC	PC	BC	FG	FS	Source
Hanunóo	Philippines	1,604	3	1,625[1]		822	890	956	Conklin 1954a
Nokopo	New Guinea	520[2]					389[3]	478	Kocher-Schmid 1991
Nuaulu	Eastern Indonesia	1,012	2–5	870	510		516	477	Ellen: NED
Tobelo	Eastern Indonesia	1,815	4	1,122	689				Taylor 1990
Tzeltal Maya	Mesoamerica	1,635	4	720			471	237	Berlin, Breedlove and Raven 1974b:30–2
Zapotec	Mesoamerica	844	10	717			452[4]	371[5]	Hunn 2008

Table 10.2 Comparative data for estimated numbers of different kinds of named category reported from ethnographies of vascular plant knowledge conducted for selected tropical and subtropical populations. In most cases it is clear that we are dealing with data for an omniscient speaker-hearer.
Key: ALL, all named categories; LF, life forms; TC, terminal categories; PC, primary categories; BC, basic categories; FG, folk generics; FS, folk specifics. Notes: 1, Conklin (1954:250): 'separate plant types'; 2. 'identified plants,' though many of those identified have same folk name; 3, 429, 40 of which are synonyms'; 4, includes 341 monotypes and 111 polytypes; 5, includes 363 monotypes and 8 polytypes.

Despite the difficulties of deriving basic categories from nomenclature (see, e.g. Taylor 1990:52–9, following Conklin), data for 'all named categories' are the easiest to assemble and arguably the most accurate and meaningful to compare, as they are simply the sum of all names at any degree of inclusiveness in a classification (including names that operate at each of the Berlinian levels). Some difficulties surrounding operationalization of the concept of life form have already been discussed in the previous section: when for example does a Berlinian 'unaffiliated generic' become a LIFE FORM? I have argued the case for distinguishing between terminal and primary categories. Terminal category, a concept adopted in some previous studies (e.g. Berlin, Breedlove and Raven 1974b:36; Taylor 1990) is used for those named categories that are shared and least inclusive. They are easy to measure because they are those categories that cannot be further subdivided. They may correspond to Berlinian SPECIFICS and VARIETALS and be expressed nomenclaturally as uninomials, binomials or trinomials. Primary named categories are the converse: the most inclusive named categories within the plant domain. In terms of the Berlinian ranking system, they are therefore sometimes LIFE FORMS, INTERMEDIATES or GENERICS. In many cases named primary and terminal categories can be identical, meaning that they present themselves simply as 'a kind of plant' without any apparent intermediate classificatory apparatus. If for Berlin a GENERIC is a category, named or not, that can be recognized perceptually, then it is the same as a basic category. But these basic-level categories for Atran (1990:6) are better described as 'generic-specieme' – combining Berlin and Bulmer's (1970) terms for the concept – as he argues that the genus-species distinction is largely irrelevant to a 'common sense' view of the world. As I have argued previously, the basic level of folk biological recognition varies between scientific genera and scientific species, between Berlin's GENERICS and his SPECIFICS, though for smaller plants and animals and other organisms the basic level may be recognized at higher scientific taxonomic levels. Conklin (1954:114) does not use versions of either generic or specific, and is careful not to predetermine the equivalence between Hanunóo and scientific rank concepts. Finally, there are VARIETALS or folk varieties, but as we have seen – and confusingly – for Berlin what others might call the folk varieties (meaning the kinds of cultivars of major crops) he calls SPECIFICS.[1]

The difficulty with the Berlinian ranks, therefore, is that they are relative, both within a classification and when comparing the folk classifications of different peoples. This is true even if we accept that the framework of conceptual ethnobiological ranks is absolute across cultures. We can never quite be sure whether a Nuaulu life form is the same thing as a Tzeltal LIFE FORM, or a Zapotec varietal the same as a Tzeltal SPECIFIC. For this reason, quantitative comparison is fraught with difficulty. If we – quite reasonably – follow Berlin (1992:96–101) and assert that everywhere people identify about 500 names for GENERIC or basic categories, we must accept that even this is an approximate number; and whether or not we are averaging out data from an entire population or picking the maximum knowledge of one key informant, and depending on a range of ecological and cultural factors, the number will alter by quite a margin. If we have difficulty with this, amongst the most reliable of cross-language or cross-cultural measures, what hope do we have with others? Much the same is true of comparisons of terminal categories. While comparative nomenclatural data are reasonably straightforward (such as ratios of uninomials to polynomials, which do appear to be diagnostic of some key cultural differences, such as between hunter-gatherers and agriculturalistis), comparison of units that require judgments of conceptual rank can be tricky.

1 Kocher-Schmid (1991:e.g. 84) uses the term 'cultivar' to refer to any cultivated plant that is treated as a discrete type, whether it is an entire genus, species, or whether a variety of a species treated as a cultivar.

Comparing ethnobotanical classifications and convergence metaphyics

The philosopher David Ludwig (2018) has recently offered a revised interpretation of the history of what he aptly calls 'convergence metaphysics'. The work of Brent Berlin, with its insistence on the idea of distinct 'chunks of biological reality' recognized pan-culturally rested intellectually on both the cognitive revolution espoused by psychologists such as George Miller (1956), and on the species realism of the 'modern evolutionary synthesis', as promoted by the likes of Ernst Mayr (1963). But Berlin's claims must also be seen in relation to his earlier work, especially that with Paul Kay on colour (Berlin and Kay 1969), which saw the first phase of a playing out of 'universal-evolutionist' hypotheses in cognitive anthropology.

The subsequent stasis was in many ways no different from what we find in other cycles of scientific and scholarly innovation. True, Scott Atran (1981) was to build upon Berlin's work impressively, and argue that converging classifications were the result of shared cognitive biases rather than shared nature, but Berlin and Atran together somehow provided us with a 'new normal'. Berlin, in particular, offered a strong model that could seemingly accommodate a growing body of empirical studies, and a basis for a highly productive research programme setting rigorous methodological standards. Its simplicity and convenience quickly established it as a common framework for description, adopted by devotees, agnostics and critics alike, and was often pressed into service as a backdrop for the exploration of new problems (e.g. Bloch 1977). As ethnobiology grew, many were content to accept this new normal unchallenged, or to sideline it. An emerging focus on 'indigenous knowledge' as an organizing framework was concerned less with how arrangements of local categories matched some universal standard, than with how these categories made local cultural (ethnoecological) and ecological sense in relation to a particular habitat. Outside the narrow world of ethnobiology, the forcefulness of the Berlin model was easier to absorb among the received truths of a certain kind of cognitive and socio-cultural anthropology. In this context the intellectualist versus utilitarian, and universalist versus relativist, debates tended to fade away 'without clear resolution'.

Ludwig is perceptive too about the missed opportunities and marginalization. Paradoxically, the growth of ethnobiology as a distinct subject separated it from the rest of anthropology at the very time when it might have become central to ongoing debates, such as those connected with the conceptualisation of nature and the 'ontological turn'. Meanwhile, there was turmoil in the world of biological taxonomy. The arrival of biomolecular barcoding, the new phylogenetics and cladistics, was giving rise to constant revision in nomenclature and taxonomic groupings, and contradictions with more conventional approaches, often with problematic practical consequences (Ereshefsky 2001). The revival of pluralism in biological taxonomy, with its recognition of discontinuity along different dimensions, reopened discussion about what species might actually be, and in turn has raised new questions as to what correspondence between scientific and folk taxa might actually entail (Hull 1992). The insights of the new pluralism were consistent with a new realism that encouraged flexibility in scientific taxonomy, and which could respond to practical concerns and social implications (as in wildlife conservation). As Lloyd (2007:49–50) remarks, there are difficulties, in terms of our present state of knowledge, in knowing what the scientific consensus on taxonomy actually is, while the number of species recognized within plant genera often closely corresponds to the intensity with which particular genera are studied. Moreover, when it comes to applying names to folk categories of plants no less than to scientific taxa, the logic of cladistics, driven by the recognition of continual binary differentiation over time, as individual organisms reproduce and modify, suggests that the level of differentiation at which we give nominal recognition to organismic difference is rather arbitrary. Indeed, the degree of difference recognized by giving a name is more likely a consequence of practicality. The problem is indicated diagrammatically in Figure 10.5.

Species definitions therefore impose divisions on ongoing evolutionary processes, as studies of the practices of taxonomists show (e.g. Kirksey 2015). Species is an unstable concept, and what makes a species visible are some very selective and arguably arbitrary decisions. Thus, for Dupré (1993:57) there are multiple plausible and defensible approaches to biological classification ('promiscuous realism') and species can be seen to emerge as conceptualized and named entities at particular moments, triggered by the coming together of different factors (scientific, economic, political), in a way somewhat similar to that in which classification itself is performed (Waterton 2003), and in a way that resembles the sociocultural processes of prehension discussed at the beginning of this chapter. And in ethnographic studies of folk classification too, quite apart from my own work among the Nuaulu, there is plenty of evidence for the intrinsic flexibility of classification systems. Thus, Sillitoe (1995:201, 206–7) warns that attempts to fit the Wola scheme into the Berlinian framework distorts their experience, and (p. 209) that 'no single classificatory scheme can comprehensively represent Wola ordering of plants', and even that 'disorder' is an essential part of the oral system (Sillitoe 1980:144).[2] Why should organized knowledge – folk or scientific, Nuaulu or Maya or Wola – exist at only one level or dimension from a realist

2 For similar comments on plant classification among other highland New Guinea peoples, see Healey (1978-9) on the Maring, and Sterly (1997) on the Simbu.

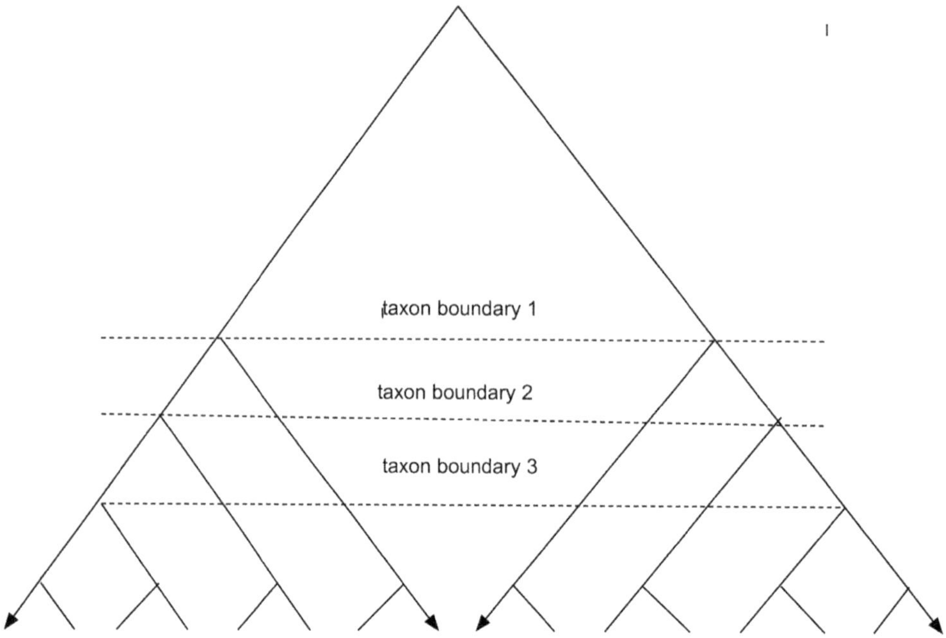

Figure 10.5 The process of continuous differentiation over time and the problem of where to mark boundaries between taxa, and which boundaries to name.

perspective? Why does there have to be THE classification shared by a single community of speakers or cultural population?

In this light, Berlin's assertion that one single pattern will 'stand out' from the rest begins to look shaky. If taxonomies used in biological systematics do not necessarily converge, why should we expect it to happen in folk classification? Modern scientific taxonomy arose in a literate culture with all the advantages of the new technologies of sharing, but also with the limitations arising from two-dimensional textual representations, the metaphor of hierarchy and with an emphasis on 'classifications' as static entities, as so well exemplified in the Linnaean scheme. Claims for convergence work better in written cultures, where authoritative dominant classifications can be more easily disseminated and enforced. By comparison, the extent to which classifications are shared in oral cultures is always 'work in progress'. Although Berlin's framework still has many virtues, the overarching intellectual case is problematic if it is seen to require the existence of some intrinsic structure of objective discontinuity in nature that folk classifications and scientific taxonomies must inevitably reflect. Neo-Whorfian developments in linguistics have led to a re-examination of Berlin's work on colour (e.g. Lalumera 2014), and may find analogues too when looking at folk classifications of biota. Similarly, notions of convergence look rather different when examined from the perspective of developments in cognitive science related to neural plasticity (Lende and Downey 2012).

But there is another reason why the reassessment of the common features of different ethnobiological classifications has been slow in coming, and this is that despite the debate appearing to be about conceptual readjustment independent of empirical findings, the real issues underlying claims for universality in the recognition of natural categories depend very much on transparency and comparability in elicitation methods, and in the availability of large datasets collected under exacting conditions of linguistic competence, supported by sympathetic expertise in relevant areas of scientific taxonomy. Such programmes of research consume enormous amounts of research time, and are logistically complex and difficult to fund; and then – as we saw in the preceeding section – grand comparison is limited by both the number of interfering variables and the conceptual assumptions and judgments of different researchers regarding ethnobiological units of measurement.

In my own long-term project among the Nuaulu people of the Indonesian island of Seram, in order to make sense of ethnobiological classification processes and degrees of correspondence between scientific taxa and folk categories, not only am I continually frustrated in my efforts by the ways in which local people vary the linguistic expression of biological difference and relationships, but equally by how the measurement of correspondence is complicated by the way scientific taxonomists constantly revise the phylogenetic 'reality' to which folk classification is supposed to be corresponding. Of course we cannot expect cognitive convergence of the kind advocated by

Atran to match phylogenies based on molecular data. Any evidence of co-evolution is going to be shaped by bodily sensory equipment for prehending the biological world unassisted by modern technologies, but we do need to recognize that sufficient semantic convergence is a necessary and universal cognitive process within social populations in order for pragmatic discourse about plants and animals to take place at all. Social sharing is always going to be to some extent an 'eclectic muddle' (Douglas 1993:161–5), both amongst ordinary folk combining morphological, ecological, utilitarian and symbolic features in their 'thinking through' of natural categories, and, with their continuing disagreements, within the community of scientific taxonomists, and the taxonomists and users of taxonomic products in the modern world. No wonder, when both ethnographers and taxonomists seek to simplify their data, and to share the same metaphorical resources in translating between folk and scientific domains, there are so many artefacts and ambiguities.

I thus reject the 'strong' model of universal-formalism, and find in a 'permissive' model a much more persuasive account, one in which cross-cultural similarities can be largely explained by the convergence and overlap of biogeographies, universal processes of cognitive economy and symmetry, and neural plasticity. It may well be that certain language families, ways of life or sociocultural configurations are less amenable to what the Berlinian scheme predicts than others, for example Papuan languages (Healey 1978–9; Sillitoe 1980; Sterly 1997), or hunter-gatherer peoples as a group. In many languages trinomials are rare, but in some, such as Zapotec, they are common (Hunn 2008:79). What Berlin calls INTERMEDIATES vary widely in their significance between linguistic cultures, and among the Hanunóo are virtually absent. Systems also vary in other features of nomenclature, such as the ratio of binomials to uninomials. For example, Nuaulu and Hanunóo are alike in having a system in which about sixty-six per cent of all named taxa are described using binomials (Conklin 1954:128); while for Hill Pandaram (Morris 1976) and !Kõ Bushmen (Heinz and Maguire 1974) hunter-gatherers, binomials are rare. Moreover, the logical and cognitive imperative to think 'taxonomically' (i.e. in terms of contrast, class inclusion and hierarchy) is hugely constrained by ecological and other practical factors that often make such an expectation unrealistic (Taylor 1990:62).

Biotaxonomy as a community of practice: Rumphius and his modern counterparts

The anthropologist who has done most to integrate the study of folk classification and the history of scientific taxonomy as it developed in Europe is Scott Atran (1990). He has shown how the schemes and concepts of different taxonomists have converged or diverged historically from what we now understand about how contemporary folk classification works in diverse cultural groups, how scientific natural history became intellectually disassociated from folk biology (see also Ellen and Harris 2000), how the concepts of scientific species and genus arose (the latter to cope with the variety of similar forms discovered as a result of European expansion into the rest of the world), and how the assumptions made by modern commentators on the history of biology, concerning say artificiality and naturalness, relate to the conceptual foundations of biological taxonomy (Atran 1985).

One of my concerns in the present book has been rather different: to establish and explore the similarities and differences in the working practices of modern biological taxonomists and the Nuaulu. As part of this endeavour, I have drawn on the writings of Rumphius, a seventeenth-century student of Ambonese natural history, not because he is in any way representative of modern taxonomic practice, but because he represents one possible beginning point in understanding the historical process by which modern practices emerged. Moreover, throughout, I have used the insights and representations of Rumphius as a way of mediating between the Nuaulu vision of their plant world and that of modern taxonomic biology. Rumphius took notes and drew illustrations (until his blindness prevented this), and may even have had herbarium specimens before all his records were lost in the combination of earthquake and tsunami (1674) and then fire (1687) that also destroyed much of the colonial town of Ambon in which he lived for the later part of his life (Beekman 2011:68–73).

Despite the possible existence of a herbarium, we may reasonably assume that Rumphius identified plants with reference to general abstractions embodied in his written descriptions, and presumably in illustrated notes never preserved. By comparison, and as I have demonstrated, Nuaulu identify plants with reference to abstract gestalts reinforced by subsequent recourse to distinctive features, all stored orally and checked where necessary against the oral knowledge of others. By contrast, botanists since Linnaeus have worked with the concept of a unique type specimen to which all new material is referred and checked. A 'type' in plant taxonomy is therefore 'an anchor to the identity of a name; 1. an element, usually a herbarium specimen, on which a species name is based; 2. species on which a genus name is based; [and] 3. a genus on which a family name is based' (Beentje 2010:125). While Nuaulu do not have a concept of 'type' in the sense of a single physical specimen to which similar specimens are compared, we have already (in Chapter 8 and above in this chapter) examined how the virtual analogue of type might work.

Sometimes the connections between Nuaulu names, and the scientific identifications for specimens given these names, do not seem to make sense, even when (or perhaps especially when) we try to find clues that might clarify the position in the work of Rumphius. Consider the case

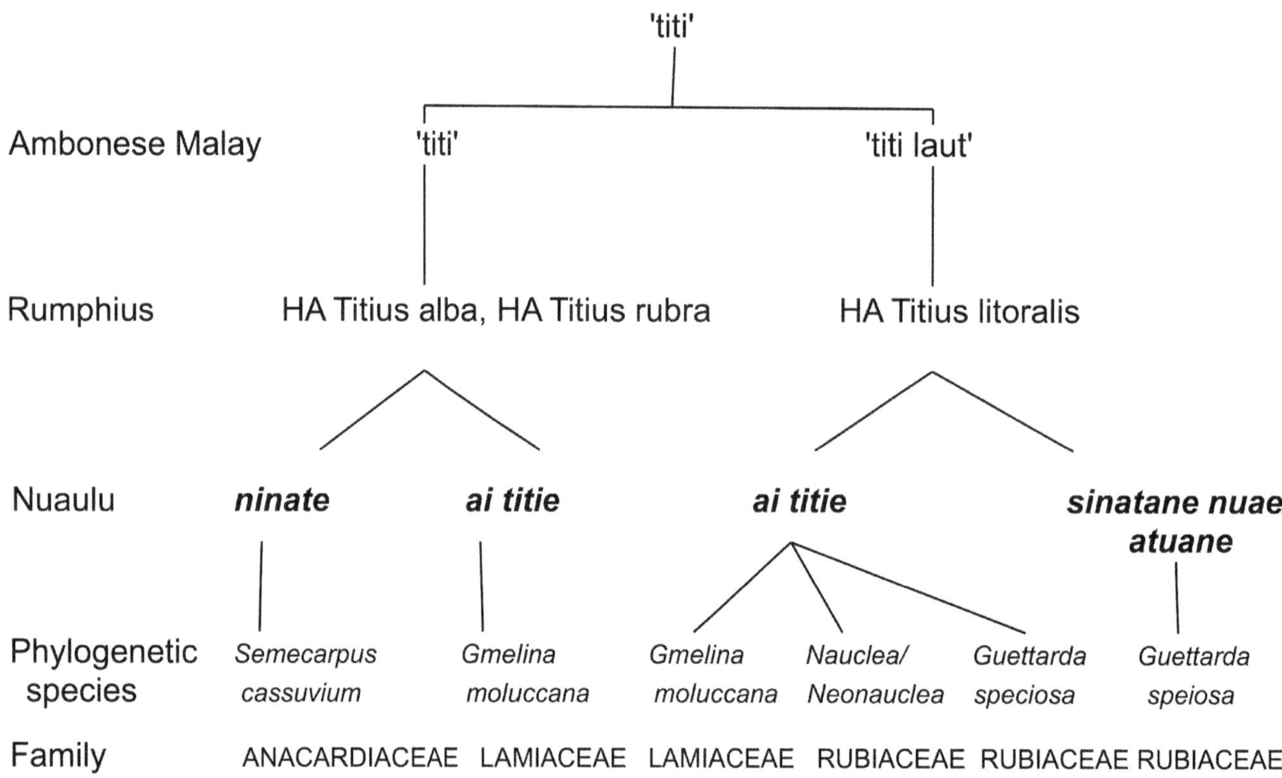

Figure 10.6 The relationship between Ambonese Malay 'kayu titi', Rumphian names, Nuaulu terminal categories and phylogenetic taxa.

of the Nuaulu names ***ninate***, ***ai titie***, ***sinatane nuae*** and ***atuane***, as applied to trees (Figure 10.6). Fortunately, there is enough evidence from Nuaulu naming practices to see clearly that the last two terms (***sinatane nuae*** and ***atuane***) work as synonyms for the same tree, in accordance with clan-based proscriptions on the use of one or other of the names. What links the three categories together is that Nuaulu translate all of these using the Ambonese Malay term 'kayu titi' (literally 'hot wood'), but distinguishing on the one hand 'kayu titi' (default) and on the other 'kayu titi laut'. Rumphius, in deferring in his usual way to Ambonese nomenclatural practice, renames both as 'Titius'. However, while the second is called 'Titius litoralis' (a literal translation of the Malay), the first he divides into two: 'Titius alba' and 'Titius rubra'. As we have discovered from Nuaulu nomenclature, a marked term (such as 'kayu titi') does not always mean that it is a kind of 'titi' in a genetic sense, so much as one distinguished from it in some crucial respect while having some features in common. We can see from Figure 10.6 that Nuaulu use the overarching AM term 'titi' to describe three Nuaulu categories (***ninate***, ***ai titie*** and ***sinatane nuae/atuane***), which in turn are applied to four or more phylogenetic taxa: ***ninate*** is unambiguously *Semecarpus cassuvium* (Anacardiaceae), ***sinatane nuae*** is unambiguously *Guettarda speciosa* (Rubiaceae), while ***ai titie*** contains specimens of *G. speciosa*, but also *Gmelina moluccana* and *Nauclea/Neonauclea* species (a mixture of Rubiaceae and Lamiaceae). The association between these various species and AM 'titi' goes some way to explaining some of the difficulties in the contemporary scientific taxonomy (see, e.g. Kok 2008, 2012).

In *The Cultural Relations of Classification* (1993) I attempted to make a fundamental distinction between identification and classification as empirical ethnobiological practices: the first being the placing of an identity (usually through naming)[3] on a particular instance of a natural kind (e.g. specimen X is a kind of Y), whereas classification involves the placement of either X or Y in any one or more abstract categories.[4] X in this latter case might, for example, be a herbarium voucher specimen consistently identified as *Syzygium*, even though taxonomists may periodically change their mind as to whether it should be placed in

3 It is of course possible to identify a plant without having a name for it. All that is required is that the person involved is satisfied that a particular plant is of the same kind that has been identified on a previous occasion, or on successive occasions. What the application of a name does is detach the identification from particular individual experiences and share the accompanying knowledge more widely, and thus limit further the risks that come with an inaccurate identification.

4 Professional botanists may distinguish between identification, nomenclature and classification slightly differently (see e.g. Bridson and Forman 1998 [1989]:2).

MELIACEAE or MYRTACEAE. Similarly, we might always regard Y as a kind of Z (e.g. a beech as a kind of tree), but we may often mistake tree Y_1 for tree Y_2. These distinctions and the errors that accompany the technical and cognitive processes involved are essentially the same whether we are dealing with Nuaulu folk biologists, ethnobotanists such as myself trying to make sense of their system, early modern naturalists with gifted observational skills such as Rumphius, or professional modern plant taxonomists working from living specimens, herbarium vouchers or even biomolecular data (e.g. Goodwin *et al.* 2015). Thus, when Nuaulu describe specimen X to me as a kind of Y, I can only test the validity of that statement by having an independent second identification that X is Y, and the more independent statements I receive that X is Y, the more I am likely to accept that there is sufficient culturally shared agreement on the matter to conclude that the identification is 'correct'. But when one informant says that X is Y and another that X is Z, what am I to make of it? In certain cases, it will be clear from the context that we are dealing with synonyms (as where gender or another social relationship mandate the use of one word rather than another), but in other cases it might be that there is no shared 'correct' view, and that different people identify and name the plant differently. I do not quite recognize in Nuaulu practice the 'surprising degree of dissent in naming plants' claimed by Sillitoe (1995:206–7) for the Wola, but we can still wonder (Sillitoe 1980:145) given the evidence for intra-cultural variation, at what point 'we move from classification to chaos' (Sillitoe 1980:145)?

Now if this latter were the standard picture in local ethnobotanical knowledge systems, then they would quickly implode, as the point of naming is to share reliable identifications. In some cases, we might agree that an informant's reponse is just incorrect: it might for example be the untutored attempt of a young child who had yet to learn how to distinguish one plant from another, or it might be a deliberate untruth, or it might be a guess by a non-expert. In the village group meetings I held in 1996 after our daily hikes into the forest to collect specimens, we began by placing specimens identified with the same name in separate piles on the floor of the ***numa insi***, the government house or 'rumah pemerintah'. Despite the specimens having been identified *in situ* by those accompanying me when the collections were made, and often by several people, who took into account a multitude of ecological factors only knowable to those who were there with me, people at the village meeting would still argue about correct names and identifications during the pile-sorting exercize, and in some instances specimens would move from pile to pile or be seen as distinct piles as a result of arguments – sometimes quite loud and vociferous – that eventually resulted in some kind of shared agreement. Similar processes take place in a scientific herbarium receiving plants for identification. Quite apart from errors arising from mislabelling or misreading labels, of which, as was noted in Chapter 1, there were many examples of in the creation of the NED, plant taxonomists can readily disagree on identifications, both within herbaria on different occasions, and between herbaria, as between Bogor and Kew in my data (Appendix 1). For example, Damon (2017:119) describes how Stevens changed the identifications of those *Calophylum* initially determined by generalists at Lae, and in many cases I have had to form judgments about which identification to accept, given competing determinations from different herbaria for specimens that I know to have come from the same plant. Even within the same herbarium, different taxonomists on different occasions will decide that a name attached to a specimen needs to be changed, sometimes several times over a period of years, not on the basis of information from new specimens or data, but simply based on a reinterpretation of the original specimen. In addition, nomenclature may change in accordance with the International Code of Botanical Nomenclature (ICBN) or International Code of Nomenclature of Cultivated Plants (ICNCP) (Turland *et al.* 2018), rules agreed after the herbarium sheets were initially prepared and the specimens named. And nomenclature can change for other reasons. As an example of this source of potential confusion, consider the name changes for Nuaulu specimens of the fern ***ahane*** in the Kew Herbarium since 1970 (Figure 10.7). These seek incremental clarity as to the 'correct' identification of a plant given a fairly tight set of rules operating within an institutional and professional system that is relatively closed, rules invented to ensure that information about plant names can be freely exchanged on a global scale without ambiguity. This, however, creates problems for local and practical end-users of taxonomic data and 'accepted' appellations, such as farmers, horticulturalists, home gardeners (e.g. Whitehouse 2014:64), nutritionists (e.g. Nesbitt *et al.* 2010) and even ethnobotanists, who find a virtue in consistency and in stable names. Over the period that I have been collecting voucher specimens with Nuaulu co-researchers, and seeking determinations from taxonomists working in international herbaria, plant names, including family names, have changed significantly.

In the case of Nuaulu plant identifiers and classifiers, there is built-in flexibility enabled by the absence of a particular ultimate physical reference point, or indeed a tight detailed permanent description that must be consulted to reach an accurate identification. Nuaulu operate as if names are stable, and some have clearly been so for many years, but in effect the relationship between name and plant is only as good as the last identification, and the frequency with which that name is applied. For both Nuaulu and professional taxonomists, plant identification is a work of constant revision, partly because biodiversity itself changes as plants continue to vary genetically and are selected

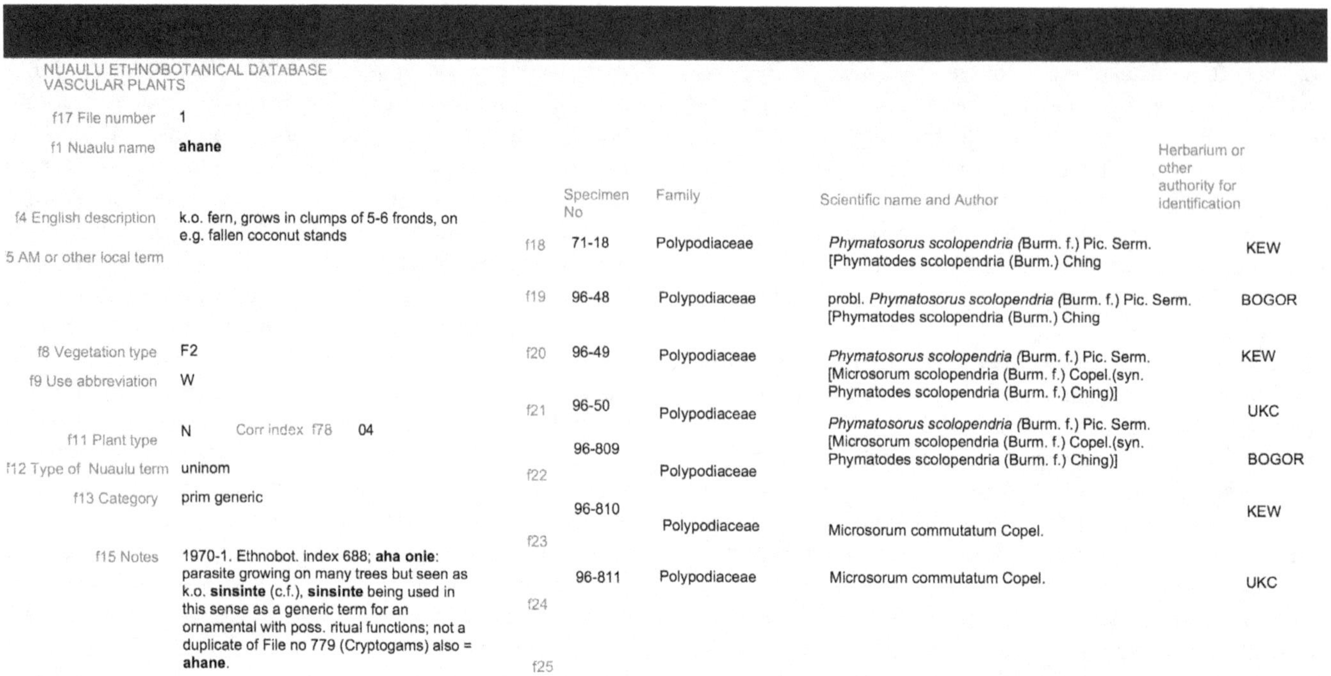

NUAULU ETHNOBOTANICAL DATABASE
VASCULAR PLANTS

f17 File number 1

f1 Nuaulu name **ahane**

f4 English description k.o. fern, grows in clumps of 5-6 fronds, on e.g. fallen coconut stands

5 AM or other local term

f8 Vegetation type F2

f9 Use abbreviation W

f11 Plant type N Corr index f78 04

f12 Type of Nuaulu term uninom

f13 Category prim generic

f15 Notes 1970-1. Ethnobot. index 688; **aha onie**: parasite growing on many trees but seen as k.o. **sinsinte** (c.f.), **sinsinte** being used in this sense as a generic term for an ornamental with poss. ritual functions; not a duplicate of File no 779 (Cryptogams) also = **ahane**.

	Specimen No	Family	Scientific name and Author	Herbarium or other authority for identification
f18	71-18	Polypodiaceae	*Phymatosorus scolopendria (*Burm. f.) Pic. Serm. [Phymatodes scolopendria (Burm.) Ching	KEW
f19	96-48	Polypodiaceae	probl. *Phymatosorus scolopendria (*Burm. f.) Pic. Serm. [Phymatodes scolopendria (Burm.) Ching	BOGOR
f20	96-49	Polypodiaceae	*Phymatosorus scolopendria (*Burm. f.) Pic. Serm. [Microsorum scolopendria (Burm. f.) Copel.(syn. Phymatodes scolopendria (Burm. f.) Ching)]	KEW
f21	96-50	Polypodiaceae	*Phymatosorus scolopendria (*Burm. f.) Pic. Serm. [Microsorum scolopendria (Burm. f.) Copel.(syn. Phymatodes scolopendria (Burm. f.) Ching)]	UKC
f22	96-809	Polypodiaceae		BOGOR
f23	96-810	Polypodiaceae	Microsorum commutatum Copel.	KEW
f24	96-811	Polypodiaceae	Microsorum commutatum Copel.	UKC
f25				

Figure 10.7 Various and successive identifications of voucher specimens in the NED of the fern Nuaulu call ***ahane****.*

for in different ways in particular environments, but also because the boundaries of those entities we call 'species' or 'varieties' change over time depending on the criteria we select to determine them. The scientific taxonomist is faced with the same practical and cognitive problems as the ethnotaxonomist. However, botanical systematists are experts who for much of their professional life specialize in only several genera or families, and while they accumulate much detailed knowledge of plants in these groups, they use a more abstract knowledge to cope with the rest. Nuaulu are on the whole generalists, as was Rumphius. Whereas professional taxonomists are constantly referring back to earlier physical specimens and descriptions, the Nuaulu process is more fluid, informed by what memories they have established from previous experience, and as confirmed or revised with benefit of other people's personal experiences. To say that the scope for error in plant identification is prodigious is not simply to make a technical point in the context of discussion about methods relevant to experts, but is an observation of theoretical relevance about the tentativeness of much plant identification.

The notion of identification as contrasted with classification has become more distinct in the context of the practices of scientific taxonomy, and as naturalists have sought to deliberately discard 'artificial' and 'practical' classifications and replace them with forms dependent on a logic internal to the practice and theory of science, whether Linnaean dependency on the primacy of sexual organs or the Darwinian model based on common ancestry, as revealed in morphology or DNA. Such classifications are often described – and some might think strangely – as 'natural' in contrast to 'artificial' classifications. Identification as a cognitive process also became more separate from that of 'real' classification through the pragmatic devices that botanists, in particular, could use to 'identify' plants, such as field keys (e.g. Clapham, Tutin and Warburg 1962).

Another major and obvious difference between Nuaulu as identifiers of plants and Rumphius and modern taxonomists, is that the Nuaulu do not write down descriptions of plants to which they can refer, or create images of their diagnostic characteristics. As argued in Chapter 9, the invention of writing – the idea of text and the physical medium on which text is created – has clearly had a massive impact on how we think about and execute the classifying process. For one thing, something as definite as 'a classification' could hardly exist before it was written down. Much has been written about the consequences of the shift from orality to literacy. Some of the claims for the impact of subsequent consequences of literacy, such as the invention of the printing press, have been shown to be false, the oral mode being capable of forms of thinking that were once imagined to be only possible with the appearance of writing. One example is the 'mental map' required to enable the Carolinian 'etak' system of navigation (Akimichi 1996). Also, it is sometimes difficult to distinguish between the impact of writing as a practice, the consequences of using paper or some other medium through which it is executed, and specialist modes of thinking and practice that have followed, such as science. And as Goody explains (1982:111):

> As with many other products of early literacy, the 'urge to classify' overwhelms reasonable human purposes and produces the over-systematisation that influences the cosmologies and religions of the book so profoundly that their prescriptions can sometimes be fulfilled only by a small elite.

The 'small elite' that Goody had in mind comprised priests and nobles, but the phrase might equally refer to scientists and scholars of the Renaissance and the Enlightenment.

The historical process of science has subsequently led to the extraction of individual plants from their contexts, their reclassification on the basis of abstract features and their grouping into a new mega-category: Plantae. For Ingold (2011), science is precisely that process by which 'storied' knowledge is turned into classificatory knowledge, or we might say, data intially processed through episodic memory becomes data encoded in mimetic memory. This process was well underway by the time of Rumphius, as was the establishment of a scholarly tradition in which the world was reimagined metaphorically as a book to read, in this particular case 'the book of nature'. But whether due to the conventions of the written word more generally, or to scientific conventions more specifically, knowledge absorbed or deliberately acquired for and from oral reservoirs thus often became distorted and simplified, not simply as a result of 'translation errors' but due to the characteristics of the new media and models associated with them.

For example, with written accounts the visual sense has become increasingly dominant in the process of plant identification, suppressing other senses such as hearing, touch, smell and taste. Writing can only capture many of the prosodic features of speech, such as tone of voice, with difficulty, and a fraction of what is ever said (Olson 1994:7). This may sometimes render otherwise distinct names as apparent homophones. Indeed, the background conventions of literacy have had implications for the recognition and reconceptualization of names. Text, after all, is a physical medium, and words, once they have been written down (in this case plant names), become more discrete and obvious 'things', with spaces between and sometimes within them (Olson 1994:76). Ad hoc descriptions are more likely to become accepted names once they are archived, even if they are not shared amongst all members of a speech community.

Though writing necessarily simplifies sensory reality and results in the encoding of mistakes, it can permit an 'archival' tradition, allowing the comparison of various versions of events, while ways of reading text became a model for verbal form. Moreover, it was only with the arrival of written texts that lists, tables and certain kinds of lexically annotated diagrams became possible (Goody 1977). Images themselves, of course, could be part of traditions of visual graphic art, and we know that otherwise non-literate peoples are able to invent and understand diagrams to explain abstract ideas to an ethnographer, as in Conklin's (1964) illustration of how Hanunóo used pictures of plants to explain kinship relations, or in the way Nuaulu employ painted or dyed motifs on shields, bark-cloth and pandanus leaves to illustrate cosmography. Before the invention of printing, pictures were not easily copied; but by the time of Rumphius, pictures were a common feature of herbals and other descriptions of natural history. Rumphius drew pictures or had pictures drawn for him to his own specifications. By the time of Linnaeus, it was widely accepted that the fundamental task of natural history was 'arrangement and designation' based on visually recognizable features (Olson 1994:226), and with Linnaeus utility and significance were abandoned. In much work in ethnobiological classification there has been a tendency to retrofit the operations of oral folk classification within the conventions of the written mode. There is nothing intrinsically wrong with this, if those conventions are being used deliberately and knowingly to shed light on how the classifications work, but problems arise when the analyst simply assumes that the conventions of the written mode must also apply to the oral. We can see this particularly well in relation to notions of hierarchy and level, when Randall (1976) cheekily asks us to speculate on how tall a 'taxonomic tree' might be, in overlap of categories, and in the suppression of dimensionality in two-dimensional representations on paper, whether using dendrograms, Venn diagrams or any other graphical device. Such conventions have no necessary place in oral representations unless they are modelled on experience of the literate mind, and serve to constrain the flexibility enjoyed by orality.

11

Nuaulu ethnobotany and theories of cultural cognition

Plants as a domain of thought and practice

Plants (or indeed other biological taxa) are fundamentally no different from other material objects perceived, organized and used by humans, in the sense that they can be analysed as a cultural domain emphasizing combinations of cognitive, semantic and linguistic criteria. The main difference with obvious cognitive implications is that as a species and genus we humans have co-evolved with plants and animals for much longer than we have with the many technical products of human culture, such as nails and hair-dryers, and therefore tend to use the cognitive resources developed for handling the biological world to handle the world of cultural artefacts, rather than the other way around. Moreover, many of the same devices used to organize the biological world have been found to apply to social cognition. A very simple lexical example of this, is the way in which binomials are used in both kinship and ethnobotanical terminologies for classificatory purposes. However, one crucial factor determining the differences in the way one domain is handled compared with another is the distinctive materialities of the objects included in those domains, for example in terms of stability of form, change over time, and the extent to which the constituent objects are 'life-like', as when comparing the abstracted qualities of things (e.g. colour), relationships (e.g. kinship) and things themselves (e.g. biota).

Nuaulu have no word for plants as a whole, but they do have a concept (Chapter 2). In Nuaulu experience the entire terrestrial world is apprehended through its vegetal properties, as these are made known visually, aurally, through taste, olfaction and touch, with each mode combining in different ways to constitute a sensorily distinctive phyto-materiality. No place or space can be other than vegetal in some of its dimensions and characteristics. Plants grow out of the ground, and the soil is composed of both living and decaying plant matter; seeds and leaves blow in the wind, plants grow in and out of water, on mountains and on the sea-shore: in many ways they define the limits of the Nuaulu world. Nuaulu apprehend the plant world through the totality of its qualities: what is vegetal is encountered phenomenologically – physical appearance, other sensory properties, utility and symbolism combined – though in any one instantiation a particular quality may come to prominence and determine the relevant classification. So, to distinguish 'plants' as either a semantic, cognitive, sensory or lexical domain is not always helpful, as each of these ways of dividing up what is perceived or conceived may reinforce any other, and in some instances contradict another way of defining the domain. Words in particular can be a problem, as their histories become embedded through language in contexts that subsequently change, and as lexemes move independently from their meanings. Considering plants as a 'knowledge domain' is even more complex, raising issues about where boundaries can be drawn. Thus, knowledge of seed dispersal and pollination inevitably involves knowledge of animal behaviour; and knowledge of nitrogen fixation of the not so obvious functions of plant roots, soils and, implicitly, microorganisms. Specifically ethnobotanical knowledge inexoribly becomes ethno-ecological knowledge. In other words, the real world is multidimensional. Our organs of sensory perception can generally cope with four dimensions. Our visual ability can manage three-dimensional space, aided by binocular vision. When we perceive the natural world, we do not do so in just two dimensions, neither is this the way in which data appear to be stored in the brain. Our obsession with two-dimensionality is a historial outcome of the evolution of our ability to inscribe representations on physical substrates, most obviously writing and other graphic arts, a secondary feature of human invention and cultural adaptation.

The overriding consensus amongst those who study ethnobiological classification today, is that the basic principles that organize the domains of plants and animals (or we might add fungi and other Eukaryota) are basically the same. But within this consensus is a group of scholars who have persuasively argued that folk biology is an innate cognitive module. Atran (1990:190, 1994) and Sperber (1994) have argued that this innateness is not theory-like (but rather 'pre-theoretical' common sense: see Chapter 3), while Keil (1979) has argued that while folk biology is an innate module it is also characterized by intuitive theory from the beginning of child development. By contrast Carey

(1987) argues that folk biology is not an innate module, but rather a framework theory that develops during the first decade of life, and which may be abandoned, or its assumptions severely modified, as increasing mastery of a domain is acquired. However, most of this discussion is in relation to animal knowledge rather than plant knowledge, and is of adult knowledge rather than that of young children. Carey also argues that while taxonomic organization and essentialism might be innate, they are unlikely to be specific to folk-biological classification. This is a view – of the limits of modularity – that I largely share, but before returning to it we must explore the possibility of differences between the folk classification of animals and plants in the Nuaulu case.

The fact that the biology of vascular plants and multicellular animals is in some important respects different suggests that this is likely to be reflected in naming and classificatory practices. Nuaulu know many more plants than they know animals, but a greater proportion of their animal world is named and classified than their plant world. The total number of plants that Nuaulu name excludes many others that have no Nuaulu name but which are found elsewhere, or for which they have no use, but are nevertheless aware that other people do (see Chapter 8, p. 124). Although some plants have features that invite resemblance to some animals (for example, with regard to the motility of some of their parts), the position of plants in the physical environment differs hugely from that of animals, partly because of their relative immobility and absolute dominance in terms of biomass, which inevitably means they are experienced as backdrop and context, rather than as objects moving around 'in an environment'. The experience of vegetation as a visually continuous phenomenon, together with ecological complexity involving the physical proximity and mixing of multiple plant species, means that, perceptually, one natural kind in the landscape can seem to morph into another. Trees, in particular, somehow stand midway between herbs and rocks as (semi)-permanent features in the landscape.

Moreover, the reproductive and life cycles of plants and animals differ in ways that influence their classification. For example, animals in many of their characteristics resemble humans more than plants do, in terms of body organization, symmetry and fundamental life processes. Even Atran (1990:74) seems to concede the point that 'because humans and animals are adjacent overlapping domains, then one might expect children to borrow from knowledge of humans to organize animals and plants, and vice versa'. This experience and knowledge influences Nuaulu in their use of human-like qualities when describing plants. Thus, while many plants may be attributed with agency and have spirit, no actual plant or type of plant that humans engage with is considered to have ever been human, and no human has even been a plant as such, unlike animals and humans, who readily interchange in Nuaulu myth and everyday experience. The closest we find are plants with totemic functions associated with myths in which humans emerge from them (e.g. sago), or different body parts in myth growing into different species of useful plants. Prominent amongst these, is the myth of origin in which certain Nuaulu clans successively emerge from the internodes of the bamboo ***tomone***. Notwithstanding, Nuaulu do not see spirits of bamboo in the way that they claim to see spirits of pythons or of certain birds. Such beliefs provide cultural reinforcement for the conceptual constituation of plants as a classificatory entity separate from animals, while challenging the idea of its simple psychological innateness. Although Nuaulu are aware that animal species can cross-breed, something they recognize in, for example, birds, the gaps between animal species are greater than those between related types of plant, where hybridization, especially in domesticed crops such as *Musa*, *Citrus* and the cucurbits, is recognized as intrinsic to modes of reproduction, and where sometimes knowing where one species begins and ends is very unclear, both in science and folk science.

This discussion brings us back to a core issue in cultural cognition: whether or not we can recognize patterns in different semantic domains to sustain the underlying notion of domain specificity. For example, are there differences between domains constituted largely through physical objects (such as plants and animals) and those that are more abstract and ideational, such as religion? As Whitehouse (2004:58) has shown, there may be several modes of transmission operating simultaneously in the domain of religion, with different cognitive architectures and emotional glue, whether liturgical or experiential. Humans absorb certain kinds of knowledge more readily than others, even within the same cultural domain. Thus, the transmission of cognitively costly aspects of religion may be very different from that of language or minimally counter-intuitive concepts. Children learn different animal and plant names, learn how to use personal names (see e.g. Hunn 2008:78), develop recognition routines, and know that plants, animals and humans constitute separate domains, at an early age, before they have a notion about, say, colour differentiation. Hunn (2002, 2008:225–6) reports such a case of 'precocious acquisition' for a seven-year-old Zapotec girl, who could identify correctly seventy-five per cent of the plants in a plant-trail experiment, and suggests that by the age of twelve children have a complete mastery of plant names. Such domain identification also occurs well before notions of religiosity. Indeed, some of the content of religiosity – belief in spirits – is often subsumed under natural-history knowledge, which provides a template for elaboration of different spirit entities. What we find is that the composition of different cultural domains, as they are conventionally and emically defined, depends on mixed strategies of transmission, although in urban-centred, literate and globalized society words and texts (whether written, spoken or electronically mediated) increasingly come to dominate. Moreover, any plant (or animal) that humans engage with inevitably becomes a cultural object rather than simply being a 'natural object' (see also Hsu 2010), caught up in a complex network of ideas transmitted socially. In other words, an organism becomes its representation, influenced by the relations a perceiver or user establishes between this entity and other similar and contrasting entities in their field of experience, and how this information is shared and transmitted with other individuals. This will influence how domain boundaries are drawn.

The mutuality of sociality, body imagery and plant knowledge

Biological-knowledge systems do not stand outside society, but are culturally and socially embedded. At the same time, the mutually constituted cognition of the natural and social worlds is part of how the human mind and body have co-evolved (Ellen 2017a:67). Plants as a group provide humans in all environments with a fundamental model of how life, beyond that which is distinctively human, works. Animals of course are phylogenetically closer to humans and easier to anthropomorphize, but because the macro life processes of animals (their growth and reproduction) are as a rule slower, and because they move, are less easy to observe and manipulate, they are less accessible to detailed and repeated scrutiny compared to plants, which are more compliant and offer less resistance. Plants provide a convenient model of organic transformation, of fructification, growth, death and decay. Although animals are closer to humans in their overall body plan, physiology and development, plants are still sufficiently close to allow the sharing of some common vocabulary. Thus, some Nuaulu terms used for plant morphology also occur as body terms (leaf and earlobe or auricle: ***tinai totue***; skin and epidermis: ***unte***; human heart and male banana 'flower': ***usue***). Among the Lio of central Flores (Sugishima 1994), trees, like humans, have hair, feet and blood. Fox (1971; see also Forth 2001) famously developed the notion of botanical idiom, firstly in relation to the Rotinese, and secondly (1980:333) as a distinctive metaphor used by peoples of eastern Indonesia when talking about marriage and kinship and social relations in general, with the notions of 'tip' and 'trunk' (or 'root'), or 'flow' of sap with regard to blood or life force. In Chapter 2 we introduced the analogical use by Nuaulu of the terms ***atue*** (tip), ***aikune*** (trunk), ***tinaie*** (stem) and ***namte*** (root), and these are often used in various combinations when talking about descent. Nuaulu will say, for example, we belong to a common trunk (Ellen 2018a:53). Similarly, Molnar (2000:90) shows how the Hoga Sara of Flores employ trunk, branch and tip and root metaphors to elucidate clan structure. Comparable plant idioms are described by Geirnaert-Martin (1992) for the Laboya of west Sumba, Barraud (1979:154–5) for Tanebar-Evav in the Kei islands, Clamagirand (1980) for the Ema of Timor, Platenkamp (1990:76) for the Tobelo of Halmahera, and by others. Moreover, plants and their growth patterns and physiological cycles enter into social life in ways that mark and measure the passing of time, from the phases of the moon that determine sap production and thereby the harvesting of bark-cloth, through the social implications of the growing season of crops in swiddens, the swidden cycle itself, to the long-term growth patterns of trees. The time to maturity of many significant plants (especially managed species such as sago) is well known and provides a fixed period that can be used to coordinate social events. Different species are linked to different social cycles: the annual cycle of the swidden, the fifteen years it takes for sago to mature, the twenty-five years for a timber-producing tree in a ***sin wesie***, or the sixty years for a coconut in a grove.

The prevalence of the botanical idiom is hardly surprising given the intimate relations all eastern Indonesian peoples have with plants and their parts (cf. Barnes 1974:231). Sometimes the idiom is plant-type specific, as when the Tana Ai of Flores (Lewis 1988:75) describe the significance of the 'eye' of a bamboo, that is the point from which a branch will emerge from a node, in contrast to the base of a branch (emphasizing substance and durability), and recapitulate this in ritual; and how the cleavage between brother and sister that comes with marriage needs to be clean, like splitting rattan without breaking it (p. 207). In the case of the Nuaulu, apart from generic plant metaphors such as ***tinaie*** ('trunk' or 'stem') to refer to the dominant line of a clan, it is the sago palm, with its numerous suckers and phenotypic continuity over generations, that is most widely used to talk about descent lines. Thus, clan segmentation is compared to the removal of suckers from a parent tree, and siblinghood to the relationship between suckers from the same palm, a figurative language widely found amongst the sago peoples of lowland Melanesia (e.g. Gell 1975:144). Sometimes the metaphors work the other way around, and instead, as amongst the Wola of highland Melanesia (Sillitoe 1995:208), social terms are used to talk about plants. This practice is exemplified here in the way Nuaulu use basic kinship (and human anatomical) terms as plant partonyms and as ways of describing plant groupings (Chapter 2, pp. 25–34).

Plants are fundamentally known through the uses that are made of them, or through their potential for such uses, and we must examine critically the claim that plants or any other organism might be perceived 'directly', intrinsically and innately as 'natural' entities, unmediated by cultural foreknowledge or context. They are in this sense what Gibson (1979), and Ingold (1986:2–7, 2000:166–8) following Gibson, mean by 'affordances' or action possibilities. So what Berlin has called special-purpose groupings are often in part categories, the salience of which is accentuated by them being affordances. This observation connects with the idea developed above, that plants entering human knowledge systems are no longer simply 'natural' objects, but are already part of material culture. Until relatively recently, virtually the entire manufactured world of the Nuaulu was made of plant materials that they had themselves collected and processed, in an economy where plant and technical knowledge are widely and evenly distributed. Even where there were long histories of trade and exchange involving acquisition of materials of other kinds (such as textiles and metal objects), there were generally good local alternatives: for cooking, bamboo internodes instead of ceramic pots; for cutting, bamboo and stone knives; for tying, lianas rather than polypropyline rope; for weaving, pandan leaves; for wearing, bark-cloth. And if the perception and identification of plants entails recognition of them as affordances, how plants transform through growth, also resembles the process by which they are transformed through making. As Ingold (2000:85–8) reminds us, in many cases growing and making are consubstantial processes, and the conceptual distinctions between them fluid. Some words are the same for the plant as for the objects that are made from them (as in the various forms of cordage described in Chapter 5); while in other cases, when transformed in the context of use the plant species is given a different name (***akunin*** > ***asinokoie***: *Actinodaphne* probl. *moluccana*); or the name of the product (say a fruit) is metonymically given the same name as the whole growing plant (***pala***: *Myristica fragrans*). As we have explored in Chapter 7 more generally, utility can profoundly influence how (particularly) more-inclusive groups of plants are conceptualized and named.

We may still wish to argue that there are special features in ethnobiological classifications that distinguish them from the organization of other domains, and that different domains require classifications with different properties, part of which is the extent to which the taxonomic model might apply (Holman 2005:83). A quantitatively ambitious attempt at such a comparison is Holman's measuring of the 'shape' of classification (breadth versus depth) with a view to testing the extent to which the natural-history domain shares organizational features with non-biological domains. Nevertheless, I would argue that any differences have less to do with the cognitive apparatus brought to bear, than with the pattern of empirical discontinuities found in a particular ecological context. For example, the mind cognizes tools in the same way as living organisms, partly because of a shared cognitive architecture but also fundamentally because our models of apprehending and thinking about the world derive from our own bodily experience, and from the social worlds of which we are part (Ellen 2017b). Indeed, it has long been recognized that at the core of human cognition is a necessary duality and tension whereby humans understand the natural world through their experience of society, and the social world through their experience of nature (e.g. Lévi-Strauss 1962). This is why, despite repeated attempts to counter naive dualism and challenge the culture-nature divide, the divide keeps on re-emerging (Astuti 2001; Leach 2000:340). Thus, there is a general tendency in human relations with the inanimate world to attribute and represent that world in organic terms, and to attribute inanimate objects with the properties of living things (Ellen 1988). It happens because we are bound to model our world directly on our experience of our own body (Mauss 1973), and we employ this selfsame model as a source of labels and concepts to interpret the world outside the body. The lexicon of animal parts is, after all, for the most part that of human anatomy. As we have seen, botanical nomenclature is less concordant, and that of inanimate objects still less, but body terms – or at least terms that appear concurrently in anatomical lexicons – are still crucial (Ellen 1977).

More than this, we tend to 'thingify' or 'entify' parts of a living system. Thus, phenomena that have life are turned into objects, only be reanimated in turn (Ellen 1988:223). The organic models we use vary along a continuum from general organic analogies (organomorphs), plant analogies (phytomorphs), animal analogies (zoomorphs), and general human analogies (anthropomorphs) to the attribution of particular personalities (personification) (ibid.:224). It is as if the mind progressively 'enlivens' non-living entities, and humanizes other entities in which it recognizes life. I have elsewhere (Ellen 1990, 2018c) illustrated this point in relation to the personification of sacred shields manufactured from ***kawasa*** (*Archidendron clypearia*) that Nuaulu attribute with soul (***matiunu***, ***nemati***). ***Kawasa*** wood is treated anthropomorphically at every stage in the production process and throughout the subsequent social lives of the shield. The first chip to be cut from the tree when making a shield is treated as the breath and the soul of the wood, and is reunited with the completed shield in the storage loft of a sacred house where it normally resides.

Because the role of the social world in the way we organize and understand nature is so entrenched in ordinary thought and discourse, this has been taken by some as good evidence that the faculties of social cognition have evolved in humans to eclipse other modules of knowledge (Mithen 1996): at some point in our evolutionary history the barriers between hitherto specialized intelligences distributed as quasi-separate neural networks in the brain – natural-history cognition, intuitive physics and others – have broken down and merged with the hegemonic module of social cognition. Regardless of whether or not such claims can be upheld, if we are to understand those cultural behaviours we describe as 'animistic' we must take into account a fundamental fact of human cognition: that we use social intelligence to make sense of the natural world. What we call 'animism' is in a way the reification of multiple instances of such thinking, of repeated observations reinforced by cultural elaboration into something approaching a coherent set of connected beliefs.

How knowledge practices are transmitted depends on their form. While all cultural transmission relies on some kind of bodily interface, what varies is the extent to which, language, and different forms of language, are involved in the process. Many bodily practices are learned through self-discovery or copying, reinforced by parental actions or those of other significant persons. In this sense, much knowledge is what I have called elsewhere (Ellen 1999) 'substantive', meaning that it may be quite complex and extensive, but is not in itself ordinarily committed to language, though it may emerge through performance. This is the case with much folk-botanical knowledge, such as that concerning plant maturation and ecology. Think for example of knowledge of weeding, in which gardeners are well aware of the properties of individual weeds and the need to remove them, but do not always know their names. People acquire this knowledge through a combination of long-term experience and interaction with plants, and occasional social interaction; but while bound by implicit rules or scripts, the knowledge is seldom systematically organized linguistically. By contrast, 'lexical' knowledge is that part encoded in language, or where the language provides a key for accessing substantive knowledge that is not itself lexicalized.

Plant classification and the meaning of life

Nuaulu ideas about animacy and agency (Ellen 2017b) are – like our own – generally fuzzy and contingent. We can easily see this by combining data from different kinds of ethnographic context and by using different elicitation procedures, when a more complex picture always emerges. However, if we look at scientific definitions, there have been repeated attempts to define the boundaries of living matter, and what this might imply. These approaches come from philosophy, developmental/cognitive psychology, recent anthropology and, particularly, ethnobiology or ethnoscience. Let us start with the ethnoscience model.

Berlin (1992) describes the concept of 'living things' as a UNIQUE BEGINNER in a general taxonomy of biological organisms. For example, Taylor (1990) in his work on Tobelo ethnobiological classification uses a strict componential form of semantic analysis. Quite reasonably, he begins by indicating that Tobelo distinguish living from non-living, defining the former as entities with the characteristic of being able to die. But rather than talk of living things, he uses the term 'biotic forms', which he then says are subdivided by Tobelo into 'sexual biotic forms' that contrast with four other groups at the same taxonomic level: coral, sponges, fungi and (a single category) 'moss, mould, bryozoans, small algae'. 'Sexual biotic forms' are in turn divided into 'breathers' and 'non-breathers', the latter exemplified by seaweed and black coral, and the former by fauna and flora.

Edmund Leach (1964) approached the issue in a different way, but with much the same result. He makes no reference to the ethnoscience literature, and would probably have been scathing in his dismissal of it if he had. Instead, he adopts a kind of logical formalism derived from Lévi-Strauss and Mary Douglas. The main problem with Leach's work, however, is not his underpinning theory, but his methods, or lack of them. Leach (1964; see also Ellen 2017b:249) draws a diagram based on his own imagination, augmented by his knowledge of the role of certain oppositions and notions from English popular culture and sacred texts. In doing so, he uses the category 'nature' in a way that he has presumably derived from Lévi-Strauss, though he is sceptical of naive dualism and ambiguous on the universal proclivity to a nature-culture opposition (e.g. Leach 1972, 2000). His 'nature' is, therefore, not the same as 'life' or 'living organisms', or even less Taylor's 'biotic forms'. He divides it into animate and inanimate, though we are left to speculate whether this means that the category 'inanimate' includes some living organisms, most obviously plants. His category 'animate' (lexico-logically) applies only to animals, given that it is then subdivided into warm-blooded and cold-blooded. However, although I have often used Leach's folk English scheme as a useful teaching example, it has always struck me as highly problematic. Not only does it suspiciously provide a convenient taxonomy of all living things that fits the logical conventions of a taxonomic approach, it conflates discourses that we know to be separate, and flies in the face of ethnographic evidence. Unfortunately, while 'living things' is a category that is easily yielded using formal elicitation techniques, it is less obvious once you aggregate data from different ethnographic contexts using different research procedures.

Possibly because of the pitfalls in defining it, or in order to avoid the problems of using it as a unique beginner in a taxonomic description, many accounts of ethnobiological classification simply avoid considering what might be meant by a 'living thing', and get on with the business of describing categories within the separate plant and animal domains. This approach is adopted by Hunn (2008) in his recent Zapotec natural history, and not even Berlin (1992) – while adopting the overriding rubric of 'ethnobiological classification' and dwelling on the formal criteria for establishing FOLK KINGDOMS of (often nomenclaturally covert) plants and animals – is interested in considering the unique beginner that might define the domain of 'living things' overall. Such approaches, however, logically (that is inferentially or syllogistically) still assume that 'living things' must exist as a categorical phenomenon. Thus, if plants are living things and animals are living things, then there must be a superordinate or more encompassing category that both belong to. But this definition assumes that we are only dealing with what might be conventionally understood as 'biological taxa'.

Nuaulu distinguish entities that have many of the qualities of life, including material objects, such as certain tools, but also other phenomena, which in the ordinary way have no corporeal existence. Thus, Nuaulu – like many peoples – think of spirits as a kind of pseudo-organism and represent them as species-like entities that can be ordered in terms of family resemblances (Ellen 1993:176–90), even if, as Boyer (1994:97) argues, their conception as such violates the intuitive principles that some psychologists (e.g. Keil 1979) have demonstrated for the domain specificity of living kinds. Because spirits have these qualities, they are sometimes also attributed with corporeal manifestations: for Nuaulu they become birds or lizards, or indeed certain animals may always be regarded as the physical manifestations of spirits, such as scarab and longhorn beetles, or the death adder (*Acanthophis antarcticus*). In a way this is not surprizing, as we can only imagine the spirit world through our experience of the physical and social world.

It was difficulties of this kind – both ontological and epistemological – that preoccupied Scott Atran in much of his *Cognitive Foundations of Natural History* (1990). For Atran there is only a superficial similarity between biological and non-biological domains, 'living things [... being ...] everywhere ranked into transitively structured taxonomies, with no other natural-object domain so structured' (ibid.:47). In concluding thus, he draws on the support of field experiments in child psychology (ibid.:50) and the work of Keil (1979). Keil has argued that children possess an ontological category of living things that includes animals and plants, allowing Atran (1990:73–4) to claim that young children 'categorically distinguish artefacts from living things' and come to presume that only the latter constitute 'natural kinds' with underlying essences, while limiting certain concepts (such as growth) to living things. For Atran (ibid.:5) there is a significant contrast both within the 'ordinary categorization of artefacts and the extraordinary scientific classification of living kinds'. Thus science is in this reading based on a kind of universal ethnoscience in which 'our universally held conception of the living world is both historically prior to, and psychologically necessary for, any scientific – or symbolic – elaboration of that world' (ibid.:13).

As well as using experimental data, Atran (1990:55–6) argues his case on formal logical grounds, assuming this to be a universal grammar. Thus, 'furniture' cannot be part of the definition of 'chair', although 'tree' is part of the definition of 'oak'. Transitive hierarchy, he therefore reasons, works for living kinds but not for artefacts, because the domain of artefacts fails to meet the inductive and deductive requirements of ranked taxonomies. Biota, like artefacts, are often placed in different categories (whether we call these taxa or not), and although he recognizes that people often 'confound' artefacts with living kinds and 'confuse' plants with things made from plants, the underlying domain structure is quite different. But one reason that artefacts are not conveniently organized taxonomically, is that Atran has defined taxonomy in such a way that it can only be used for biota (1990:57). If we are to argue on the basis of logical formalism, then we might forgive ordinary people from deviating from it in the practices of their ordinary lives. No wonder Atran (1990:71) dismisses that work in developmental psychology (e.g. Carey 1987; see also Gelman 2003) that shows that children will spontaneously attribute a common invented property, a kind of underlying nature, to dogs, flowers and inanimate objects. Such theories and data tend to support the idea of the integration of domains of knowledge, and of fundamental ambiguity in the concept of 'natural kind'.

Thus, through the working out of the recent evolutionary history of the embodied brain, and polythetic linkage between different cognitive and semantic domains in any particular cultural population, we will find a group of things that are regarded as like life, or life-like. But in attributing the notion of 'life', and more narrowly 'animacy', what features does the mind latch on to? The evidence of the previous section would suggest that we start with physical resemblance. Given our dominant sense of vision, it is visually salient attributes that first register and that are most easily encoded in memory. In other species it might be other senses that occupy this position, and even in humans these other senses can be critical in determining whether something is likely to be living. Imagine a world in which life is mainly apprehended and registered through the senses of sound or tactility (which we can perhaps imagine, especially if we are blind) or through smell (which most people can only imagine with difficulty). In reality, of course, our cognition of life is multi-sensorial and synaesthetic, but we must start with physical form. Nuaulu adult perceptions of plants are fully multi-sensorial – touch, smell, taste (and even sound) being quite as important in identification as visual clues, though not evenly so. I say 'adult perceptions' because early learning routines abhore ambiguity and uncertainty; but as humans acquire more knowledge, so the range of means to identify X as Y increases, along with a more nuanced and circumspect understanding of difficult cases.

Some of the classic boundary problems of the category 'living matter' are raised through an engagement with forms of life that do not fit easily into the cognitively universal (but not lexically universal) prototypical categories of 'plant' or 'animal', and which are peripheral to biological life as most ordinary people experience it. When introducing what Nuaulu understand by a plant in Chapter 2, I reviewed some of these issues in relation to fungi, algae, lichens, slime molds and some marine invertebrates. For Nuaulu, all these forms are undeniably living, even though some seem to lack movement, others much evidence of growth, and others still any capacity for reproduction. Movement in the narrow sense is easily observable, growth less so, but reproduction in many cases has to be inferred. In most folk biologies, while sexual biotic forms clearly reproduce, entities such as seaweed, lichen and sponges merely endure.

Our different senses register different qualities on the life/non-life continuum, but combined offer a sufficiently discriminating instrument. So, by deliberately selecting liminal forms we can test the extent to which the notion of 'life' applies to them, and if that test is passed, whether such entities conform best to animal or to plant prototypes. But we also need to recognize that both these prototypes are in semantic tension, simultaneously sharing features and contrasting them. All living matter could be said to have both a vegetal and animalistic aspect, there being a 'vegetal quality' especially found in plants, and an 'animal quality' especially found in animals. This is why the placement of certain peripheral organisms in classificatory space varies between cultures, compared with the greater regularities reported for core plants and animals. But it is also reflected in the evolutionary convergence of forms that we might otherwise have no problem is assigning to one or other of the taxonomic kingdoms. Thus, for Nuaulu, certain plants have attributes of animals, such as the insectivorous pitcher plants (***koitipi***: *Nepenthes* spp.) found in the Manusela National park, and certain animals the attributes of plants, as in various phasmid stick and leaf insects (***kau ai otoe***, e.g. *Platycrana viridana*). But in addition, we are constantly primed to look out for what is potentially human in other parts of the living world. We try to make things anthropomorphic, and while the basic attributes of morphology may make this more promising in the case of, say, the Nuaulu marsupial cuscus, with its superficially simian anatomy and multiple symbolic resonances (Ellen 1972), conservationists struggle to convince us of the humanity in the hapless pearl mussel (Carrithers, Bracken and Emery 2011).

In the attribution of life, and even more so of animacy in the sense of 'animality' (Reed 1988), morphological resemblance is not enough, 'motion' is often necessary, though seldom sufficient, in all or any of its manifestations, and is often accompanied by multi-sensorial – often synaesthetic – characteristics that are shared with biological life, for example the expenditure of heat and the emission of sound. Everywhere liminal biological forms are attributed with animacy, it is because in some way or another they display a characteristic that is semantically rooted in movement or its metaphorical extension, such as 'locomotion', 'growth', 'reproduction', 'fission', 'fusion', 'fragmentation'. Even 'eruption' and 'erosion' are recognizably, for all humans, 'kinds of motion'. When we talk about the 'living' landscape we have in mind the idea that it is dynamic, whether vegetally, animalistically or geomorphologically so. Time, change and cause are all described in various contexts as if they were like motion. But to qualify for the condition of life, the source of motion has to be independent of any interlocutor.

We now come to agency in the strict sense. As Gell (1998:72) puts it, 'agency implies the possession of a mind which intends actions prior to performing them' and 'what matters is where an animated object stands in a network of social relations' (ibid.:123). For Gell, there is only a slight dividing line between the intentionality of humans and of anthropomorphic objects (ibid.:132). The problem with this is that, conceptually, it mixes up agency not only with intention, but also with animacy and sociality. The term agency has been much stretched in recent anthropological discussions of animism and technology, but in this context I think we need to shrink it back to its core meaning: an intervention to produce a particular result, or something with 'the faculty of an agent' (SOED 1973:37).

A quality that Nuaulu commonly associate with the essentials of animal and human life is breath (***nahai***), and breath too is a kind of motion. When Nuaulu refer to ***nahai*** they are first and foremost thinking of human life, but this does not mean that they do not also readily extend the notion to other biota and non-biota. At birth, the moment at which the child breathes independently, when the chest begins to move and the lungs ventilate, is the beginning of autonomous human life; while death is the expiration of breath and the end of movement, and therefore is also (paradoxically) a characteristic of life. Breath as a concept is recurrent in ritual, and is attributed to physical entities that are anthropomorphized, such as sacred houses or ritual shields. So, when cicadas sing it is evidence of breath; even when the wind blows through certain rocks to produce eery sounds it may be taken as evidence of life, recalling somewhat the remarks of Herodotus on the Colossi of Memnon at Thebes.

Finally, we come to intentionality. In the Nuaulu world, humans, some animals, spirits, the godhead, animals and some plants that have become coterminous with spirits might be said to act intentionally. Their actions in the world do not simply have consequences independent of the humans that interact with them, they have minds that permit them to make plans and act in the world in ways that are deliberate responses to human and other behaviour, and which are often contrary.

I am hardly the first to argue that the attribution of life is necessarily gradual, contextual and, from a biological perspective, sometimes inconveniently deviant. Bloch (1998:53) calls this the 'more or less' character of life. In other words, it is a matter of degree. Ingold (2006) argues that we can make anything seem alive, though we do not always choose to do so. Moreover, there is much ethnographic evidence that people do not agree about what life is and no universal distinctions as to what is alive and what not. Ingold (2011:29) further speaks of 'bringing things to life', of 'things in life, rather than life in things', suggesting that life is less a property of individual entities than a phenomenon of which those entities are part. In this model, the animacy of the 'life-world' is 'ontologically prior to their differentiation' (Ingold 2006:10). What I have tried to do here is to unpack those conditions for the recognition of life and to show how they might constitute a series of progressive cognitive steps that when aggregated are more likely to prompt the attribution of life. These are: recognition of morphological resemblance, recognition of motion, and recognition of independent motion.[1] As we have seen, these are each reflected in a pattern of conceptual attribution that follows a broadly phylogenetic progession: we apprehend physical entities as being plant-like, animal-like, human-like and – ultimately – like individual human persons (Ellen 1988).

Part of the problem in the literature is that the discourses on bio-cognition, life and animism begin from different starting points and have different intellectual histories. While analytically we need to separate these, and certainly not confuse and conflate the concepts and terms employed, in particular ethnographic cases it is unlikely that we will ever discover a convenient congruence. It may be that we need to separate vital from symbolic forces – that we need both 'vital and symbolic ontologies' – just as Mauss suggested we separate technical from symbolic classifications, and Berlin general-purpose 'natural' schemes from special-purpose schemes. Unfortunately, real-life examples do not give us much hope that this is possible. In some cases we may find evidence of two life forces (spiritual and biological), as Rival (2012) suggests for Makushi cassava, but my informed guess is that because vital and technical processes are already situated in relational complexes when we experience them, in practical everyday life organisms and things are treated as though they are motivated by a single underlying force, unless otherwise prompted by a particular tradition of theological casuistry. This would be my reading of the Nuaulu data.

While all cultural objects are potentially regarded in this way, in practice some objects are more alive, that is, are more likely to evoke the characteristics of agency or indeed intentionality than others. Part of the problem with existing anthropological treatments of the category 'living things' is that they are either logical extrapolations through polythetic extension, based on formal ethnotaxonomic deduction/induction, rooted in observations of how children under artificial conditions perform in field experiments of a very abstract kind; or – in contrast – assume that life is recognized phenomenologically prior to its differentiation. Nuaulu ideas as to what is animate and agentive are always fuzzy and contingent. By combining data from different kinds of ethnographic context, using different elicitation procedures, a more complex picture emerges.

The same applies to animism, with which the attribution of life is too readily conflated. If, following Descola (2005:183–4), animism is 'the granting by humans to non-humans of an interiority identical to theirs, an attribution that humanizes animals (and plants)' (Rival 2012:70), I would say that it is virtually impossible to separate ethno-theories of life processes from what is sometimes described as animism. However, not every object in nature is animated, and animism is not totalizing. Recent work (including my own) suggests much selectivity with which animism is evoked ontologically, and limits to the extension of personhood as a humanlike category. Praet (2014) even argues that the outstanding feature of animism is its peculiar restrictiveness.

Methodological reflexivity and ontological friction

Arguably, this book has been a work of theoretical deconstruction rather than one of construction, though not I think one of destruction. Reverting to its central theses, I am in favour of a context-rich approach to understanding how ethnobotanical classification works. I would contend that 'natural classification', as endorsed by Berlin, Atran and Boster (e.g. Boster 1996) is better seen as a fallback mode of organization, a classification of last resort rather than of first resort, hidden in the background of day-to-day perception, only becoming apparent and necessary occasionally in the working practices and ontologies of non-scientific systems of practical knowledge. Otherwise, it is visible only when stripped of historical accretions in circumstances where plant classifications begin to be simplified to achieve the objectives of science, or formal natural history, as we begin to see in the work of Rumphius. The domains that emerge as a consequence, do so aided by the perception of physical difference and similarity in a particular ecological context, but only due to the social and technical forces impinging on them, rather than because they are inherently innate. The 'radical simplification' (Ellen 2019a) entailed in producing classifications such as those of Linnaeus, based on reproductive-organ morphology, cladistics, DNA bar-coding, or even Berlin's universal formalist model, can only be achieved by removing content and organizational features that are important for people using classification on a routine basis in most non-scientific circumstances. While helpfully consistent with the wider aims of theoretical evolutionary biology, the claims for the existence of 'natural classification' can introduce distorting

1 Compare this with a similar progression noted by Rival (2012:71) in the contributions from Ingold and others in the animism debate regarding the logical implicatory relationships between communication, intentionality, consciousness, life and movement. The problem, as Rival observes, is in a 'hazardous slippage' between concepts.

artefacts when we attempt to use them to understand folk schemes. The simplifications required of Atran's cross-cultural approach, and especially the assumptions necessary in order to quantify it, are so radical that they arguably present a picture that knocks out much of the operational meaning for local people.

Recent developments in the anthropology of environmental perception and engagement have tended to move beyond cognitivism, and implicitly to disparage it. These new approaches – for example studies of individual people/plant entanglement, analyses using actor-network theory (ANT), an emphasis on plant materiality and agency, and a consideration of plants as an emotive domain (e.g. Calvet-Mir and Salpeteur 2016; Ryan 2012; van der Veen 2014; and for a summary Ellen 2016) – provide valuable additional tools for the better understanding of classification, but with a much greater accent on methodological reflexivity and the wider context of political ecology. Tsing (1999:186–7), for example, has looked at how nature is made into a 'utopian space of collaboration' through the practice of naming trees. She notes that naming nature in Mankiling in south Kalimantan is collaborative, and critiques 'botanizing' as an insidious imperial practice resulting in the creation of texts in which the collaborating local people are ignored, and which nevertheless leaves a subaltern trace. Using Mankiling botanizing texts that focus on a list of plants, she observes, as others have done before her (Ellen 1996a), that the tree lists of scientists contain both local and scientific names, but are simplistic, often giving a poor idea of how local people really name trees.

The ecology of ways in which people interact with their world is the same for every cultural population on Earth. People see, feel, smell, hear and taste plants. Nuaulu see and cognize the shapes of plants: they feel the difference between the rough bark of a *Tectona grandis* and the fine stinging hairs on a nettle; they smell the pungent aroma of durian and the sweet smell of *Cananga* wafting from a freshly plucked flower placed behind the ear of a lover; they hear the cracking sound of a dead dipterocarp as it crashes through the canopy after being hit by lightning, and the thunderous thud of a 16-metre tall sago palm that, several years after flowering and rotten to the core, finally falls against surrounding palms and sinks into the swamp; they taste the astringent bitterness of *Alstonia scholaris* wood used in an anti-malarial cocoction, the sourness of the lime used to flavour gravy that makes sago porridge palatable, the piquancy of chillis and the sweetness of mango. What makes the Nuaulu different in their engagement with their plant world, is that all these sensory inputs are interpreted in a particular ecological context: the physical appearance of a particular species is inevitably compared with other plants in a given field of vision or in a landscape travelled through over time; the tactile features of plant materials are inevitably interpreted through the range of all other materials experienced, as are sounds and tastes. So, the acousticality of a lowland tropical rainforest comprising a particular composition and diversity is different from the acoustical features, say, of a prairie or an arid zone, or of an English country garden. But in addition, the perception and interpretation of plant species is instrumentalized and particularized, and their juxtaposition in a landscape is often the deliberate or inadvertent consequence of human interaction over many hundreds of years. The meanings and uses of plants are encoded in language, and how they are used in combination with other plants gives them new complexified references that are understood in relation to those imagined worlds of myth, spirits and ritual that give structure to a totality of experience. Nuaulu see the world in terms of the relations they establish with individual types of plants and in terms of the relations connecting those plants.

If we look at folk understandings of plants and how they are organized as part of a system of knowledge in a particular place or cultural tradition, there is seldom evidence of the kind of coherent and self-conscious overarching ontology that we associate with a Linnaean taxonomic framework, though there is plenty of evidence of inferential causation based on biological resemblance of the kind we associate with science (Atran 1998). Thus, any 'scientific common sense' is located less in a shared 'pre-theoretical' framework of names and arrangements (ethnobotanical classification itself), than in the the universal cognitive apparatus of experimental thinking and inference. Moreover, in many cases the descriptions ethnobotanists provide are partial, and simply reflect the questions they ask of their data; and what's more are bound to reflect and refract the scientific ontology that they themselves operate within. Unless there is a reflexive tradition of science among the subjects being researched, the very idea of 'plant classification' as a separate department of knowledge is likely to seem very strange. But local people may come to endorse new ways of (scientifically) representing their plant knowledge, as these are valorized as legitimate official representations and become recognized as a practical means of storing and transmitting knowledge (e.g. Miller 2016). My interest, and that of other scientists, in Nuaulu knowledge has certainly made some Nuaulu more conscious of something called 'plant knowledge', reifying it as a distinct subject; whereas previously plants were just embedded and diffused through everything else they know and do. During the rainy months of 2015 as I sat in the shelter of the Rouhua ***numa insi*** with Iane, Saleh and others discussing the 'wants list' I had brought with me in order to complete my herbarium collections, I became an irrelevance in the conversation. They had learned to 'botanize' just as they had also learned to write down the names of plants I was seeking. They discussed amongst themselves the different names, their properties and where to find them, organized them into lists and groups, and had thus become instrumental in transferring knowledge from the older illiterate generation to the younger less knowledgeable but competently literate generation of self-conscious bearers of traditional knowledge in a modern Indonesia.

APPENDIX 1

List of plant taxa and supporting vouchers for Nuaulu plant terms discussed in the text, arranged phylogenetically

Class, order and family are provided (from left to right) under 'taxonomic information' in the following table. The arrangement of these higher taxa follows APG (Angiosperm Phylogeny Group) IV (Stevens 2001). Genera and species are listed alphabetically within families, except for rattans (which are grouped together within ARECACEAE), and except where (in some cases) this would separate species labelled by the same Nuaulu name. Sometimes old family names are included as synonyms, such as PALMAE and COMPOSITAE, where these are still widely used. The first genus and species names are those given in THE PLANT LIST (an online collaborative project by the Royal Botanic Gardens at Kew and the Missouri Botanic Gardens in St Louis) at the time the appendix was completed. Names in square brackets are those provided at the time of the original determination by the identifying institution, or any subsequent revisions provided by those institutions where they differ from THE PLANT LIST name. Organization of pteridophyte taxa follows the revision of Christenhusz and Chase (2014). In the absence of voucher specimens, the following additional sources have been consulted (with abbreviation used placed in brackets): Bolton and Matoke 2005 (BOLTON & MATOKE), Coursey 1967 (COURSEY), Dransfield *et al.* 2008 (DRANSF), Heyne 1913–17 (HEYNE), Matthews 2014 (MATTHEWS), Mercy Morris (MORRIS), PROSEA 1992–2003 (PROSEA), Rumphius 2011 (RUMPH) and Whitmore *et al.* 1989 (WHITM). For a discussion of identification procedures adopted see Chapter 1, and for other abbreviations used here see separate list in Appendix 5.

Taxonomic information	Voucher code or other source of identification		Nuaulu name
Lycopodiophyta			
Lycopodiales			
Lycopodiaceae			
Lycopodiella cernua (L.) Pic. Serm. [*Lycopodium cernuum* L.]	96-819	KEW	***soka-sokae numte*** syn. ***mono numte***
	96-818	BOGOR	
Lycopodiella cernua (L.) Pic. Serm. [*Lycopodium cernuum* L.]	96-820	UKC	
	96-992	BOGOR	
Tapeinidium amboynense (Hook.) C.Chr.	96-993	KEW	
Tapeinidium amboynense (Hook.) C.Chr.	96-994	UKC	
Lycopodiella cernua (L.) Pic. Serm. [*Lycopodium cernuum* L.]	71-55	KEW	***numte***
Lycopodiella cernua (L.) Pic. Serm. [*Lycopodium cernuum* L.]	71-75	KEW	
Selaginellaceae			
Selaginella P. Beauv. sp.	71-28	KEW	***soka-sokae***
Selaginella P. Beauv. sp.	71-28	KEW	***soka-sokae msinae***
Selaginella willdenowii (Desv.) Baker	96-980	BOGOR	***soka-sokae onate***
Selaginella willdenowii (Desv.) Baker	96-981	UKC	
Selaginella willdenowii (Desv.) Baker	96-982	KEW	
Selaginella P. Beauv. sp.		PROSEA	***soka-sokae mnatinae***

Taxonomic information	Voucher code or other source of identification		Nuaulu name
Pteridophyta			
Ophioglossales			
Ophioglossaceae			
Helminthostachys zeylanica (L.) Hook	96-1132	BOGOR	***utatone***
Helminthostachys zeylanica (L.) Hook.	96-905	BOGOR	***naene (totu) onate***
	96-1085	BOGOR	
Psilotales			
Psilotaceae			
Psilotum nudum (L.) Griseb	96-513	BOGOR	***soka-sokae marae***
Psilotum nudum (L.) P. Beauv.	96-514	KEW	
Psilotum nudum (L.) P. Beauv.	96-515	UKC	
Marattiales			
Marattiaceae			
Angiopteris evecta (G. Forst.) Hoffm. *sensu lato*	96-70	KEW	***muku(ne) sonone*** syn. ***sonone***
Angiopteris evecta (G. Forst.) Hoffm. *sensu lato*	96-71	UKC	
Angiopteris evecta (G. Forst.) Hoffm. *sensu lato*	96-71a	UKC	
Schizaeales			
Schizaeaceae			
Lygodium flexuosum (L.) Sw.	71-20	KEW	***isane***
Lygodium scandens (L.) Sw.	96-64	KEW	***sipoi***
Lygodium circinatum (Burm. f.) Sw.	96-65	UKC	
Lygodium scandens (L.) Sw.	96-63	BOGOR	
Cyatheales			
Cyatheaceae			
Cyathea Sm. sp.	96-66	BOGOR	***(ai) mukune***
Cyathea Sm. sp.	96-1374	UKC	***mukune***
Cyathea Sm. sp.	96-852	KEW	***muku(ne) putie***
Cyathea Sm. sp.	96-851	BOGOR	
Cyathea Sm. sp.	96-853	UKC	
Cyathea Sm. sp.	96-854	BOGOR	
Cyathea Sm. sp.	96-69	BOGOR	***muku(ne) metene***
Cyathea Sm. sp.	96-66	BOGOR	***muku(ne) msinae***
Polypodiales			
Pteridaceae			
Acrostichum aureum L.	96-1341	BOGOR	***naene (totu) onate***
	96-1342	KEW	
Acrostichum aureum L.	96-1343	UKC	
Pityrogramma calomelanos (L.) Link	71-36	KEW	***sina-sina wane***
Pteris semipinnata L. [*Pteris ensiformis* Burm. f.]	71-83	KEW	***pupu onate***
Pteris semipinnata L.	96-849	KEW	***mukune***
Pteris semipinnata L.	96-850	UKC	***mukune***
	96-1372	KEW	***mukune***
	96-1373	UKC	***mukune***
Pteris vittata L.	71-31	KEW	***sapana***
Pteris. L. sp		KEW	***kau-kau***
Pteris. L. sp	71-68	KEW	***kopane***
Aspleniaceae			
Asplenium nidus L.		PROSEA	***nite (unue)***
Chingia ferox (Blume) Holttum	96-762	BOGOR	***pupu hanaie***
Diplazium sp. prob. *esculentum* (Retz.) Sw.	96-1321	KEW	***one***

Taxonomic information	Voucher code or other source of identification		Nuaulu name
Diplazium sp. prob. *esculentum* (Retz.) Sw.	96-1322	UKC	
Pneumatopteris auctipinna Holttum	96-1320	BOGOR	***one***
	96-1321	BOGOR	
	96-60	BOGOR	***pupu***
Sphaerostephanos unitus var. *mucronatus* (Christ) Holttum	96-61	KEW	
Sphaerostephanos unitus var. *mucronatus* (Christ) Holttum	96-62	UKC	
Sphaerostephanos unitus var. *mucronatus* (Christ) Holttum	96-763	UKC	***pupu hanaie***
Sphaerostephanos unitus var. *mucronatus* (Christ) Holttum	96-764	KEW	
Stenochlaena palustris (Burm.) Bedd.		PROSEA	***naene***
Stenochlaena palustris (Burm. f.) Bedd.	96-1311	BOGOR	***nae(ne) tinse***
Stenochlaena palustris (Burm. f.) Bedd.	96-1312	KEW	
Stenochlaena palustris (Burm. f.) Bedd.	96-1313	UKC	
Polypodiaceae			
Colysis macrophylla (Blume) C. Presl	71-67	KEW	***sapuane, sapua(n) unue***
Drynaria sparsisora (Desv.) T. Moore	15-67	BOGOR	
	15-68	KEW	
	96-808	BOGOR	
Drynaria quercifolia (L.) Smith	96-1214	BOGOR	***naene putie***
Tectaria repanda (Willd.) Holttum	96-1215	UKC	
Tectaria repanda (Willd.) Holttum	96-1216	KEW	
Microsorum commutatum (Blume) Copel	71-30	KEW	***oune***
Nephrolepis biserrata (Sw.) Schott	96-663	BOGOR	***pupu pina***
Nephrolepis biserrata (Sw.) Schott	96-664	UKC	
Nephrolepis biserrata (Sw.) Schott	96-665	KEW	
Nephrolepis hirsutula (G. Forst.) C. Presl	71-32	KEW	***pupu kaia***
Nephrolepis biserrata (Sw.) Schott	96-1104	BOGOR	
Nephrolepis biserrata (Sw.) Schott *sensu lato*	96-1105	KEW	
Nephrolepis biserrata (Sw.) Schott *sensu lato*	96-1106	UKC	
Phymatosorus membranifolium (R. Br.) S.G. Lu [*Phymatodes nigrescens* (Blume) J. Sm.]	71-71	KEW	***kinekane***
Phymatosorus scolopendria (Burm. f.) Pic. Serm. [*Phymatodes scolopendria* (Burm. f.) Ching., *Microsorum scolopendria* (Burm. f.) Copel.]	71-18	KEW	***ahane***
	96-48	BOGOR	
Phymatosorus scolopendria (Burm. f.) Pic. Serm. [*Phymatodes scolopendria* (Burm. f.) Ching., *Microsorum scolopendria* (Burm. f.) Copel.]	96-49	KEW	
Phymatosorus scolopendria (Burm. f.) Pic. Serm. [*Phymatodes scolopendria* (Burm. f.) Ching., *Microsorum scolopendria* (Burm. f.) Copel.]	96-50	UKC	
	96-809	BOGOR	
Microsorum commutatum Copel.	96-810	KEW	
Microsorum commutatum Copel.	96-811	UKC	
Tectaria leuzeana (C. Presl) Copel. [*Pleocnemia leuzeana* (Gaudich.) C. Presl]	71-72	KEW	***muku msinae***
Tectaria leuzeana (C. Presl) Copel. [*Pleocnemia leuzeana* (Gaudich.) C. Presl]	96-67	KEW	
Tectaria leuzeana (C. Presl) Copel. [*Pleocnemia leuzeana* (Gaudich.) C. Presl]	96-68	UKC	
Tectaria leuzeana (C. Presl) Copel. [*Pleocnemia leuzeana* (Gaudich.) C. Presl]	96-67	KEW	***mukune***

Taxonomic information	Voucher code or other source of identification		Nuaulu name
Tectaria leuzeana (C. Presl) Copel. [*Pleocnemia leuzeana* (Gaudich.) C. Presl]	96-421	BOGOR	
Tectaria leuzeana (C. Presl) Copel. [*Pleocnemia leuzeana* (Gaudich.) C. Presl]	96-422	KEW	
Tectaria leuzeana (C. Presl) Copel. [*Pleocnemia leuzeana* (Gaudich.) C. Presl]	96-423	UKC	
	96-848	BOGOR	
other Polypodiaceae		PROSEA	***aha totu onate***
OTHER FERNS		PROSEA	***aha onie***
		PROSEA	***senene***
		PROSEA	***kaie***
Cycadophyta			
Cycadales			
Cycadaceae			
Cycas rumphii Miq. / *C. revoluta* Thunb.		PROSEA	***punen***
Pinophyta			
Pinales			
Araucariaceae			
Agathis Salisb. sp.		PROSEA	***kamane metene***
Agathis Salisb. sp.		PROSEA	***kamane putie***
Agathis sp. probl. *dammara* (Lamb.) Rich.	96-991	BOGOR	***kama(ne) wae***
Agathis dammara (Lamb.) Rich.	96-988	BOGOR	***ai nante***
Agathis Salisb. sp.	96-989	UKC	
Agathis Salisb. sp.	96-990	KEW	
Podocarpaceae			
Podocarpus rumphii Blum probl.		PROSEA	***ai yoko***
Gnetophyta			
Gnetales			
Gnetaceae			
Gnetum gnemon L.		PROSEA	***utatam***
Gnetum gnemon L. var.	96-394	BOGOR	***uta(tam) marane*** syn. ***utatam wesie***
	96-395	KEW	
Gnetum gnemon L. var.	96-396	UKC	
Gnetum poss. *latifolium* Blume	96-506	UKC	***ai tinae***
	96-505	KEW	
	96-504	BOGOR	

Taxonomic information	Voucher code or other source of identification		Nuaulu name
Magnoliophyta			
Magnoliales			
Myristicaceae			
Horsfieldia bivalvis Merr.	96-47	UKC	***ananione***
	96-48	KEW	
	15-83	KEW	
	96-246	BOGOR	
Horsfieldia bivalvis Merr.	95-540	BOGOR	
	96-541	KEW	
Horsfieldia bivalvis Merr	96-542	UKC	
Knema cinerea Warb.	15-88	BOGOR	***ai oi***
Knema cinerea Warb.	15-84	BOGOR	
Knema cinerea Warb., probl.	96-247	UKC	
	96-248	KEW	
	15-83	KEW	
	96-246	BOGOR	
Myristica Gronov. sp.	71-43	KEW	***asaherane***
	15-20	KEW	
Myristica Gronov. sp.	96-295	KEW	***nanu***
Myristica Gronov. sp.	96-296	BOGOR	
	96-815	KEW	
	96-816	KEW	
Myristica insipida R. Br.	96-817	UKC	
Myristica Gronov. sp.	96-834a	BOGOR	
Myristica fragrans Houtt.		PROSEA	***pala***
Virola surinamensis (Rol. ex Rottb.) Warb. [*Myristica* sp. probl. *fatua* Houtt.]	15-26	BOGOR	***ai unene, asaherane***
	96-201	BOGOR	
	96-202	UKC	
	96-203	KEW	
	96-518	BOGOR	
	96-1166	BOGOR	
	96-1167	KEW	
	96-1168	UKC	
Annonaceae			
Annona muricata L.		PROSEA	***tunene***
Annona muricata L.		PROSEA	***tune*** (*wa*)***rata***, ***tune***(*ne*) ***warata***
Annona reticulata L.		PROSEA	***ai pinamo***
Annona reticulata L., probl.		PROSEA	***ai pina ikine***
	96-238	KEW	
	96-239	UKC	
	96-237	BOGOR	
Cananga odorata (Lam.) Hook. f. & Thomson	96-324	KEW	***kupaleno***
	96-213	UKC	
	96-214	KEW	
	96-215	UKC	
	96-322	KEW	
	96-323	BOGOR	
Polyalthia glauca Boerl.	96-1217	BOGOR	***sakaputi*** syn. ***sakana***
Polyalthia glauca Boerl.	96-1218	UKC	
	96-1219	KEW	

Taxonomic information	Voucher code or other source of identification		Nuaulu name
Polyalthia lateriflora Kurz	96-1033a	BOGOR	***ai namte***
	96-1034	KEW	
Polyalthia lateriflora Kurz	96-1035	UKC	
Xylopia malayana Hook. f. & Thoms.	96-620	BOGOR	***anae ate***
Laurales			
Lauraceae			
Actinodaphne Nees	96-507	UKC	***akunin*** syn. ***asinokoie akunin hanaie***
Actinodaphne Nees.	96-508	KEW	
Actinodaphne probl. *moluccana* Blume	96-509	UKC	
Actinodaphne probl. *rumphii* Blume		PROSEA	***akunin pina***
Cinnamomum iners Reinw. ex Blume		PROSEA	***weio***
Cinnamomum verum J. Presl.		PROSEA	
Cryptocarya R. Br. sp.	15-123	BOGOR	***aunasate***
	15-124	KEW	
	15-125	KEW	
	96-562	KEW	
Litsea resinosa Blume	96-561	BOGOR	***ana naine***
Litsea resinosa Blume	96-563	UKC	
Persea americana Mill.		PROSEA	***advokat***
other Lauraceae	96-487	KEW	***ai mpiae***
	96-486	BOGOR	
	96-488	UKC	
	96-389	KEW	***ai numa***
	96-120	KEW	
	96-121	UKC	
	96-388a	BOGOR	
	96-388b	UKC	
	96-119	BOGOR	
	96-390	UKC	
	96-873	BOGOR	
	96-874	KEW	
	96-875	UKC	
			ai numa hanaie
			ai numa pina
	96-481	KEW	***nakunatae***
	96-480	BOGOR	
	96-482	UKC	
	96-482	KEW	
	96-496	KEW	***ai unene***
	96-247	UKC	
	96-248	KEW	
	96-495	BOGOR	
	96-246	BOGOR	
	96-497	UKC	
	96-1054	BOGOR	
	96-1055	KEW	
	96-1056	UKC	
	15-46	BOGOR	

Taxonomic information	Voucher code or other source of identification		Nuaulu name
Monimiaceae			
Steganthera hirsuta Perkins	15-89	BOGOR	***ai weti***
	15-90	KEW	
Piperales			
Piperaceae			
Piper L.		PROSEA	***kam(u)***
Piper aduncum L.	96-723	BOGOR	***ai tohu***
Piper L.sp.	96-724	KEW	
Piper aduncum L.	96-725	UKC	
Piper aduncum L.	96-282	KEW	
Piper aduncum L.	15-51	BOGOR	
	15-52	KEW	***ai kam***
Piper betle L.		PROSEA	***kam niane***
Piper probl. *caninum* Blume		PROSEA	***kam hanate***
Piper corylistachyon (Miq.) C. DC.	96-1317	BOGOR	***kam hau***
Piper L. sp.	96-1318	KEW	
Piper probl. *amboinense* (Miq.) C. DC. L. sp.	96-1319	UKC	
Piper corylistachyon (Miq.) C. DC.	96-751	BOGOR	***kam tunne***
Piper corylistachyon (Miq.) C. DC.	96-752	UKC	
Piper L. sp.	96-753	KEW	
Piper corylistachyon (Miq.) C. DC.	96-1326a	BOGOR	
	96-1327a	KEW	
	96-1328a	UKC	
Piper corylistachyon (Miq.) C. DC.	96-1326	BOGOR	***punara wane***
Piper L. sp.	96-1327	KEW	
Piper L. sp.	96-1328	UKC	
Piper L. sp.	96-372	BOGOR	***kam (w)apane, kamu apane***
Piper L. sp.	96-373	KEW	
Piper L. sp.	96-374	UKC	
	96-680	label only	
	96-681	label only	
	96-682	label only	
Piper forstenii C. DC.	96-768	UKC	
	96-769	BOGOR	
Piper forstenii C. DC.	96-770	KEW	
Piper L. sp.	96-526	UKC	***kam inate***
Piper L. sp.		PROSEA	***kam putie***
	96-1383	KEW	
Piper nigrum L.		PROSEA	***lada***
Piper L. sp.	96-1116	BOGOR	***kam sisinate***
Piper L. sp.	96-1117	KEW	
Piper L. sp.	96-1118	UKC	
Piper reinwardtianum Miq.		PROSEA	
Piper L. sp.		PROSEA	***kam wanae ukune***
Piper probl. *retrofractum* Vahl.			***kam malaka***
Piper probl. *sarmentosum* Roxb.			***kam(u) wane***
Piper L. sp.			***kam minne***
Piper L. sp.	96-940	KEW	
Piper caninum Blume	96-938a	BOGOR	
Piper sundaicum Blume	15-69	BOGOR	
	15-70	KEW	

Taxonomic information	Voucher code or other source of identification		Nuaulu name
Piper sundaicum Blume	15-7	BOGOR	
	96-938a	KEW	
Alismatales			
Araceae			
Aglaonema simplex Blume	96-1066	BOGOR	***kanna***
Aglaonema simplex Blume	96-1067	BOGOR	
	96-1068	KEW	
Aglaonema simplex Blume	96-1344	BOGOR	***wetune***
	96-1345a	KEW	
	96-1345b	KEW	
	96-1346a	UKC	
Aglaonema simplex Blume	96-1346b	BOGOR	
Colocasia esculenta (L.) Schott.	96-695	BOGOR	***punetene, pun***
	15-36	BOGOR	
Colocasia esculenta (L.) Schott. var.		MATTHEWS	***punetene metene***
Colocasia esculenta (L.) Schott. var.		MATTHEWS	***punetene putie***
Colocasia esculenta (L.) Schott. var.		MATTHEWS	***punetene taraga***
Colocasia esculenta (L.) Schott.		PROSEA	***sikeue***
Colocasia esculenta (L.) Schott.		PROSEA	***sikeue hatusoke***
Colocasia esculenta (L.) Schott.		PROSEA	***sikeue nakatua***
Colocasia esculenta (L.) Schott.		PROSEA	***sikeue tekene***
Colocasia esculenta (L.) Schott.		PROSEA	***sikeue tombelu***(*e*)
Colocasia esculenta (L.) Schott.		PROSEA	***sikeue hau rihue***
Colocasia esculenta (L.) Schott.		PROSEA	***sikeue kunie***
Colocasia esculenta (L.) Schott.		PROSEA	***sikeue masikunne***
Colocasia esculenta (L.) Schott.		PROSEA	***sikeue msinae***
Colocasia esculenta (L.) Schott.		PROSEA	***sikeue nikate***
Colocasia esculenta (L.) Schott.		PROSEA	***sikeue saparuane***
Epipremnum pinnatum (L.) Engl. [*Rhaphidophora pinnata* (L.) Schott.]	96-899	BOGOR	***kupawane***
Epipremnum pinnatum (L.) Engl. [*Rhaphidophora pinnata* (L.) Schott.]	96-900	BOGOR	
	96-901a	KEW	
	96-901b	UKC	
	96-1381	KEW	
	96-35	BOGOR	
	96-36	KEW	
	96-37	UKC	
Homalomena cordata Schott. var.	96-882	BOGOR	***puni msinae***
Homalomena cordata Schott. var.	96-16	BOGOR	
Homalomena cordata Schott. var.	96-883	UKC	
	96-884	KEW	
Homalomena pendula (Blume) Bakh. f.	96-643	BOGOR	***pun***i
	96-644	KEW	
Homalomena pendula (Blume) Bakh. f.	96-645	UKC	
Homalomena cordata Schott.	15-16	BOGOR	***puni msinae***
Homalomena cordata Schott.	96-777	BOGOR	***puni hatae***
Pothos L. sp.	71-37	KEW	***sunene***
	71-69	KEW	
Scindapsus Schott sp.	96-1057	BOGOR	***monone wane***
	96-1058	KEW	

Taxonomic information	Voucher code or other source of identification		Nuaulu name
Scindapsus Schott sp.	96-1059	UKC	
Xanthosoma sagittifolium (L.) Schott.		PROSEA	***yohoru***
Xanthosoma sagittifolium (L.) Schott.		PROSEA	***yohoru msinae***
Xanthosoma sagittifolium (L.) Schott.		PROSEA	***yohoru putie***
other Araceae			***pina totue***
Dioscoreales			
Dioscoreaceae			
Dioscorea alata L.		PROSEA	***hueni***
Dioscorea alata L. var.		PROSEA	***hueni (huae) hatasia***
Dioscorea alata L. var.	75-03	UKC	
Dioscorea alata L. var.		PROSEA	***hueni marae***
Dioscorea alata L. var.		PROSEA	***hueni msinae***
Dioscorea alata L. var.		PROSEA	***hueni nante***
Dioscorea alata L. var.		PROSEA	***hueni nakatua***
Dioscorea alata L. sp.	96-1302	BOGOR	***huen(i) puti(e)***
Dioscorea L. sp.	96-1303	KEW	
Dioscorea L. sp.	96-1304	UKC	
Dioscorea alata L. var.		PROSEA	***hueni suneminete***
Dioscorea alata L. var.		PROSEA	***hueni sune totue***
Dioscorea alata L. var.		PROSEA	***hueni tumata***
Dioscorea esculenta (Lour.) Burkill		PROSEA	***siahue***
Dioscorea esculenta (Lour.) Burkill		PROSEA	***siahue kawei***
Dioscorea esculenta (Lour.) Burkill		PROSEA	***siahue loloeno***
Dioscorea esculenta (Lour.) Burkill		PROSEA	***siahue masikunne***
Dioscorea esculenta (Lour.) Burkill		PROSEA	***siahue msinae***
Dioscorea esculenta (Lour.) Burkill		PROSEA	***siahue nusa ini***
Dioscorea esculenta (Lour.) Burkill	75-02	UKC	
Dioscorea esculenta (Lour.) Burkill		PROSEA	***siahue putie***
Dioscorea esculenta (Lour.) Burkill		PROSEA	***siahue sapatu***
Dioscorea esculenta (Lour.) Burkill	73-04	UKC	***siahue siaie***
Dioscorea esculenta (Lour.) Burkill		PROSEA	***siahu tekene***
Dioscorea hispida Dennst.	96-668	BOGOR	***kawasine***
Dioscorea hispida Dennst.	96-669	UKC	
Dioscorea hispida Dennst. var.	96-1293	BOGOR	***kawasine masikunne***
	96-1294	KEW	
Dioscorea hispida Dennst. var.	96-1295	UKC	
Dioscorea hispida Dennst. var.		PROSEA	***kawasine putie***
Dioscorea L. sp.	96-670	KEW	
Dioscorea pentaphylla L.	15-02	BOGOR	***loloeno***
Dioscorea L. sp.	96-00	KEW	
Dioscorea L. sp.	96-1393	UKC	
Dioscorea pentaphylla L.		COURSEY	***akae***
Dioscorea L. sp.	15-64	BOGOR	
Dioscorea pentaphylla L. var.	96-726	BOGOR	***aka hunuhunue, akae hunuhue***
Dioscorea L.	71-03b	KEW	
Pandanales			
Pandanaceae			
	96-352	BOGOR	***sina-sina wane***
Freycinetia Gaudich. sp.	96-353	KEW	
Freycinetia prob. *ceramensis* Martelli	96-354	UKC	

Taxonomic information	Voucher code or other source of identification		Nuaulu name
Pandanus Parkinson sp.	96-896	KEW	***koae koio*** syn. ***kokoio***
Pandanus Parkinson sp.	96-897	UKC	
Pandanus Parkinson sp.	96-898	UKC	
Pandanus Parkinson sp.	96-895	UKC	
Pandanus amaryllifolius Roxb			***koae***
Pandanus amaryllifolius Roxb. ♀		PROSEA	***koae metene***
Pandanus amaryllifolius Roxb. ♂		PROSEA	***koae putie***
Pandanus conoideus Lam.		PROSEA	**(*koae*) *saune*** syn. ***saune, koai'a***
Pandanus conoideus Lam.		PROSEA	***saune hehue***
Pandanus Parkinson sp.			***koae kinoe***
Pandanus krauelianus K. Schum.[*P. amboinensis* Warb.]	96-1179	UKC	***paone***
Pandanus Parkinson sp.	96-1178	KEW	
Pandanus Parkinson sp.	96-1180	KEW	
Pandanus probl. *polycephalous* Lam.		PROSEA RUMPH	***kokoio nuae***
Pandanus tectorius Sol. ex Balf. f.	96-1275	BOGOR	***sau*(*ne*) *hahu***
Pandanus Parkinson sp.	96-1276	KEW	
Pandanus Parkinson sp.	96-1277	UKC	
Pandanus Parkinson sp.		PROSEA	***sanoma***
Liliales			
Smilacaceae			
Smilax zeylanica Wight.	96-748	BOGOR	***kane-kane wane* (*inie*)** syn. ***wane unie***
Smilax L. sp.	96-351	UKC	
Smilax L. sp.	96-749	UKC	
Smilax L. sp.	96-750	KEW	
Smilax L. sp.	96-350	KEW	
	96-349	BOGOR	
Asparagales			
Xanthorrhoeaceae			
Dianella javanica Kunth	96-1323	BOGOR	***kokohahau***
Dianella sp., probl.	96-1325	UKC	
Amaryllidaceae			
Allium L.			***pahane***
Allium cepa L. (Aggregatum Group)		PROSEA	***paha msinae, pahane msina***
Allium sativum L.		PROSEA	***paha putie, pahane puti***
Crinum asiaticum Blanco	96-659	BOGOR	***napu nane***
Crinum L.	96-660	KEW	
Orchidaceae			
Dendrobium anosmum Lindl. [*D. superbum* Rchb. f.]	71-65	KEW	***koriwane***
Phreatia secunda (Blume) Lindl.	71-76	KEW	***kopuni***
			sipane
Robiquetia gracilistipes J.J. Sm.	96-519	BOGOR	***sipane hanaie***
Pomatocalpa Breda. sp.	96-520	KEW	
	96 521	UKC	
Oberonia padangensis Schltr. [*O.* nr. *spathulata* Lindl.]	71-53a	KEW	***sipa nione***
Grammatophyllum scriptum (L.) Blume	71-53b	KEW	
Dendrobium nr. *crumenatum* Sw. [*D. ephemerum* J.J. Sm.]	71-53c	KEW	
Dendrobium nr. *crumenatum* Sw. [*D. ephemerum* J.J. Sm.]	71-86	KEW	
Phalaenopsis amabilis (L.) Blume	71-29	KEW	unknown 12

Taxonomic information	Voucher code or other source of identification		Nuaulu name
Vanilla planifolia Jacks. ex Andrews		PROSEA	***vanili***
Asparagaceae			
Agave cantala (Haw.) Roxb. ex Salm-Dyck		PROSEA	***kinawane***
Cordyline fruticosa (L.) A. Chev. [*C. terminalis* Kunth]		PROSEA	***sone***
Cordyline fruticosa (L.) A. Chev., var. [*C. terminalis* Kunth]		PROSEA	***sone totu onate***
Cordyline fruticosa (L.) A. Chev., var. [*C. terminalis* Kunth]		PROSEA	***sone totu ikine***
Cordyline fruticosa (L.) A. Chev., var. [*C. terminalis* Kunth]	96-1039	BOGOR	***sone msinae***
	96-1040	KEW	
Cordyline fruticosa (L.) A. Chev. [*C. terminalis* Kunth]	96-1041	BOGOR	
Cordyline probl.	96-324	KEW	***kokohau***
Arecales			
Arecaceae [Palmae]			
Areca catechu L.		PROSEA	***kanai***
Areca catechu L.	96-1274	UKC	***kanai ia nanae, kanai iane anae*** syn. ***kanai ia nenane***
Areca catechu L.		PROSEA	***kanai kakaue***
Areca catechu L.		PROSEA	***kanai makuana***
Areca catechu L.	96-1152	UKC	***kanai nopane***
	96-1273	UKC	
Areca catechu L.		PROSEA	***kanai pona***
Areca catechu L.	96-1151	UKC	***kanai puti(e)***
Areca catechu L.		PROSEA	***kanai puti asaehene***
Areca catechu L.		PROSEA	***kanai sopite***
Areca probl. *glandiformis* Lam.		PROSEA	***(kanai) apane***
Arenga pinnata Merr. (syn. *A. saccharifera* (Wurmb.))		PROSEA	***naha***
Arenga pinnata Merr. (syn. *A. saccharifera* (Wurmb.)) var.		PROSEA	***naha niane***
RATTANS			***meute***
Calamus L. sp.	96-747	KEW	***meu(te) nunte***
Calamus L. sp.	96-746a	UKC	
Calamus L. sp.	96-976	UKC	***meu(te) hahu kikae***
Calamus L. sp.	96-975	KEW	
Calamus L. sp.	96-974	BOGOR	
Calamus L. sp.	96-146	UKC	***meu(te) wa ukune***
Calamus L. sp.	96-147	BOGOR	
Calamus L. sp.	96-148	KEW	
Calamus L. sp.	96-546	BOGOR	***meu(te) hehue***
Calamus L. sp.	96-547	KEW	
	96-548	KEW	
Calamus L.sp.	96-1078	BOGOR	
	96-216	BOGOR	***meu(te) munate, meu(te) wasaura***
Calamus L.sp.	96-217	UKC	
Calamus L.sp.	96-218	KEW	
Calamus L.sp.	96-1261	BOGOR	***meute werane, meu werane***
Calamus L. sp.	96-1262	KEW	
Calamus L. sp.	96-1263	UKC	
Calamus L. sp.	96-586	KEW	
Calamus L. sp.	96-587	UKC	
Daemonorops Blume. ex Schult f. sp.	96-585	BOGOR	***meu(te) patona***
OTHER RATTANS		PROSEA	***meu onate*** syn. ***meu kute, meu on***
	96-973	UKC	***meu peune, meute peun***
	96-971	BOGOR	
	96-972a	KEW	

Taxonomic information	Voucher code or other source of identification		Nuaulu name
STANDING PALMS			
Calyptrocalyx spicatus (Lam.) Blume	96-1031	KEW	***tepine***
	96-267	BOGOR	
	96-267a	UKC	
	96-267b	KEW	
	96-1032	KEW	
Pinanga rumphiana (Mart.) J. Dransf. & Govaerts [*P. punicea* (Zipp. ex Blume) Merr.]	96-1033a	BOGOR	
	96-1033b	KEW	
Caryota rumphiana Mart.	96-759	BOGOR	***panuke*** syn. ***tepine panuke*** syn. ***katue panuke***
Caryota rumphiana Mart.	96-760	UKC	
Caryota rumphiana Mart.	96-761	KEW	
Cocos nucifera L.		PROSEA	***nione***
Cocos nucifera L.		PROSEA	***nione ibreda***
Cocos nucifera L.		PROSEA	***nione inai***
Cocos nucifera L.		PROSEA	***nione koine***
Cocos nucifera L.		PROSEA	***nione kori***
Cocos nucifera L.		PROSEA	***nione marae***
Cocos nucifera L.		PROSEA	***nione msinae***
Cocos nucifera L.		PROSEA	***nione naha***
Cocos nucifera L.		PROSEA	***nione nene sanae***
Cocos nucifera L.		PROSEA	***nione nione***
Cocos nucifera L.		PROSEA	(***nione***) ***nuelo***
Cocos nucifera L.		PROSEA	***nione warane*** syn. ***nione putie***
Cocos nucifera L.		PROSEA	***nio***(***ne***) ***weke hutu-hutu*** syn. ***wekete***
Drymophloeus oliviformis (Giseke) Mart.	96-1249	BOGOR	***piasoi***
Drymophloeus oliviformis (Giseke) Mart.	71-77	KEW	
Drymophloeus oliviformis (Giseke) Mart.	96-1250	KEW	
Drymophloeus oliviformis (Giseke) Mart.	96-1251	UKC	
Hydriastele costata F.M. Bailey (PL)		DRANSF	***mapua***
Hydriastele microcarpa (Scheff.) W.J. Baker & Loo [*Gronophyllum microcarpum* Scheff.]	96-986	BOGOR	***inene***
Hydriastele microcarpa (Scheff.) W.J. Baker & Loo [*Gronophyllum microcarpum* Scheff.]	96-987	KEW	
Metroxylon sagu Rottb.	96-159	BOGOR	***hatane***
Metroxylon sagu Rottb.	96-160	KEW	
Metroxylon sagu Rottb.		UKC	
Metroxylon sagu Rottb.		PROSEA	***hata***(***ne***) ***apane***
Metroxylon sagu Rottb.		PROSEA	***hata metene, hata nuni metene, hata***(***ne***) ***metene, hata numetene***
Metroxylon sagu Rottb.	96-1248	UKC	***hata nun***(***i***) ***msinae, hata***(***ne***) ***msinae, hata***(***ne***) ***nuni msinae***
Metroxylon sagu Rottb.		PROSEA	***hata napaune, hata nuni anapaune, hatane anapaune***
Metroxylon sagu Rottb.	96-159	BOGOR	***hata***(***ne***) ***nuni mane***
Metroxylon sagu Rottb.	96-160	KEW	
Metroxylon sagu Rottb.		PROSEA	***hata***(***ne***) ***nuni nasinana, hata***(***ne***) ***nun nasinanae***
Metroxylon sagu Rottb.		PROSEA	***hata***(***ne***) ***nuni sekane, hata***(***ne***) ***sekane***

Taxonomic information	Voucher code or other source of identification		Nuaulu name
Metroxylon sagu Rottb.		PROSEA	***hata(ne) nuni tamno, hata nuni tamnone, hatane numtamnon***
Metroxylon sagu Rottb.		PROSEA	***hata(ne) nuni wakane***
Metroxylon sagu Rottb.		PROSEA	***hatane nunuari, hata(ne) nuni weri***
Metroxylon sagu Rottb.		PROSEA	***hata(ne) warata*** syn. ***hata(ne) ai***
Metroxylon sagu Rottb.		PROSEA	***hatane kainune***
Metroxylon sagu Rottb.		PROSEA	***hatane kaiohatai***
Metroxylon sagu Rottb.		PROSEA	***(hata) muena, hata mena, hata(ne) umena***
Nypa fruticans Wurmb.		PROSEA	***tepine huae ina***
Oncosperma filamentosa Blume [*O.tigillarium* (Jack) Ridley]		PROSEA	***katue***
Orania sylvicola (Griff.) H.E. Moore [*O. macrocladus* Mart.]		PROSEA	***korane***
	96-1380	UKC	
	96-1379	KEW	
Pigafetta filaris (Giseke) Becc.		DRANSF	***makoi*** syn. ***ai samine***
Rhopaloblaste ceramica (Miq.) Burret	96-564	BOGOR	***huaena***
Salacca zalacca (Gaertn.) Voss		PROSEA	***salak***
Saribus rotundifolius (Lam.) Blume [*Livistona rotundifolia* (Lam.) Mart.]		DRANSF	***nasinana***
Poales			
Bromeliaceae			
Ananas comosus (L.) Merr.		PROSEA	***sesene***
Ananas comosus (L.) Merr.		PROSEA	***sese(ne) bogor*** syn. ***sese pokoro***
Ananas comosus (L.) Merr.		PROSEA	***sese nusa inae, sesene nusa inae***
Ananas comosus (L.) Merr.		PROSEA	***sese(ne) potoni***
Ananas comosus (L.) Merr.		PROSEA	***sese(ne) warata***
Ananas comosus (L.) Merr.		PROSEA	***sesene gula***
Ananas comosus (L.) Merr.		PROSEA	***sesene kamruku***
Ananas comosus (L.) Merr.		PROSEA	***sese(ne) msinae*** syn. ***sesene kapane***
Cyperaceae			
Cyperus compressus L.	71-41	KEW	***anaupua, ana upa***
Cyperus diffusus Vahl.	71-47	KEW	***hahuseite***
Scleria P.J. Bergius	96-44a	UKC	
	96-45	BOGOR	
Scleria P.J. Bergius	96-46	KEW	
Scleria P.J. Bergius	96-47	UKC	
Cyperus elatus L.	96-1338	BOGOR	***tamkari (waene)***
Cyperus L.	96-1339	KEW	
Cyperus L.	96-1340	UKC	
Flagellariaceae			
Flagellaria indica L.	96-59	KEW	***makanitu (wane)***
	96-58	BOGOR	
	96-57	BOGOR	
	96-742	KEW	
Flagellaria indica L.	96-742	BOGOR	
Flagellaria indica L.	96-743	UKC	
Flagellaria indica L.	96-744	KEW	

Taxonomic information	Voucher code or other source of identification		Nuaulu name
Poaceae			
Bambusa bambos var. *spinosa* (Buch.-Ham.) S.S. Jain & S. Biswas		PROSEA	***tomo kainun***
Bambusa vulgaris var. *striata* (yellow culm group) Schrader ex Wendland		PROSEA	***wana tomo, tomo seane***
Centotheca lappacea (L.) Desv.	1971-50	KEW	***ekene***
Setaria palmifolia (J. Koenig) Stapf.	15-17	BOGOR	
		PROSEA	***nesa-nesa***
Chloris barbata Sw	96-275	KEW	***nesa-nesa hanaie***
Chloris barbata Sw	96-276	UKC	
	96-274	BOGOR	
Coix lacryma-jobi L.		PROSEA	***kaihoke***
Cymbopogon rectus A. Camus	96-1110	BOGOR	***hisae***
	96-1111	KEW	
	96-1112	UKC	
Dendrocalamus Nees.	96-909	BOGOR	(***wana***) ***ona***
	96-910	KEW	
Dendrocalamus asper (Scult.) Backer, probl.	96-911	UKC	
Dendrocalamus Nees.	96-171	BOGOR	***wana hatu***
	96-172	KEW	
Dendrocalamus Nees.	96-173	UKC	
	96-714	BOGOR	
	96-715	KEW	
	96-716a	BOGOR	
Dendrocalamus Nees.	96-716b	UKC	
Dendrocalamus Nees.	96-171	BOGOR	***wanate*** syn. ***wana hatu***
	96-172	KEW	
Dendrocalamus Nees.	96-173	UKC	
Dendrocalamus Nees., sp., probl		PROSEA	***wana apane***
Digitaria setigera Roth ex Roem. & Schult.	96-278	KEW	***nesa-nesa pina***
Digitaria setigera Roth ex Roem. & Schult.	96-279	UKC	
	96-277	BOGOR	
Gigantochloa apus (Schult.) Kurz, probl.	96-1014	UKC	***tenne***
	96-1013	BOGOR	
	96-1015	KEW	
Gigantochloa atter (Hassk.) Kurz, probl.	96-1018	KEW	***wana tunne***
	96-1016	BOGOR	
	96-1017	UKC	
k.o. bamboo	03-23	UKC	***tomo temine***
Imperata exaltata Brongn.	71-84	KEW	***oni***
Imperata cylindrica (L.) Beauv.		PROSEA	
Leptaspis urceolata (Roxb.) R. Br.	96-696	BOGOR	***saenanane, saenonane***
Leptaspis urceolata (Roxb.) R. Br.	96-697	KEW	
Leptaspis urceolata (Roxb.) R. Br.	96-698	UKC	
Miscanthus floridulus (Labill.) Warb. ex K. Schum. & Lauterb.		PROSEA	***niune***
Miscanthus Anderson, sp.		PROSEA	***niu tunne***
Miscanthus Anderson, sp.		PROSEA	***niune ikine***
Oryza sativa L.	96-1148	UKC	***ana msinae***
Oryza sativa L.		PROSEA	***ana*** syn. ***hotone***
Oryza sativa L.		PROSEA	***ana nusa inae***

Taxonomic information	Voucher code or other source of identification		Nuaulu name
Pennisetum macrostachyum (Brongn.) Trin.	96-55	KEW	***asu etute***
	96-54	BOGOR	
Pennisetum macrostachyum (Brongn.) Trin.	96-56	UKC	
Phragmites karka (Retz.) Trin. ex Steude., poss.		PROSEA	***seate***
Saccharum officinarum L.		PROSEA	***tohu***
Saccharum officinarum L. var.		PROSEA	***tohu hehue***
Saccharum officinarum L. var.		PROSEA	***tohu ito***
Saccharum officinarum L. var.		PROSEA	***tohu manteri***
Saccharum officinarum L. var.		PROSEA	***tohu metene***
Saccharum officinarum L. var.		PROSEA	***tohu mitane***
Saccharum officinarum L. var.		PROSEA	***tohu putie***
Saccharum officinarum L. var.		PROSEA	***tohu tona***
Saccharum officinarum L. var.		PROSEA	***tohu tona metene***
Saccharum officinarum L. var.		PROSEA	***tohu tona putie***
Saccharum spontaneum L.	96-1290	BOGOR	***niune tunne***
Saccharum spontaneum var. *edulis* (Hassk.) K. Schum.		PROSEA	***topikane***
Saccharum L. sp.		PROSEA	***to(hu) pikane***
	96-1291	KEW	
	96-1292	UKC	
Schizostachyum Nees.	96-268	BOGOR	***suenie***
	96-269	KEW	
	96-270	UKC	
Schizostachyum brachycladum Kurz, probl.	96-1092	BOGOR	***tapine***
	96-1093	KEW	
Schizostachyum brachycladum Kurz, probl.	96-1094	UKC	
Schizotachyum lima (Blanco) Merr., probl.	96-271	BOGOR	***tomone***
	96-272	KEW	
	96-273	UKC	
Schizotachyum lima (Blanco) Merr., probl.		PROSEA	***tomo(ne) onate*** syn. ***tomo kehane***
Setaria italica (L.) Pal, probl.	96-1149	UKC	***hotone*** syn. ***ana hotone (manipa)***
	96-1150	UKC	
Sorghum bicolor (L.) Moench. [*Sorghum vulgare* Pers.], probl.	96-1238	UKC	***salea***
Sorghum propinquum (Kunth) Hitchc.	96-90	KEW	***niwapane, niu wapane, niune apane***
	96-89	BOGOR	
Sorghum Moench.	96-91	UKC	
	96-297	BOGOR	
Thysanolaena maxima (Roxb.) Kuntze	96-298	KEW	***niwapane, niu wapane***
Thysanolaena maxima (Roxb.) Kuntze	96-299	UKC	
Triticum aestivum L.		PROSEA	***manae***
Zea mays L.		PROSEA	***(ana) manipa*** syn. ***yakon(i), manipa***
Zea mays L. var.	96-1153	UKC	***yakon(i) nikate***
Zea mays L. var.		PROSEA	***yakon msinae***
Zea mays L. var.		PROSEA	***yakon putie***
other Poaceae		PROSEA	***huetune***
		PROSEA	***munanea***
		PROSEA	***uta numa***

Taxonomic information	Voucher code or other source of identification		Nuaulu name
Commelinales			
Commelinaceae			
Amischotolype hispida A. Rich.	96-661	KEW	***hono-hono wesie***
Amischotolype hispida A. Rich.	96-662	UKC	
	96-821	BOGOR	
	96-822	KEW	
	96-823	UKC	
Aneilema R. Br. sp.	71-07	KEW	***hono-hono hanaie***
	96-149	BOGOR	
Aneilema R. Br. sp.	96-624	KEW	
	96-625	UKC	
Aneilema sp., probl.	96-825	KEW	***hono-hono hanaie msinae***
	96-824	BOGOR	
	96-826	UKC	
Pollia Miq.	96-624	KEW	***hono-hono***
Pollia secundiflora (Blume) Bakh. f. [*Pollia sorzogonensis* Miq.]	96-623	BOGOR	***hono-hono hanaie putie***
Pollia secundiflora (Blume) Bakh. f. [*Pollia sorzogonensis* Miq.]	96-1133	BOGOR	***hono-hono pina***
Pollia Miq.	96-1134a	KEW	
Pollia Miq.	96-1134b	KEW	
Pollia Miq.	96-1144	UKC	
Zingiberales			
Musaceae			
Musa L.		PROSEA	***uri***
Musa L. x var.		PROSEA	(***uri***) ***kainonone***
Musa L. x var.		PROSEA	(***uri***) ***hanoate***
Musa L. x var.		PROSEA	***uri hunan itane, uri hunan***(***e***) ***ntane, uri hunanitane***
Musa L. x var.		PROSEA	***uri ibreda***
Musa L. x var.		PROSEA	***uri kapananoi***
Musa L. x var.		PROSEA	***uri kokune***
Musa L. x var.		PROSEA	***uri nuetetue***
Musa L. x var.		PROSEA	***uri pesa, uri geser***
Musa L. x var.		PROSEA	***uri poitai***
Musa L. x var.		PROSEA	***uri punau***
Musa L. x var.		PROSEA	***uri suaie***
Musa L. x var.		PROSEA	***uri tonamune***
Musa L. x var.		PROSEA	***uri wanate***
Musa L. x var.		PROSEA	***uri warakai***
Musa L. x var.		PROSEA	***uri warata*** syn. ***uri ikine***
Musa acuminata Colla. (AA) var.		PROSEA	***uri numanaite***
Musa acuminata Colla. (AA) var.		PROSEA	***uri tope***
Musa acuminata Colla. (AA) var.		PROSEA	***uri empatpuluh hari***
Musa acuminata Colla. (AA) var.		PROSEA	***uri hutu ate***
Musa acuminata Colla. (AA) var.		PROSEA	***uri onana hutu ete***
Musa acuminata Colla. (AA, Sucrier subgroup) var.		PROSEA	***uri hunahane***
Musa acuminata Colla. (AA, Sucrier subgroup) var.		PROSEA	***moe-moe***
Musa acuminata Colla. (AA, Sucrier subgroup) var.		PROSEA	***uri hunananane***
Musa acuminata Colla. (AA, Sucrier subgroup) var.		PROSEA	***uri kuako***
Musa acuminata Colla. (AA, Sucrier group) var.		PROSEA	***uri pinate***

Taxonomic information	Voucher code or other source of identification		Nuaulu name
Musa acuminata Colla. (AAA, Gros Michel subgroup) var.		PROSEA	(***uri***) ***moane***
Musa poss. *acuminata* subsp. *banksii* (F. Muell.) Simmonds, or *M. lolodensis* Cheesman	96-1254	UKC	***uri poi***
	96-1252	BOGOR	
	96-1253	KEW	
Musa x *paradisiaca* L. (AAB, Silk subgroup) var.		PROSEA	***uri pelau hua metene***
Musa x *paradisiaca* L. (AAB, Silk subgroup) var		PROSEA	***uri pelau hua putie***
Musa x *paradisiaca* L. (AAB, Silk subgroup) var.		PROSEA	***uri tamno***
Musa x *paradisiaca* L. (AAB, Silk subgroup) var.		PROSEA	***uri pelau***
Musa x *paradisiaca* L. (AAB, Silk subgroup); non-Cavendish AAA var.		PROSEA	***uri susu***
Musa x *paradisiaca* L. (ABB) var.		PROSEA	***uri abu-abu***
Musa x *paradisiaca* L. (ABB) var.		PROSEA	***uri hua puti*** syn. ***uri makuana***
Musa x *paradisiaca* L. (ABB) var.		PROSEA	***uri lewaka***
Musa x *paradisiaca* L. (ABBB) var.		PROSEA	***uri pau***
Musa x *paradisiaca* L. (AAB) or *M. acuminata* Colla. AAA) var.		PROSEA	***uri kasupa***
Musa x *paradisiaca* var. (ABB)		PROSEA	***uri makuana*** syn. ***uri hua putie***
Musa troglodytarum L. (Callimusa section)		PROSEA	***uri marinanante***
Heliconiaceae			
Heliconia bihai (L.) L.		RUMPH	***kokine***
Cannaceae			
Canna indica L.	71-56	KEW	***kasipehe***
Canna indica L.	96-1211	KEW	
	96-1212	KEW	
Marantaceae			
Donax canniformis (G. Forst.) K. Schum. [*Donax canaeformis* Roffe]	96-705	BOGOR	***nonanae***
Donax canniformis (G. Forst.) K. Schum.	96-706	UKC	
Donax canniformis (G. Forst.) K. Schum.	96-707	KEW	
Donax canniformis (G. Forst) K. Schum.	96-1389	KEW	
Hornstedtia Retz. sp.		PROSEA	***kikun***
Hornstedtia Retz. sp.		PROSEA	***kikun*** (***hua***) ***nawe***
Hornstedtia Retz. sp.	03-05	UKC	***kikun tonamune***
Phrynium macrocephalum K. Schum.	96-1212b	BOGOR	***kasipehe***
Phrynium macrocephalum K. Schum.	96-1213	UKC	
Phrynium Willd.	96-1391	KEW	***wainite***
Phrynium pubinerve Blume, var.	96-678	BOGOR	(***wainite***) ***makanotue***
Phrynium Willd. sp.	96-650	KEW	
Phrynium Willd. sp.	96-679	KEW	
Phrynium Willd. sp.	96-1122	KEW	
Phrynium Willd. sp.	96-1392	KEW	
Phrynium pubinerve Blume	15-87	BOGOR	
	15-88	KEW	
Phrynium pubinerve Blume [*Phrynium capitatum* Willd.], var.	96-1028	BOGOR	***wainite*** (***totu***) ***onate***
	96-1029	KEW	
Phrynium pubinerve Blume [*Phrynium capitatum* Willd.], var.	96-1030	UKC	
Phrynium Willd. sp.	96-1391a	KEW	***wainite totu nawe***
Phrynium Willd. sp.	96-1391b	KEW	***wainite manuaia***

Taxonomic information	Voucher code or other source of identification		Nuaulu name
Phrynium Willd. sp., probl.	15-34	KEW	***wainite putie***
Zingiberaceae			
Alpinia Roxb. sp.	96-306	BOGOR	***makakoii***
Alpinia Roxb. sp.	96-307	KEW	
Alpinia Roxb. sp.	96-308	UKC	
Alpinia Roxb. sp.	96-381	UKC	
Alpinia Roxb. sp.		PROSEA	***makakoii pina***
Alpinia Roxb. sp.	96-517	BOGOR	
Alpinia Roxb. sp.	71-610	SING	***tonamune***
	96-78	BOGOR	
Alpinia Roxb. sp	96-79	KEW	
	96-313	BOGOR	
Alpinia Roxb. sp	96-314	KEW	
	96-315	UKC	
	96-380	UKC	
Alpinia gigantea Blume	15-57	BOGOR	
	15-58	KEW	
Alpinia galanga (L.) Willd. [*Languas galanga* (L.) Stuntz]	96-689	BOGOR	***nakuasi***
	96-690	KEW	
	96-691a	UKC	
Alpinia galanga (L.) Willd. [*Languas galanga* (L.) Stuntz]	96-691b	BOGOR	
Alpinia nutans (L.) Roscoe.	03-12	UKC	***kamkari***
Amomum aculeatum Roxb.	15-35	BOGOR	***katou totu ikine***
Amomum aculeatum Roxb.	96-1082	BOGOR	***kikun marae*** syn. ***kikun*****(*****i*****)** ***werane***
	96-1083	KEW	
Amomum aculeatum Roxb.	96-1084	UKC	
	15-19	KEW	
Amomum compactum Sol. ex Maton [*Amomum cardamomum* Willd.]	96-711	BOGOR	***kikun*** **(*****hua*****)** ***pukune***
Amomum compactum Sol. ex Maton [*Amomum cardamomum* Willd.]	96-712	UKC	
	96-713	KEW	
Cheilocostus speciosus (J. Koenig)	96-424	BOGOR	***kamamuri***
	96-82	KEW	
	96-426	BOGOR	
	15-10	KEW	
	15-11	BOGOR	
Curcuma Roxb.		PROSEA	***kunie***
Curcuma aurantiaca Zijp	96-683	BOGOR	***kunie apane***
	96-684	KEW	
Curcuma aurantiaca Zijp	96-685	UKC	
Curcuma viridiflora Roxb.	96-1138	BOGOR	
	96-1139	KEW	
	96-1140	UKC	
Curcuma longa L.	96-1382a	UKC	***kun tun, kunie tunn***
	96-1382b	KEW	
Globba marantina L. [*Globba strobilifera* Zoll. et Mor.]	96-1308	BOGOR	***soi mutu putie***
	96-1309	KEW	
	96-1310	UKC	
Hornstedtia minor Valeton	96-1220	BOGOR	**(*****kikuni*****)** ***manipa***

Taxonomic information	Voucher code or other source of identification		Nuaulu name
Hornstedtia Retz.	96-1221	KEW	
Hornstedtia Retz.	96-1222	UKC	
Kaempferia galanga L.	96-1086	BOGOR	***sukue***
	96-1087	KEW	
	96-1088	UKC	
Riedelia Oliver. sp.	15-18	BOGOR	***nakuasi***
	15-33	BOGOR	
Zingiber gramineum Noronha	96-1230a	BOGOR	***kuni wekatisie***
	96-1230b	UKC	
	96-1229	KEW	
	96-1231	KEW	
Zingiber montanum (J. Koenig) Link ex A. Dietr. [*Z. cassumunar* Roxb.]		PROSEA	***siaune***
Zingiber montanum (J. Koenig) Link ex A. Dietr. [*Z. cassumunar* Roxb.]	96-666	BOGOR	***soi kukini***
Zingiber officinale Roscoe.		PROSEA	***soi tunne*** syn. ***soi msinae***
Zingiber papuanum Val.	96-676	BOGOR	***soi mutune***
Zingiber zerumbet (L.) Roscoe ex Sm. [*Z. americans* Blume]	15-08	BOGOR	***kaiihoke***
Zingiber Boehmer sp.	96-677	KEW	
Zingiber Boehmer and Globba			***soi***
Zingiber Boehmer or Globba		PROSEA	***soi mutu(ne) marae***
Zingiber Boehmer or Globba		PROSEA	***soi mutu(ne) msinae***
other Zingiberaceae	96-667	KEW	***soi putie***
Ranunculales			
Menispermaceae			
Arcangelisia flava (L.) Merr.	71-106	UKC	***awane kunie***
	71-437	UKC	
	96-429	BOGOR	
	96-493	KEW	
	96-494	UKC	
Pericampylus glaucus (Lam.) Merr.	96-1101	BOGOR	***awane wasi***
	96-1102	KEW	
	96-1103	UKC	
Ranunculaceae			
Hepatica Mill.	71-82	KEW	***numte***
Dilleniales			
Dilleniaceae			
Dillenia sp.		PROSEA	***asora***
Tetracera scandens Merr.	96-1226	BOGOR	***sapate wane***
Tetracera scandens Merr.	96-1227	UKC	
	96-1228	KEW	
Vitales			
Vitaceae			
Cissus javana DC.	96-605	BOGOR	***mahosu wane***
Cissus javana DC.	96-606	UKC	
	96-607	KEW	
Cissus repens A. Chev.	96-657	BOGOR	***mataetuae***
Cissus repens A. Chev.	96 658	UKC	

Taxonomic information	Voucher code or other source of identification		Nuaulu name
Leea Royen ex L.	96-85	KEW	***anapakane***
Leea Royen ex L.	96-84	BOGOR	
	96-258	BOGOR	
	96-259	UKC	
Leea Royen ex L.	96-260	KEW	
	15-75	KEW	
Leea indica (Burm. f.) Merr.	15-76	BOGOR	
	96-41	BOGOR	***auhaune***
Leea Royen ex L. sp.	96-250	BOGOR	
Leea Royen ex L. sp.	96-241	KEW	
Leea Royen ex L. sp.	96-242	KEW	
Leea Royen ex L. sp.	96-385	BOGOR	
Leea Royen ex L. sp.	96-386	KEW	
Leea Royen ex L. sp.	96-387	UKC	
	15-30	KEW	
Leea indica (Burm. f.) Merr. sp.	15-31	BOGOR	
Leea Royen ex L.			***auhaune hanaie***
Leea Royen ex L. sp.		PROSEA	***auhaune pina***
Tetrastigma lanceolarium Planch.	96-1119	BOGOR	***ken awane***
	96-1120	KEW	
Tetrastigma lanceolarium Planch.	96-1121	UKC	
Tetrastigma (Miq.), probl. *papillosum* Planch.	96-453	UKC	**(*ni*)*mentinse***
	96-452	BOGOR	
Tetrastigma (Miq.), probl. *papillosum* Planch.	96-451	UKC	
	96-525	UKC	
Oxalidales			
Oxalidaceae			
Averrhoa bilimbi L		PROSEA	***taulela maninue***
Averrhoa carambola L.		PROSEA	***taulela***
	96-846	KEW	
	96-847	UKC	
	96-845	BOGOR	
Averrhoa carambola L.		PROSEA	***taulela marimto***
Elaeocarpaceae			
Elaeocarpus augustifolius Blume	96-335	KEW	***riane***
Elaeocarpus augustifolius Blume	96-336	UKC	
	96-334	BOGOR	
Elaeocarpus L.	96-329	KEW	***ai nasate***
	96-328	BOGOR	
Elaeocarpus probl. *serratus* L.	96-330	UKC	
Malpighiales			
Erythroxylaceae			
Erythroxylum ecarinatum Hochr.	15-117	BOGOR	***anine***
	96-476	KEW	
	96-475	BOGOR	
	96-477	UKC	
	15-118	KEW	
	15-119	KEW	
Rhizophoraceae			
Bruguiera gymnorhiza (L.) Lam.		PROSEA	***hana*** syn. ***ai unate***

Taxonomic information	Voucher code or other source of identification		Nuaulu name
Rhizophora L.		PROSEA	
Calophyllaceae			
Calophyllum L.		PROSEA	***ahutau wesie***
Calophyllum L.	96-447	KEW	***akenue***
	96-446	BOGOR	
	96-478a	BOGOR	
	96-478b	KEW	
Calophyllum L.	96-128	BOGOR	***ahutau(ne) numte*** syn. ***ahutau mkauke***
Calophyllum articulatum P.F. Stevens, probl.	96-129a	UKC	
	96-129b	KEW	
	96-130	KEW	
Calophyllum L.		PROSEA	***ahutau(ne) hanaie***
Calophyllum L.		PROSEA	***ahutau(ne) pina***
Calophyllum inophyllum L.	96-862	KEW	***ahutau(ne) nuae*** syn. ***hitaio***
	96-863	UKC	
	96-861	BOGOR	
incl. *Calophyllum inophyllum* L.	96-31	KEW	***ahutaune***
	96-30	BOGOR	
	96-29	BOGOR	
Calophyllum probl. *C. soulattri* Burm. f.		PROSEA	***ahutau totu onate*** syn. ***ahutau metene***
Clusiaceae			
Garcinia L.		PROSEA	***nerune***
Garcinia L. sp.		PROSEA	***neru(ne) putie***
Garcinia L. sp.	96-198	BOGOR	
	96-199	KEW	
	96-200	UKC	
	96-985	UKC	***neru(ne) msinae*** syn. ***neru(ne) metene***
Garcinia L. sp.	96-984	KEW	
	96-983	BOGOR	
Garcinia celebica L.	96-1157	BOGOR	***komine***
	96-1158	KEW	
Garcinia celebica L.	96-1159	UKC	
Garcinia x *mangostana* L.		PROSEA	***mangosta, mankositaa***
Garcinia tetrandra Pierre, probl.		PROSEA	***setu***
Dichapetalaceae			
Dichapetalum probl. *Pa* (Becc.) Boerl.	96-245b	UKC	***awane metene***
	96-243	BOGOR	
	96-244	KEW	
	96-245a	KEW	
Passifloraceae			
Passiflora foetida L.	71-21	KEW	***pitiri***
	96-406	BOGOR	
Passiflora foetida L.	96-407	KEW	

Taxonomic information	Voucher code or other source of identification		Nuaulu name
Passiflora foetida L.	96-408	UKC	
Salicaceae			
Casearia glabra Roxb. [syn. *Casearia moluccana* Blume]	15-95	BOGOR	***kasauni apane***
	96-510	BOGOR	
	96-511	KEW	
	96-512	UKC	
	15-96	KEW	
Flacourtia rukam Zoll. & Moritzi	96-757	BOGOR	***tom-tom*** syn. ***tom-tome***
Flacourtia inermis Roxb. (det. S. Zmarzty)	96-758	KEW	
Homalium foetidum (Roxb.) Benth	96-113	UKC	***pono***
	96-114	BOGOR	
	96-115	KEW	
Euphorbiaceae			
Acalypha amentacea Roxb.	96-699	BOGOR	***pita(r)***
Acalypha caturus Blume	96-700	KEW	
Acalypha caturus Blume	96-701	UKC	
Aleurites moluccannus Willd.		PROSEA	***kahea***
Aleurites moluccanus Willd.	96-1010	BOGOR	***kahea pukune***
Aleurites moluccanus Willd.			***kahea nawe***
Croton L. sp. poss.	96-1011	KEW	
Aleurites moluccana Willd., probl.	96-1012	UKC	
Claoxylon A. Juss. sp	96-1114	KEW	***kaiohite***
Claoxylon A. Juss. sp.	96-1115	UKC	
Codiaeum variegatum (L.) Rumph. ex A. Juss.		PROSEA	***sinsinte***
Codiaeum variegatum (L.) Rumph. ex A. Juss. var.	96-1080	KEW	***sinsin(te) amasen***
Codiaeum A. Juss	96-1079	BOGOR	
Codiaeum variegatum (L.) Rumph. ex A. Juss.	96-1081	UKC	
Codiaeum variegatum (L.) Rumph. ex A. Juss. var.		PROSEA	***sinsin(te) masikunne***
Codiaeum variegatum (L.) Rumph. ex A. Juss. var.		PROSEA	***sinsin matopae, sinsinte matopai***
	15-41	KEW	***sinsin(te) msinae***
Codiaeum variegatum (L.) Rumph. ex A. Juss. var.	15-42	BOGOR	
Codiaeum variegatum (L.) Rumph. ex A. Juss. var.	15-43	BOGOR	
Codiaeum variegatum (L.) Rumph. ex A. Juss. var.		PROSEA	***sinsin(te) totu nawe***
Codiaeum variegatum (L.) Rumph. ex A. Juss. var.	96-302	UKC	***sinsin(te) (totu) onate***
	96-301	BOGOR	
	96-300	KEW	
Codiaeum variegatum (L.) Rumph. ex A. Juss. var.		PROSEA	***sinsin(te) totu pukune***
Croton cf. *oblongus* Burm. f.	96-1359	KEW	unknown 13
Croton cf. *oblongus* Burm. f.	96-1361	KEW	unknown 14
	96-1360	BOGOR	
Croton cf. *oblongus* Burm. f.	96-1362	UKC	
Endospermum moluccanum (Teijsm. & Binn.) Kurz	96-534	BOGOR	***au oi(ne)***
Endospermum moluccanum (Teijsm. & Binn.) Kurz	96-535	KEW	
Endospermum moluccanum (Teijsm. & Binn.) Kurz	96-536	UKC	
Euphorbia hirta L.	71-4	KEW	***mono nute***
Hevea brasiliensis (Willd. ex A. Juss.) Müll. Arg. [or x with *H. spruceana* (Benth.) Müll. Arg.]		PROSEA	***kataa onate***
Hevea brasiliensis (Willd. ex A. Juss.) Müll. Arg. [or x with *H. spruceana* (Benth.) Müll. Arg.]		PROSEA	***kataa ikine***
Hevea brasiliensis (Willd. ex A. Juss.) Müll. Arg. [or x with *H. spruceana* (Benth.) Müll. Arg.]		PROSEA	***kataa***

Taxonomic information	Voucher code or other source of identification		Nuaulu name
Hancea penangensis (Müll. Arg.) S.E.C. Sierra, Kulju & Welzen [*Mallotus echinatus* Elmer.]	96-617b	KEW	***ai hatu hatae***
	96-617a	BOGOR	
	96-618	UKC	
	96-619	KEW	
Jatropha L.	96-27	KEW	***ai kamane***
	96-26	BOGOR	
	96-28	BOGOR	
Jatropha curcas L.		PROSEA	
Macaranga Thouars.	96-745a	BOGOR	***ini***
	96-746	KEW	
	96-747a	BOGOR	
Macaranga Thouars.	96-747b	UKC	
	96-790	BOGOR	
	96-791	KEW	
	96-792	UKC	
Macaranga Thouars., probl.	96-876	KEW	***kupatonae***
	96-568	KEW	
	96-569	UKC	
	96-567	BOGOR	
	96-877a	BOGOR	
	96-877b	KEW	
	96-878	UKC	
Macaranga involucrata (Roxb.) Baill.	96-420	UKC	***hun*(*an*)*ori*** syn. ***nori-nori***
Macaranga involucrata (Roxb.) Baill.	96-419	KEW	
Macaranga trichocarpa Müll. Arg.	96-1019	BOGOR	
	96-418	BOGOR	
	96-1020a	KEW	
Macaranga trichocarpa Müll. Arg.	96-1020b	BOGOR	
	96-1021	UKC	
Cheilocostus speciosus (J. Koenig) C. Specht	15-12	BOGOR	
Macaranga involucrata (Roxb.) Baill.	15-13	BOGOR	
Macaranga involucrata (Roxb.) Baill.	15-15	BOGOR	
Macaranga hispida (Blume) Müll. Arg.	96-210	UKC	***ranoi***
	96-211	BOGOR	
Macaranga hispida (Blume) Müll. Arg.	96-212	KEW	
	96-793	BOGOR	
Macaranga hispida (Blume) Müll. Arg.	96-794	KEW	
	96-795	BOGOR	
Macaranga mappa A. Juss.	96-472	KEW	***nihuhune, nihihune***
Macaranga Thouars. sp.	96-473	KEW	
Macaranga Thouars. sp.	96-474	UKC	
	96-1353	BOGOR	
	96-1354	KEW	
	96-1355	BOGOR	
Macaranga sp. nr. *tanarius* (L.) Müll. Arg.	96-223	KEW	***hunua***
	96-222	BOGOR	
Macaranga sp. nr. *tanarius* (L.) Müll. Arg.	96-224	UKC	
	96-842	BOGOR	
Macaranga Thouars. sp.	96-843	KEW	
Macaranga Thouars. sp.	96-844	UKC	

Taxonomic information	Voucher code or other source of identification		Nuaulu name
Mallotus Lour., probl.		PROSEA	***ai hatu***
	96-104a	KEW	
	96-105	UKC	
	96-106	KEW	
	96-375	BOGOR	
	96-376	KEW	
	96-377	UKC	
	96-104b	BOGOR	
Mallotus paniculatus (Lam.) Müll. Arg.	96-1205	BOGOR	***hanu hatae***
Mallotus tiliifolius (Blume) Müll. Arg.	96-1206	KEW	
Mallotus tiliifolius (Blume) Müll. Arg.	96-1207	UKC	
Mallotus mollissimus (Geiseler) Airy Shaw	15-77	BOGOR	***kaha pukune***
Mallotus mollissimus (Geiseler) Airy Shaw	15-78	BOGOR	
Manihot Mill.		PROSEA	***kasipii***
Manihot cf. *carthaginensis* subsp. *glaziovii* (Müll. Arg.) Allem [*Manihot* cf. *glaziovii* Müll. Arg.]	96-220	KEW	***kasipii sombar***
Manihot cf. *carthaginensis* subsp. *glaziovii* (Müll. Arg.) Allem [*Manihot* cf. *glaziovii* Müll. Arg.]	96-219a	UKC	
Manihot cf. *carthaginensis* subsp. *glaziovii* (Müll. Arg.) Allem [*Manihot* cf. *glaziovii* Müll. Arg.]	96-219b	BOGOR	
Manihot cf. *carthaginensis* subsp. *glaziovii* (Müll. Arg.) Allem [*Manihot* cf. *glaziovii* Müll. Arg.]	96-221	UKC	
Manihot cf. *carthaginensis* subsp. *glaziovii* (Müll. Arg.) Allem [*Manihot* cf. *glaziovii* Müll. Arg.]	96-319	BOGOR	
Manihot cf. *carthaginensis* subsp. *glaziovii* (Müll. Arg.) Allem [*Manihot* cf. *glaziovii* Müll. Arg.]	96-320	KEW	
Manihot cf. *carthaginensis* subsp. *glaziovii* (Müll. Arg.) Allem [*Manihot* cf. *glaziovii* Müll. Arg.]	96-321	UKC	
Manihot cf. *carthaginensis* subsp. *glaziovii* (Müll. Arg.) Allem [*Manihot* cf. *glaziovii* Müll. Arg.]	09-R1	UNPAD	
Manihot cf. *carthaginensis* subsp. *glaziovii* (Müll. Arg.) Allem [*Manihot* cf. *glaziovii* Müll. Arg.]	09-R3	UNPAD	
Manihot esculenta Crantz var.	09-R5	UNPAD	***kasipii honue*** syn. ***kasipii paru***
Manihot esculenta Crantz var.	09-R10	UNPAD	***kasipii kanapua***
Manihot esculenta Crantz var.	09-R4	UNPAD	***kasipii masikunne***
	09-R9	UNPAD	
Manihot esculenta Crantz var.		PROSEA	***kasipii mentega***
Manihot esculenta Crantz var.		PROSEA	***kasipii msinae*** syn. ***kasipii postel***
Manihot esculenta Crantz var.		PROSEA	***kasipii puna***
Manihot esculenta Crantz var.	09-R2	UNPAD	***kasipii putie*** syn. ***kasipii wasaena***
Manihot esculenta Crantz var.		PROSEA	***kasipii sinsinte*** syn. ***kasipii makuana***
Manihot esculenta Crantz var.		PROSEA	***kasipii unuhutu***
Manihot esculenta Crantz var.	09-R11	UNPAD	***kasipii wanate***
Manihot esculenta Crantz var.	09-R6	UNPAD	***kasipii posteli***
Manihot esculenta Crantz var.	09-R7	UNPAD	
Manihot esculenta Crantz var.		PROSEA	***posteli marae***
Manihot esculenta Crantz var.	09-R8	UNPAD	***posteli metene*** syn. ***kasipii hatumetene***
Manihot esculenta Crantz var.		PROSEA	***posteli msinae***
Manihot esculenta Crantz var.		PROSEA	***posteli putie***

Taxonomic information	Voucher code or other source of identification		Nuaulu name
Manihot esculenta Crantz var.		PROSEA	***posteli upone***
Mallotus miquelianus Pax & K. Hoffm.		PROSEA	***ai menie***
Melanolepis multiglandulosa (Blume) Reich. f. and Zoll.	71-85	KEW	***noha***
Melanolepis multiglandulosa (Blume) Reich. f. and Zoll.	96-778	UKC	
Melanolepis multiglandulosa (Blume) Reich. f. and Zoll.	96-779	KEW	
Melanolepis multiglandulosa (Blume) Reich. f. and Zoll.	96-780	UKC	
Ricinus communis L.	96-856	BOGOR	***jarak (jepang)***
	95-855	KEW	
Ricinus communis L.	96-857	UKC	
	96-857	BOGOR	
Sauropus androgynus (L.) Merr.	71-13	KEW	***karatu, kastu, karata***
Sauropus androgynus (L.) Merr.	96-1223	KEW	
Sauropus androgynus (L.) Merr.	96-1224	BOGOR	
Sauropus androgynus (L.) Merr.	96-1224	UKC	
Sauropus androgynus (L.) Merr.	96-1225	BOGOR	
	03-22	UKC	
Triadica cochinchinensis Lour.	96-549	BOGOR	***sioe***
	96-549	KEW	
Homalanthus novoguineensis (Warb.) K. Schum.	96-550	UKC	
Homalanthus novoguineensis (Warb.) K. Schum.	96-551	KEW	
other Euphorbiaceae	96-131	UKC	***aniono, anionoi***
	96-132a	BOGOR	
	96-132b	KEW	
	96-133	KEW	
	96-437	KEW	
	96-438	UKC	
	96-439	BOGOR	
	96-527a	BOGOR	
	96-527b	KEW	
	96-997	KEW	***asihata(ia)***
	96-996a	BOGOR	
	96-996b	KEW	
Phyllanthaceae			
Actephila excelsa var. *javanica* (Miq.) Pax & K. Hoffm. [*Actephila javanica* Miq.]	96-1160	BOGOR	***puane nimusie***
	96-196	KEW	
	96-197	UKC	
	96-195	BOGOR	
	96-1161	KEW	
	96-1162	UKC	
Antidesma stipulare Blume (det. P. Hoffmann)	96-1076	KEW	***ahau uni***
	96-1075	BOGOR	
	96-1077	BOGOR	
Antidesma stipulare Blume (det. P. Hoffmann)	96-537	BOGOR	***ahauni, ahau un(i), ahau unu***
Antidesma stipulare Blume (det. P. Hoffmann)	96-538	UKC	
Antidesma stipulare Blume (det. P. Hoffmann)	96-539	KEW	
Antidesma montanum Blume	15-85	BOGOR	
	15-86	KEW	
Breynia cernua (Poir.) Müll. Arg.	71-39	KEW	***ai kihoke***

Taxonomic information	Voucher code or other source of identification		Nuaulu name
Glochidion Forster & Forster f. sp.	96-552	KEW	
	96-553a	KEW	
Glochidion rubrum Blume	96-553b	BOGOR	
Glochidion Forster & Forster f. sp.	96-554	KEW	
Glochidion macrocarpum Blume	15-108	BOGOR	***ai mane***
Glochidion macrocarpum Blume	15-109	KEW	
Glochidion macrocarpum Blume	15-110	KEW	
Phyllanthus borneensis Müll. Arg. [*Glochidion borneense* Boerl.]	96-1069	BOGOR	***tom-tom hatae*** syn. ***tom-tom wesie***
	96-135	KEW	
	96-136	UKC	
	96-134	BOGOR	
Glochidion Forster & Forster f. sp.	96-1070	KEW	
Glochidion Forster & Forster f. sp.	96-1071	UKC	
Phyllanthus borneensis Müll. Arg. [*Glochidion borneense* Boerl.]	96-600	BOGOR	***sonae***
	96-600	UKC	
	96-601	KEW	
Fabales			
Fabaceae			
Acacia Mill. sp.	96-785	KEW	***akasie***
	96-784	BOGOR	
Acacia probl. *mangium* Willd.	96-786	UKC	
Arachis hypogaea L.		PROSEA	***kaisan***
Arachis hypogaea L. var.		PROSEA	***kaisan msinae***
Arachis hypogaea L. var.		PROSEA	***kaisan putie***
Archidendron clypearia (Jack) I.C. Nielson [*Pithecellobium clypaeria* Benth.]		PROSEA	***kawasa***
Archidendron clypearia (Jack) I.C. Nielsen [*Pithecellobium clypaeria* Benth.]		PROSEA	***kawasa hanaie***
	96-208	KEW	
	96-209	UKC	
	96-207	BOGOR	
Archidendron clypearia (Jack) I.C. Nielsen [*Pithecellobium clypaeria* Benth.]		PROSEA	***kawasa pina*** syn. ***ananaine***
Bauhinia semibifida Roxb. [*Phanera semibifida* (Roxb.) Benth.]	96-614	BOGOR	***hihane wane, kihane wane***
Bauhinia L. sp.	96-615	KEW	
Bauhinia L. sp.	96-616	UKC	
Phanera kockiana (Korth) Benth. [*Bauhinia kockiana* Benth.]	96-1239	BOGOR	
	96-1240	KEW	
	96-1241	UKC	
Caesalpinia L. sp.	71-541a	UKC	***kapikute***
Caesalpinia L. sp.	71-542a	UKC	***kapikute ikine***
Caesalpinia L. sp.	71-542b	SING	
Caesalpinia L. sp.	71-541b	SING	***kapikute mainae***
Caesalpinia pulcherrima (L.) Sw.	71-6	KEW	***kupa ruso***
Callerya atropurpurea (Wall.) Schot [*Millettia atropurpurea* (Wall) Benth.]		PROSEA	***ai kasitena***
Chamaecrista (L.) Moench. (syn. *Cassia*) sp.	96 1386	KEW	***saumahu pinaie***

Taxonomic information	Voucher code or other source of identification		Nuaulu name
Canavalia maritima Thouars	96-859a	BOGOR	***kasitena nuae***
	96-859b	KEW	
Canavalia maritima Thouars	96-860	UKC	
Cassia fistula L.		PROSEA	***kihue putie***
Crotalaria retusa L.	71-22	KEW	unknown 7
Cynometra cauliflora L.		PROSEA	***namu-namu***
Delonix regia (Hook.) Raf.		PROSEA	***ai nuelo***, (***ai***) ***yainuelo***
Dendrolobium umbellatum (L.) Benth. [*Desmodium umbellatum* (L.) DC.]	71-57	KEW	***ai moti***
Derris spp.			(***awane***) ***munu***
Derris probl. *elliptica* (Wall.) Benth.		UKC	(***awane***) ***munu tunne***
Derris trifoliata Lour.	96-1072	BOGOR	(***awane***) ***munu apane*** syn. (***awane***) ***munu wesie***
Derris trifoliata Lour.	96-1073	UKC	
	96-1074	KEW	
Desmodium heterocarpon (L.) DC.	71-63	KEW	***mono nuae***
Desmodium heterocarpon (L.) DC.	71-64	KEW	
	96-635	no spec	
	96-636	no spec	
	96-637	no spec	
Desmodium sequax Wall.	71-16	KEW	(***mono***) ***panu-panu***
	73-3	UKC	
	96-88	BOGOR	
Leptaspis urceolata (Roxb.) R. Br.	15-97	BOGOR	
	15-98	KEW	
Erythrina variegata L. [*Erythrina indica* Lam.]	96-915	BOGOR	***rene***
Erythrina L. sp.	96-916	KEW	
Erythrina L. sp.	96-917	UKC	
Falcataria moluccana (Miq) Barneby & J.W. Grimes		PROSEA	***nisoae*** syn. ***onsoai***
Falcataria moluccana (Miq) Barneby & J.W. Grimes, var., probl.		PROSEA	***nisoae putie*** syn. ***onsoai putie***
Falcataria moluccana (Miq) Barneby & J.W. Grimes, var., probl.	96-1395	KEW	
Falcataria moluccana (Miq) Barneby & J.W. Grimes, var., probl.		PROSEA	***nisoae msinae*** syn. ***onsoai msinae***
Glycine max (L.) Merr.		PROSEA	***kedelai***
Inocarpus fagifer (Parkinson) Fosberg [*Inocarpus edulis* Forst.]		PROSEA	***kaine***
Intsia bijuga (Colebr.) Kuntze	96-558	BOGOR	***tonate***
	96-559	UKC	
Intsia Thouars. sp.	96-560	KEW	
	96-919	BOGOR	
Intsia bijuga (Colebr.) Kuntze	96-920	KEW	
	96-921a	BOGOR	
	96-921b	KEW	
Koompassia excelsa (Becc.) Taub.	96-116	UKC	***ima***
	96-117	KEW	
	96-118	UKC	
Mastersia bakeri Backer ex Koord.	96-771	BOGOR	(***tuha***) ***wane***
Mastersia bakeri Backer ex Koord.	96-772	BOGOR	
Mastersia bakeri (Koord.) Back.	96-773	KEW	

Taxonomic information	Voucher code or other source of identification		Nuaulu name
Millettia sericea Wight & Arn.	96-1335	BOGOR	***awane senane***
	96-432	KEW	
	96-431a	BOGOR	
	96-1336	KEW	
	96-1337	UKC	
Mimosa pudica L.	96-8	KEW	***pitiri malu***
Mimosa pudica L.	96-7	BOGOR	
Mimosa pudica L.	96-9	UKC	
Mimosa pudica L.	15-24	BOGOR	
Mimosa pudica L.	15-29	BOGOR	
Neptunia oleracea Lour. [*Neptunia prostrata* (Lam.) Baill.]		PROSEA	***masikunne***
Parkia speciosa Hassk.			***muku hanaie***
Phaseolus vulgaris L.		PROSEA	***buncis***
Pongamia pinnata (L.) Pierre		PROSEA	
Pongamia pinnata (L.) Pierre	96-1193	BOGOR	***namte***
	96-1194	KEW	
Pongamia pinnata (L.) Pierre	96-1195	UKC	
Pterocarpus Jacq.		PROSEA	***nana kunini***
Pterocarpus Jacq.		PROSEA	***nana minete***
Pterocarpus indicus Willd.	96-1196	BOGOR	***atipanate***
Vitex cofassus Reinw. ex Blume [see also Verbenaceae]	96-1197a	KEW	
	96-1197b	BOGOR	
	96-1198	KEW	
Pterocarpus indicus Willd.	71-239	SING	***nana***
Pterocarpus indicus Willd.	15-126	BOGOR	
	15-127	KEW	
	15-128	KEW	
	15-132	KEW	
Pterocarpus indicus Willd.	96-1127	BOGOR	***nana tope***
Pterocarpus indicus Willd.	96-1126	BOGOR	
Pterocarpus indicus Willd.	96-1127	UKC	
	96-1128	KEW	
Senna alata (L.) Roxb. [*Cassia alata* L.]	96-736	BOGOR	***ai turi***
Senna alata (L.) Roxb. [*Cassia alata* L.]	96-737	KEW	
Senna alata (L.) Roxb. [*Cassia alata* L.]	96-738	UKC	
Senna occidentalis (L.) Link.	96-3	KEW	***monote***
	96-2	BOGOR	
Senna occidentalis (L.) Link.	96-4	UKC	
Senna sophera (L.) Roxb. [*Cassia sophera* L.]	71-8	KEW	***ai saja***
Tamarindus indica L.		PROSEA	***asamjawa***
Vigna Savi.	73-04	UKC	***kihue***
Vigna radiata L. R. Wilczek		PROSEA	
	96-1375	UKC	
	96-1376	UKC	
	96-1377	UKC	
Vigna radiata L. R. Wilczek		HEYNE	***manora, kihue manora, kihue leka***
Vigna unguiculata (L.) Walp. grp. *sesquipedalis*		PROSEA	***kacang panjang***

Taxonomic information	Voucher code or other source of identification		Nuaulu name
other Fabaceae			***kihue marae***
	96-76	KEW	***mono maianane***
	96-75	BOGOR	
	96-77	UKC	
	96-907	KEW	***naka-naka naene*** syn. ***naka-naka nuae***
	96-906	BOGOR	
	96-908	UKC	
	71-60	KEW	***rama(ne) wane***
	96-164	UKC	
	96-264	BOGOR	
	96-265	KEW	
	96-266	UKC	
Rosales			
Rhamnaceae			
Colubrina Rich. ex Brongn. sp.	96-927	KEW	***amnae ai anawasi*** syn. ***(n)amnae nuae***
Rosaceae			
Prunus L. sp.	96-957	KEW	***ai ane***
	96-956	BOGOR	
Prunus arborea (Blume) Kalkman or *Prunus wallaceana* Kalkman	96-958	UKC	
	96-958	KEW	
Prunus javanica (Teijsm. & Binn.) Miq., probl.	96-1098	BOGOR	***aunasate***
	96-1099	KEW	
Prunus javanica (Teijsm. & Binn.) Miq., probl.	96-1100	UKC	
Rubus L. sp.	96-401	KEW	***awane putie***
	96-400	BOGOR	
Rubus L. sp.	96-402	UKC	
Rubus fraxinifolius Poir.	96-977	BOGOR	***(a)wane unie***
Rubus L. sp.	96-978	KEW	
Rubus L. sp.	96-979a	UKC	
Rubus L. sp.	96-979b	BOGOR	
Rubus L. sp.	03-06	UKC	
Rubus moluccanus L.	71-03c	KEW	***kupa rosi***
Cannabaceae			
Celtis L. sp.	96-262	KEW	***sosone***
	96-261	BOGOR	
Celtis probl. *philippinensis* Blanco	96-263	UKC	
Ulmaceae			
Trema orientalis (L.) Blume	96-1258	BOGOR	***sapa(ne)***
Trema Lour. sp.	96-410	KEW	
Trema Lour. sp.	96-411	UKC	
	96-409	BOGOR	
Trema Lour. sp.	96-1259	KEW	
Trema Lour. sp.	96-1260	UKC	
Moraceae			
Antiaris prob. *toxicaria* Lesch		PROSEA	***ai polo***
	96-479	BOGOR	
Artocarpus J.R. Forst. & G. Forst. sp.		PROSEA	***ai une***
Artocarpus J.R. Forst. & G. Forst. sp.			***sune***

Taxonomic information	Voucher code or other source of identification		Nuaulu name
Artocarpus J.R. Forst. & G. Forst. sp.		PROSEA	***sune nimasae***
Artocarpus altilis (Parkinson ex F.A. Zorn) Fosberg		PROSEA	***sune niane***
Artocarpus altilis (Parkinson ex F.A. Zorn) Fosberg [*Artocarpus communis* J.R. Forst. & G. Forst.]	96-1130a	BOGOR	***une***
Artocarpus J.R. Forst. & G. Forst. sp.	96-1130b	UKC	
	96-1131	KEW	
Artocarpus J.R. Forst. & G. Forst. sp.	96-1129	KEW	
	96-1369	BOGOR	
	96-1370	KEW	
	96-1371	UKC	
Artocarpus elasticus Reinw. ex Blume		PROSEA	***nimasi***
Artocarpus heterophyllus Lam.	96-1329	BOGOR	***utatila***
Ficus L. sp.	96-1330	KEW	
Ficus L. sp.	96-1331	UKC	
Artocarpus integer (Thunb.) Merr.		PROSEA	***tahenae***
Artocarpus integer (Thunb.) Merr.		PROSEA	
Artocarpus integer (Thunb.) Merr.		PROSEA	***tahenae msinae***
Artocarpus integer (Thunb.) Merr.	96-1089	BOGOR	***tahenae nusa ina***
	96-1090	UKC	
	96-1091	KEW	
Artocarpus integer (Thunb.) Merr.		PROSEA	
	96-1330a	KEW	
	96-1331a	UKC	***tahenae warata***
	96-1329a	UKC	
Broussonetia papyrifera (L.) Vent.	96-1107	BOGOR	***hisa onate***
	96-1108	KEW	
Broussonetia papyrifera (L.) Vent.	96-1108	BOGOR	
	96-1109	KEW	
Ficus L. sp.	96-797	KEW	***amnae wesie***
	96-796	BOGOR	
Ficus L. sp.	96-798	UKC	
Ficus adenosperma Miq.	96-702	BOGOR	***nusu***
Ficus L. sp.	96-703	UKC	
Ficus L. sp.	96-704	KEW	
Ficus L. sp.		PROSEA	***nunu(e) ina aia***
Ficus L. sp.	96-455	KEW	***nunu inane***
	96-141	KEW	
Ficus L. sp.	96-142	UKC	
	96-454	BOGOR	
	96-140	BOGOR	
Ficus L. sp.	96-456	UKC	
Ficus L. sp.		PROSEA	***nunu mkauke***
Ficus L. sp.		PROSEA	***nunu(e) potone***
Ficus L. sp.	96-1192	UKC	***nunue tinse***
	96-1191	KEW	
Ficus L. sp.	96-1190	UKC	
Ficus L. sp., probl.	96-913	KEW	***pate***
	96-912	BOGOR	
	96-914	UKC	
Ficus L. sp.		PROSEA	***tatane***
Ficus L. sp.	96-369	UKC	***nunu(e)***

Taxonomic information	Voucher code or other source of identification		Nuaulu name
Ficus benjamina L., probl.		PROSEA	***nunu(e) putie***
Ficus botryocarpa subsp. *hirtella* C.C. Berg [*Ficus conora* King]	71-679	SING	***kaiohite***
Ficus callosa Willd.	96-566	KEW	
Ficus glandulifera Wall.	96-641	BOGOR	
Ficus glandulifera Wall.	96-642	UKC	
Ficus copiosa Steud.	96-1113	BOGOR	
Ficus botryocarpa Miq.	15-37	BOGOR	***kaiohite***
Ficus botryocarpa subsp. *hirtella* C.C. Berg [*Ficus conora* King]	71-679	SING	***nana ai***
	96-1188	KEW	
Ficus L. sp.	96-1189	UKC	
Ficus pungens Reinw.ex Blume	96-745	BOGOR	
	96-368	KEW	
	96-367	BOGOR	
	96-1005	BOGOR	
Ficus lepicarpa Blume	96-754	BOGOR	***kako hatae***
Ficus L. sp.	96-755	KEW	
Ficus lepicarpa Blume	96-756	UKC	
Ficus nervosa subsp. pubinervis (Blume) C.C. Berg [*Ficus pubinervis* Blume]	96-573	BOGOR	***waronone***
Ficus L. sp.	96-574	UKC	
Ficus L. sp.	96-575	KEW	
Ficus nervosa subsp. pubinervis (Blume) C.C. Berg	15-105	BOGOR	
	15-106	KEW	
	15-107	KEW	
	15-133	KEW	
	96-1378	KEW	***waronone rapa***
Ficus L. sp.		PROSEA	
Ficus L. sp.		PROSEA	***waronone kahuite***
Ficus polyantha Warb. ex K. Schum. & Lautertb.	96-582	BOGOR	***kaenuri, kaenuni***
Ficus polyantha Warb. ex K. Schum. & Lautertb.	96-583	UKC	
	96-584	KEW	
	96-1141	BOGOR	
	96-1142	KEW	
	96-1143	UKC	
Ficus cf. *septica* Burm f.	96-13	KEW	***tahola***
	71-70	KEW	
	96-14	KEW	
Ficus cf. *septica* Burm f.	96-15	KEW	
Ficus tinctoria G. Forst.	96-1305	BOGOR	***(ai) supa***
Ficus L. sp.	96-490	KEW	
Ficus L. sp.	96-491	UKC	
Ficus tinctoria G. Forst.	96-686	BOGOR	
Ficus tinctoria G. Forst.	96-687	UKC	
Ficus callosa Willd.	96-688	KEW	
	96-489	BOGOR	
Ficus L. sp.	96-1306	KEW	
Ficus L. sp.	96-1307	UKC	

Taxonomic information	Voucher code or other source of identification		Nuaulu name
Ficus tinctoria G. Forst.	15-120	BOGOR	
	15-121	KEW	
	15-122	KEW	
Ficus variegata Blume	15-129	BOGOR	***waruae, uarue***
	15-130	KEW	
	15-131	KEW	
	96-433	BOGOR	
Poikilospermum suaveolens (Blume) Merr.	96-1051	BOGOR	***kinoke wane*** (***inae***)
	96-1052	KEW	
Poikilospermum suaveolens (Blume) Merr.	96-1053	UKC	
Urticaceae			
NETTLES			***sinatane***
Dendrocnide microstigma (Gaudich. ex Wedd) Chew	96-1025	BOGOR	***sinatane urone***
	96-1026	KEW	
Dendrocnide microstigma (Gaudich. ex Wedd) Chew	96-1027	UKC	
Dendrocnide stimulans (L. f.) Chew. [*Laportea stimulans* Miq.]	96-1022	BOGOR	***sinata***(***ne***) ***onate***
	96-1023	KEW	
Dendrocnide stimulans (L. f.) Chew. [*Laportea stimulans* Miq.]	96-1024	UKC	
Elatostema sinuatum (Blume) Hassk.	96-1281	BOGOR	***pasate***
Elatostema sinuatum (Blume) Hassk.	96-1282	KEW	
Elatostema sinuatum (Blume) Hassk.	96-1283	UKC	
	03-01	UKC	
Laportea interrupta (L.) Chew	71-78	KEW	***hahusina***
Fleurya interrupta (L.) Gaudich.	96-717	BOGOR	
Laportea interrupta (L.) Chew	96-718	KEW	
Laportea interrupta (L.) Chew	96-719	UKC	
Pipturus argenteus Wedd.	96-739	BOGOR	***karosine***
	96-740	KEW	
	96-741	UKC	
Pipturus argenteus Wedd.	96-1169	BOGOR	
	96-1170	UKC	
Pipturus argenteus Wedd.	96-1171	KEW	
Procris frutescens Blume, probl.	71-44	KEW	***sinatane urone***
Urticastrum decumanum (Roxb.) Kuntze [*Laportea decumana* Wedd.]	96-1296	BOGOR	***sina*** (***manusia***)
	96-1297	KEW	
Urticastrum decumanum (Roxb.) Kuntze [*Laportea decumana* Wedd.]	96-1298	UKC	
	96-1351	BOGOR	
Cucurbitales			
Anisophylleaceae			
Anisophyllea disticha Baill. (Heyne)		PROSEA	***ai anawasi***
Cucurbitaceae			
Benincasa hispida (Thunberg ex Murray) Cogniaux		PROSEA	***rorong***
Citrullus lanatus (Thunberg) Matsum. & Nakai		PROSEA	(***tumno***) ***pateka***
Curcubita probl. *maxima* Duchesne	03-11	UKC	***awane munie***
Cucurbita L. sp.		PROSEA	***awane munie huae***
Cucurbita L. sp.		PROSEA	***napu***
Cucurbita probl. *moschata* Duchesne		PROSEA	***napu sina***

Taxonomic information	Voucher code or other source of identification		Nuaulu name
Cucurbita moschata Duchesne		PROSEA	***napu onate***
			tumno
Cucurbita pepo L.		PROSEA	***tumno nawe***
Cucumis sativus L.		PROSEA	***papino*** syn. ***temon***
Lagenaria siceraria (Molina) Standl. [*Cucurbita lagenaria* L.]		PROSEA	***napu potone***
Luffa acutangula (L.) Roxb., *L. cylindrica* (L.) M. Roem.		PROSEA	***sayor patola***
Trichosanthes cucumerina L.		PROSEA	***napu tekene***
Zehneria japonica (Thunb.) H.Y. Liu [*Melothria indica* L.]	71-59	KEW	***awane kin-kin***
Tetramelaceae			
Octomeles mollucana Teijsm. & Boerl., poss.	96-732	UKC	***ai punara***
	96-731	KEW	
	96-730	BOGOR	
Octomeles sumatrana Miq.	96-327	KEW	***punara noene***
Octomeles sumatrana Miq.	96-325	BOGOR	
Begoniaceae			
Begonia isoptera Dryand.	96-1202	BOGOR	***pasite***
Begonia L. sp.	96-1203	KEW	
Begonia L. sp.	96-1204	UKC	
Begonia isoptera Dryand.	96-225	BOGOR	***rorie***
Begonia L. sp.	71-26	KEW	
Begonia L. sp.	96-226	KEW	
Begonia L. sp.	96-227	UKC	
	96-1356	BOGOR	
Begonia L. sp.	96-1357	KEW	
Begonia L. sp.	96-1358	UKC	
Begonia L. sp.		PROSEA	***rori(e) msinae***
Fagales			
Casuarinaceae			
Casuarina equisetifolia L.	96-782	KEW	***neune***
	96-781	BOGOR	
Casuarina rumphiana Miq.	96-835a	UKC	***asuan(i) mata hunue***
Gymnostoma rumphianum (Miq.) L.A.S. Johnson	96-783	UKC	***neune anae***
Fagaceae			
Castanopsis sp., probl.	PROSEA		***ai nasa tiane***
Myrtales			
Combretaceae			
Terminalia catappa L.	96-33	KEW	***kaeane***
	96-32	BOGOR	
Terminalia catappa L.	96-34	UKC	
	96-595	BOGOR	
Terminalia catappa L.	96-33	KEW	***kaeane msinae*** syn. ***kaea wesie***
Terminalia catappa L.	96-00	KEW	***kaeane putie***
Terminalia poss. *copelandii* Elmer.		PROSEA	***kaea(ne) nuae***
Onagraceae			
Ludwigia L. sp.	96-20	BOGOR	***mono totu mkauke***
	96-19	KEW	
Ludwigia L. sp.	96-21	UKC	
Lythraceae			
Duabanga moluccana Blume	71-73	KEW	***niotune***
Duabanga sonneratioides Buch.-Ham	73-19	UKC	

Taxonomic information	Voucher code or other source of identification		Nuaulu name
Lawsonia inermis L.	96-24	UKC	
Lawsonia inermis L.	96-25	KEW	
	96-23	BOGOR	***naka-naka***
Punica granatum L.	PROSEA		***dalima***
Myrtaceae			
Decaspermum fruticosum J.R. Forst. & G. Forst., probl.	96-999	KEW	***kupanaha(ne) wesie*** syn. ***ai huae kauke*** syn. ***kupanaha ukune***
Eucalyptus deglupta Blume		PROSEA	***ai iene***
Eugenia sp.			***hukila***
Melaleuca cajuputi Powell [syn. *M. leucadendra* L.]	96-1175	KEW	***sakaputi*** syn. ***sakana***
Melaleuca cajuputi Powell [syn. *M. leucadendra* L.]	96-1176	UKC	
	96-1177	KEW	
	96-998	BOGOR	
	96-1000	UKC	
Psidium guajava L.	96-1332	BOGOR	***koiawas***
Psidium guajava L.	96-1333	KEW	
Psidium guajava L.	96-1334	UKC	
Rhodamnia cinerea Jack.	71-11	KEW	***ai kau***
Rhodamnia cinerea Jack.	96-543	BOGOR	
	96-543	KEW	
	96-544	UKC	
	96-545	KEW	
Rhodamnia cinerea Jack.	96-950b	BOGOR	***katinoie numte***
Syzygium aromaticum (L.) Merr. & L.M. Perry		PROSEA	***kupanahane***
Syzygium aromaticum (L.) Merr. & L.M. Perry		PROSEA	***kupanahane hua ikine***
Syzygium aromaticum (L.) Merr. & L.M. Perry		PROSEA	***kupanahane hua mainae***
Syzygium Gaertn. sp.	96-469	KEW	***aninoi***
Syzygium Gaertn. sp.	96-470	UKC	
	96-471	BOGOR	
	96-959a	BOGOR	
	96-959b	KEW	
	96-960	BOGOR	
Syzygium Gaertn.		PROSEA	***apaine***
Syzygium Gaertn.	96-98	BOGOR	
	96-99	KEW	
	96-100	UKC	***apaine numte***
			apaine totu onate
Syzygium Gaertn.	96-348	KEW	***apaine totu mkauke*** syn. ***apaine totu pukune***
	96-345	KEW	
	96-346	BOGOR	
Syzygium Gaertn. sp.	96-347	UKC	
	96-348	UKC	
Syzygium acuminatissimum (Blume) DC. [*Eugenia acuminatissima* Kurz]	96-1042	BOGOR	***wasone***
Syzygium Gaertn. sp.	96-1043	KEW	
Syzygium Gaertn. sp.	96-1044	UKC	
Syzygium aqueum (Burm. f.) Alston [*Eugenia aquea* Burm. f.]	96-591	BOGOR	***atupikae***
Syzygium Gaertn. sp.	96-592	KEW	
Syzygium Gaertn. sp.	96-593	UKC	

Taxonomic information	Voucher code or other source of identification		Nuaulu name
Syzygium Gaertn. sp.	96-889	KEW	
Syzygium Gaertn. sp.	96-890	UKC	
Syzygium aqueum (Burm. f.) Alston		PROSEA	***popote***
	96-86	BOGOR	
Syzygium aqueum (Burm. f.) Alston [*Eugenia aqueua* Burm. f.]			***popo(te) msinae***
Syzygium aqueum (Burm. f.) Alston [*Eugenia aqueua* Burm. f.]	96-671	BOGOR	
Syzygium aqueum (Burm. f.) Alston [*Eugenia aqueua* Burm. f.]	96-672	BOGOR	
Syzygium aqueum (Burm. f.) Alston [*Eugenia aqueua* Burm. f.]	96-673	UKC	
	15-61	BOGOR	
	15-62	KEW	
Syzygium Gaertn. sp.	15-63	BOGOR	
Syzygium aqueum (Burm. f.) Alston [*Eugenia aqueua* Burm. f.]	96-1245	BOGOR	***popote putie***
Syzygium Gaertn. sp.	96-1246	KEW	
Syzygium aqueum (Burm. f.) Alston	96-1247	UKC	
Syzygium boerlagei (Merr.) Govaerts			***asapopote***
	15-114	BOGOR	
	15-115	KEW	
	15-116	KEW	
Syzygium Gaertn. sp.	96-101	UKC	***asapop(o) manea***
	96-102	BOGOR	
Syzygium Gaertn. sp.	96-103	KEW	
Eugenia L. sp.	96-1255	BOGOR	
Syzygium Gaertn. sp.	96-1256	KEW	
Syzygium Gaertn. sp.	96-1257	UKC	
Syzygium Gaertn. sp.	96-166	KEW	***asapopo makae***
	96-165	BOGOR	
	96-167	UKC	
Syzygium boerlagei (Merr.) Govaerts	15-21	BOGOR	
	15-25	KEW	
	15-26a	KEW	
Syzygium Gaertn. sp.	96-953	UKC	***auuhone***
	96-954	BOGOR	
	96-954	KEW	
Syzygium Gaertn. sp.	96-955	KEW	
Syzygium Gaertn. sp., probl.	96-963	KEW	***ene***
	96-962	BOGOR	
Syzygium Gaertn. sp., probl.	96-964	UKC	
	96-964	KEW	
Syzygium Gaertn. sp., probl.	96-332	KEW	***hino***
	96-331	BOGOR	
	96-333	BOGOR	
Syzygium malaccense (L.) Merr. & L.M. Perry [*Eugenia malaccensis* L.]	96-1048	BOGOR	***kanoke***
Syzygium malaccense (L.) Merr. & L.M. Perry [*Eugenia malaccensis* L.]	96-1049	UKC	

Taxonomic information	Voucher code or other source of identification		Nuaulu name
	96-1050	KEW	
Syzygium malaccense (L.) Merr. & L.M. Perry [*Eugenia malaccensis* L.]	96-1289b	BOGOR	***nutune***
Syzygium malaccense (L.) Merr. & L.M. Perry [*Eugenia malaccensis* L.]	96-1287	BOGOR	***nutu(ne) msinae***
	96-1288	KEW	
Syzygium malaccense (L.) Merr. & L.M. Perry [*Eugenia malaccensis* L.]	96-1289a	UKC	
Syzygium malaccense (L.) Merr. & L.M. Perry [*Eugenia malaccensis* L.]	96-1289b	BOGOR	
Syzygium malaccense (L.) Merr. & L.M. Perry [*Eugenia malaccensis* L.]		PROSEA	***nutu(ne) putie***
Syzygium rumphii Govaerts.		HEYNE	***ai msinae***
Syzygium Gaertn. sp.	96-789	KEW	
	96-788	BOGOR	
	96-788	KEW	
	96-787	BOGOR	
Tristaniopsis whiteana (Griff.) Peter G. Wilson & J.T. Waterh. [*Tristania whitiana* Griff.]	96-1184	BOGOR	***hino numte***
Tristaniopsis whiteana (Griff.) Peter G. Wilson & J.T. Waterh. [*Tristania whitiana* Griff.]	96-1185	UKC	
	96-1186	KEW	
Xanthostemon verus (Roxb.) Peter G. Wilson [*Metrosideros vera* Roxb.]	96-365	KEW	***nahane***
	96-364	BOGOR	
Xanthostemon verus (Roxb.) Peter G. Wilson [*Metrosideros vera* Roxb.]	96-366	UKC	
Melastomataceae			
Clidemia hirta (L.) D. Don	96-73	KEW	***mono karanate***
	96-72	BOGOR	
	96-74	BOGOR	
Melastoma L. sp.	96-950a	UKC	***katinoie numte***
Melastoma L. sp.	96-951	KEW	
Melastoma L. sp.	96-952	KEW	
Melastoma imbricatum Wall. ex C.B. Clarke	96-310	KEW	***katinoie***
	96-309	BOGOR	
Melastoma L. sp.	96-311	UKC	
	96-830	BOGOR	
Melastoma L. sp.	96-831	KEW	
Melastoma L. sp.	96-832	UKC	
Melastoma malabathricum L.	96-1285	KEW	***katinoi wan anoi***
	96-440	BOGOR	
	96-441	KEW	
	96-442	UKC	
Melastoma malabathricum L.	15-59	BOGOR	***kasinoi waene anoi***
	15-60	KEW	
Melastoma affine D. Don	96-1284	BOGOR	
	71-80	KEW	
Melastoma malabathricum L.	96-1286	UKC	
Melastoma polyanthum Blume	96-655	BOGOR	***kamnatine***

Taxonomic information	Voucher code or other source of identification		Nuaulu name
Melastoma moluccanum Blume	96-656	KEW	
	03-07	UKC	
Pternandra cordata Baill.	96-528	BOGOR	***popote wesie***
Pternandra caerulescens Jack	96-529	UKC	
Pternandra caerulescens Jack	96-530	KEW	
Sapindales			
Burseraceae			
Canarium sp.	15-134	KEW	***kamane***
Canarium asperum Benth.	15-102	BOGOR	***kamane kamine***
	15-103	KEW	
	15-104	KEW	
	96-1045	BOGOR	
	96-1046	KEW	
	96-1047a	BOGOR	
	96-1047b	UKC	***(kama) kamine, kama minne***
Canarium decumanum Gaertn.	96-1047c	UKC	***kamine***
Canarium hirsutum Willd.	96-1270	BOGOR	***au mamine***
	96-1271	KEW	
	96-1272	UKC	
Canarium L. sp.	96-1347	UKC	***iane***
Canarium indicum L.	15-99	BOGOR	
Canarium L. sp.	15-100	KEW	
Canarium L. sp.	15-101	KEW	
Canarium L. sp.	96-894	BOGOR	
Canarium L. sp.	96-835	BOGOR	
Canarium L. sp.	96-936	KEW	
Canarium L. sp.	96-937	KEW	
Canarium L. sp.	96-891	BOGOR	***iane kauke, ia(ne) mkauke*** syn. ***ia yapono*** syn. ***ia haruku*** syn. ***ia pukune*** syn. ***ia katanepane***
Canarium indicum L. [*Canarium commune* L.]		PROSEA	***ia(ne) hanaie***
Canarium indicum L. [*Canarium commune* L.]	96-1006	BOGOR	***ia(ne) hanate*** syn. ***ia nawe*** syn. ***kahiaue***
Canarium indicum L.	15-01	BOGOR	
	15-136	BOGOR	
Canarium indicum L. [*Canarium commune* L.]		PROSEA	***ia(ne) nawe***
Canarium indicum L. [*Canarium commune* L.]		PROSEA	***ia(ne) pukune***
Canarium L. sp.	96-892	KEW	
Canarium L. sp.	96-893	UKC	
Canarium sylvestre Gaertn.	15-05	BOGOR	***ananate***
	15-06	KEW	
Canarium sylvestre Gaertn.	15-07	BOGOR	
Canarium sylvestre Gaertn.	15-137	BOGOR	
Canarium L. sp.	96-192	BOGOR	***ananate, ia ananate***
	96-193	KEW	
	96-194	KEW	
	96-358	BOGOR	
Canarium L. sp.	96-359	KEW	
Canarium L. sp.	96-360	UKC	
	96-621	KEW	
Canarium sylvestre Gaertn.	96-622	UKC	

Taxonomic information	Voucher code or other source of identification		Nuaulu name
Canarium sylvestre Gaertn.	15-91	BOGOR	
Anacardiaceae			
Anacardium occidentale L.		PROSEA	***bigi keluar***
Bouea macrophylla Griff.		PROSEA	***gandaria***
Campnosperma auriculata Hook. f.	96-599	BOGOR	***sonae***
	96-944	BOGOR	***sonae totu ikin**e* syn. ***sono o manue***
Campnosperma Thwaites sp.	96-945	KEW	
Campnosperma Thwaites sp.	96-946	UKC	
Dracontomelon dao (Blanco) Merr. & Rolfe	96-887	KEW	***karane, kerane***
	96-886	KEW	
	96-885	BOGOR	
Mangifera foetida Lour.		PROSEA	***aune***
Mangifera indica L.		PROSEA	***apalam***
Mangifera indica L. var.		PROSEA	***apalam hua papae***
Mangifera indica L. var.		PROSEA	***apalam hua pukune***
Mangifera indica L. var.		PROSEA	***apalam tunne***
Mangifera indica L. var.		PROSEA	(***apalam***) ***kesi***
Mangifera odorata Griff.		PROSEA	***kuine***
Semecarpus cassuvium Roxb.	96-611a	BOGOR	***ninate***
Semecarpus cassuvium Roxb.	96-611b	UKC	
Semecarpus cassuvium Roxb.	96-612	UKC	
	96-613	KEW	
Spondias cytherea Sonn.		WHITM	***nihasa***
Spondias novoguineensis Kosterm.	96-1267a	BOGOR	
	96-1267b	KEW	
	96-1268	KEW	
Spondias novoguineensis Kosterm	96-1269	UKC	
Spondias pinnata (L. f.) Kurz		WHITM	
Sapindaceae			
Allophylus L. sp.	96-838	KEW	***pasa natae, pasan atae, pasane hata***
	96-837	BOGOR	
Nephelium lappaceum L.		PROSEA	***rambutan***
Simaroubaceae			
Brucea javanica (L.) Merr. [*Picrasma javanica* Blume]	96-284	KEW	***amnae***
	96-283	BOGOR	
Brucea javanica (L.) Merr. [*Picrasma javanica* Blume]	96-285	UKC	
	96-391	BOGOR	
	96-392	BOGOR	
Brucea javanica (L.) Merr.	96-393	KEW	
	96-926	BOGOR	
Brucea javanica (L.) Merr.	96-928	UKC	
	15-44	BOGOR	
Brucea javanica (L.) Merr.	71-79	KEW	***amnae napotone***
Meliaceae			
Aglaia sapindina Harms.	96-720	BOGOR	***apunue***
	96-721	KEW	
Aglaia sapindina Harms.	96-722	UKC	

Taxonomic information	Voucher code or other source of identification		Nuaulu name
Chisocheton ceramicus Miq. [*Chisocheton sandoricocarpus* Koord. & Valeton]	96-1299	BOGOR	***ai osi*** syn. ***tahena ukuna***
Aglaia parviflora DC.	96-1300	KEW	
	96-257	BOGOR	
	96-382	KEW	
	96-383	BOGOR	
	96-384	KEW	
	96-255	BOGOR	
	96-256	KEW	
	96-1301	UKC	
	15-83	KEW	
	96-1390	BOGOR	
Aglaia sapindina (F. Muell.) Harms	15-81	BOGOR	
Aglaia sapindina (F. Muell.) Harms	15-82	BOGOR	
	96-483	BOGOR	***asu hunne***
	96-484	KEW	
Dysoxylum caulostachyum Miq., poss.	96-485	UKC	
Heynea trijuga Roxb. ex Sims			***menie yatanunuta*** syn. ***mamasu***
Lansium parasiticum (Osbeck) K.C. Sahni & Bennet [*Lansium domesticum* Correa], var.		PROSEA	***nasate***
Lansium parasiticum (Osbeck) K.C. Sahni & Bennet [*Lansium domesticum* Correa], var.		PROSEA	***duku***
Walsura Roxb. sp., probl.	96-961	KEW	***aninoi***
Rutaceae			
Citrus L.		PROSEA	***nusi***
Citrus L. sp.		PROSEA	***moni***
Citrus aurantifolia (Christm.) Swingle		PROSEA	***nusi sina***
Citrus aurantium L.		PROSEA	***nusi*** (***ruei***) ***hun***(***i***)
Citrus hystrix DC.		PROSEA	***nusi ramatae***
Citrus limon (L.) Burm. f.		PROSEA	***nusi moune***
Citrus maxima (Burm.) Merr.		PROSEA	(***nusi***) ***peka-peka***
Citrus medica L.		PROSEA	(***nusi***) ***kapane***
Citrus nobilis Lour.		PROSEA	(***nusi***) ***mananai***
Citrus reticulata Blanco		PROSEA	***nusi une***
Citrus sinensis (L.) Osbeck		PROSEA	(***nusi***) ***koromase, nusi masine***
Citrus sinensis (L.) Osbeck		PROSEA	***nusi masine, nusi koramase***
Lunasia amara Blanco	96-1172	BOGOR	***atuane***
	96-931	UKC	
	96-930	BOGOR	
	96-1173	KEW	
Lunasia amara Blanco	96-1174	UKC	
Murraya paniculata (L.) Jack.		PROSEA	***kemunin***
Zanthoxylum nitidum (Roxb.) DC.	96-1350	BOGOR	***awane seite***
	96-1146	BOGOR	
	96-1147	KEW	
Zanthoxylum L. sp.	96-1348	KEW	
Zanthoxylum nitidum (Roxb.) DC.	96-1349	BOGOR	
	96-1145	BOGOR	
other Rutaceae	96-1004	KEW	***topi-topi***
	96-1003	UKC	

Taxonomic information	Voucher code or other source of identification		Nuaulu name
	96-1002	BOGOR	
Malvales			
Thymelaeaceae			
Aquilaria Lam., or *Gyrinops versteegii* (Gilg) Domke; incl. *Phaleria capitata* Jack.		WHITM	***alalasie***
	96-429	KEW	
	96-430	UKC	
	96-428a	BOGOR	
Aquilaria probl. *audate* (Oken) Merr., or *Gyrinopsis cumingiana* (Decne.) Ridl		PROSEA	***alalasie pina***
Phaleria capitata Jack.	15-111	BOGOR	***alalasie hanaie***
	15-112	KEW	
	15-113	KEW	
	03-10	UKC	
Phaleria perrottetiana (Deccne.) Fern.-Vill.	96-629	BOGOR	***atupikae***
Phaleria Jack. sp.	96-630	KEW	
Phaleria Jack. sp.	96-631	UKC	
	96-888	BOGOR	
Phaleria perrottetiana (Decne.) Fern.-Vill.	96-576	BOGOR	***ai weti***
	96-597	KEW	
	96-598	KEW	***ai weti atue***
Bixaceae			
Bixa orellana L.		PROSEA	***kasupa***
Dipterocarpaceae			
Shorea selanica (Lam.) Blume		WHITM	(***kamane***) ***onie***
	96-338	KEW	
	96-339	UKC	
	96-337	BOGOR	
Shorea Roxb. ex Gaertn. f. sp.		PROSEA	***oni***(***e***) ***msinae***
Shorea Roxb. ex Gaertn. f. sp.		PROSEA	***oni***(***e***) ***putie***
Shorea Roxb. ex Gaertn. f. sp.		PROSEA	***onie hatae*** syn. ***punara hatae***
Shorea selanica (DC.) Blume		PROSEA	***ai ama***
Malvaceae			
Abelmoschus moschatus Medik.	96-626	BOGOR	***pionenane***
	71-38	KEW	
Abelmoschus moschatus Medik.	96-627	UKC	
Abelmoschus moschatus Medik.	96-628	KEW	
Abroma augusta (L.) L. f [*Abroma* (syn. *Ambroma*) *mollis* DC.]	96-1210	KEW	***saumahu***
Abroma fastuosa R. Br.	96-1208	BOGOR	
Abroma augusta (L.) L. f [*Abroma* (syn. *Ambroma*) *mollis* DC.]	96-1209	UKC	
	71-333	KEW	
Ceiba pentandra (L.) Gaertn.		PROSEA	***kapu*** syn. ***kanapua***
Commersonia bartramia (L.) Merr.	96-444	KEW	***nopa putie***, ***nopoi putie*** syn. ***nopoi waene***
	96-443	BOGOR	
Commersonia bartramia (L.) Merr.	96-445	UKC	
Durio L.		PROSEA	***tunene***
Durio zibethinus Murr. var.		PROSEA	***tunene hua ikine***
Durio zibethinus Murr. var.		PROSEA	***tunene hua mainae***

Taxonomic information	Voucher code or other source of identification		Nuaulu name
Durio zibethinus Murr. var.		BOLTON & MATOKE	***hutiene***
Gossypium barbadense L. [*Gossypium acuminatum* Roxb.]	96-1036	BOGOR	***kanapua***
Gossypium barbadense L. [*Gossypium acuminatum* Roxb.]	96-1037	BOGOR	
Gossypium barbadense L. [*Gossypium acuminatum* Roxb.]	96-1038	UKC	
Heritiera littoralis Aiton	96-1187	BOGOR	***nanae***
Heritiera littoralis Aiton	71-637	SING	
Heritiera littoralis Aiton	96-594	BOGOR	
	96-595	KEW	
Heritiera littoralis Aiton	96-595	BOGOR	
Heritiera Dryand.	96-596	KEW	
Hibiscus L. sp.			***pioi***
Hibiscus probl. *mutabilis* L.	71-19	KEW	***pioi nenane***
Hibiscus rosa-sinensis L.		PROSEA	***huka hukai***
Hibiscus tiliaceus L.	96-579	BOGOR	***hanu***
Hibiscus tiliaceus L.	96-580	KEW	
	96-581	UKC	
	96-1366	BOGOR	
	96-1367	UKC	
	96-1368	UKC	
Kleinhovia hospita L.	96-803	UKC	***kinae***
Kleinhovia hospita L.	96-804	KEW	
Kleinhovia hospita L.	96-802	BOGOR	
Microcos ceramensis Burret	15-74	BOGOR	***ai pina onate***
	15-73	KEW	
Sterculia L.		PROSEA	***pomane***
Sterculia foetida L.		PROSEA	***pomane msinae***
Sterculia foetida L.		PROSEA	***pomane putie***
	96-692	BOGOR	
	96-693	KEW	
No identification	96-694	UKC	
Theobroma cacao L.		PROSEA	***sakalate***
Theobroma cacao L. var.	96-1156	UKC	(***sakalate***) ***kau-kau***
	96-1155	KEW	
	96-1154	BOGOR	
Theobroma cacao L. var.	96-922	UKC	***sakalate msinae***
Urena lobata L.	96-304	KEW	***kaote, kaota***
Urena lobata L.	96-871	KEW	
	96-303	BOGOR	
	96-870	BOGOR	
Urena lobata L.	96-305	UKC	
Urena lobata L.	96-1314	BOGOR	
Urena lobata L.	15-03	BOGOR	
Urena lobata L.	96-1315	KEW	
Urena lobata L.	96-1316	UKC	
Urena lobata L.	73-04	KEW	
Urena lobata L.	73-05	KEW	
Urena lobata L.	96-872	UKC	
	15-04	KEW	
other Malvaceae	71-40	KEW	***mono nione sanene*** (***senene***)
	71-24	KEW	***kaute***

Taxonomic information	Voucher code or other source of identification		Nuaulu name
Brassicales			
Caricaceae			
Carica papaya L. var.		PROSEA	***papae hua pukune*** syn. ***papae nusa ina***
Carica papaya L. var.		PROSEA	***papae wakane***
Brassicaceae			
Brassica juncea (L.) Czern.		PROSEA	***sesawi***
Brassica napus L.			***sesawi***
Brassica oleracea L.		PROSEA	***kol***
Rorippa indica (L.) Hiern			***sesawi***
Tropaeolum majus L.,			***sesawi***
Moringaceae			
Moringa oleifera Lam.		PROSEA	***kelor***
Caryophyllales			
Nepenthaceae			
Nepenthes maxima Reinw. var.	03-04	UKC	***koitipi putie***
Nepenthes maxima Reinw. var.	03-03	UKC	***koitipi sakalate***
Nepenthes mirabilis (Lour.) Druce	03-02	UKC	***koitipi***
Polygonaceae			
Polygonum L. sp.	96-766	KEW	***manahu wan anoi***
Polygonum L. sp.	96-767	UKC	
Amaranthaceae			
Celosia argentea L.	71-14	KEW	***manahute, manahu(te) msinae***
Amaranthus tricolor L.	96-674	BOGOR	
Amaranthus tricolor L.		KEW	***manahute marae***
Amaranthus viridis L.	73-01	KEW	***mono manuate, mono manahutane, mono manhutanane, mono mana hutanane***
Amaranthus viridis L.	73-06	KEW	
Celosia argentea L.	96-765	BOGOR	***manahu wan anoi***
Celosia L. sp.	96-169	KEW	
Celosia L. sp.	96-170	UKC	
	96-168	BOGOR	
Cyathula argentica L.	96-169	KEW	***manahute putie*** syn. ***manahu wan anoi, manahute waene anoi***
Amaranthus spinosus L.	73-03d	KEW	
Cyathula prostrata (L.) Blume	73-08	KEW	***mono monatohu***
Cyathula prostrata (L.) Blume	96-185	UKC	***panu-panu, panu-panu totu ikine***
	96-184	BOGOR	
Cyathula Blume sp.	96-183	KEW	
	96-184	BOGOR	
Cyathula prostrata (L.) Blume	96-183	KEW	
Cyathula prostrata (L.) Blume	96-185	UKC	***panu-panu totu onate***
	96-184	BOGOR	
Cyathula prostrata Blume	96-183	KEW	
other Amaranthaceae		PROSEA	***manuhutune***
Portulacaceae			
Portulaca oleracea L.	71-62	KEW	***mono senane***
Cactaceae			
Pilocereus plachyclades F. Ritter, poss.		MORRIS	***kupa unie***

Taxonomic information	Voucher code or other source of identification		Nuaulu name
Cornales			
Cornaceae			
poss. *Alangium* if so then probl. *uniloculare* (Griff.) King [*Alangium griffithii* (C.B. Clarke) Harms]		WHITM	***ai nona***
Homalium foetidum Benth. [see also Saliaceae]		PROSEA	
	96-137	BOGOR	
	96-138	KEW	
	96-139	UKC	
Ericales			
Sapotaceae			
Manilkara zapota (L.) P. Royen		PROSEA	***sauporune***
Palaquium obovatum (Griff.) Engl.		PROSEA	***katou totu onate***
Palaquium obovatum (Griff.) Engl.		PROSEA	***katou totu ikine***
	15-23	KEW	
	15-32	KEW	
Palaquium obovatum (Griff.) Engl.		PROSEA	
Ebenaceae			
Diospyros elliptifolia Merr.	71-660	SING	***ai yarane***
Diospyros maritima Blume	96-608	BOGOR	
	96-609	KEW	
	96-610	UKC	
Diospyros ebenaster Retz.		PROSEA	
Diospyros sp. *poss. pilosanthera* Blanco, and/or *D. ulo* Merr.		WHITM	***kamakunte***
	96-231	UKC	
	96-232	BOGOR	
	96-233	KEW	
	96-92	BOGOR	***anametene***
	96-93	KEW	
Diospyros rostrata (Merr.) Bakh.	96-94	UKC	
	96-448	BOGOR	
	96-449	KEW	
	96-450	UKC	
	96-675	BOGOR	
Diospyros rostrata (Merr.) Bakh.	15-79	BOGOR	
	15-80	BOGOR	
Lecythidaceae			
Barringtonia J.R. Forst. & G. Forst. sp.	96-317	KEW	***hanasane***
	96-253	KEW	
Barringtonia J.R. Forst. & G. Forst. sp.	96-254	UKC	
	96-316	BOGOR	
	96-252	BOGOR	
Barringtonia J.R. Forst. & G. Forst. sp.	96-318	UKC	
	96-839	BOGOR	
Barringtonia J.R. Forst. & G. Forst. sp.	96-840	KEW	
Barringtonia J.R. Forst. & G. Forst. sp.	96-841	UKC	
Barringtonia asiatica Kurz	96-923	BOGOR	***hutune***
Barringtonia, probl. *racemosa* (L.) Spreng.	96-924a	KEW	
Primulaceae			
Ardisia Sw. sp.	96-467	KEW	***msikene***
	96-466	BOGOR	
	96-468	UKC	

Taxonomic information	Voucher code or other source of identification		Nuaulu name
Ardisia humilis Vahl.	96-708	BOGOR	***msikene hatae***
	96-709	UKC	
	96-710	KEW	
Tapeinosperma Hook. f. sp.	96-939	UKC	***anakoine***
	96-941	KEW	
	96-942	BOGOR	
Tapeinosperma Hook. f. sp.	96-943	KEW	
	96-379	BOGOR	
Actinidiaceae			
Saurauia ceramica Miq.	15-33	BOGOR	***kikun(i) werane***
Saurauia tristyla DC.	96-531	BOGOR	***katoe***
Saurauia Willd. sp.	96-532	UKC	
Saurauia Willd. sp.	96-533	KEW	
Gentianales			
Rubiaceae			
Coffea L.		PROSEA	***kopi***
Coffea arabica L. var.	71-17	KEW	***kopi jawa***
Coffea arabica L. var.		PROSEA	***kopi mkauke***
Coffea canephora Pierre ex A. Froehner var. *robusta* (L. Linden) A. Chev.		PROSEA	***kopi onate***
Guettarda speciosa L.	96-651	BOGOR	***ai titie***
Guettarda L. sp.	96-865	KEW	
	96-866	UKC	
Guettarda speciosa L.	96-930	BOGOR	***sinatane nuae*** syn. ***atuane***
	96-903	KEW	
	96-929	BOGOR	
	96-930	KEW	
	96-902	BOGOR	
Hedyotis costata (Roxb.) Kurz.	71-25	KEW	unknown 11
Ixora L. or nr. sp.	71-87	KEW	***sinunane, sainunan***
	96-516	KEW	
Ixora L. sp.	96-425	UKC	***sinuri***
Ixora L. sp.	96-427	KEW	
	96-428	BOGOR	
	96-431	BOGOR	
Morinda citrifolia L.	71-45	KEW	***nenu***
	96-42	BOGOR	
Morinda citrifolia L.	96-577	BOGOR	
Morinda citrifolia L.	96-578	KEW	
Nauclea orientalis L.	96-1123	BOGOR	***menene***
Nauclea Merr. sp.	96-1124	KEW	
Nauclea Merr. sp.	96-1125	UKC	
Nauclea Merr. sp.	96-363	KEW	***panne***
Nauclea Merr. sp.	96-361	UKC	
	96-362	BOGOR	
Nauclea Merr. sp.	96-368	UKC	
Nauclea or *Neonauclea* or nr. sp.	96-652	KEW	***ai titie***
	96-864	BOGOR	
Neolamarckia macrophylla (Roxb.) Bosser [*Anthocephalus macrophyllus* Havil.]		PROSEA	

Taxonomic information	Voucher code or other source of identification		Nuaulu name
Neolamarckia cadamba (Roxb.) Bosser [*Anthocephalus cadamba* Miq.]	96-555	BOGOR	***hana(panne)***
Nauclea sp., probl.	96-556	KEW	
Nauclea sp., probl.	96-557	UKC	
Neonauclea lanceolata subsp. *gracilis* (Vidal) Ridsdale [*Nauclea philippensis* Havil]	96-251	UKC	unknown 4
Neolamarckia macrophylla (Roxb.) Bosser [*Anthocephalus macrophyllus* Havil.]		PROSEA	
	96-903	KEW	***sinatane wane***
	96-902	BOGOR	
	96-904	UKC	
Ophiorrhiza canescens Blume	96-1060	BOGOR	***kamruku***
	96-1061	KEW	
Ophiorrhiza L. sp.	96-1062	UKC	
Ophiorrhiza L. sp.	96-1364	KEW	
	96-1363	BOGOR	
Ophiorrhiza L. sp.	96-1365	UKC	
Paederia verticillata Blume	96-174	BOGOR	***awane notute***
	96-1387	KEW	***awane notute***
Timonius timon (Spreng.) Merr. [*Timonius* probl. *sericeus* (Desf.) K. Schum.]		PROSEA	***kokune pina***
Uncaria Schreb. sp.	96-340	BOGOR	***kapine wane*** syn. ***awane kapine***
	96-312	KEW	
	96-341	KEW	
Uncaria Schreb. sp.	96-342	UKC	
	96-370	BOGOR	
	96-371a	KEW	
	96-371b	UKC	
Uncaria probl. *cordata* (Lour.) Merr.	96-1181	UKC	***kapine wane msinae***
Uncaria Schreb. probl.	96-1182	UKC	
	96-1183	KEW	
Uncaria probl. *gambir* (Hunter) Roxb.		PROSEA	***kapine wane putie***
other Rubiaceae	96-800	UKC	***ansaha***
	96-800	BOGOR	
	96-799	BOGOR	
	96-801	KEW	
	96-175	KEW	
	96-176	UKC	
	96-868	KEW	***hana nuae***
	96-867	BOGOR	
	96-869	UKC	
	96-250	KEW	***nosate***
	96-111	KEW	
	96-112	UKC	
	96-249	BOGOR	
	96-110	BOGOR	
Gentianaceae			
Fagraea tacapala Leenh., probl.		WHITM	***kamkaune***
Apocynaceae			
Alstonia R. Br. sp.		WHITM	***sene, saenae***

Taxonomic information	Voucher code or other source of identification		Nuaulu name
Alstonia scholaris (L.) R. Br.	96-1007	BOGOR	***ito***
Alstonia scholaris (L.) R. Br.	96-1008	UKC	
	96-1009	KEW	
Alstonia scholaris (L.) R. Br.		PROSEA	***pompule, pompulai, pule***
Alstonia spectabilis R. Br.	96-1264	BOGOR	***urone***
	96-1265	KEW	
Alstonia spectabilis R. Br., poss.	96-1266	UKC	
Catharanthus roseus (L.) G. Don.	71-9	KEW	unknown 10
Cerbera manghas L.	96-38	UKC	***mataponane***
	96-39	KEW	
Cerbera manghas L.	96-39	BOGOR	
Cerbera L. sp.	96-40	KEW	
	96-653a	KEW	
Rauwolfia sumatrana Jack	96-653b	BOGOR	
Rauwolfia sumatrana Jack	96-654	UKC	
Hoya sp probl. *dischidia*	71-42	KEW	***timne isie, timnisie***
	71-52	KEW	
	96-412	BOGOR	
	96-413	KEW	
	96-414	UKC	
Plumeria rubra L.		PROSEA	***kupa kupero hanae, kupanaia, kupahaia*** syn. ***kupa kubur***
Solanales			
Convolvulaceae			
Ipomoea aquatica Forssk.		PROSEA	***tanko***
Ipomoea batatas (L.) Lam.		PROSEA	***kasitena***
Ipomoea batatas (L.) Lam. var.		PROSEA	***kasitena masikunne*** syn. ***sakitena***
Ipomoea batatas (L.) Lam. var.		PROSEA	***kasitena msinae***
Ipomoea batatas (L.) Lam. var.		PROSEA	***kasitena papae***
Ipomoea batatas (L.) Lam. var.		PROSEA	***kasitena putie***
Ipomoea pes-caprae (L.) R. Br.	96-1	BOGOR	***kasitena nuae***
Ipomoea pes-caprae (L.) R. Br.	96-5	KEW	
Ipomoea pes-caprae (L.) R. Br.	96-6	UKC	
	96-858	BOGOR	
Ipomoea quamoclit L., probl.	96-163	UKC	(***k***)***upa tonae***
	96-161	BOGOR	
	96-162	KEW	
Merremia peltata Merr.	96-1135	BOGOR	***hunane***(***ne***) ***wane***
	96-1136	KEW	
Merremia peltata Merr.	96-1137	BOGOR	
Solanaceae			
Capsicum L.		PROSEA	***karatupa***
Capsicum annuum L. var.		PROSEA	***karatupa hua metene***
Capsicum annuum L. [*Capsicum frutescens* L.] var.	96-1237a	KEW	***karatupa konopu***
Capsicum annuum var. *cordiforme* Edwall	96 1237b	BOGOR	
Capsicum annuum var. *cordiforme* Edwall	96-1239a	UKC	
Capsicum annuum L. [*Capsicum frutescens* L.] var.	96-1063	BOGOR	(***karatupa***) ***manesane*** syn. ***karatupa*** (***hua***) ***marae***
Capsicum annuum L. [*Capsicum frutescens* L.] var.	96-1064	KEW	
Capsicum annuum L. [*Capsicum frutescens* L.] var.	96-1065	UKC	

Taxonomic information	Voucher code or other source of identification		Nuaulu name
Capsicum annuum L. [*Capsicum frutescens* L.] var.	96-646	UKC	***karatupa mkauke***
Capsicum annuum L. [*Capsicum frutescens* L.] var.		PROSEA	***karatupa mnaha etute***
Capsicum annuum L. var.		PROSEA	***karatupa onate***
Capsicum annuum L. [*Capsicum frutescens* L.] var.		PROSEA	***karatupa poko tuene***
Capsicum annuum L. var.		PROSEA	***karatupa utane***
Capsicum annuum L. [*Capsicum frutescens* L.] var.		PROSEA	***karatupa walon***
Capsicum annuum L. [*Capsicum frutescens* L.] var.		PROSEA	***karatupa warata, karatupa putie***
Capsicum annuum L. [*Capsicum frutescens* L.] var.	96-724a	UKC	
Lycopersicum esculentum Mill.		PROSEA	***tamati***
Lycopersicum esculentum Mill. var.		PROSEA	***tamati hua ikine***
Lycopersicum esculentum Mill. var.		PROSEA	***tamati hua onate***
Nicotiana tabacum L.		PROSEA	***tapako***
Nicotiana tabacum L. var.		PROSEA	***tapako malole***
Nicotiana tabacum L. var.		PROSEA	***tapako pane***
Nicotiana tabacum L. var.		PROSEA	***tapako sinsinte***
Nicotiana tabacum L. var.		PROSEA	***tapako tahenae***
Solanum melongena L.		PROSEA	***tonne***
Solanum melongena L. var.		PROSEA	***tonne marae***
Solanum melongena L. var.		PROSEA	***tonne metene***
Solanum melongena L. var.		PROSEA	***tonne msinae***
Solanum melongena L. var.		PROSEA	***tonne nusa***
Solanum melongena L. var.		PROSEA	***tonne pina***
Solanum melongena L. var.		PROSEA	***tonne putie***
Solanum torvum Sw.	96-287	KEW	***ton(ne) hanaie*** syn. ***ton(ne) wesie***
	96-286	BOGOR	
Solanum L. sp.	96-288	UKC	
	96-288	BOGOR	
	96-834	BOGOR	
Solanum torvum Sw.	96-835	UKC	
Solanum torvum Sw.	96-836	KEW	
Solanum torvum Sw.	96-1384	KEW	
Solanum ansuivi Lam. [*S. indicum* L.]	15-38	BOGOR	
	15-39	BOGOR	
	15-40	KEW	
Solanum tuberosum L.		PROSEA	***kentang***
Boraginales			
Boraginaceae			
Cordia myxa Thwaites	96-570	BOGOR	***kanapone***
	96-571	KEW	
Cordia myxa Thwaites	96-572	UKC	
Lamiales			
Oleaceae			
	96-343	KEW	***anahata katapa***
	96-344	BOGOR	

Taxonomic information	Voucher code or other source of identification		Nuaulu name
	96-965	UKC	
	96-966	KEW	
	96-967	BOGOR	
	96-967	KEW	
Gesneriaceae			
Cyrtandra sp. nr. *C. decurrens*, poss. *C. pilosa* de Vries Blume.	71-81	KEW	***rorie waene***
Acanthaceae			
Andrographis paniculata (Burn.) Nees.	96-1234a	KEW	***ai turi nuae***
Andrographis paniculata (Burn.) Nees.	96-1232	BOGOR	***hono-hono nuae***
Andrographis paniculata (Burn.) Nees.	96-1233	UKC	
Andrographis paniculata (Burn.) Nees.	96-1233a	KEW	
Asystasia australasia F.M. Bailey	96-1385	KEW	***kamruku pina***
Asystasia gangetica (L.) T. Anders., s.l.	71-46	KEW	***matapete onone***
Asystasia gangetica (L.) T. Anders., s.l.	71-12	KEW	
Dicliptera Juss. sp.	71-74	KEW	***moune (hatae) putie***
	96-150	BOGOR	
	96-151	BOGOR	
Dicliptera Juss. sp.	96-152	KEW	
Justicia gendarussa Burm. f. [*Gendarussa vulgaris* Nees]		PROSEA	***mosone***
Justicia gendarussa Burm. f. [*Gendarussa vulgaris* Nees]	96-1278	BOGOR	***moso(ne) putie***
Justicia gendarussa Burm. f. [*Gendarussa vulgaris* Nees]	96-1279	UKC	
	96-1280	KEW	
Justicia gendarussa Burm. f. [*Gendarussa vulgaris* Nees]	96-727	BOGOR	***moso(ne) msinae***
	96-728	KEW	
Justicia gendarussa Burm. f. [*Gendarussa vulgaris* Nees]	96-729	UKC	
Justicia gendarussa Burm. f., or nr.		PROSEA	***moso(ne) metene***
Peristrophe bivalvis (L.) Merr. [*P. roxburghiana* (Schult.) Bremek.]	15-53	BOGOR	***moune***
	71-66	KEW	
	15-54	KEW	
nr *Peristrophe bivalvis* (L.) Merr. [*P. roxburghiana* (Schult.) Bremek.]		PROSEA	***moune hatae marae***
	71-66	KEW	***moune (hatae) msinae***
Pseuderanthemum Radik. sp.	71-48	KEW	***kamruku***
Pseuderanthemum Radlk. sp.		PROSEA	***kamruku metene***
Pseuderanthemum Radik. sp.		PROSEA	***kamruku sororione***
Pseuderanthemum Radik. sp.	71-10	KEW	***kamruku wane*** syn. ***korosine wane, kamruku awane***
Pseuderanthemum Radik. sp.		PROSEA	***kamruku notane***
Pseuderanthemum Radik. sp		PROSEA	***kamruku korisine***
Pseuderanthemum Radik. sp.	71-33	KEW	***mo wapane***
other Acanthaceae	71-51	KEW	***mono maune***
	71-49	KEW	***ohone***
Bignoniaceae			
Crescentia cujete L.		PROSEA	***kalapasa***
Crescentia cujete L.		PROSEA	***kalapasa ikine***
Crescentia cujete L.		PROSEA	***kalapasa mainae*** syn. ***kalapasa hua onate***
Jacaranda obtusifolia Bonpl. [*Jacaranda filicifolia* D. Don]		PROSEA	***makasanane***
Oroxylum indicum (L.) Benth. ex Kurz		PROSEA	***koitipa***

Taxonomic information	Voucher code or other source of identification		Nuaulu name
Verbenaceae [now Lamiaceae]			
Premna nitada K. Schum.	15-93	BOGOR	***pasa nuae***
	15-94	KEW	
Premna serratifolia L. [*Premna integrifolia* L.]	71-54	KEW	***ai suto***
	96-879	BOGOR	
Premna serratifolia L.	96-880	KEW	
Premna L. sp.	96-881	UKC	
Premna serratifolia L. [*Premna obtusifolia* R. Br.]	96-588	BOGOR	***akioae***
Premna serratifolia L. [*Premna integrifolia* L.]	96-589	KEW	
Premna serratifolia L. [*Premna integrifolia* L.]	96-590	UKC	
Premna serratifolia L. [*Premna obtusifolia* R. Br.]	96-638	BOGOR	
Premna serratifolia L.	96-639	KEW	
Premna serratifolia L.	96-640	UKC	
Stachytarpheta aff. *indica* (L.) Vahl	71-1	KEW	***manahu(te) msinae***
Tectona grandis L. f.		PROSEA	***salimuri***
Vitex cofassus Reinw. ex Blume		PROSEA	***pasane***
Vitex cofassus Reinw. ex Blume	96-1095	BOGOR	***pasa tope***
Vitex cofassus Reinw. ex Blume	96-1096	UKC	
	96-1097	KEW	
Vitex cofassus Reinw. ex Blume	96-1195	BOGOR	
Vitex negundo L.	96-1236	KEW	***ai turi nuae***
Vitex trifolia L.	96-1234b	BOGOR	
Vitex L. sp.	96-1235	UKC	
Lamiaceae			
Callicarpa longifolia Lam.		PROSEA	***notone***
	96-126	UKC	
	96-127	KEW	
	96-523	BOGOR	
	96-125	BOGOR	
Callicarpa longifolia Lam.		PROSEA	***kokune***
	96-126	UKC	
	96-127	KEW	
	96-523	BOGOR	
	96-125	BOGOR	
	15-55	BOGOR	
Callicarpa longifolia Lam.	96-647	BOGOR	***koku(ne) hanaie***
	96-398	KEW	
	96-399	UKC	
	96-397	BOGOR	
Callicarpa longifolia Lam.	96-648	KEW	
Callicarpa L. sp.	96-649	UKC	
Callicarpa longifolia Lam.	96-924	KEW	
Callicarpa longifolia Lam.	96-925	UKC	
Callicarpa longifolia Lam.	96-647	BOGOR	
Callicarpa L. sp.	96-648	KEW	
Callicarpa L. sp.	96-649	UKC	
Clerodendrum L.			***unuhutu*** syn. ***atu inate***
Clerodendrum blumeanum var. *blumeanum* Schauer	71-5a	KEW	
Clerodendron colebrookianum Walp.	71-5b	KEW	***unuhutu hanaie***
	96-22	BOGOR	
	96-87	BOGOR	

Taxonomic information	Voucher code or other source of identification		Nuaulu name
	96-228	BOGOR	
	96-229	KEW	
	96-230	UKC	
Clerodendrum rumphianum de Vriese	96-81b	KEW	***unuhutu msinae***
	96-80b	BOGOR	
Clerodendrum speciosissimum Drapiez	96-81a	KEW	***unuhutu pina***
	96-80a	BOGOR	
Gmelina moluccana Backer (ex K. Heyne)			
Hyptis capitata Jacq.	96-186	KEW	***kananuku hanaie***
	96-187	BOGOR	
Hyptis capitata Jacq.	96-188	UKC	
Mentha arvenensis L.		PROSEA	***mono hau mesie***
Ocimum americanum L., *O. basilicum* L. & poss. *O. gratissum* L		PROSEA	***kananukue***
Ocimum sp.		PROSEA	
Orthosiphon aristatus (Blume) Miq. [*Orthosiphon stamineus* Benth.]		PROSEA	***manapesi mau***
Orthosiphon aristatus (Blume) Miq. [*Orthosiphon stamineus* Benth.]	03-13	UKC	
Petraeovitex multiflora (Sm.) Merr.	71-58	KEW	***papate wane*** syn. ***awane papate***
	96-95	BOGOR	
	96-96	KEW	
	96-97	BOGOR	
	96-463	BOGOR	
	96-464	UKC	
	96-465	KEW	
Asterales			
Goodeniaceae			
Scaevola frutescens K. Krause	96-632	BOGOR	***hoka***
Scaevola taccada (Gaertn.) Roxb.	96-633	KEW	
Scaevola taccada (Gaertn.) Roxb.	96-634a	UKC	
Scaevola frutescens K. Krause	96-932	BOGOR	***okar***
Scaevola taccada (Gaertn.) Roxb	96-933	UKC	
Scaevola taccada (Gaertn.) Roxb	96-934b	KEW	
Asteraceae [Compositae]			
Ageratum conyzoides L.	71-23	KEW	***mono mara osu, mono marosu***
Cyanthillium cinereum (L.) H. Rob. [*Vernonia cinera* (L.) Less.]	71-2	KEW	unknown 5
Eupatorium sp., probl.	96-417	UKC	***mono seane***
	96-415	BOGOR	
	96-416	KEW	
Gynura procumbens Cass.	71-03a	KEW	***cintar manis***
Helianthus annuus L.		PROSEA	***kupa rani*** syn. ***bunga matahari***
Melanthera biflora (L.) H. Wild	71-15	KEW	***nimunone***
	96-177	BOGOR	
	96-179	UKC	
	96-178	KEW	
Mikania cordata (Burm. f.) B.L. Roxb.	15-50	BOGOR	***papari, paperi***
	15-49	KEW	
Synedrella nodiflora (L.) Gaertn.	71-34a	KEW	***mono monatohu***
Synedrella nodiflora (L.) Gaertn.	71-34b	KEW	***mono nuetetue***
Crassocephalum crepidioides (Benth.) S. Moore	71-35	KEW	

Taxonomic information	Voucher code or other source of identification		Nuaulu name
Ageratum conyzoides L.	71-61	KEW	
Synedrella nodiflora (L.) Gaertn.	73-02	KEW	
	96-10	BOGOR	
	96-11	UKC	
	96-12	UKC	
Synedrella nodiflora (L.) Gaertn.	71-27a	KEW	***mono tohu***
Synedrella nodiflora (L.) Gaertn.	71-27b	KEW	**(*mono*) *une-une taie***
other Asteraceae	71-03e	KEW	***mono mahusine***
Apiales			
Pittosporaceae			
Pittosporum ferrugineum Dryand. ex W.T. Aiton	96-1242	BOGOR	***kam mnaha***
	96-1243	KEW	
Pittosporum ferrugineum Dryand ex W.T. Aiton	96-1244	UKC	
Araliaceae			
Osmoxylon nr. *boerlagei* Warb. Philipson [*Boerlagidendron* sp.]	96-51	BOGOR	**(*ai*) *mkoue***
Osmoxylon nr. *boerlagei* Warb. Philipson [*Boerlagidendron* sp.]	96-52	KEW	
Osmoxylon nr. *boerlagei* Warb. Philipson [*Boerlagidendron* sp.]	96-53	UKC	
Osmoxylon nr. *boerlagei* Warb. Philipson [*Boerlagidendron* sp.]	96-43	UKC	
Panax L.		PROSEA	***jinsen***
Polyscias cumingiana Fern.-Vill.	96-1199	BOGOR	***masahatune***
	96-1200	KEW	
	96-1201	UKC	
	15-09	KEW	
Apiaceae [Umbelliferae]			
Apium graveolens L.		PROSEA	***selederi***
Coriandrum sativum L.		PROSEA	***ketumbar***
Daucus carota L.		PROSEA	***worteli***
Leucas zeylanica R. Br.		PROSEA	***anahuai***
Mackinlaya celebica (Harms.) Philipson	15-27	BOGOR	***masahahute***
Petroselinum crispum (Miller) Nyman ex A.W. Hill		PROSEA	***peterseli***
k.o. unidentified dicotyledonous tree	96-812	UKC	***kinekane***
	96-813a	BOGOR	
	96-813b	KEW	
	96-814	KEW	
	03-09	UKC	

APPENDIX 2

Alphabetical index of Nuaulu plant names with scientific identifications

Scientific synonyms, superseded names and other taxonomic information are provided in Appendix 1 in phylogenetic order. Page numbers provide references made in the text, in the tables and in Appendix 1.

Nuaulu name	Scientific identification	Pages
advokat	*Persea americana*	82, 130–1, 136, 175
ahane	k.o. fern incl. *Phymatosorus scolopendria*	23n3, 51–5, 103, 125, 157–8, 172
aha onie	k.o. fern	23, 51, 53, 125, 173
aha totu onate	k.o. POLYPODIACEAE	51, 53–4, 125, 173
ahau uni	*Antidesma stipulare, A. montanum*	37, 114, 194
ahesa	k.o. tree	
ahorane	k.o. tree; 96-501 BOGOR, 96-502 KEW, 96-503 UKC; 1996 plot 10	
ahutau metene syn. ***ahutau totu onate***	*Calophyllum* probl. *C. soulattri*	77–8, 190
ahutau mkauke syn, ***ahutau(ne) numte***	*Calophyllum articulatum*, probl.	77–8, 190
ahutau numte syn. ***ahutau mkauke***	*Calophyllum articulatum*, probl.	77–8, 190
ahutau totu onate syn. ***ahutau metene***	probl. *C. soulattri*	77–8, 190
ahutau wesie	*Calophyllum* sp.	78, 190
ahutaune	incl. *C. inophyllum*	77–8, 81, 115, 127, 129, 144, 190
ahutau(ne) nuae syn. ***hitaio***	*Calophyllum inophyllum*	77–8, 190
ahutau(ne) hanaie	*Calophyllum* sp.	38–9, 77–8, 190
ahutau(ne) pina	*Calophyllum* sp.	38–9, 77–8, 190
ai	tree	75–7
ai ama	*Shorea selanica*	43, 209
ai anawasi	*Anisophyllea disticha*	51, 201
ai ane	*Prunus arborea, P. wallaceana*	113–14, 198
ai emane	k.o. tree	
ai hatu	probl. *Mallotus* sp.	192
ai hatu hatae	*Hancea penangensis*	193
ai iene	*Eucalyptus deglupta*	203
ai inate	k.o. tree	
ai itine	k.o. tree	
ai kam	*Piper aduncum*	106, 149, 176
ai kamane	*Jatropha curcas*	108, 192
ai kau	*Rhodamnia cinerea*	44, 203
ai kihoke	*Breynia cernua, Glochidion rubrum*	44, 116, 194
ai mane	*Glochidion macrocarpum*	195
ai menie	*Mallotus miquelianus*	194
(***ai***) ***mkoue***	*Osmoxylon* nr. *boerlagei*	44, 220
ai moti	*Dendrolobium umbellatum*	196
ai mpiae	k.o. LAURACEAE	175
ai msinae	*Syzygium rumphii*	75, 86, 205
(***ai***) ***mukune***	*Cyathea* spp., *Tectaria leuzeana, Pteris semipinnata*	37, 52–3, 75, 117, 171
ai naine	k.o. tree	
ai naka(***-naka***)	k.o. tree	44
ai namte	*Polyalthia lateriflora*	45, 175
ai nante	*Agathis dammara*	173
ai nasa tiane	*Castanopsis* sp., probl.	202
ai nasate	*Elaeocarpus* probl. *serratus*	189

Nuaulu name	Scientific identification	Pages
ai nona	poss. *Alangium* if so then probl. *uniloculare*, or *Homalium foetidum*	212
ai nuelo, (ai) yainuelo	*Delonix regia*	45, 104, 196
ai numa	k.o. LAURACEAE	42, 175
ai numa hanaie	k.o. LAURACEAE	39
ai numa pina	k.o. LAURACEAE	39
ai numte	k.o. tree	
ai oi	*Knema cinerea*	174
ai osi syn. ***tahena ukuna***	*Chisocheton ceramicus, Aglaia parviflora*	16, 75, 116, 208
ai parae	96-968 UKC, 96-969 KEW, 96-970 UKC	
ai pina ikine	*Annona reticulata*, probl.	117, 174
ai pina onate	*Microcos ceramensis*	210
ai pinamo	*Annona reticulata, Microcos ceramensis*	43, 82, 117, 174
ai polo	*Antiaris* probl. *toxicaria*	66, 198
ai punara	*Octomeles moluccana*	45, 136, 150, 202
ai purae	k.o. tree; 1996 plot 10	
ai rapa	96-203 UKC, 96-205 BOGOR, 96-206 KEW, 96-522 BOGOR	
ai saja	*Senna sophera*	104, 197
ai sote	k.o. tree; 1996 plot 9	44
(***ai***) ***supa***	*Ficus tinctoria, F. callosa*	17, 75, 81, 114, 200
ai suto	*Premna serratifolia*	44, 50, 75–6, 86, 110, 218
(***ai***) ***tahane***	96-403 BOGOR, 96-404 KEW, 96-405 UKC	
ai tihane	96-1001 BOGOR	
ai tinae	*Gnetum* probl. *latifolium*	173
ai titie	*Guettarda speciosa, Nauclea* or *Neonauclea*	156, 213–14
ai tohu	*Piper aduncum*	106, 176
ai totu onate	k.o. tree	
ai turi	*Senna alata*	76, 86, 128–9, 136, 197
ai turi nuae	*Andrographis paniculata, Vitex trifolia, V. negundo*	117, 217–18
ai unate syn. ***hana***	*Bruguiera gymnorrhiza; Rhizophora* sp.	75, 118, 189
ai une	*Artocarpus* sp.	198
ai unene	*Virola surinamensis;* also k.o. LAURACEAE	174–7
ai weti	*Phaleria perrottetiana, Steganthera hirsuta*	75, 117, 176, 209
ai yarane	*Diospyros elliptifolia, D. maritima, D. ebenaster*	113, 143, 212
ai yoko	k.o. Podocarpaceae	45, 173
akae	*Dioscorea pentaphylla*	60–1, 133, 137, 178
aka hunuhunue	*Dioscorea pentaphylla* var.	43, 61, 178
akahante	k.o. tree	
	k.o. tree; 96-107 BOGOR, 96-108 UKC, 96-109 KEW, 96-355 BOGOR,	
akahonte	k.o. tree; 96-356 KEW, 96-357 UKC, 96-1163 BOGOR, 96-1164 KEW, 96-1165 KEW	
akahon tonu-tonu	k.o. tree; 96-602 KEW, 96-603 UKC, 96-604 KEW; 1996 plot 10	40
akaokoe	k.o. tree	
akasie	*Acacia* probl. *mangium* Willd.	195
akenue	*Calophyllum* sp.	77–8, 190
akioae	*Premna serratifolia*	
akunin syn. ***asinokoie***	*Actinodaphne* spp.	107, 110, 163, 175
akunin hanaie	*Actinodaphne moluccana*	39, 107, 175
akunin pina	*Actinodaphne rumphii*	39, 107, 175
alalasie	Incl. *Aquilaria* sp., or poss. *Gyrinops versteegii; Phaleria capitata*	39, 209
alalasie hanaie	*Phaleria capitata*	209
alalasie pina	*Aquilaria* probl. *audate* or *Gyrinopsis cumingiana*	209
aleha	k.o. tree	
amnae	incl. *Brucea javanica*	58, 81, 207
amnae ai anawasi, syn. (***n***)***amnae nuae***	*Colubrina* sp.	198
amnae nuae, (***n***) ***amnae nuae***, syn. ***amnae ai anawasi***	*Colubrina* sp.	198
amnae napotone	*Brucea javanica*	207
amnae wesie	*Ficus* sp.	58–9, 81, 199
ana (hotone) syn. ***hotone***	*Oryza sativa*	39, 42, 45, 66, 115, 127, 129, 131, 184
(***ana***) ***manipa*** syn. ***yakon***(***i***)	*Zea mays*	41, 45, 66, 131, 184

Nuaulu name	Scientific identification	Pages
ana msinae	*Oryza sativa* var.	66, 183
ana naine	*Litsea resinosa*	175
ana nusa inae	*Oryza sativa* var.	66, 183
anae ate	*Xylopia malayana*	86, 175
anahata katapa	k.o. tree; probl. OLEACEAE; 1996 plots 8 and 10	217
anahuai	*Leucas zeylanica*	
anakoine	*Tapeinosperma* sp.	14, 118, 213
anametene	*Diospyros rostrata*	85–6, 118, 128, 143, 212
ananaine syn. **kawasa pina**	*Archidendron clypearia*	77, 195
ananaine pina	k.o. tree	
ananaine totu nawe	k.o. tree	
ananaine totu pukune	k.o. tree	
ananate	*Canarium* incl. *C. sylvestre*	78–80, 206
ananinoi	k.o. tree; 1996 plot 10	
ananione	*Horsfieldia bivalvis*	86, 174
anapakane	*Leea indica*	188
anaupua, ana upa	*Cyperus compressus*	40, 45
aniate	k.o. tree	
anine	*Erythroxylum ecarinatum*	118, 189
aninoi	*Syzygium* sp., *Walsura* sp.	203, 208
aniono	k.o. EUPHORBIACEAE	86, 194
ansaha	k.o. RUBIACEAE	214
apaine	*Syzygium* spp.	203
apaine numte	*Syzygium* sp.	203
apaine totu mkauke syn. **apaine totu pukune**	*Syzygium* sp.	203
apaine totu onate	*Syzygium* sp.	203
apaine totu pukune syn. **apaine totu mkauke**	*Syzygium* sp.	203
apalam	*Mangifera indica*	82, 115, 129, 136, 207
apalam hua papae	*Mangifera indica* var.	42, 45, 207
apalam hua pukune	*Mangifera indica* var.	207
(**apalam**) **kesi**	*Mangifera indica* var.	207
apa(**lam**) **olo**	*Mangifera indica* var.	
apalam tunne	*Mangifera indica* var.	42, 207
apetine	96-498.1 BOGOR, 96-498.2 KEW, 96-499 UKC, 96-500 KEW	

Nuaulu name	Scientific identification	Pages
apunue	*Aglaia sapindina*	86, 208
aputie	96-460 BOGOR, 96-461 KEW, 96-462.1 UKC, 96-462.2 KEW	
asaherane	*Virola surinamensis, Syzygium boerlagei*	76, 86, 113, 117, 174
asamjawa	*Tamarindus indica*	82, 197
asapopote	*Syzygium* incl. *S. boerlagei*	19, 77, 82, 204
asapopo makae	*Syzygium boerlagei*	204
asapopo manea	*Syzygium* sp., *Eugenia* sp.	204
asihata(**ia**)	probl. EUPHORBIACEAE	109, 194
asinokoie syn. **akunin**	*Actinodaphne* spp.	163, 175
asora	*Dillenia* sp., probl.	118, 188
asuan(**i**) **mata hunue**	*Casuarina rumphiana*	44, 202
asu etute	*Pennisetum macrostachyum*	15, 39, 41, 184
asu hunne	*Dysoxylum caulostachyum*, probl.	208
atipanate	*Pterocarpus indicus, Vitex cofassus*	197
atuane	*Lunasia amara*	156, 208, 213
atuate	k.o. tree; 1996 plot 9	
atu inate syn. **unuhutu**	*Clerodendrum* spp.	218
atupikae	*Syzygium aqueum, Phaleria perrottetiana*	117, 204, 209
au mamine	*Canarium hirsutum*	206
au oi(**ne**)	*Endospermum moluccanum*	86, 191
auhaune	*Leea indica*	189
auhaune hanaie	*Leea* sp.	39,189
auhaune pina	*Leea* sp., probl.	39, 189
aunasate	*Prunus javanica* probl., *Cryptocarya* sp.	117, 175, 198
aune	*Mangifera foetida*	136, 207
auuhone	*Syzygium* sp.	204
awane	vine	68, 83–4, 100–1, 107, 149–50
awane apane	k.o. vine	42, 59
awane kin-kin	*Zehneria japonica*	40, 50, 68, 83, 133, 202
awane kunie	*Arcangelisia flava*	45, 188
awane metene	*Dichapetalum* probl. *papuanum*	190
awane minne	k.o. vine	

Nuaulu name	Scientific identification	Pages
hono-hono hanaie	*Aneilema* sp., *Pollia* spp., probl.	39–40, 116, 185
hono-hono hanaie msinae	*Aneilema* sp., probl.	40, 185
hono-hono hanaie putie	*Pollia secundiflora*	40, 185
hono-hono nuae	*Andrographis paniculata*	39–40, 217
hono-hono pina	*Pollia secundiflora*	39–40, 185
hono-hono wesie	*Amischotolype hispida*	40, 59, 184
hotone syn. ***ana*** (***hotone***)	*Oryza sativa*	42, 45, 127, 131, 184
hotone manipa	*Setaria italica*	41
huaena	*Rhopaloblaste ceramica*	91, 144, 182
hueni	*Dioscorea alata*	37, 46, 60–1, 105, 115, 127, 137, 148, 178
huen(***i***) ***puti***(***e***)	*Dioscorea alata* var.	37, 61, 178
hueni (***huae***) ***hatasia***	*Dioscorea alata* var.	43, 61, 178
hueni marae	*Dioscorea alata* var.	61, 178
hueni msinae	*Dioscorea alata* var.	61, 178
hueni nakatua	*Dioscorea alata* var.	61, 178
hueni nante	*Dioscorea alata* var.	61, 178
hueni sune totue	*Dioscorea alata* var.	61, 178
hueni suneminete	*Dioscorea alata* var.	61, 178
hueni tumata	*Dioscorea alata* var.	61, 178
huetune	k.o. POACEAE	184
huka hukai	*Hibiscus rosa-sinensis*	210
hukila	*Eugenia* sp.	203
hun(***an***)***ori***, ***nori-nori***	*Macaranga involucrata*, *M. trichocarpa*, *Cheilocostus speciosus*	39–40, 76, 106, 117, 192
hunane(***ne***) ***wane***	*Merremia peltata*	41, 83, 215
hunua	*Macaranga* sp. nr. *tanarius*	26–7, 76, 103, 192
hutiene	*Durio zibethinus* var.	115, 136, 210
hutune	*Barringtonia* probl. *racemosa*	212
iane	*Canarium*	37–8, 78–81, 114–15, 118, 136, 144, 206
(***ia***) ***ananate***	*Canarium sylvestre*	78–80, 206
ia(***ne***) ***hanaie***	*Canarium indicum*	38, 78–9, 206
ia(***ne***) ***hanate*** syn. ***ia nawe*** syn. ***kahiaue***	*Canarium indicum*	37–8, 78–9, 206
ia(***ne***) ***nawe*** syn. ***ia***(***ne***) ***hanate*** syn. ***kahiaue***	*Canarium indicum*	78–9, 206
ia(***ne***) ***pukune***	*Canarium indicum*	78–9, 206
iane kauke, ***ia***(***ne***) ***mkauke*** syn. ***ia yapono*** syn. ***ia haruku*** syn. ***ia pukune*** syn. ***ia katanepane***	*Canarium sylvestre*	37–8, 40, 43, 78–9, 206
ima	*Koompassia excelsa*	196
inene	*Hydriastele microcarpa*	91, 143, 181
ini	*Macaranga* sp.	192
ipina onate	k.o. tree	
isane	*Lygodium flexuosum*	53–4, 83, 171
ito	*Alstonia scholaris*	110, 215
jarak (***jepang***)	*Ricinus communis*	45, 194
jinsen	*Panax* sp.	220
kacang panjang	*Vigna unguiculata* grp. *sesquipedalis*	50, 67, 105, 135, 197
kaeane, ***kaieane***	*Terminalia catappa*	118, 202
kaea(***ne***) ***nuae***	*Terminalia* poss. *copelandii*	202
kaeane msinae syn. ***kaea wesie***	*Terminalia catappa* var.	202
kaeane putie	*Terminalia catappa* var.	202
kaenuri, ***kaenuni***	*Ficus polyantha*	81, 200
kaea wesie syn. ***kaeane msinae***	*Terminalia catappa* var.	202
kaha pukune	*Mallotus mollissimus*	193
kahea	*Aleurites moluccanus*	115, 191
kahea nawe	*Aleurites moluccanus*	191
kahea pukune	*Aleurites moluccanus*	191
kahiaue	*Canarium indicum*	37, 78–9, 206
kaie	k.o. fern	53, 173
kaihoke	*Coix lacryma-jobi*	66, 183
kaiihoke	*Zingiber zerumbet*	188
kaine	*Inocarpus fagifer*	128, 196
kaiohite	*Claoxylon* sp.; *Ficus botryocarpa* subsp. *hirtella*, *F. callosa*, *F. glandulifera*, *F. copiosa*	81, 113–14, 129, 191, 200
kaisan	*Arachis hypogaea*	67, 105, 115, 195
kaisan msinae	*Arachis hypogaea* var.	67, 195
kaisan putie	*Arachis hypogaea* var.	195
kakeimate	k.o. tree; 03-06 UKC	

Nuaulu name	Scientific identification	Pages
kakeimate totu onate	k.o. tree; 03-14 UKC	
kako hatae	*Ficus lepicarpa*	200
kalapasa	*Crescentia cujete*	115, 217
kalapasa hua onate syn. ***kalapasa mainae***	*Crescentia cujete* var.	218
kalapasa ikine	*Crescentia cujete* var.	217
kalapasa mainae syn. ***kalapasa hua onate***	*Crescentia cujete* var.	218
kamkari	*Alpinia nutans*	102, 187
kam(u)	*Piper*	34, 106, 111, 136, 149, 176
kam (w)apane, kamu apane	*Piper* sp.	37, 59, 106, 176
kam hanate	*Piper* probl. *caninum*	43, 106, 176
kam hau	*Piper corylistachyon, P. amboinense*	106, 176
kam inate	*Piper* sp.	106, 176
kam malaka	*Piper* probl. *retrofractum*	45, 106, 176
kam minne	*Piper sundaicum*	106, 176
kam mnaha	*Pittosporum ferrugineum*	220
kam niane	*Piper betle*	58–9, 106, 176
kam putie	*Piper* sp.	106, 176
kam sisinate	*Piper* sp. probl. *reinwardtianum*	106, 176
kam tunne	*Piper corylistachyon*	42, 58, 106, 176
kam wanae ukune	*Piper* sp.	106, 179
kam(u) wane	*Piper* probl. *sarmentosum*	106, 176
kamakunte	*Diospyros* sp. *poss. pilosanthera* Blanco, and/or *D. ulo* Merr.	212
kamane	resin-producing tree	108, 119, 173, 206
(***kamane***) ***onie***	*Shorea selanica*	108, 173
kama(ne) wae	*Agathis dammara*	108, 173
kamamuri	*Cheilocostus speciosus*	187
kamane metene	*Agathis* sp.	108, 173
kamane putie	*Agathis* sp.	108, 173
kamine, (***kama***) ***kamine, kama minne***	*Canarium asperum, Canarium decumanum*	78–9, 80, 108, 206
kamkari	2003–12 UKC	187
kamkaune	probl. *Fagraea tacapala*	118, 215
kamnatine	*Melastoma polyanthum*	40, 206

Nuaulu name	Scientific identification	Pages
kamruku	*Pseuderanthemum* incl. *Ophiorrhiza canescens*	102–3, 214, 217
kamruku korisine	*Pseuderanthemum* sp.	217
kamruku metene	*Pseuderanthemum* sp.	217
kamruku notane	*Pseuderanthemum* sp., probl.	44, 217
kamruku pina	*Asystasia australasia*	39, 217
kamruku sororione	*Pseuderanthemum* sp.	103, 217
kamruku wane syn. ***korosine wane***	*Pseuderanthemum* sp.	103, 217
kanai	*Areca catechu*	26–7, 34, 91, 106, 111, 114–15, 127, 180
(***kanai***) ***apane***	*Areca* probl. *glandiformis*	58–9, 91, 180
kanai ia nanae, kanai iane anae syn. ***kanai ia nenane***	*Areca catechu* var.	180
kanai kakaue	*Areca catechu* var.	180
kanai makuana	*Areca catechu* var.	180
kanai nopane	*Areca catechu* var.	180
kanai pona	*Areca catechu* var.	180
kanai puti asaehene	*Areca catechu* var.	180
kanai puti(e)	*Areca catechu* var.	180
kanai sopite	*Areca catechu* var.	44, 180
kananae	k.o. tree	
kananukue	*Ocimum americanum, O. basilicum, O. gratissum*	219
kananuku hanaie	*Hyptis capitata*	39, 219
kananuku putie	*Ocimum* sp.	250
kanapone	*Cordia myxa*	118, 216
kanapua	*Gossypium barbadense*	45, 67,76, 209–10
kane-kane wane (***unie***)	*Smilax zeylanica*	39–40, 179
kanna	*Aglaonema simplex*	102, 177
kanoie	k.o. tree; 96-434 BOGOR, 96-435 KEW, 96-436 UKC	
kanoke	*Syzygium malaccense*	82, 205
kanunte	k.o. vine [Matoke & Bolton 2005]	
kaote	*Urena lobata*	16, 210
kapenene	k.o. tree	
kapikute	*Caesalpinia* sp.	195
kapikute ikine	*Caesalpinia* sp.	195
kapikute mainae	*Caesalpinia* sp.	195

Nuaulu name	Scientific identification	Pages
kapine wane* syn. *awane kapine	*Uncaria* sp.	214
kapine wane msinae	*Uncaria* probl. *cordata*	214
kapine wane putie	*Uncaria* probl. *gambir*	214
kapu	*Ceiba pentandra*	209
karane, kerane	*Dracontomelon dao*	207
karatu, kastu, karata	*Sauropus androgynus*	66, 194
karatupa	*Capsicum annuum*	67, 115, 134–5, 215
karatupa* (*hua*) *marae syn. (***karatupa***) ***manesane***	*Capsicum annuum* var.	135, 216
karatupa hua metene	*Capsicum annuum* var.	135, 215
karatupa konopu	*Capsicum annuum* var. *cordiforme*	135, 215
(***karatupa***) ***manesane*** syn. ***karatupa* (*hua*) *marae***	*Capsicum annuum* var.	xiii, 135, 216
karatupa mkauke	*Capsicum annuum* var.	135, 216
karatupa mnaha etute	*Capsicum annuum* var.	44, 135, 216
karatupa onate	*Capsicum annuum* var.	135, 216
karatupa poko tuene	*Capsicum annuum* var.	135, 216
karatupa utane	*Capsicum annuum* var.	135, 216
karatupa walon	*Capsicum annuum* var.	135, 216
karatupa warata	*Capsicum annuum* var.	43, 135, 216
karosine	*Pipturus argenteus*	201
kasauni apane	*Casearia glabra*	59, 191
kasipehe	*Canna indica, Phrynium macrocephalum*	102, 117–18, 186
kasipii	*Manihot*	60–1, 104–5, 114–15, 133–4, 136, 151, 193
kasipii hatumetene	*Manihot esculenta* var.	132, 134. 193
kasipii honue syn. ***kasipii paru***	*Manihot esculenta* var.	132, 138, 193
kasipii kanapua	*Manihot esculenta* var.	45, 132, 193
kasipii kasitena	*Manihot esculenta* var.	132, 193
kasipii makuana syn. ***kasipii sinsinte***	*Manihot esculenta* var.	42, 45, 132, 193
kasipii masikunne	*Manihot esculenta* var.	42, 132, 134, 193
kasipii mata rase	*Manihot esculenta* var.	132, 193
kasipii mentega	*Manihot esculenta* var.	132, 193
kasipii msinae syn. ***kasipii postel***	*Manihot esculenta* var.	132, 193
kasipii paru syn. ***kasipii honue***	*Manihot esculenta* var.	43, 50, 132, 134–5, 138, 193
kasipii puna	*Manihot esculenta* var.	132, 193
kasipii putie	*Manihot esculenta* var.	132, 134, 193
kasipii sinsinte syn. ***kasipii makuana***	*Manihot esculenta* var.	110, 132, 193
kasipii sombar	*Manihot* cf. *carthaginensis* subsp. *glaziovii*	50, 60, 134, 193
kasipii super mie	*Manihot escuenta* var.	132, 135, 193
kasipii tiga bulan	*Manihot esculenta* var.	132, 193
kasipii unuhutu	*Manihot esculenta* var.	132, 193
kasipii wanate	*Manihot esculenta* var.	132, 193
kasipii wasaena	*Manihot esculenta* var.	132, 134, 193
kasitena	*Ipomoea batatas*	60–1, 105, 115, 134, 136, 215
kasitena masikunne syn. ***sakitena***	*Ipomoea batatas* var.	215
kasitena msinae	*Ipomoea batatas* var.	215
kasitena nuae	*Ipomoea pes-caprae, Canavalia maritima*	42, 60, 117, 196, 215
kasitena papae	*Ipomoea batatas* var.	215
kasitena putie	*Ipomoea batatas* var.	215
kasupa	*Bixa orellana*	45, 86, 107, 118–19, 209
kataa	*Hevea brasiliensis*	93, 115, 191
kataa ikine	*Hevea brasiliensis* var.	191
kataa onate	*Hevea brasiliensis* var.	191
katinoie	incl. *Melastoma imbricatum*	118, 143, 205
katinoi wan anoi syn. ***kasinoi waene anoi***	*Melastoma malabathricum, M. affine*	143, 205
katinoie numte	*Melastoma* sp., *Rhodamnia cinerea*	117, 203, 205
katoe	*Saurauia tristyla*	118, 213
katou	*Palaquium obovatum*	115, 212
katou totu ikine	*Palaquium obovatum* var., *Amomum aculeatum*	117, 187, 212
katou totu onate	*Palaquium obovatum* var.	212
katue	*Oncosperma filamentosa*	37, 90–1, 144, 150, 181–2

Nuaulu name	Scientific identification	Pages
pasite	*Begonia isoptera*	102, 202
pate	*Ficus* sp., probl.	81, 128, 199
piasoi	*Drymophloeus oliviformis*	91, 181
pina totue	k.o ARACEAE	59, 178
pioi	*Hibiscus* sp.	210
pioi nenane	*Hibiscus mutabilis*	210
pionenane	*Abelmoschus moschatus*	209
pita(r)	*Acalypha amentacea, A.caturus*	191
pitiri	*Passiflora foetida*	118, 190
pitiri malu	*Mimosa pudica*	14, 102, 197
pomane	*Sterculia foetida*	115, 210
pomane msinae	*Sterculia* sp. probl. *foetida*	210
pomane putie	*Sterculia* sp. probl. *foetida*	210
pom (pule)	*Alstonia scholaris*	215
pono	*Homalium foetidum*	191
popote	*Syzygium aqueum*	19, 82, 115, 129, 136, 204
popo(te) msinae	*Syzygium aqueum* var.	42, 204
popote putie	*Syzygium aqueum* var.	128, 204
popote wesie	*Pternandra cordata, P. coerulescens*	59, 77, 150, 206
postel	*Manihot esculenta* var.	114–15, 131, 135, 151, 193
postel marae	*Manihot esculenta* var.	193
postel metene	*Manihot esculenta* var.	151, 193
postel msinae	*Manihot esculenta* var.	193
postel putie	*Manihot esculenta* var.	133, 151, 193
postel upone	*Manihot esculenta* var.	133, 193
puane nimusie	*Actephila excelsa* var. *javanica*	44, 194
punara hatae syn. ***onie hatae***	*Shorea* sp.	209
punara noene	*Octomeles sumatrana*	150, 202
punara wane	*Piper corylistachyon*	83, 150, 176
punen	*Cycas rumphii / C. revoluta.*	92, 118, 173
punetene, pun	*Colocasia esculenta* var.	59, 105, 115, 148, 177
punetene metene	*Colocasia esculenta* var.	177
punetene putie	*Colocasia esculenta* var.	177
punetene taraga	*Colocasia esculenta* var.	177
puni	*Homalomena* incl. *H. pendula*	105, 115, 132, 177
puni hatae	*Homalomena* probl. *pendula*	177
puni msinae	*Homalomena cordata* var.	177
puni putie	*Homolomena cordata* var.	177
pupu	various ferns incl. *Sphaerostephanos unitus* var. *mucronatus*	51, 53, 55–6, 118, 172
pupu hanaie	*Chingia ferox, Sphaerostephanos unitus* var. *mucronatus*	39, 53, 55, 116, 171–2
pupu kaia	*Nephrolepis hirsutula, N. biserrata*	53–5, 172
pupu onate	*Pteris semipinnata*	53, 55, 103, 171
pupu pina	*Nephrolepis biserrata*	39, 53, 55, 172
rama(ne) wane	k.o. FABACEAE (vine)	198
rambutan	*Nephelium lappaceum*	82, 131, 136, 207
ranoi	*Macaranga hispida*	76, 192
rapa	k.o. tree	45
rene	*Erythrina variegata*	76, 86, 196
riane	*Elaeocarpus augustifolius*	86, 118, 189
roa	k.o. tree	157
rorie	*Begonia* incl. *B. isoptera*	102, 136, 202
rori(e) msinae	*Begonia* sp.	102, 202
rorie waene	*Cyrtandra* sp. nr. *C.decurrens* poss. *C. pilosa*	102, 217
rorong	*Benincasa hispida*	68, 201
saenae	*Alstonia* sp.	215
saenonane	*Leptaspis urceolata*	183
saiaune	k.o. tree; 1996 plot 7	
sakalate	*Theobroma cacao*	40, 42, 45, 93, 115, 133, 210
(***sakalate***) ***kau-kau***	*Theobroma cacao* var.	40, 45, 210
sakalate msinae	*Theobroma cacao* var.	210
sakaputi syn. ***sakana***	*Melaleuca cajuputi, Polyalthia glauca*	203, 174
sakitena syn. ***kasistena masikunne***	*Ipomoea batatas* var.	215
sakuala	k.o. tree	
salak	*Salacca zalacca*	91, 136, 182
salea	*Sorghum bicolor*	66, 184
salimuri	*Tectona grandis*	218
sanoma	*Pandanus* sp.	88, 90, 107, 179
sapa(ne)	*Trema orientalis*	103, 128, 198

Nuaulu name	Scientific identification	Pages
sapana	*Pteris vittata*	53, 55, 103, 171
sapate wane	*Tetracera scandens*	83, 188
sapuane,* *sapua(n) unue	*Colysis macrophylla,* *Drynaria sparsisora*	53, 55, 116, 172
saune syn. (***koae***) ***saune***	*Pandanus* various spp. incl. *conoideus*	87–8, 90, 106, 127, 136
sau(ne) hahu	*Pandanus tectorius*	44, 88, 179
saumahu	*Abroma augusta,* *A. fastuosa*	66, 76, 86, 209
saumahu pinaie	*Chamaecrista* sp.	39, 195
saune hehue	*Pandanus conoideus*	88, 179
sauporune	*Manilkara zapota*	212
sayor patola	*Luffa acutangula,* *L. cylindrica*	202
seate	k.o. Poaceae	184
selederi	*Apium graveolens*	220
senene	k.o. fern	53, 173
sesawi	*Brassica juncea, B. napus; Rorippa indica, Tropaeolum majus*	66, 135, 138, 211
sesene	*Ananas comosus*	67, 83, 115, 118, 136, 182
sese(ne) bogor syn. ***sese pokoro***	*Ananas comosus* var.	45, 182
sese(ne) gula	*Ananas comosus* var.	182
sese(ne) kamruku	*Ananas comosus* var.	182
sese(ne) msinae syn. ***sesene kapane***	*Ananas comosus* var.	182
sese(ne) nusa inae	*Ananas comosus* var.	45, 182
sese(ne) pokoro syn. ***sese(ne) bogor***	*Ananas comosus* var.	182
sese(ne) potoni	*Ananas comosus* var.	182
sese(ne) warata	*Ananas comosus* var.	182
setu	*Garcinia tetrandra,* probl.	190
siahu tekene	*Dioscorea esculenta* var.	178
siahue	*Dioscorea esculenta*	60–1, 105, 115, 137, 178
siahue kawei	*Dioscorea esculenta* var.	61, 178
siahue loloeno	*Dioscorea esculenta* var.	45, 178
siahue masikunne	*Dioscorea esculenta* var.	61, 178
siahue msinae	*Dioscorea esculenta* var.	61, 178
siahue nusa ina	*Dioscorea esculenta* var.	61, 178
siahue putie	*Dioscorea esculenta* var.	61, 178
siahue sapatu	*Dioscorea esculenta* var.	61, 178
siahue siaie	*Dioscorea esculenta* var.	61, 178
siaune	*Zingiber montanum*	188
sikeue	*Colocasia esculenta*	46, 60, 105, 110, 127, 177
sikeue hatusoke	*Colocasia esculenta* var.	60, 177
sikeue haurihue	*Colocasia esculenta* var.	43, 60
sikeue kunie	*Colocasia esculenta* var.	60, 177
sikeue masikunne	*Colocasia esculenta* var.	60, 177
sikeue msinae	*Colocasia esculenta* var.	60, 177
sikeue nakatua	*Colocasia esculenta* var.	44, 60, 177
sikeue nikate	*Colocasia esculenta* var.	43, 60, 177
sikeue saparuane	*Colocasia esculenta* var.	45, 60, 177
sikeue tekene	*Colocasia esculenta* var.	44, 60, 177
sikeue tombelu(e)	*Colocasia esculenta* var.	45, 60, 177
sina (manusia)	*Urticastrum decumanum*	58, 129, 201
sina-sina wane	*Pityrogramma calomelanos,* *Freycinetia* sp.	40, 53, 55, 83, 90, 117, 171
sinata(ne) onate	*Dendrocnide stimulans*	201
sinatane	nettles	129, 201
sinatane nuae syn. ***atuane***	*Guettarda speciosa*	156, 213
sinatane urone	*Dendrocnide microstigma,* poss. *Procris frutescens*	116, 201
sinatane wane	Rubiaceae: k.o. vine	214
sinepane	no ident.	
sinsinte	*Codiaeum variegatum*	37, 45, 51, 54, 58, 94, 104, 106, 110, 115, 191
sinsin(te) amasen	*Codiaeum variegatum* var.	191
sinsin(te) masikunne	*Codiaeum variegatum* var.	110, 191
sinsin(te) matopae	*Codiaeum variegatum* var.	43, 191
sinsin(te) msinae	*Codiaeum variegatum* var.	191
sinsin(te) totu nawe	*Codiaeum variegatum* var.	37, 42, 191
sinsin(te) (totu) onate	*Codiaeum variegatum* var.	41, 191
sinsin(te) totu pukune	*Codiaeum variegatum* var.	191
sinunane	*Ixora* or nr. sp.	213
sinuri	*Ixora* sp.	213
sioe	*Triadica cochinchinensis,* *Homalanthus novoguineensis*	116, 194

Nuaulu name	Scientific identification	Pages
wainite totu nawe	*Phrynium* sp.	186
wana apane	*Dendrocalamus* sp.	59, 95–6, 183
wana hatu	*Dendrocalamus* sp.	43, 95–7, 183
(***wana***) ***ona***	*Dendrocalamus asper*, probl.	41, 94–6, 150, 183
wana tomo syn. ***tomo seane***	*Bambusa vulgaris* var. *striata* (yellow culm group)	94–6, 183
wanate	Various bamboo incl. *Dendrocalamus*	45, 64, 94–7, 144, 150, 183
wanate tun syn. ***wana tunne***	*Gigantochloa atter*	95–7, 183
waronone	*Ficus*, incl. *F. nervosa* subsp. *pubinervis*	200
waronone kahuite	*Ficus* sp.	81, 200
waronone rapa	*Ficus* sp.	81, 200
waruae syn. ***uraue***	*Ficus variegata*	81, 201
wasone	*Syzygium acuminatissimum*	203
wasusu	k.o. vine; 96-827 BOGOR, 96-828 KEW, 96-829 UKC	
weio	*Cinnamomum iners, C. verum*	175
wekete	*Cocos nucifera* var.	26, 151, 181
wetune	*Aglaonema simplex*	177
worteli	*Daucus carota*	220
yakon(***i***) syn. (***ana***) ***manipa***	*Zea mays*	26, 66, 115, 131, 184
yakon(***i***) ***msinae***	*Zea mays* var.	184
yakon(***i***) ***nikate***	*Zea mays* var.	184
yakon(***i***) ***putie***	*Zea mays* var.	184
yana	k.o. tree; Bolton & Makote 2005	63–4
yohoru	*Xanthosoma sagittifolium*	60–1, 115, 133–4, 178
yohoru msinae	*Xanthosoma sagittifolium* var.	178
yohoru putie	*Xanthosoma sagittifolium* var.	178

APPENDIX 3

Alphabetical index of scientific plant taxa with Nuaulu names

Scientific identification	Nuaulu name	Pages
Abelmoschus moschatus	***pionenane***	209
Abroma augusta, A. fastuosa	***saumahu***	86, 209
Acacia probl. *mangium* Willd.	***akasie***	195
Acalypha amentacea, A. caturus	***pita(r)***	191
Acrostichum aureum	***naene (totu) onate***	53, 117, 171
Actephila excelsa var. *javanica*	***puane nimusie***	194
Actinodaphne moluccana	***akunin hanaie***	163
Actinodaphne rumphii	***akunin pina***	175
Agathis dammara	***kama(ne) wae***	4, 7, 108, 173
Actinodaphne spp.	***akunin*** syn. ***asinokoie***	107, 163, 175
Actinodaphne spp.	***asinokoie*** syn. ***akunin***	107, 163, 175
Agathis dammara	***ai nante***	4, 7, 108, 173
Agathis sp.	***kamane metene***	173
Agathis sp.	***kamane putie***	173
Agave cantala	***kinawane***	88, 90, 130, 180
Ageratum conyzoides	***mono mara osu, mono marosu, mono nuetetue***	17, 44, 103, 116, 219
Aglaia parviflora	***asi osi*** syn. ***tahena ukuna***	75, 208
Aglaia sapindina	***apunue***	16, 86, 116, 208
Aglaonema simplex	***kanna***	102, 129, 177
Aglaonema simplex	***wetune***	102, 129, 177
Alangium probl. *uniloculare, Homalium foetidum*	***ai nona***	212 191, 212
Aleurites moluccanus	***kahea***	7, 115, 127, 191
Aleurites moluccanus	***kahea nawe***	7, 115, 127, 191
Aleurites moluccanus	***kahea pukune***	7, 115, 127, 191
Allium	***pahane***	19, 124, 179
Allium cepa (Aggregatum Group)	***paha(ne) msinae***	67, 135, 179
Allium sativum	***paha(ne) putie***	37, 42, 179
Allophylus sp.	***pasa(ne) hatae***	109, 207
Alpinia galanga	***nakuasi***	128, 187
Alpinia gigantea	***tonamune***	187
Alpinia nutans	***kamkari***	102, 187
Alpinia sp.	***makakoii***	39, 106, 116, 187
Alpinia sp.	***makakoii pina***	39, 106, 116, 187
Alstonia scholaris	***ito***	110, 168, 215
Alstonia scholaris	***pom(pule)***	110, 168, 215
Alstonia spectabilis	***urone, urune***	215
Alstonia sp.	***saenae***	215
Amaranthus spinosus	***manahute waene anoi***	66–7, 148, 211
Amaranthus tricolor	***manahute marae***	42, 66, 148, 211
Amaranthus tricolor	***manahu(te) msinae***	42, 66, 148, 211
Amaranthus viridis	***mono manuate, mono manahutane, mono manhutanane, mono mana hutanane***	50, 66–7, 103, 148, 211

Scientific identification	Nuaulu name	Pages
Musa x var.	***uri wanate***	45, 63–4, 185–6
Musa x var.	***uri warakai***	45, 63–4, 185–6
Musa x var.	***uri warata*** syn. ***uri ikine***	63–4, 185–6
Myristica fragrans	***pala***	163, 174
Myristica insipida	***nanu***	174
Nauclea orientalis	***menene***	213–14
Nauclea sp.	***panne***	86, 113, 116, 156, 213–14
Nauclea sp.	***ai titie***	213–14
Neolamarckia cadamba, Nauclea sp.	***hana*(*panne*)**	86, 113, 116, 156, 213–14
Neolamarckia macrophylla	unknown4	214
Neonauclea lanceolata subsp. *gracilis*	***nosate***	72, 214
Neonauclea sp.	***ai titie***	116, 156
Nepenthes maxima var.	***koitipi putie***	38, 118, 166, 211
Nepenthes maxima var.	***koitipi sakalate***	42, 45, 211
Nepenthes spp. incl. *mirabilis*	***koitipi***	38, 118, 166, 211
Nephelium lappaceum	***rambutan***	82, 131, 136
Nephrolepis biserrata	***pupu pina***	53–5, 172
Nephrolepis hirsutula, N. biserrata	***pupu kaia***	53–5, 172
Neptunia oleracea	***masikunne***	197
Nicotiana tabacum	***tapako***	67, 115, 216
Nicotiana tabacum var.	***tapako malole***	45, 131, 216
Nicotiana tabacum var.	***tapako pane***	216
Nicotiana tabacum var.	***tapako sinsinte***	40, 45, 216
Nicotiana tabacum var.	***tapako tahenae***	216
Nypa fruticans	***tepine huae ina***	90–3, 144, 182
Oberonia padangensis	***sipa nione***	116, 179
Ocimum americanum, O. basilicum, O. gratissum	***kananukue***	219
Ocimum sp.	***kananuku putie***	219
Octomeles moluccana	***ai punara***	202
Octomeles sumatrana	***punara noene***	45, 136, 150, 202
Oncosperma filamentosa	***katue***	90–1, 93, 144, 182
Ophiorrhiza canescens, O. sp.	***kamruku***	55, 103, 214
Orania sylvicola	***korane***	90–1, 182

Scientific identification	Nuaulu name	Pages
Oroxylum indicum	***koitipa***	218
Orthosiphon aristatus	***manapesi mau***	102, 136, 219
Oryza sativa	***ana* (*hotone*)** syn. ***hotone***	45, 66–7, 115, 127, 129, 131, 183
Oryza sativa	***hotone*** syn. ***ana* (*hotone*)**	45, 66–7, 115, 127, 129, 131, 183
Oryza sativa var.	***ana msinae***	66–7, 183
Oryza sativa var.	***ana nusa inae***	66–7, 183
Osmoxylon nr. *boerlagei*	**(*ai*) *mkoue***	220
Paederia verticillata	***awane notute***	43, 214
Palaquium obovatum	***katou***	115, 117, 212
Palaquium obovatum var	***katou totu ikine***	115, 117, 212
Palaquium obovatum var.	***katou totu onate***	115, 117, 212
Panax sp.	***jinsen***	220
Pandanus		87, 107, 110, 118, 127, 159
Pandanus conoideus	**(*koae*) *saune*** syn. ***saune, koaia***	88, 90, 106, 136, 179
Pandanus conoideus	***saune hehue***	88, 90, 179
Pandanus krauelianus	***paone***	88–90, 179
Pandanus tectorius	***sau*(*ne*) *hahu***	88, 90, 179
Pandanus sp.	***koae kinoe***	87–8, 90, 118, 179
Pandanus sp.	***koae koio*** syn. ***kokoio***	88, 90, 118, 179
Pandanus sp.	***koae metene***	88, 90, 118, 179
Pandanus sp.	***koae putie***	88, 90, 118, 179
Pandanus sp.	***kokoio*** syn. ***koae koio***	88–9, 90, 179
Pandanus sp.	***kokoio nuae***	88–9, 90, 179
Pandanus sp.	***sanoma***	88, 90, 179
Pandanus spp. incl. *amaryllifolius*	***koae***	87–90, 106, 179
Pandanus various spp. incl. *conoideus*	***saune*** syn. **(*koae*) *saune***	88, 90, 179
Parkia speciosa	***muku hanaie***	39, 197
Passiflora foetida	***pitiri***	118, 130, 190
Pennisetum macrostachyum	***asu etute***	39–40, 184
Pericampylus glaucus	***awane wasi***	59, 188
Peristrophe bivalvis	***moune***	102, 217
Peristrophe sp.	***moune msinae***	102, 217

Scientific identification	Nuaulu name	Pages
Premna serratifolia	**ai suto**	50, 75, 86, 110, 218
Premna serratifolia	**akioae**	218
Procris frutescens, probl.	**sinatane urone**	116, 201
Prunus arborea, *P. wallaceana*	**ai ane**	113–14, 198
Prunus javanica	**aunasate**	86, 117, 198
Prynaria quercifolia	**naene putie**	172
Pseuderanthemum sp.	**kamruku**	102–3, 217
Pseuderanthemum sp.	**kamruku korisine**	217
Pseuderanthemum sp.	**kamruku metene**	217
Pseuderanthemum sp.	**kamruku notane**	217
Pseuderanthemum sp.	**kamruku sororione**	103, 217
Pseuderanthemum sp.	**korosine wane** syn. **kamruku wane**	103, 217
Pseuderanthemum sp.	**kamruku wane** syn. **korosine wane**	103, 217
Pseuderanthemum sp.	**mo wapane**	103, 217
Psidium guajava	**koiawas**	130, 203
Psilotum nudum	**soka-sokae marae**	40, 51, 103, 171
Pteris sp.	**kau-kau**	40, 45, 53, 55, 171
Pteris ensiformis, *P. semipinnata*	**pupu onate**	53, 55, 103, 171
Pteris semipinnata	(**ai**) **mukune**	53, 117, 171
Pteris vittata	**sapana**	53, 55, 103, 171
Pteris sp	**kopane**	54–5, 103, 171
Pternandra cordata, *P. coerulescens*	**popote wesie**	59, 77, 155, 206
Pterocarpus incl. *P. indicus*	**nana**	128, 197
Pterocarpus indicus	**nana tope**	128, 197
Pterocarpus indicus	**atipanate**	117, 128, 197
Pterocarpus sp.	**nana kunini**	197
Pterocarpus sp.	**nana minete**	197
Punica granatum	**dalima**	118, 203
Rauwolfia sumatrana	**mataponane**	116, 215
Rhizophora sp.	**hana** syn. **ai unate**	189
Rhodamnia cinerea	**ai kau**	44, 86, 203
Rhodamnia cinerea	**katinoie numte**	117, 203
Rhopaloblaste ceramica	**huaena**	90–1, 182
Ricinus communis	**jarak** (**jepang**)	45, 194
Riedelia sp.	**nakuasi**	106, 188
Robiquetia gracilistipes	**sipane hanaie**	39, 104, 179
Rorippa indica	**sesawi**	138, 211
Rubus fraxinifolius	(**a**)**wane unie**	43, 83, 149, 198
Rubus moluccanus	**kupa rosi**	103–4, 198
Rubus sp.	**awane putie**	198
Saccharum officinarum var.	**tohu hehue**	184
Saccharum officinarum var.	**tohu ito**	184
Saccharum officinarum var.	**tohu manteri**	184
Saccharum officinarum var.	**tohu metene**	184
Saccharum officinarum var.	**tohu mitane**	44, 184
Saccharum officinarum var.	**tohu pikane**	184
Saccharum officinarum var.	**tohu putie**	184
Saccharum officinarum var.	**tohu tona**	43, 184
Saccharum officinarum var.	**tohu tona metene**	42, 184
Saccharum officinarum var.	**tohu tona putie**	184
Saccharum spontaneum	**niune tunne**	40, 67, 184
Saccharum spontaneum var. *edulis*	**topikane**	40, 67, 184
Saccharum spp. incl. *S. officinarum*	**tohu**	67, 95, 98–9, 107, 115, 184
Salacca zalacca	**salak**	91, 136, 182
Saribus rotundifolius	**nasinana**	91, 182
Saurauia ceramica	**kikun**(**i**) **werane**	118, 213
Saurauia tristyla	**katoe**	118, 213
Sauropus androgynus	**karatu, kastu, karata**	66, 194
Scaevola taccada	**okar**	118, 219
Scaevola taccada, *S. frutescens*	**hoka**	118, 219
Schizostachyum brachycladum	**tapine**	95, 98, 184
Schizotachyum lima	**tomone**	95–6, 97, 144, 184
Schizotachyum lima	**tomo**(**ne**) **onate**, syn. **tomo kehane**	95–6, 97, 144, 184
Schizotachyum sp.	**suenie**	94–5, 184
Scindapsus sp.	**monone wane**	177
Scleria sp.	**hahuseite**	18, 103, 116, 182
Selaginella willdenowii	**soka-sokae onate**	51, 170
Selaginella sp.	**soka-sokae mnatinae**	51, 170

Scientific identification	Nuaulu name	Pages
k.o. tree	***niosane***	
k.o. tree	***nopoi***	
k.o. tree	***oi'i***	
k.o. tree	***rapa***	45
k.o. tree	***roa***	157
k.o. tree	***taha iuene***	
k.o. tree	***taha unene***	
k.o. tree	***taha waene***	
k.o. tree	***una sinatane***	
k.o. tree	***yana***	63–4
k.o. tree; 03-06 UKC	***kakeimate***	
k.o. tree; 03-14 UKC	***kakeimate totu onate***	
k.o. tree; 1996 plot 7	***saiaune***	
k.o. tree; 1996 plot 7	***tanasane***	
k.o. tree; 1996 plot 8	***muna hau***	
k.o. tree; 1996 plot 8	***nanae ane***	
k.o. tree; 1996 plot 8	***sukae***	
k.o. tree; 1996 plot 9	***ai sote***	44
k.o. tree; 1996 plot 9	***atuate***	
k.o. tree; 1996 plot 10	***ai purae***	
k.o. tree; 1996 plot 10	***ananinoi***	
k.o. tree; 96-107 BOGOR, 96-108 UKC, 96-109 KEW, 96-355 BOGOR, 96-356 KEW, 96-357 UKC, 96-1163 BOGOR, 96-1164 KEW, 96-1165 KEW	***akahonte***	
k.o. tree; 96-122 BOGOR, 96-123 Kew, 96-124 UKC	***manane***	
k.o. tree; 96-143 BOGOR, 96-144 KEW, 96-145 UKC	***siri***	
k.o. tree; 96-203 UKC, 96-205 BOGOR, 96-206 KEW, 96-522 BOGOR	***ai rapa***	
k.o. tree; 96-367 BOGOR, 96-368 KEW, 96-369	***manene***	
k.o. tree; 96-403 BOGOR, 96-404 KEW, 96-405 UKC	(***ai***) ***tahane***	
k.o. tree; 96-434 BOGOR, 96-435 KEW< 96-436 UKC	***kanoie***	
k.o. tree; 96-501 BOGOR, 96-502 KEW, 96-503 UKC; 1996 plot 10	***ahorane***	
k.o. tree; 96-602 KEW, 96-603 UKC, 96-604 KEW; 1996 plot 10	***akahon tonu-tonu***	40
k.o. tree; 96-947.1 BOGOR, 96.947.2 KEW, 96-948 KEW, 96-949 KEW	***msine***	
k.o. tree; 96-968 UKC, 96-969 KEW, 96-970 UKC	***ai parae***	
k.o. tree; 96-1001 BOGOR	***ai tihane***	
k.o. tree; MALVACEAE [TILIACEAE]	***kaute***	102, 211
k.o. tree; probl. OLEACEAE; 1996 plots 8 and 10	***anahata katapa***	217
resin-producing tree	***kamane***	108, 119, 173, 206
k.o. vegetable	***topa***	
vine	***awane***	68, 83–4, 100–1, 107, 149–50
k.o. vine	***awane apane***	45, 59
k.o. vine	***awane mine***	
k.o. vine	***awane nuae***	
k.o. vine	***hau uroe waene***	
k.o. vine	***kanunte***	
k.o. vine	***sosone wane***	
k.o. vine; 96-827 BOGOR, 96-828 KEW, 96-829 UKC	***wasusu***	
k.o. vine: 96-1330 UKC	***utatila***	199
[Bolton & Matoke 2005]	***nahune***	
96-460 BOGOR, 96-461 KEW, 96-462.1 UKC, 96-462.2 KEW	***aputie***	
96-498.1 BOGOR, 96-498.2 KEW, 96-499 UKC, 96-500 KEW	***apetine***	
96-805 BOGOR, 96-806 KEW, 96-867 UKC	***moite wane***	
NO IDENTIFICATION		
	lea mena	
k.o. tree	***manene haniae***	
no ident.	***manu unue***	
no ident.	***mkoue muaie***	
k.o. herb/weed	***mono makae***	103
k.o. tree	***nahani nani***	
k.o. tree	***sakuala***	
no ident.	***sinepane***	
no ident.	***taunaki***	

APPENDIX 4

Eleven plot surveys of forest in south Seram traversed and used by Nuaulu living in Rouhua, 1996

Plot 1: Nahati: near river Makoihiru adjacent Sekanima's

nisi ahue

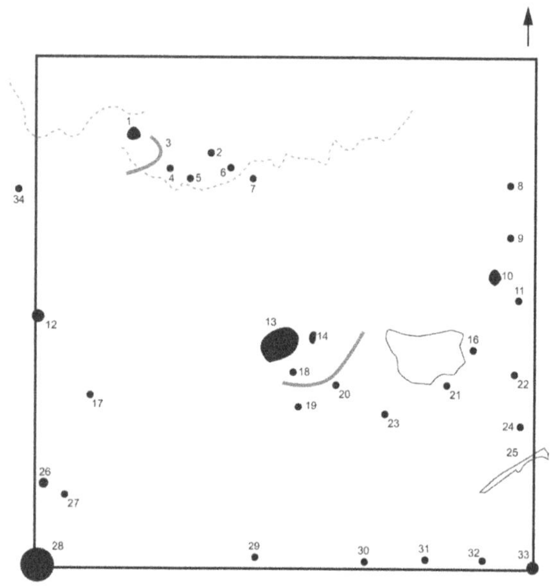

Map location: 129 10' E 03 21' S
Date: 06-02-96
Description: on coralline limestone, not known to have been used for gardens. 35 trees over dbh 10.
Slope: 30-40
Altitude in meters: 200
Notes:
▬ = rock edges
- - - - = stream bed

Key:

1	ai hatu	*Mallotus*	EUPHORBIACEAE
2	notone	*Callicarpa longifolia*	LAMIACEAE
3	awane papate	*Petraeovitex*	LAMIACEAE
4	siri		
5	akahonte		
6	asapopote	*Syzygium boerlagei*	MYRTACEAE
7	aniono		EUPHORBIACEAE
8	aniono		EUPHORBIACEAE
9	aniono		EUPHORBIACEAE
10	akahonte		
11	nosate	*Neonauclea lanceolata*	RUBIACEAE
12	nunu inane	*Ficus*	MORACEAE
13	ima	*Koompassia excelsia*	FABACEAE
14	ai hatu	*Mallotus* sp.	EUPHORBIACEAE
15	wane	(k.o. liana)	
16	ai hatu	*Mallotus* sp.	EUPHORBIACEAE
17	ai hatu	*Mallotus* sp.	EUPHORBIACEAE
18	pono	probl. *Homalium foetidum*	SALICACEAE
19	nosate	*Neonauclea lanceolata*	RUBIACEAE
20	aniono		EUPHORBIACEAE
21	ai hatu	*Mallotus* sp.	EUPHORBIACEAE
22	nosate	*Neonauclea lanceolata*	RUBIACEAE
23	anametene	*Diospyrus rostrata*	EBENACEAE
24	aniono		EUPHORBIACEAE
25	asapopo mane	*Syzygium*	MYRTACEAE
26	apaine totu onate	*Syzygium*	MYRTACEAE
27	tom-tom wesie	*Glochidion*	PHYLLANTHACEAE
28	anametene	*Diospyrus rostrata*	EBENACEAE
29	aniono		EUPHORBIACEAE
30	apaine	*Syzygium*	MYRTACEAE
31	au oine	*Endospermum moluccanum*	EUPHORBIACEAE
32	ai numa		LAURACEAE
33	asapopote	*Syzygium boerlagei*	MYRTACEAE
34	ahutau numte	*Callophylum* probl. *articulatum*	CLUSIACEAE

Plot 2: Mon River, 1 km E Rouhua

Map location: 129 10' E 03 20' S
Date: 09-02-1996
Description: depleted forest on sandstone; open canopy, some cut stands, ground cover of young saplings, ferns and *Selaginella*.
Slope: 40-50
Altitude in meters: 150
Notes: 30 = 3.7m dbh
═ = rock edges

Key:

1	puane musie	*Actephila excelsa*	PHYLLANTHACEAE
2	kamakunte	probl. *Diospyros pilosanthera*	EBENACEAE
3	katue	*Oncosperma filamentosa*	ARECACEAE
4	katue	*Oncosperma filamentosa*	ARECACEAE
5	ai pina onate	*Microcos ceramensis*	MALVACEAE
6	ranoi	*Macaranga hispida*	EUPHORBIACEAE
7	kupaleno	*Cananga odorata*	ANNONACEAE
8	huaena	*Rhopaloblaste ceramica*	
9	rapa	probl. *Ficus*	MORACEAE
10	?		
11	kawasa	*Archidendron clypearia*	FABACEAE
12	?		
13	kupatonae	*Macaranga*	EUPHORBIACEAE
14	anapakane	*Leea indica*	VITACEAE
15	anapakane	*Leea indica*	VITACEAE
16	asaherane	*Virola surinamensis*	MYRISTICACEAE
17	nosate	*Neonauclea lanceolata*	RUBIACEAE
18	awane metene	*Dichapetalum papuanum*	DICHAPETALACEAE
19	kamakunte probl.	*Diospyros pilosanthera*	EBENACEAE
20	manane		
21	meu wa ukune	*Calamus*	ARECACEAE
22	sosone	*Celtis*	CANNABACEAE
23	ai unene		LAURACEAE
24	anapakane	*Leea indica*	VITACEAE
25	asaherane	*Virola surinamensis*	MYRISTICACEAE
26	pono	probl. *Homalium foetidum*	SALICACEAE
27	sosone	*Celtis*	CANNABACEAE
28	ai osi	*Aglaia parviflora*	MELIACEAE
29	nakunatae		LAURACEAE
30	nosate	*Neonauclea lanceolata*	RUBIACEAE

Plot 3: Nahane Ukune (beyond Sama), South Seram

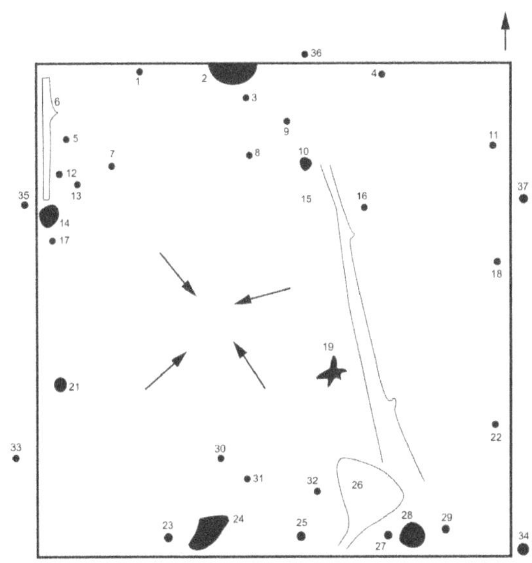

Map location: 129 10' E 03 19' S
Date: 11-02-1996
Description: k.a. Wae sana ukune (trib. of Upa), 25 metres from road, created by Philippine logging company. Mature forest with much *Shorea* adjacent logging concession. An old village site (Samalo) is located in the vicinity.
Slope: 30
Altitude in meters: 200
Notes: 14 = 2.5m dbh, 2 = 7.5m dbh

Key:

1	anahata katapa		OLEACEAE
2	onie	*Shorea selanica*	DIPTEROCARPACEAE
3	ahutau numte	*Callophylum* probl. *articulatum*	CLUSIACEAE
4	rapa	poss. *Ficus*	MORACEAE
5	ai numa hanaie		LAURACEAE
6	ananate (fallen)	*Canarium sylvestre*	BURSERACEAE
7	kawasa	*Archidendron clypearia*	FABACEAE
8	onie	*Shorea selanica*	DIPTEROCARPACEAE
9	aunasate	*Prunus javanica*	ROSACEAE
10	ananate	*Canarium sylvestre*	BURSERACEAE
11	ahutau numte	*Callophylum* probl. *articulatum*	CLUSIACEAE
12	hanasane	*Barringtonia asiatica*	LECYTHIDACEAE
13	rapa	poss. *Ficus*	MORACEAE
14	onie	*Shorea selanica*	DIPTEROCARPACEAE
15	onie	*Shorea selanica*	DIPTEROCARPACEAE
16	ahutau numte	*Callophylum* probl. *articulatum*	CLUSIACEAE
17	ahutaune	*Callophylum inophyllum*	CLUSIACEAE
18	aunasate	*Prunus javanica*	ROSACEAE
19	apaine totu mkauke	*Syzygium*	MYRTACEAE
20	onie	*Shorea selanica*	DIPTEROCARPACEAE
21	ahutaune	*Callophylum inophyllum*	CLUSIACEAE
22	aunasate	*Prunus javanica*	ROSACEAE
23	anametene	*Diospyrus rostrata*	EBENACEAE
24	hino	*Syzygium*	MYRTACEAE
25	nanu	*Myristica insipida*	MYRISTICACEAE
26	onie (fallen)	*Shorea selanica*	DIPTEROCARPACEAE
27	riane	*Elaeocarpus augustifolius*	ELAEOCARPACEAE
28	onie (dead)	*Shorea selanica*	DIPTEROCARPACEAE
29	ananate	*Canarium sylvestre*	BURSERACEAE
30	hino	*Syzygium*	MYRTACEAE
31	onie	*Shorea selanica*	DIPTEROCARPACEAE
32	ananate	*Canarium sylvestre*	BURSERACEAE
33	onie	*Shorea selanica*	DIPTEROCARPACEAE
34	nanu	*Myristica insipida*	MYRISTICACEAE
35	anahata katapa		OLEACEAE
36	nahane	*Xanthostemon verus*	MYRTACEAE
37	onie	*Shorea selanica*	DIPTEROCARPACEAE

Plot 4: Agroforest (nisi ahue) on Ratipanisa, a branch of the Yoko river, near Rouhua

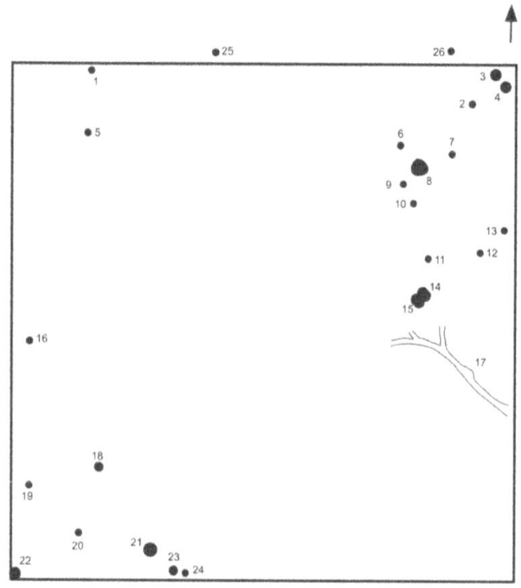

Map location: 129 08' E 03 20' S
Date: 13-02-96
Description : Land last cleared for garden by Numapena, perhaps 20 years ago.Now owned by Saete and Pela'uma Soumori. Now mostly chempedak(*A. integer*, Sonawe - ainakahata) These are self-set rather than planted
Slope: 30
Altitude in meters:100
Notes:
Ground cover is oni (*Imperata exaltata*), soka-sokae (*Selaginella*), a lot of young tahenae saplings, mukune (*Cyathea*), unuhutu msinae (*Clerodendron*) [monne for Soumori], timnisie (*Asclepias daceae*), nite unue (*Asplenium nidus*), makakoihi, buane nimusie (*Carallia brachiata*) and pono (*Eugenia palembarica*)
18 = 80cm dbh

Key:

1	karane		BURSERACEAE
2	tahenae	*Artocarpus integer*	MORACEAE
3	tahenae	*Artocarpus integer*	MORACEAE
4	tahenae	*Artocarpus integer*	MORACEAE
5	ai numa		LAURACEAE
6	tahenae	*Artocarpus integer*	MORACEAE
7	tahenae	*Artocarpus integer*	MORACEAE
8	tahenae	*Artocarpus integer*	MORACEAE
9	tahenae	*Artocarpus integer*	MORACEAE
10	tahenae	*Artocarpus integer*	MORACEAE
11	iane	*Canarium indicum*	BURSERACEAE
12	tahane		
13	ahauni	*Antidesma stipulare*	PHYLLANTHACEAE
14	tahenae	*Artocarpus integer*	MORACEAE
15	tahenae	*Artocarpus integer*	MORACEAE
16	kokune	*Callicarpa longifolia*	LAMIACEAE
17	tahenae	*Artocarpus integer*	MORACEAE
18	tahenae	*Artocarpus integer*	MORACEAE
19	kamamuri	*Cheilocostus speciosus*	ZINGIBERACEAE
20	tahenae	*Artocarpus integer*	MORACEAE
21	tahenae	*Artocarpus integer*	MORACEAE
22	tahenae	*Artocarpus integer*	MORACEAE
23	tahenae	*Artocarpus integer*	MORACEAE
24	tahenae	*Artocarpus integer*	MORACEAE
25	kupanahane	Myrciathes fragrans	MYRTACEAE
26	nana	*Pterocarpus indicus*	FABACEAE

Plot 5: Benteng (AM 'fort'): old settlement site of Besi people, located by Iana Ikine and Iana Onate rivers, both tributaries of Upa.

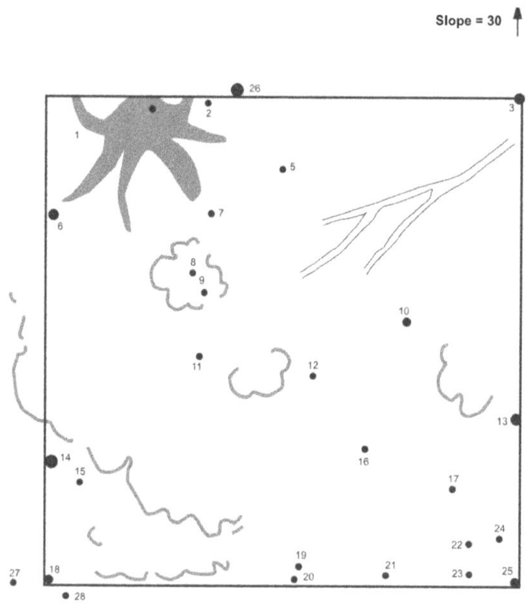

Map location: 129 09' E 03 18' S
Date: 14-02-1996
Description: Coralline limestone outcrop, some evidence of human fortification (walling), dominated by large *Ficus*. Liana growth mainly mentinse (*Tetrastigma*), awane kunie and papate wane (*Petraeovitex*). Old settlement site (called 'Hatu Sin' by Nuaulu), perhaps 200 years since last occupation.
Slope: Indeterminate
Altitude in meters 300
Notes: Site associated with war between Sepa and Besi. After retreat to Benteng, the Nuaulu clans Sonaue-ainakahata and Soumori acted as guides to Besi people, showing them the way to the North.
Ground cover mainly ferns
1 = 7m diam.
= rock edges
= stream bed

Key:

1	nunu inane	*Ficus*	MORACEAE
2	apaine	*Syzygium*	MYRTACEAE
3	ai hatu	probl. *Mallotus*	EUPHORBIACEAE
4	apetine		
5	apetine		
6	apaine totu mkauke	*Syzygium*	MYRTACEAE
7	apetine		
8	apaine	*Syzygium*	MYRTACEAE
9	apetine		
10	apetine		
11	asaherane	*Virola surinamensis*	MYRISTICACEAE
12	apetine		
13	akahonte		
14	anametene	*Diospyrus rostrata*	EBENACEAE
15	apetine		
16	apetine		
17	apetine		
18	apaine	*Syzygium*	MYRTACEAE
19	ai hatu	probl. *Mallotus*	EUPHORBIACEAE
20	ai hatu	probl. *Mallotus*	EUPHORBIACEAE
21	asaherane	*Virola sunnamensis*	MYRISTICACEAE
22	ai hatu	probl. *Mallotus*	EUPHORBIACEAE
23	ai hatu	probl. *Mallotus*	EUPHORBIACEAE
24	ai osi	*Aglaia parviflora*	MELIACEAE
25	apaine	*Syzygium*	MYRTACEAE
26	apaine	*Syzygium*	MYRTACEAE
27	apetine		
28	apetine		

Plot 6: Agroforest (nisi ahue) on Iana Onate Matai

Map location: 129 09' E 03 19' S
Date: 19-02-1996
Description: old garden site (**nisi ahue**). Lohia Sonaue-ainakahata had a garden here 30 years ago: planted with langsat, coconut, betelnut, durian, clove, coffee and banana.
Slope: 20°
Altitude in meters: 83
Notes: 32 = 110 dbh

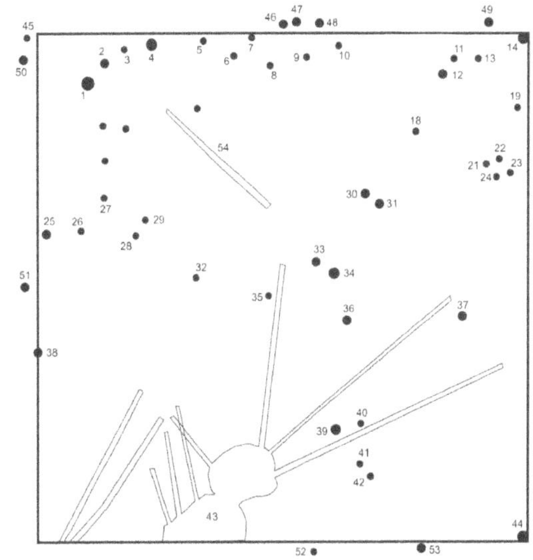

Key:

1	kinae	*Kleinhovia hospita*	MALVACEAE
2	kinae	*Kleinhovia hospita*	MALVACEAE
3	nana	*Pterocarpus indicus*	FABACEAE
4	kinae	*Kleinhovia hospita*	MALVACEAE
5	nana	*Pterocarpus indicus*	FABACEAE
6	msikene	*Ardisia humilis*	PRIMULACEAE
7	kokoio	*Pandanus*	PANDANACEAE
8	kokoio	*Pandanus*	PANDANACEAE
9	ai nasate	*Elaeocarpus* probl. *serratus*	ELAEOCARPACEAE
10	kanai	*Areca catechu*	ARECACEAE
11	kanai	*Areca catechu*	ARECACEAE
12	kanai	*Areca catechu*	ARECACEAE
13	kanai	*Areca catechu*	ARECACEAE
14	kanai [ia nanae]	*Areca catechu*	ARECACEAE
15	kanai	*Areca catechu*	ARECACEAE
16	kanai	*Areca catechu*	ARECACEAE
17	anametene	*Diospyrus rostrata*	EBENACEAE
18	puane musie	*Actephila excelsa*	PHYLLANTHACEAE
19	kanai	*Areca catechu*	ARECACEAE
20	urone	*Alstonia villosa*	APOCYNACEAE
21	kanai	*Areca catechu*	ARECACEAE
22	kanai	*Areca catechu*	ARECACEAE
23	kanai	*Areca catechu*	ARECACEAE
24	kanai	*Areca catechu*	ARECACEAE
25	asaherane	*Virola surinamensis*	MYRISTICACEAE
26	kanoie		
27	nana	*Pterocarpus indicus*	FABACEAE
28	ananate	*Canarium sylvestre*	BURSERACEAE
29	taha waene		
30	msikene	*Ardisia humilis*	PRIMULACEAE
31	msikene	*Ardisia humilis*	PRIMULACEAE
32	nione	*Cocos nucifera*	ARECACEAE
33	kaiohitae	*Claoxylon*	EUPHORBIACEAE
34	kaiohitae	*Claoxylon*	EUPHORBIACEAE
35	msikene	*Ardisia humilis*	PRIMULACEAE
36	anapakane	*Leea indica*	VITACEAE
37	nione	*Cocos nucifera*	ARECACEAE
38	kananae		
39	kanoie		
40	msikene	*Ardisia humilis*	PRIMULACEAE
41	asaherane	*Virola surinamensis*	MYRISTICACEAE
42	anapakane	*Leea indica*	VITACEAE
43	wana hatu	*Dendrocalamus*	POACEAE
44	kanai	*Areca catechu*	ARECACEAE
45	tunene	*Annona muricata*	ANNONACEAE
46	panuke	*Caryota rumphiana*	ARECACEAE
47	panuke	*Caryota rumphiana*	ARECACEAE
48	panuke	*Caryota rumphiana*	ARECACEAE
49	kanai ia nanae	*Areca catechu*	ARECACEAE
50	asahune	*Dysoxylum* sp.	MELIACEAE
51	nasate	*Lansium parasiticum*	MELIACEAE
52	anapakane	*Virola surinamensis*	VITACEAE
53	nione	*Cocos nucifera*	ARECACEAE
54	kanai	*Areca catechu*	ARECACEAE

Plot 7: Part of sin wesie (protected forest) belonging to clan Penisa at Mon sanae

Map location : 129 10 E 03 20 S '
Date : 21-02-1996
Description : Edge of protected area
Slope : 20
Altitude in meters : 300
Notes :
The boundaries of this *sin wesie* go some kilometres inland and include large areas of sago (*Metroxylon sagu*)

—— = rock edges

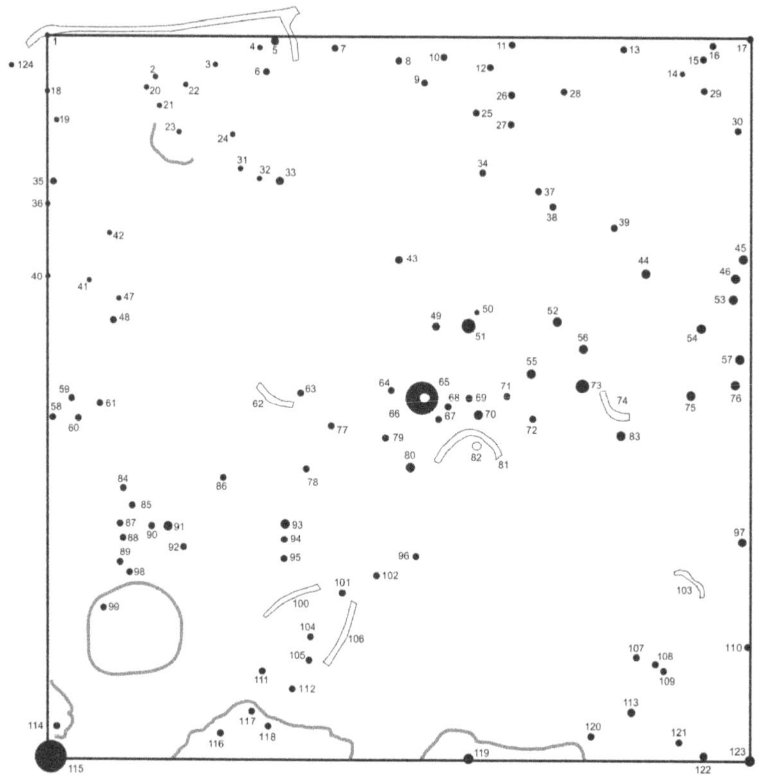

Key:

1	pono	probl. *Homalium foetidum*	SALICACEAE
2	ananione	*Horsfieldia bivalvis*	MYRISTICACEAE
3	apetine		
4	anametene	*Diospyrus rostrata*	EBENACEAE
5	mataponane	*Cerbera manghas*	APOCYNACEAE
6	akahon tonu-tonu		
7	anametene	*Diospyrus rostrata*	EBENACEAE
8	ahutaune	*Calophylum inophyllum*	CLUSIACEAE
9	akahon tonu-tonu		
10	ai pina onate	*Microcos ceramensis*	MALVACEAE
11	kawasa	*Archidendron clypearia*	FABACEAE
12	ai pina onate	*Microcos ceramensis*	MALVACEAE
13	hanapanne	probl. *Nauclea*	RUBIACEAE
14	ranoi	*Macaranga hispida*	EUPHORBIACEAE
15	waruae	*Ficus variegata*	MORACEAE
16	waruae	*Ficus variegata*	MORACEAE
17	ranoi	*Macaranga hispida*	EUPHORBIACEAE
18	ai numa		LAURACEAE
19	anametene	*Diospyrus rostrata*	EBENACEAE
20	hihane wane	*Phanerla kockiana*	FABACEAE
21	ai msinae	*Syzygium* probl. *rumphii*	MYRTACEAE
22	rapa		
23	ai msinae	*Syzygium* probl. *rumphii*	MYRTACEAE
24	panuke	*Caryota rumphiana*	ARECACEAE
25	tanasane		
26	aunasate	probl. *Prunus javanica*	ROSACEAE
27	ai pina onate	*Microcos ceramensis*	MALVACEAE
28	ananate	*Canarium sylvestre*	BURSERACEAE
29	waruae	*Ficus variegata*	MORACEAE
30	ranoi	*Macaranga hispida*	EUPHORBIACEAE
31	anametene	*Diospyrus rostrata*	EBENACEAE
32	sene	probl. *Alstonia*	APOCYNACEAE
33	au oine	*Endospermum moluccanum*	EUPHORBIACEAE
34	ai nona	probl. *Homalium foetidum*	SALICACEAE
35	akenue	*Calophyllum*	CLUSIACEAE
36	ai pina onate	*Microcos ceramensis*	MALVACEAE
37	ai unene		LAURACEAE
38	ai pina onate	*Microcos ceramensis*	MALVACEAE
39	waruae	*Ficus variegata*	MORACEAE
40	rapa		
41	ananate	*Canarium sylvestre*	BURSERACEAE
42	asapopo makae	*Syzygium boerlaegi*	MYRTACEAE
43	tanasane		
44	ananione	*Horsfieldia bivalvis*	MYRISTICACEAE
45	kam natine	*Melastoma polyanthum*	MELASTOMATACEAE
46	auhaune	*Leea indica*	VITACEAE
47	ai msinae	*Syzygium* probl. *rumphii*	MYRTACEAE
48	aunasate	probl. *Prunus javanica*	ROSACEAE
49	pono	*Homalium foetidum*	SALICACEAE
50	nunue tinse	*Ficus*	MORACEAE
51	tahane		
52	panuke	*Caryota rumphiana*	ARECACEAE
53	ai pina onate	*Microcos ceramensis*	MALVACEAE
54	kamakunte	probl. *Melia excelsa*	MELIACEAE
55	ananate	*Canarium sylvestre*	BURSERACEAE
56	waruae	*Ficus variegata*	MORACEAE
57	hukila	probl. *Litsea*	LAURACEAE
58	ai weti atue	*Tectona grandis*	VERBENACEAE
59	rapa		
60	kamakunte	probl. *Melia excelsa*	MELIACEAE
61	asapopo mane	*Syzygium*	MYRTACEAE
62	mahosu wane	*Cissus javana*	VITACEAE
63	ananione	*Horsfieldia bivalvis*	MYRISTICACEAE
64	ai msinae	*Syzygium* probl. *rumphii*	MYRTACEAE
65	nunue	*Ficus*	MORACEAE
66	anametene	*Diospyrus rostrata*	EBENACEAE
67	nunue tinse	*Ficus*	MORACEAE
68	ananate	*Canarium sylvestre*	BURSERACEAE
69	kanoie		
70	ananate	*Canarium sylvestre*	BURSERACEAE
71	panuke	*Caryota rumphiana*	ARECACEAE
72	neru putie	*Garcinia*	CLUSIACEAE
73	ai nona	*Alangium*	ALANGIACEAE
74	kapine wane	*Uncaria*	RUBIACEAE
75	auhaune	*Leea indica*	VITACEAE
76	kawasa	*Archidendron clypearia*	FABACEAE
77	akahon tonu-tonu		
78	mukune	*Cyathea*	CYATHEACEAE
79	ai msinae	*Syzygium* probl. *rumphii*	MYRTACEAE
80	akenue	*Calophyllum*	CLUSIACEAE
81	awane papate	*Petraeovitex*	LAMIACEAE
82	hukila	probl. *Eugenia*	MYRTACEAE
83	asaherane	*Virola ceramensis*	MYRISTICACEAE
84	mukune	*Cyathea*	CYATHEACEAE
85	anametene	*Diospyrus rostrata*	EBENACEAE
86	tahane	*Neonauclea lanceolata*	
87	nosate	*Neonauclea lanceolata*	RUBIACEAE
88	ananate	*Canarium sylvestre*	BURSERACEAE
89	ai unene		LAURACEAE
90	ranoi	*Macaranga hispida*	EUPHORBIACEAE
91	ai pina onate	*Microcos ceramensis*	MALVACEAE
92	asora	probl. Dillenia	DILLENIACEAE
93	ananate	*Canarium sylvestre*	BURSERACEAE
94	rapa		
95	huaena	*Rhopaloblaste ceramica*	ARECACEAE
96	auhaune	*Leea indica*	VITACEAE
97	saeaune		
98	pono	*Homalium foetidum*	SALICACEAE
99	anametene	*Diospyrus rostrata*	EBENACEAE
100	asapopo mane	*Syzygium*	MYRTACEAE
101	asaherane	*Virola ceramensis*	MYRISTICACEAE
102	auhaune	*Leea indica*	VITACEAE
103	kapine wane	*Uncaria*	RUBIACEAE
104	ai pina onate	*Microcos ceramensis*	MALVACEAE
105	hukila	probl. *Eugenia*	MYRTACEAE
106	asapopo mane	*Syzygium*	MYRTACEAE
107	ahutaune	*Calophylum inophyllum*	CLUSIACEAE
108	kawasa	*Archidendron clypearia*	FABACEAE
109	ai numa		LAURACEAE
110	asaherane	*Virola ceramensis*	MYRISTICACEAE
111	pono	probl. *Homalium foetidum*	SALICACEAE
112	anametene	*Diospyrus rostrata*	EBENACEAE
113	ai pina onate	*Microcos ceramensis*	MALVACEAE
114	ai hatu	probl. *Mallotus*	EUPHORBIACEAE
115	sonae	*Campnosperma auriculata*	ANACARDIACEAE
116	pono	*Homalium foetidum*	SALICACEAE
117	ai hatu	probl. *Mallotus*	EUPHORBIACEAE
118	ai hatu	probl. *Mallotus*	EUPHORBIACEAE
119	ai nona	probl. Alangium	ALANGIACEAE
120	ai pina onate	*Microcos ceramensis*	MALVACEAE
121	ai pina onate	*Microcos ceramensis*	MALVACEAE
122	akenue	*Calophyllum*	CLUSIACEAE
123	asapopo makae	*Syzygium boerlagei*	MYRTACEAE
124	pono	*Homalium foetidum*	SALICACEAE

Plot 8: Sokonana

Map location: 129 08' E 03 14' S
Date: 24-02-96
Description: Ground cover mainly debris of cut trees and young saplings; bracken along edge. Max D of trees = 220 cm, mean D = 60-70 cm.
Slope: Slight decline to south towards Ruatan
Altitude in meters: 100
Notes: Smaller trees cut by people from Bunara

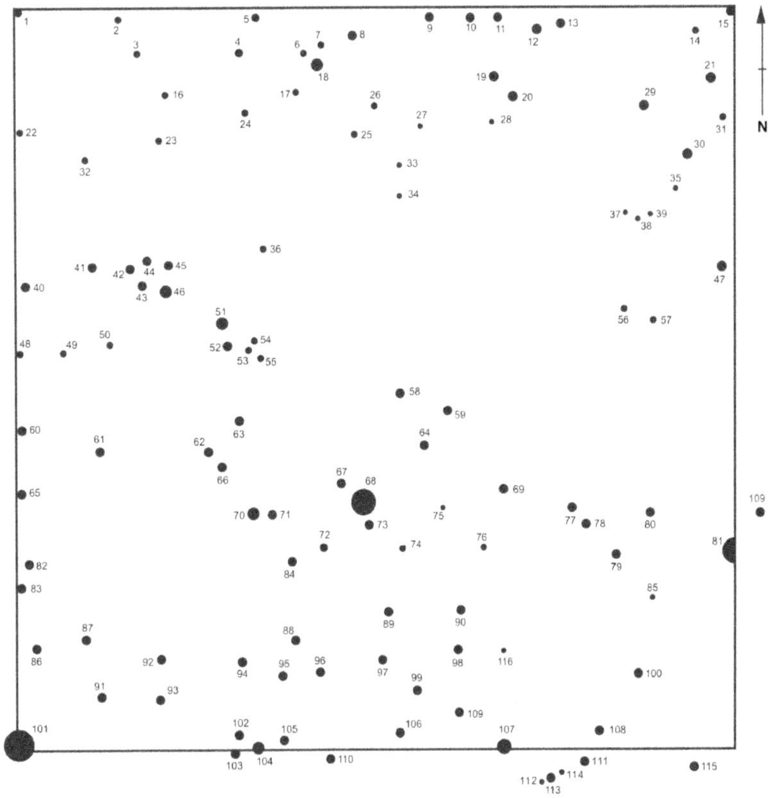

Key:

No.	Nuaulu name	Species	Family
1	anakoine	*Tapeinosperma*	PRIMULACEAE
2	msikene	*Ardisia humilis*	PRIMULACEAE
3	kinekane	*Phymatosorus membranifolium*	POLYPODIACEAE
4	rapa		
5	aleha		
6	kinekane	*Phymatosorus membranifolium*	POLYPODIACEAE
7	kinekane	*Phymatosorus membranifolium*	POLYPODIACEAE
8	akenue	*Calophyllum*	CLUSIACEAE
9	anahata katapa		OLEACEAE
10	anahata katapa		OLEACEAE
11	nanu	*Myristica insipida*	MYRISTICACEAE
12	nahane	*Xanthostemon verus*	MYRTACEAE
13	nahane	*Xanthostemon verus*	MYRTACEAE
14	rapa		
15	nanu	*Myristica insipida*	MYRISTICACEAE
16	kinekane	*Phymatosorus membranifolium*	POLYPODIACEAE
17	ai numte		
18	kinekane	*Phymatosorus membranifolium*	POLYPODIACEAE
19	hino	*Syzygium*	MYRTACEAE
20	kinekane	*Phymatosorus membranifolium*	POLYPODIACEAE
21	akenue	*Calophyllum*	CLUSIACEAE
22	kinekane	*Phymatosorus membranifolium*	POLYPODIACEAE
23	kinekane	*Phymatosorus membranifolium*	POLYPODIACEAE
24	nanu	*Myristica insipida*	MYRISTICACEAE
25	ananate	*Canarium sylvestre*	BURSERACEAE
26	hino	*Syzygium*	MYRTACEAE
27	ai mpiae		LAURACEAE
28	nanu	*Myristica insipida*	MYRISTICACEAE
29	akenue	*Calophyllum*	CLUSIACEAE
30	nanu	*Myristica insipida*	MYRISTICACEAE
31	kawasa	*Archidendron clypearia*	FABACEAE
32	kawasa	*Archidendron clypearia*	FABACEAE
33	aunasate	*Prunus javanica*	ROSACEAE
34	nanae	*Heritiera littoralis*	STERCULIACEAE
35	hino	*Syzygium*	MYRTACEAE
36	akenue	*Calophyllum*	CLUSIACEAE
37	kinekane	*Phymatosorus membranifolium*	POLYPODIACEAE
38	kinekane	*Phymatosorus membranifolium*	POLYPODIACEAE
39	kinekane	*Phymatosorus membranifolium*	POLYPODIACEAE
40	asaherane	*Virola surinamensius*	MYRISTICACEAE
41	iane	*Canarium indicum*	BURSERACEAE
42	asaherane	*Virola surinamensius*	MYRISTICACEAE
43	asaherane	*Virola surinamensius*	MYRISTICACEAE
44	asaherane	*Virola surinamensis*	MYRISTICACEAE
45	anahata katapa		OLEACEAE
46	hino	*Syzygium*	MYRTACEAE
47	anahata katapa		OLEACEAE
48	asaherane	*Virola surinamensis*	MYRISTICACEAE
49	hino	*Syzygium*	MYRTACEAE
50	hanasane	*Barringtonia asiatica*	LECYTHIDACEAE
51	ai msinae	*Syzygium rumphii*	MYRTACEAE
52	hanasane	*Barringtonia asiatica*	LECYTHIDACEAE
53	kinekane	*Phymatosorus membranifolium*	POLYPODIACEAE
54	kinekane	*Phymatosorus membranifolium*	POLYPODIACEAE
55	kinekane	*Phymatosorus membranifolium*	POLYPODIACEAE
56	msikene	*Ardisia humilis*	MYRSINACEAE
57	nanu	*Myristica insipida*	MYRISTICACEAE
58	apaine	*Syzygium*	MYRTACEAE
59	kinekane	*Phymatosorus membranifolium*	POLYPODIACEAE
60	naneane		
61	hanasane	*Barringtonia asiatica*	LECYTHIDACEAE
62	anine	*Ficus*	MORACEAE
63	kinekane	*Phymatosorus membranifolium*	POLYPODIACEAE
64	kinekane	*Phymatosorus membranifolium*	POLYPODIACEAE
65	rapa		
66	aleha		
67	ananate	*Canarium sylvestre*	BURSERACEAE
68	hino	*Syzygium*	MYRTACEAE
69	anahata katapa		OLEACEAE
70	nerune	*Garcinia*	CLUSIACEAE
71	asapopo mane	*Syzygium*	MYRTACEAE
72	ahutaune	*Calophylum inophyllum*	CLUSIACEAE
73	nanu	*Myristica insipida*	MYRISTICACEAE
74	ai numte		
75	ai mpiae		LAURACEAE
76	kawasa	*Archidendron clypearia*	FABACEAE
77	kinekane	*Phymatosorus membranifolium*	POLYPODIACEAE
78	ai msinae	*Syzygium*	MYRTACEAE
79	?		
80	nanu	*Myristica insipida*	MYRISTICACEAE
81	hino	*Syzygium*	MYRTACEAE
82	kinekane	*Phymatosorus membranifolium*	POLYPODIACEAE
83	nanu	*Myristica insipida*	MYRISTICACEAE
84	anametene	*Diospyrus rostrata*	EBENACEAE
85	katinoie	*Melastoma imbricatum*	MELASTOMATACEAE
86	hanasane	*Barringtonia asiatica*	LECYTHIDACEAE
87	hanasane	*Barringtonia asiatica*	LECYTHIDACEAE
88	nanu	*Myristica insipida*	MYRISTICACEAE
89	ahutaune	*Calophyllum inophyllum*	CLUSIACEAE
90	ai numte		
91	sukae		
92	asuani mata hunue		
93	ai pina onate	*Microcos ceramensis*	MALVACEAE
94	nanu	*Myristica insipida*	MYRISTICACEAE
95	ai pina onate	*Microcos ceramensis*	MALVACEAE
96	katinoie	*Melastoma imbricatum*	MELASTOMATACEAE
97	asuani mata hunue		
98	ai numte		
99	rapa		
100	ai numte		
101	anahata katapa		OLEACEAE
102	nahane	*Xanthostemon verus*	MYRTACEAE
103	nahane	*Xanthostemom verus*	MYRTACEAE
104	nanu	*Myristica insipida*	MYRISTICACEAE
105	ahutaune	*Calophyllum inophyllum*	CLUSIACEAE
106	kinekane	*Phymatosorus membranifolium*	POLYPODIACEAE
107	hino	*Syzygium*	MYRTACEAE
108	msikene	*Ardisia humilis*	MYRSINACEAE
109	ai pinamo	*Annona reticulata*	ANNONACEAE
110	asapopo mane	*Syzygium*	MYRTACEAE
111	nanu	*Myristica insipida*	MYRISTICACEAE
112	kinekane	*Phymatosorus membranifolium*	POLYPODIACEAE
113	kinekane	*Phymatosorus membranifolium*	POLYPODIACEAE
114	kinekane	*Phymatosorus membranifolium*	POLYPODIACEAE
115	ahutaune	*Calophyllum inophyllum*	CLUSIACEAE
116	rapa		

Plot 9: Rohnesi

Map location: 129 06' E 03 04' S
Date: 01-03-1996
Description:
Slope: steep decline towards east

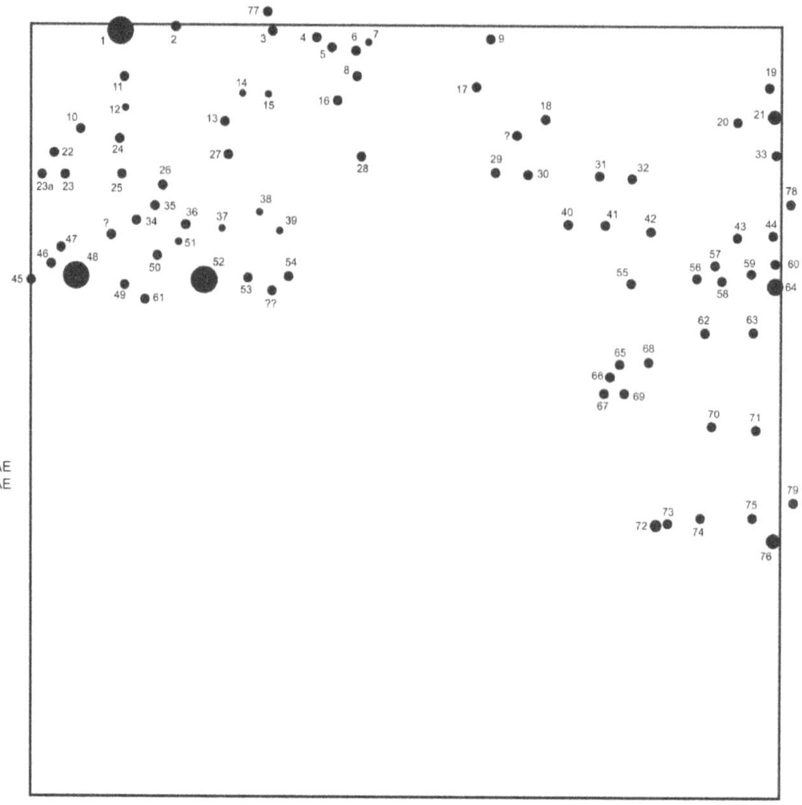

Key:

1	kama wae	*Agathis dammara*	ARAUCARIACEAE
2	ai suto	*Premna serratifolia*	LAMIACEAE
3	nahane	probl *Xanthostemon ve*	MYRTACEAE
4	ahutaune	*Calophyllum inophyllum*	CLUSIACEAE
5	ai nante	*Agathis dammara*	ARAUCARIACEAE
6	asuani mata hunue		
7	asuani mata hunue		
8	asuani mata hunue		
9	ai ane	*Prunus*	ROSACEAE
10	?		
11	ai mpiae		LAURACEAE
12	?		
13	apaine	*Syzygium*	MYRTACEAE
14	ai mpiae		LAURACEAE
15	kawasa	*Archidendron clypearia*	FABACEAE
16	ai ane	*Prunus*	ROSACEAE
17	ai ane	*Prunus*	ROSACEAE
18	asihata		probl. EUPHORBIACEAE
19	asihata		probl. EUPHORBIACEAE
20	ahutaune	*Calophyllum inophyllum*	CLUSIACEAE
21	kawasa	*Archidendron clypearia*	FABACEAE
22	muna hau		
23	ai numte		
23a	rapa		
24	apaine	*Syzygium*	MYRTACEAE
25	ai numte		
26	hino	*Syzygium*	MYRTACEAE
27	asu hunne	*Dysoxlum caulustachyu*	MELIACEAE
28	ahutaune	*Calophyllum inophyllum*	CLUSIACEAE
29	asuani mata hunue		
30	asuani mata hunue		
31	ai numte		
32	ahutaune	*Calophyllum inophyllum*	CLUSIACEAE
33	ahutaune	*Calophyllum inophyllum*	CLUSIACEAE
34	apaine	*Syzygium*	MYRTACEAE
35	hino	? *Syzygium*	MYRTACEAE
36	ananaine	*Litsea resinosa*	LAURACEAE
37	kawasa	*Archidendron clypearia*	FABACEAE
38	?		
39	ai numte		
40	hino	*Syzygium*	MYRTACEAE
41	asuani mata hunue		
42	ai nante	*Agathis dammara*	ARAUCARIACEAE
43	asihata		probl EUPHORBIACEAE
44	ahutaune	*Calophyllum inophyllum*	CLUSIACEAE
45	nakunatae		LAURACEAE
46	pasane hatae	*Allophylus*	SAPINDACEAE
47	pasane hatae	*Allophylus*	SAPINDACEAE
48	hino	*Syzygium*	MYRTACEAE
49	ai itine		
50	ahutau numte	*Calophyllum* probl *articulatum*	CLUSIACEAE
51	ai numte		
52	kama wae	*Agathis alba*	ARAUCARIACEAE
53	ahutaune	*Calophyllum inophyllum*	CLUSIACEAE
54	ai weti	*Phaleria perrottetiana*	THYMELEACEAE
55	ahutaune	*Calophyllum inophyllum*	CLUSIACEAE
56	asuani mata hunue		
57	ai ane	*Prunus*	ROSACEAE
58	inene	*Hydriastele microcarpa*	ARECACEAE
59			
60	ai numte		
61	hino	*Syzygium*	MYRTACEAE
62	ahutaune	*Calophyllum inophyllum*	CLUSIACEAE
63	inene	*Hydriastele microcarpa*	ARECACEAE
64	ai itine		
65	ahutaune	*Calophyllum inophyllum*	CLUSIACEAE
66	ahutaune	*Calophyllum inophyllum*	CLUSIACEAE
67	ahutaune	*Calophyllum inophyllum*	CLUSIACEAE
68	kawasa	*Archidendron clypearia*	FABACEAE
69	ahutaune	*Calophyllum inophyllum*	CLUSIACEAE
70	ahutaune	*Calophyllum inophyllum*	CLUSIACEAE
71	ai itine		
72	asuani mata hunue		
73	ai sote		
74	ahutaune	*Calophyllum inophyllum*	CLUSIACEAE
75	aninoi	Syzygium	MYRTACEAE
76	atuane	*Lunasia amara*	RUTACEAE
77	?		
78	asuani mata hunue		
79	ahutaune	*Calophyllum inophyllum*	CLUSIACEAE

Plot 10: Wae Pia (line transect)

Map location: 129 01' E 03 11' S
Date: 14-02-1996
Description: Lowland rainforest approx. 200m from Trans-Seram highway
Slope: 20
Altitude in meters: 200

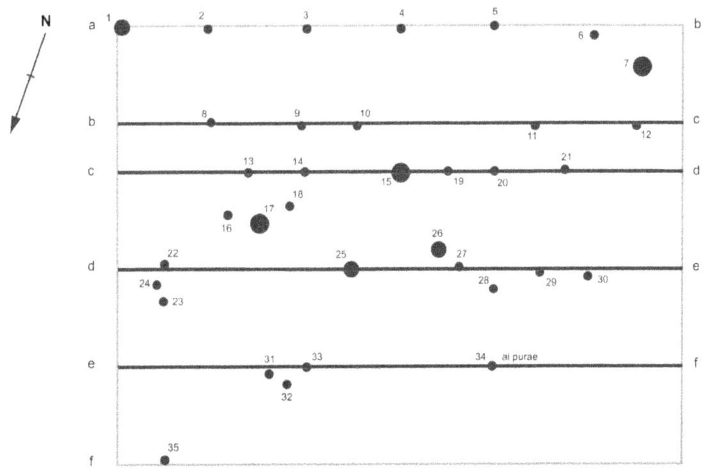

Key:

1	anine	*Ficus*	MORACEAE
2	nanae	*Heritiera littoralis*	STERCULIACEAE
3	ahutaune totu onate	*Calophyllum* probl. *soulattri*	CLUSIACEAE
4	ahutaune totu onate	*Calophyllum* probl. *soulattri*	CLUSIACEAE
5	apaine	*Syzygium*	MYRTACEAE
6	asaherane	*Virola surinamensis*	MYRISTICACEAE
7	apaine	*Syzygium*	MYRTACEAE
8	asaherane	*Virola surinamensis*	MYRISTICACEAE
9	ahutaune	*Calophyllum inophyllum*	CLUSIACEAE
10	asaherane	*Virola surinamensis*	MYRISTICACEAE
11	asaherane	*Virola surinamensis*	MYRISTICACEAE
12	asaherane	Virola surinamensis	MYRISTICACEAE
13	msine		
14	ai kihoke	*Glochidion rubrum*	PHYLLANTHACEAE
15	msine		
16	asaherane	*Virola surinamensis*	MYRISTICACEAE
17	ai msinae	*Syzygium* probl. *rumphii*	MYRTACEAE
18	ai msinae	*Syzygium* probl. *rumphii*	MYRTACEAE
19	ananinoi	*Syzgium*	MYRTACEAE
20	anakoine	*Tapeinosperma*	PRIMULACEAE
21	ahutaune	*Calophyllum inophyllum*	CLUSIACEAE
22	akahon tonu-tonu		
23	katinoie	*Melastoma imbricatum*	MELASTOMATACEAE
24	asuani mata hunue		
25	hino numte	*Tristaniopsis whiteana*	MYRTACEAE
26	rapa		
27	kupanahane	*Myrcianthes fragrans*	MYRTACEAE
28	rapa		
29	ene	*Syzygium*	MYRTACEAE
30	meu hahu kikae	*Calamus*	ARECACEAE
31	nerune	*Garcinia*	CLUSIACEAE
32	asuani mata hunue		
33	rapa		
34	ai purae		
35	anine	*Ficus*	MORACEAE

Plot 11: Amatene: old settlement site, northwest of Upa estuary

Map location: 129 07' E 03 19' S
Date: 04-03-1996
Description: Lots of evidence of previous cultivation. More open spaces covered in **sonae** grass and **soka-sokae**
Slope: negligible
Altitude in meters: 400
Notes: Identified with Sounaue clan, but also Pia associations

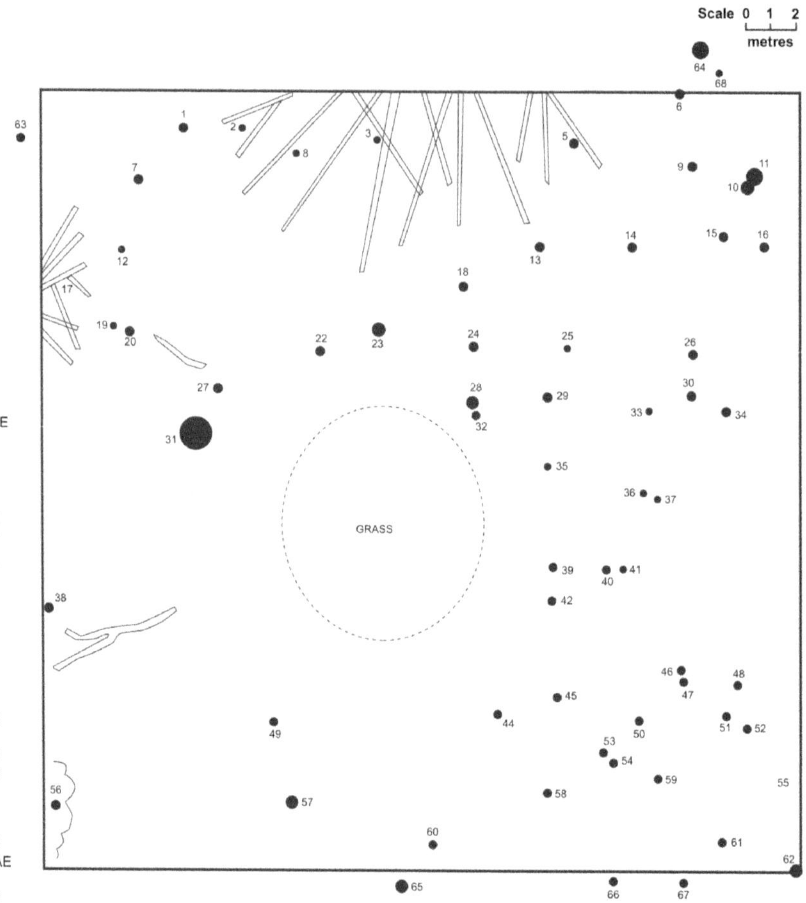

Key:

1	asapopote	*Syzygium boerlagei*	MYRTACEAE
2	kanoke	*Syzygium malaccense*	MYRTACEAE
3	pono	probl *Homalium foetidum*	SALICACEAE
4	tomone	probl. *Schizostachyum lima*	POACEAE
5	kawasa	*Archidendron clypearia*	FABACEAE
6	asu hunne	*Dysoxylum* probl. *caulostachyum*	MELIACEAE
7	kawasa	*Archidendron clypearia*	FABACEAE
8	riane	*Elaeocarpus augustifolius*	ELAEOCARPACEAE
9	nopa putie	*Commersonia bartramia*	MALVACEAE
10	nisoae	*Falcataria moluccana*	FABACEAE
11	nopa putie	*Commersonia bartramia*	MALVACEAE
12	urone	*Alstonia villosa*	APOCYNACEAE
13	hunori	*Macaranga involucrata*	EUPHORBIACEAE
14	hunori	*Macaranga involucrata*	EUPHORBIACEAE
15	asapopo msinae	*Syzygium aqueum*	MYRTACEAE
16	hunori	*Macaranga involucrata*	EUPHORBIACEAE
17	tomone	*Schizostachyum lima*	POACEAE
18	kokoio	*Pandanus*	PANDANACEAE
19	kawasa	*Archidendron clypearia*	FABACEAE
20	kawasa	*Archidendron clypearia*	FABACEAE
21	kawasa	*Archidendron clypearia*	FABACEAE
22	asapopo msinae	*Syzygium aqueum*	MYRTACEAE
23	asapopo msinae	*Syzygium aqueum*	MYRTACEAE
24	urone	*Alstonia villosa*	APOCYNACEAE
25	hunori	*Macaranga involucrata*	EUPHORBIACEAE
26	kawasa	*Archidendron clypearia*	FABACEAE
27	kupatonae	*Macaranga*	EUPHORBIACEAE
28	aune	*Mangifera foetida*	ANACARDIACEAE
29	pate	*Ficus*	MORACEAE
30	ito	*Alstonia scholaris*	APOCYNACEAE
31	ito	*Alstonia scholaris*	APOCYNACEAE
32	aune	*Mangifera foetida*	ANACARDIACEAE
33	riane	*Elaeocarpus augustifolius*	ELAEOCARPACEAE
34	au mamine	*Canarium hirsutum*	BURSERACEAE
35	hunori	*Macaranga involucrata*	EUPHORBIACEAE
36	hunori	*Macaranga involucrata*	EUPHORBIACEAE
37	hunori	*Macaranga involucrata*	EUPHORBIACEAE
38	asapopo msinae	*Syzygium aqueum*	MYRTACEAE
39	ito	*Alstonia scholaris*	APOCYNACEAE
40	kokune	*Callicarpa longifolia*	LAMIACEAE
41	asa hunne	*Dysoxylum caulostachyum*	MELIACEAE
42	urone	*Alstonia villosa*	APOCYNACEAE
43	popote	*Syzygium aqueum*	MYRTACEAE
44	asapopo mane	*Syzygium*	MYRTACEAE
45	asapopo msinae	*Syzygium aqueum*	MYRTACEAE
46	hunori	*Macaranga involucrata*	EUPHORBIACEAE
47	hunori	*Macaranga involucrata*	EUPHORBIACEAE
48	riane	*Elaeocarpus augustifolius*	ELAEOCARPACEAE
49	hata nuni tamnone	*Metroxylon sagu*	ARECACEAE
50	urone	*Alstonia villosa*	APOCYNACEAE
51	nione	*Cocos nucifera*	ARECACEAE
52	pate	*Ficus*	MORACEAE
53	urone	*Alstonia villosa*	APOCYNACEAE
54	panne	*Neolamarckia macrophylla*	RUBIACEAE
55	kokoio	*Pandanus*	PANDANACEAE
56	wana tun	*Gigantochlea atter*	POACEAE
57	hata nuni tamnone	*Metroxylon sagu*	ARECACEAE
58	kokoiyo	*Pandanus*	PANDANACEAE
59	urone	*Alstonia villosa*	APOCYNACEAE
60	asapopo msinae	*Syzygium aqueum*	MYRTACEAE
61	hanu	*Hibiscus tiliaceius*	MALVACEAE
62	hanu	*Hibiscus tiliaceius*	MALVACEAE
63	anine	*Ficus*	MORACEAE
64	urone	*Alstonia villosa*	APOCYNACEAE
65	asapopo msinae	*Syzygium aqueum*	MYRTACEAE
66	hunori	*Macaranga involucrata*	EUPHORBIACEAE
67	hanu	*Hibiscus tiliaceius*	MALVACEAE
68	asu hunne	*Dysoxylon caulostachyum*	MELIACEAE

APPENDIX 5

Abbreviations and conventions used in text and appendices

ABRI	Angkatan Bersenjata Republik Indonesia: Indonesian Armed Forces
AH	Ambonese Herbal (Rumphius)
AM	Ambonese Malay
Asl	above sea level (altitude)
BOGOR	Herbarium Bogoriense, Indonesia
BP	before present
CE	common era
circum	circumference
dbh	diameter at breast height (as in measurements of tree size)
Engl.	English (language)
HA	Herbarium Amboinense
ID	identification
incl.	including
Ind.	Indonesian
KEW	Herbarium of the Royal Botanic Gardens at Kew
k.o.	kind of
l.	length
m	metre
n	number
NED	Nuaulu Ethnobotanical Database
Nua.	Nuaulu
NW	New World
OW	Old World
pl	plural
Port.	Portuguese
poss.	possibly
probl.	probably
PROSEA	Plant Resources of Southeast Asia: an international project for documenting the plant resources of South-East Asia published in 19 volumes.
SIL	Summer Institute of Linguistics
sing	singular
SING.	Singapore Botanic Gardens
s.l.	*senso lato*: in the broad sense
sp.	species not further identified, singular
spp.	species not further identified, plural.
syn.	synonym
UKC	Herbarium of the Ethnobiology Laboratory, University of Kent at Canterbury
unident	unidentified
UNPAD	Departemen Biologi, FMIPA, Universitas Padjadjaran, Indonesia
vill	village
x	hybrid (of two plant species)

REFERENCES

Akimichi, T. 1996. Image and reality at sea: fish and cognitive mapping in Carolinean navigational knowledge. In R. Ellen and K. Fukui (eds), *Redefining Nature: Ecology, Culture and Domestication*, pp. 493–514. Oxford: Berg.

Alcántara-Salinas, G., Ellen, R.F. and Rivera-Hernández, J.E. 2016. Ecological and behavioral characteristics in grouping Zapotec bird categories in San Miguel Tiltepec, Oaxaca, Mexico. *Journal of Ethnobiology* 36(3):658–82.

Amir, H. and Wind, J. 1978. The nature reserves Way Mual and Way Nua, Central Seram, Maluku, Fo/JNS/73/013. Field Report 8. Unpublished, 14 pp.

Astuti, R. 2001. Are we all natural dualists? A cognitive developmental approach. *Journal of the Royal Anthropological Institute* 7:429–47.

Atlas Bahasa Tanah Maluku. 1996. Ambon: Pusat Pengkajian dan Pengembangan Maluku. Universitas Pattimura dan Summer Institute of Linguistics.

Atran, S. 1983. Covert fragmenta and the origins of the botanical family. *Man* 18(1):51–71.

——— 1985. The nature of folk-botanical life forms. *American Anthropologist* 87(2):298–315.

——— 1990. *Cognitive Foundations of Natural History: Towards an Anthropology of Science.* Cambridge: Cambridge University Press.

——— 1994. Core domains versus scientific theories: evidence from systematics and Itza-Maya folk-biology. In L. Hirschfeld and S. Gelman (eds), *Mapping the Mind*, pp. 316–40. Cambridge: Cambridge University Press.

——— 1998. Folk biology and the anthropology of science. *Behavioural and Brain Sciences* 21:547–611.

——— 1999. Itzaj Maya folkbiological taxonomy: cognitive universals and cultural particulars. In D.L. Medin and S. Atran (eds), *Ethnobiology*, pp. 120–203. Cambridge, MA: MIT Press.

——— 2007. Ordinary constraints on the semantics of living kinds: a commonsense alternative to recent treatments of natural-object terms. *Mind and Language* 2(1):27–63.

Atran, S. and Medin, D.L. (eds) 2010. *The Native Mind and the Cultural Construction of Nature.* Cambridge, MA: MIT Press.

Atran, S., Medin, D and Ross, N. 2004. Evolution and devolution of knowledge: a tale of two biologies. *Journal of the Royal Anthropological Institute* 10(2):395–420.

Badan Pusat Statistik, Kabupaten Maluku Tengah. 1971. *Memori Maluku Tengah 1971*. Ambon: Badan Pusat Statistik, Kabupaten Maluku Tengah.

——— 2014. *Kecamatan Amahai Dalam Angka 2014.* Amahai: Badan Pusat Statistik, Kabupaten Maluku Tengah.

Bailey, L.H. 1923. Various cultigens, and transfers in nomenclature. *Gentes Herbarum* 1(3):113–36.

Balée, W.L. 1994. *Footprints of the Forest: Ka'apor Ethnobotany – The Historical Ecology of Plant Utilization by an Amazonian People.* New York: Columbia University Press.

——— (ed.) 1998. *Advances in Historical Ecology.* New York: Columbia University Press.

Balée, W.L. and Erickson, C.L. 2006. *Time and Complexity in Historical Ecology: Studies in the Neotropical Lowlands.* New York: Columbia University Press.

Badan Pusat Statistik 2014. *Kecamatan Amahai Dalam Angka 2014.* Badan Pusat Statistik Kabupaten Maluku Tengah, Amahai.

Balick, M.J. 1994. Ethnobotany, drug development and biodiversity conservation – exploring the linkages. In G.D. Chadwick and J. Marsh (eds), *Ethnobotany and the Search for New Drugs*, pp. 4–18. Chichester: Wiley.

Barnes, R.H.. Barnes, R.H. 1974. *Kédang: A Study of the Collective Thought of an Eastern Indonesian People.* Oxford: Clarendon Press.

Barnes, R.H. and Medin, D. 2008. *The Native Mind and the Cultural Construction of Nature.* Cambridge, MA: The MIT Press.

Barrau, J. J. 1979. Coping with exotic plants in folk taxonomies. In R.F. Ellen and D.A. Reason (eds), *Classifications in Their Social Context*, pp. 139–44. London: Academic Press.

Barraud, C. 1979. *Tanêbar-Evav: Une Société de Maisons Tournée Vers le Large.* Paris: CNRS.

Bates, D.M. 1985. Plant utilization: patterns and prospects. *Economic Botany* 39(3):241–65.

Beekman, E.M. 2011. Introduction. In G.E. Rumphius, *The Ambonese Herbal* (trans. E.M. Beekman), vol. 1, pp. 1–169. New Haven, CT: Yale University Press.

Beentje, H.J. 2010. *The Kew Plant Glossary: An Illustrated Dictionary of Plant Identification Terms.* Kew: Royal Botanic Gardens.

Belharte, S. 2011. The ecological view: management of woody perennials and the transition to vegeculture in Southeast Asia and the Pacific. In G. Barker and M. Janowski (eds), *Why Cultivate: Anthropological and Archaeological Approaches to Foraging-farming Transitions in Southeast Asia*, pp. 27–46. Cambridge: MacDonald Institute for Archaeological Research.

Berlin, B. 1972. Speculations on the growth of ethnobotanical nomenclature. *Language in Society* 1:51–86.

——— 1978. Ethnobiological classification. In E. Rosch and B. Lloyd (eds), *Cognition and Categorization*, pp. 9–16. Hillsdale NJ: Lawrence Erlbaum.

——— 1992. *Ethnobiological Classification: Principles of Categorization of Plants and Animals in Traditional Societies.* Princeton, NJ: Princeton University Press.

Berlin, B., Boster, J.S. and O'Neill, J.P. 1980. *The Perceptual Bases of Ethnobiological Classification: Evidence from Aguaruna Jívaro Ornithology.* Berkeley, CA: Language Behavior Research Laboratory, University of California.

Berlin, B., Breedlove, D.E. and Raven, P.H. 1966. Folk taxonomies and biological classification. *Science* 154:273–5.

——— 1973. General principles of classification and nomenclature in folk biology. *American Anthropologist* 75:214–42.

——— 1974a. Folk systematics in relation to biological classification and nomenclature. *Annual Review of Ecology and Systematics* 4:259–71.

——— 1974b. *Principles of Tzeltal Plant Classification: An Introduction to the Botanical Ethnography of a Mayan-speaking Community in Highland Chiapas.* New York: Academic Press.

Berlin, B. and Kay, P. 1969. *Basic Color Terms.* Berkeley, CA: University of California Press.

Bernstein, J. and R. Ellen 1995. *Licuala* palms in Brunei Dusun ethnobotany. *Brunei Museum Journal* 10, 97-110.

———. R. Ellen and Bantong Antaran 1997. The use of plot surveys for the study of ethnobotanical knowledge: a Brunei Dusun example. *Journal of Ethnobiology* 17(1), 69-96.

Blench, R.M. 2016. Splitting up Proto-Malayopolynesian: new models of dispersals from Taiwan. In B. Prasetyo, T. Surti Nastiti and T. Simanjuntak (eds), *Austronesian Diaspora: A New Perspective*, pp. 77–104. Jakarta: Gadjah Mada University Press.

Bloch, M. 1977. The past and the present in the present. *Man* (NS) 12(2):278–92.

——— 1991. Language, anthropology and cognitive science. *Man* (NS) 26(2):183–98.

——— 1998. Why trees, too, are good to think with: towards an anthropology of the meaning of life. In L. Rival (ed.), *The Social Life of Trees: Anthropological Perspectives on Tree Symbolism*, pp. 39–56. Oxford: Berg.

Blust, R. 1974. Proto-Austronesian syntax: the first step. *Oceanic Linguistics* 13:1–15.

——— 1993. Central and Central-Eastern Malayo-Polynesian. *Oceanic Linguistics* 32(2):241–93.

——— 2018. *Austronesian Comparative Dictionary*: www.trussel2.com/acd (accessed 10 December 2018).

Bolton, R.A. 1990. A Preliminary Description of Nuaulu Phonology and Grammar. University of Texas at Arlington: unpublished MA thesis.

——— 1997. The social context of Nuaulu struggle for health and life. Unpublished paper: Fuller Theological Seminary, Pasadena, CAL.

——— n.d. *Hatane Sona Runa Sikenae Sagu.* Proyek Kerja Sama UNPATTI - SIL, Ambon: Universitas Pattimura dan Summer Institute of Linguistics.

Bolton, R.A and Matoke, H. 2005. *Kamus Sou Naunue – Sou Manai; Bahasa Indonesia – Nuaulu.* Ambon: Program Kerja Sama, Badan Pemberdayaan Masyarakat, Propinsi Maluku dan SIL [Summer Institute of Linguistics] International, Cabang Maluku.

Boomgaard, P. 2003. In the shadow of rice: roots and tubers in Indonesian history, 1500–1950. *Agricultural History* 77(4):582–610.

Boster, J. 1996. Human cognition as a product and agent of evolution. In R.F. Ellen and K. Fukui (eds), *Redefining Nature: Culture, Ecology and Domestication*, pp. 269–89. Oxford: Berg.

Bowker, G.C. and Star, S.L. 2000. *Sorting Things Out: Classification and its Consequences.* Cambridge, MA: MIT Press.

Boyer, P. 1994. *The Naturalness of Religious Ideas: A Cognitive Theory of Religion.* Berkeley: University of California Press.

Brennan, E. 2017. Heterogenous Cloth: An Ethnography of the Coming Into Being of Barkcloth Artefacts at the Royal Botanic Gardens, Kew, and amongst the Nuaulu of Nua Nea village, Maluku, Eastern Indonesia. PhD thesis, University College London.

Bridson, D. and Forman, L. 1998. *The Herbarium Handbook.* Kew: Royal Botanic Gardens.

Broersma, R. 1934. Koopvaardij in de Molukken. *Koloniaal Tijdschrift* 23:129–47, 320–50.

Brown, C.H. 1977. Folk botanical life-forms: their universality and growth. *American Anthropologist* 79: 317–42.

——— 1984. *Language and Living Things: Uniformities in Folk Classification and Naming.* New Brunswick, nj: Rutgers University Press.

Brown, C.H. and Chase, P.K. 1981. Animal classification in Juchitan Zapotec. *Journal of Anthropological Research* 37(1):61–70.

Bulmer, R.N.H. 1970. Which came first, the chicken or the egg-head? In J. Pouillon and P. Maranda *(eds), Echanges et Communications; Mélanges Offerts* a *Claude Lévi-Strauss, pp.* 1069–91. The Hague: Mouton.

——— 1974a. Memoirs of a small game hunter: on the track of unknown animal categories in New Guinea. *Journal d'Agriculture Tropicale et de Botanique Appliquée* 21:79–99.

——— 1974b. Folk biology in the New Guinea highlands. *Social Science Information* 13(4–5):9–28.

——— 1979. Mystical and mundane in Kalam classification of birds. In R.F. Ellen and D. Reason (eds), *Classifications in Their Social Context*, pp. 57–79. London: Academic Press.

Bulmer, R.N.H. and Tyler, M. 1968. Karam classification of frogs. *Journal of the Polynesian Society* 77:333–85.

Cain, A.J. 1971 [1954]. *Animal Species and their Evolution.* London: Hutchinson.

Calvet-Mir, L. and Salpeteur, M. 2016. Humans, plants, and networks: critical review. *Environment and Society: Advances in Research* 7:107–28.

Carey, S. 1987. *Conceptual Change in Childhood.* Cambridge, Mass: MIT Press.

Carrithers, M., Bracken, L.J. and Emery, S. 2011. Can a species be a person? A trope and its entanglements in the Anthropocene. *Current Anthropology* 52:661–85.

Casagrande, D.G. 2017. Ethnoscientific implications of classification as a socio-cultural process. In H. Kopnina and E. Shoreman-Ouimet (eds), *Routledge Handbook of Environmntal Anthropology*, pp. 55–68. London: Routledge.

Chase, M.W. 2004. Monocot relationships: an overview. *American Journal of Botany* 91:1645–56.

Christenhusz, M.J.M. and Chase, M.W. 2014. Trends and concepts in fern classification. *Annals of Botany* 113:571–94.

Christensen, H. 2002. *Ethnobotany of the Iban and the Kelabit.* Aarhus: Forest Department, Sarawak, Malaysia/University of Aarhus.

Clamagirand, B. 1980. Social organization of the Ema of Timor. In J. Fox (ed.), *The Flow of Life: Essays on Eastern Indonesia*, pp. 134–51. Cambridge, MA.: Harvard University Press.

Clapham, A.R., Tutin, T.G. and Warburg, E.F. 1962. *Flora of the British Isles* (2nd edition). Cambridge: Cambridge University Press.

Clarke, D. 1968. *Analytical Archaeology*. London: Methuen.

Collins, J.T. 1980. The historical relationship of the languages of central Maluku, Indonesia. Unpublished PhD. dissertation: University of Chicago.

——— 1984. Linguistic research in Maluku: a report on recent fieldwork. *Oceanic Linguistics* 21 (1-2), 73-146.

Conklin, H.C. 1954. The Relation of Hanunóo Culture to the Plant World. PhD thesis: Yale University.

——— 1962. Lexicographical treatment of folk taxonomies. *International Journal of American Linguistics* 28:119–41.

——— 1963. The Oceanian-African hypothesis and the sweet potato. In J. Barrau (ed.), *Plants and the Migration of Pacific People*, pp. 129–36. Honolulu: Bishop Museum Press.

——— 1964. Ethnogenealogical method, In W.H. Goodenough (ed.), *Explorations in Cultural Anthropology*, pp. 22–55. New York: McGraw-Hill.

——— 1988. Des orientements, des vents, des riz...: pour une etude lexicologique de savoirs traditionnels. *Journal d'Agriculture Traditionelle et de Botanique Appliquée* 33:3–9.

Cook, F.E.M. 1995. *Economic Botany Data Collection Standard.* Kew: Royal Botanic Gardens.

Corner, E.J.H. 1959. Taxonomic note on *Ficus* L., Asia and Australasia. I. Subgen. *Urostigma* (Gasp.) Miq. *Gardens' Bulletin Singapore* 17:368–415.

Cosiaux, A. 2012. Local Uses of Tree Species and Contribution of Mixed Tree Gardens to Livelihoods of Saleman Village near the Manusela National Park, Seram Island, Maluku (Indonesia). MA thesis: Universitairé Montpelier 2.

Coursey, J.G. 1967. *Yams: An Account of the Nature, Origins, Cultivation and Utilisation of Useful Members of the Dioscoreaceae.* London: Longmans, Green.

Crosby, A. 1972. *The Columbian Exchange: Biological and Cultural Consequences of 1492.* Westport, CT: Greenwood Press.

Crumley, C.L. (ed.) 1994. *Historical Ecology: Cultural Knowledge and Changing Landscapes.* Santa Fe, NM: School of American Research Press.

Damon, F. 2017. *Trees, Knots and Outriggers: Environmental Knowledge in the Northeast Kula Ring.* Oxford: Berghahn.

Darlington, P.J. 1957. *Zoogeography: The Geographical Distribution of Animals.* New York: Wiley.

Darwin, C.R. 1928[1859]. *The Origin of Species.* London: Dent.

Descola, P. 2005. *Par-delá Nature et Culture.* Paris: Gallimard.

Donohue, M. and Denham, T. 2010. Farming and language in Island Southeast Asia: reframing Austronesian history. *Current Anthropology* 51(2):223–56.

Douglas, M. 1993. Hunting the pangolin [correspondence]. *Man* (NS) 28(1):161–5.

Dougherty, J. 1979. Learning names for plants and plants for names. *Anthropological Linguistics* 21:298–315.

Dransfield, J. and Manokaran, N. (eds) 1994. *Plant Resources of South-East Asia. No 6: Rattans.* Bogor, Indonesia: PROSEA Foundation.

Dransfield, J., Uhl, N.W., Asmussen, C.B., Baker, W.J., Harley, M.M. and Lewis, C.E. 2008 [1987]. *Genera Palmarum: The Evolution and Classification of Palms.* Kew Publishing, Royal Botanic Gardens, Kew.

Dumont, L. 1986. *Essays on Individualism: Modern ideology in Anthropological Perspective.* Chicago: Chicago University Press.

Dupré, J. 1993. *The Disorder of Things.* Cambridge, MA: Harvard University Press.

Edison, H.S., Sutanto, A., Hermanto, C., Uji, T. and Razak N. 1996. *The Exploration of Musaceae in Maluku Island, 18 November–14 December 1996.* Montpellier, France: Research Institute for Fruits.

Edwards, I.D. 1993. Introduction. In I.D. Edwards, A.A. Macdonald and J. Proctor (eds), *Natural History of Seram, Maluku, Indonesia*, pp. 1–12. Andover: Intercept.

——— 1994. Rainforest people. *Between the Leaves: The DPI Forest Service Journal.* Brisbane: Queensland Forest Service.

Edwards, I.D., MacDonald, A.A. and Proctor, J. (eds) 1993. *Natural History of Seram, Maluku, Indonesia*. Andover: Intercept.

Edwards, I.D., Payton, R.W., Proctor, J. and Riswan, S. 1990. Altitudinal zonation of the rain forests in the Manusela National Park, Seram, Maluku, Indonesia. In P. Baas, K. Kalkman and R. Geesink (eds), *The Plant Diversity of Malesia*, pp. 161–75. Amsterdam: Kluwer.

——— 1993. Rainforest types in the Manusela National Park. In *Natural History of Seram, Maluku, Indonesia* (eds) I.D. Edwards, A.A. MacDonald and J. Proctor, pp. 63–74. Andover: Intercept.

Elias, M., Rival, L. and McKey, D. 2000. Perception and management of cassava (*Manihot esculenta* Crantz) diversity among Makushi Amerindians of Guyana (South America). *Journal of Ethnobiology* 20(2):239–65.

Ellen, R.F. 1972. The marsupial in Nuaulu ritual behaviour. *Man* (NS) 7:223–38.

——— 1977. Anatomical classification and the semiotics of the body. In J. Blacking (ed.), *The Anthropology of the Body*, pp. 343–73. London: Academic Press.

——— 1978a. *Nuaulu Settlement and Ecology: The Environmental Relations of an Eastern Indonesian Community.* The Hague: Martinus Nijhoff.

——— 1978b. Restricted faunas and ethnozoological inventories in Wallacea. In P.H. Stott (ed.), *Man and Nature in South East Asia*, pp. 1–16. London: School of Oriental and African Studies.

——— 1979a. Omniscience and ignorance: variation in Nuaulu knowledge, identification and classification of animals. *Language in Society* 8:337–64.

——— 1979b. Sago subsistence and the trade in spices: a provisional model of ecological succession and imbalance in Moluccan history. In P. Burnham and R.F. Ellen (eds), *Social and Ecological Systems*, pp. 43–74. London: Academic Press.
——— 1979c. Introductory essay. In R.F. Ellen and D.A. Reason (eds), *Classifications in Their Social Context*, pp. 1–32. London: Academic Press.
——— 1983. Semantic anarchy and ordered social practice in Nuaulu personal naming. *Bijdragen Tot de Taal-, Land- en Volkenkunde* 139:18–45.
——— 1985. Patterns of indigenous timber extraction from Moluccan rain forest fringes. *Journal of Biogeography* 12:559–87.
——— 1988. Fetishism. *Man* (NS) 23:213–35.
——— 1990. Nuaulu sacred shields: the reproduction of things or the reproduction of images? *Etnofoor* 3(1):5–25.
——— 1991a. Nuaulu betel chewing: ethnobotany, technique and cultural significance. *Cakalele: Maluku Research Journal* 2(2):97–122.
——— 1991b. Grass, grerb or weed?: a Bulmerian meditation on the category 'monote' in Nuaulu plant classification. In A. Pawley (ed.), *Man and a Half: Essays in Honour of Ralph Bulmer*, pp. 95–101. Auckland: Uniprint.
——— 1993. *The Cultural Relations of Classification: An Analysis of Nuaulu Animal Categories from Central Seram.* Cambridge: Cambridge University Press.
——— 1996a. Putting plants in their place: anthropological approaches to understanding the ethnobotanical knowledge of rainforest populations. In D.S. Edwards, W.E. Booth and S.C. Choy (eds), *Tropical Rainforest Research - Current Issues* 457–65. Dordrecht: Kluwer.
——— 1996b. Introduction. In R. Ellen and K. Fukui (eds), *Redefining Nature: Ecology, Culture and Domestication*, pp. 1–36. Oxford: Berg.
——— 1996c. The cognitive geometry of nature: a contextual approach. In P. Descola and G. Palsson (eds), *Nature and Society: Anthropological Perspectives*, pp. 103–23. London: Routledge.
——— 1998. Palms and the prototypicality of trees: some questions concerning assumptions in the comparative study of categories and labels. In L. Rival (ed.), *The Social Life of Trees*, pp. 57–79. Oxford: Berg.
——— 1999. Modes of subsistence and ethnobiological knowledge: between extraction and cultivation in Southeast Asia. In D.L. Medin and S. Atran (eds), *Folkbiology*, pp. 91–117. Cambridge, MA: MIT Press.
——— 2003a. Variation and uniformity in the construction of biological knowledge across cultures. In H. Selin (ed.), *Nature Across Cultures: Views of Nature and the Environment in Non-Western Cultures*, pp. 47–74. Dordrecht: Kluwer.
——— 2003b. *On the Edge of the Banda Zone: Past and Present in the Social Organization of a Moluccan Trading Network.* Honolulu: University of Hawai'i Press.
——— 2004a. From ethno-science to science, or 'What the indigenous knowledge debate tells us about how scientists define their project'. *Journal of Cognition and Culture* 4(3–4):409–50.
——— 2004b. Processing Metroxylon sagu (ARECACEAE) as a technological complex: a case study from south central Seram, Indonesia. *Economic Botany* 58(4):601–25.
——— 2004c. The distribution of Metroxylon sagu and the historical diffusion of a complex traditional technology. In D. Henley and P. Boomgaard (eds), *Smallholders and Stockbreeders: Histories of Food Crop Production and Animal Husbandry in Southeast Asia*, pp. 69–106. Leiden: KITLV Press.
——— 2006a. Local knowledge and management of sago palm (*Metroxylon sagu* Rottboell) diversity in South Central Seram, Maluku, eastern Indonesia. *Journal of Ethnobiology* 26(2):83–123.
——— 2006b. Introduction. In R. Ellen (ed.), *Ethnobiology and the Science of Humankind.* [Special Issue of the Journal of the Royal Anthropological Institute 12], pp. 1–22. Oxford: Blackwell.
——— 2006c. *The Categorical Impulse: Collected Essays in the Anthropology of Classifying Behaviour.* Oxford: Berghahn.
———_ 2007. Plots, typologies and ethnoecology: local and scientific understandings of forest diversity on Seram. In P. Sillitoe (ed.), *Global vs Local Knowledge*, pp. 41–74. Oxford: Berghahn.
——— 2008. Ethnomycology among the Nuaulu of the Moluccas: putting Berlin's 'General principles' to the test. *Economic Botany* 62:483–96.
——— 2009a. A modular approach to understanding the transmission of technical knowledge: Nuaulu basket-making from Seram, eastern Indonesia. *Journal of Material Culture* 14(2):243–77.
——— 2009b. Cognition, nature, and the 'nature-deficit syndrome'. Review of Scott Atran and Douglas Medin, 'The native mind and the cultural construction of nature'. *Current Anthropology* 50(3):406–7.
——— 2010. Why aren't the Nuaulu like the Matsigenka? Knowledge and categorization of forest diversity on Seram, eastern Indonesia. In L.M. Johnson and E.S. Hunn (eds), *Landscape Ethnoecology: Concepts of Biotic and Physical Space*, pp. 116–40. Oxford: Berghahn.
——— 2011. Sago as a buffer against subsistence stress and as a currency of inter-island trade networks in eastern Indonesia. In G. Barker and M. Janowski (eds), *Why Cultivate? Anthropological and Archaeological Approaches to Foraging-farming Transitions in Southeast Asia*, pp. 47–60. Cambridge: McDonald Institute Monographs.
——— 2012. *Nuaulu Religious Practices: The Frequency and Reproduction of Rituals in a Moluccan Society.* Leiden: KITLV Press.
——— 2016. Is there a role for ontologies in understanding plant knowledge systems? *Journal of Ethnobiology* 36(1):10–28.

——— 2017a. Rethinking the relationship between studies of ethnobiological knowledge and the evolution of human cultural cognition. In C. Power, M. Finnegan and H. Callan (eds), *Human Origins: Contributions From Social Anthropology*, pp. 59–83. Oxford: Berghahn.

——— 2017b. Tools, agency and the category of living things. In T. Pommerening and W. Bisang (eds), *Classification From Antiquity to Modern Times: Sources, Methods, and Theories From an Interdisciplinary Perspective*, pp. 239–62. Berlin: De Gruyter.

——— 2017c. Contingency and adaptation over five decades in Nuaulu forest-based plant knowledge. In P. Sillitoe (ed.), *Indigenous Knowledge: Enhancing its Contribution to Natural Resources Management*, pp. 28–39. Wallingford: CAB International.

——— 2018a. *Kinship, Population and Social Reproduction in the 'New Indonesia': A Study of Nuaulu Cultural Resilience*. Abingdon: Routledge.

——— 2018b. A comment on 'Revamping the metaphysics of ethnobiological classification' by David Ludwig. *Current Anthropology* 59(4):424–26.

——— 2019a. Transmitting symbolic concepts from the perspective of cultural cognition – the acquisition and transfer of folk-biological knowledge. In J. Althoff, D. Berrens and T. Pommerening (eds), *Finding, Inheriting or Borrowing? The Construction and Transfer of Knowledge in Antiquity and the Middle Ages*, pp. 41–70. Bielefeld: Transcript Verlag.

——— 2019b. Contrasting acoustical signatures and the power in ritual: Nuaulu spirit healing in the Moluccan islands. In N. Porath (ed.), *Hearing Southeast Asia: Sounds of Hierarchy and Power in Context*, pp. 174–201. Copenhagen: NIAS Press.

——— 2019c. Ritual, landscapes of exchange, and the domestication of *Canarium*: a Seram case study. *Asian Perspectives* 58(4):261–86.

——— 2019d. Understanding geometrical features of Nuaulu shield design. *Journal of Material Culture* 24(2):210–231.

——— in press.. Nuaulu use and management of Culturally salient polymorphisms in *Codiaeum variegatum*: explaining the biocultural dimensions of leaf variegation in a Southeast Asian ornamental. *Journal of Ethnobiology*.

Ellen, R.F. and Glover, I.C. 1979. A further note on flaked stone material from Seram, Eastern Indonesia. *Asian Perspectives* 20(2):236–4.

Ellen, R.F. and Harris, H. 2000. Introduction. In R. Ellen, P. Parkes and A. Bicker (eds), *Indigenous Environmental Knowledge and its Transformations: Critical Anthropological Perspectives*, pp. 1–33. Amsterdam: Harwood.

Ellen, R.F. and Puri, R.K. 2016. Conceptualising 'core' medicinal floras: a comparative and methodological study of phytomedical resources in related Indonesian populations. *Conservation and Society* 14(4):345–58.

Ellen, R.F. and Soselisa, H.L. 2012a. A comparative study of the socio-ecological concomitants of cassava (*Manihot esculenta*) diversity, local knowledge and management in Eastern Indonesia. *Ethnobotany Research and Applications* 10:15–35.

——— 2012b. Cassava diversity and toxicity in relation to environmental degradation: a feature of food security in the Moluccas, Indonesia. In A.-K. Hornidge and C. Antweiler (eds), *Environmetal Uncertainty and Local Knowledge: Southeast Asia as a Laboratory of Global Ecological Change*, pp. 215–42. Bielefeld: Transcript Verlag.

Ellen, R.F., Soselisa, H.L. and Wulandari, A.P. 2012. The biocultural history of *Manihot esculenta* in the Moluccan islands of eastern Indonesia: assessing the evidence for the movement and selection of cassava germplasm. *Journal of Ethnobiology* 32(2):157–84.

Ereshefsky, M. 2001. *The Poverty of Linnaean Hierarchy: A Philosophical Study of Biological Taxonomy*. Cambridge: Cambridge University Press.

Etkin, N.L. 1994. Consuming a therapeutic landscape: a multicontextual framework for assessing the health significance of human-plant interactions. In J. Flagler and R.P. Poincelot (eds), *People-plant Relationships: Setting Research Priorities*, pp. 61–81. Philadelphia, PA: Haworth Press.

Flenley, J.R. 1979. *The Equatorial Rainforest: A Geological History*. London: Butterworths.

Florey, M. 1991. Shifting patterns of language allegiance: a generational perspective from eastern Indonesia. In H. Steinhauer (ed.), *Papers in Austronesian Linguistics* 1:39–47. Canberra: Pacific Linguistics.

——— 1993. The reinterpretation of knowledge and its role in the process of language obsolescence. *Oceanic Linguistics* 32(2):295–309.

——— 1997. Skewed performance and structural variation in the process of language obsolescence. In C. Ode and W. Stokhof (eds), *Proceedings of the Seventh International Conference on Austronesian Linguistics*, pp. 639–60. Amsterdam/Atlanta: Editions Rodopi B.V.

——— 1998. Alune incantations: continuity or discontinuity in verbal art? *Journal of Sociolinguistics* 2(2):205–31.

——— 2001. Threats to indigenous knowledge: a case study from eastern Indonesia. In L. Maffi(ed.), *On Biocultural Diversity: Linking Language, Knowledge and the Environment*, pp. 325–42. Washington DC: Smithsonian Institution Press.

——— 2005a. Language shift and endangerment. In A. Adelaar and N.P. Himmelmann (eds), *The Austronesian Languages of Asia and Madagascar*, pp. 43–64. London: Routledge.

——— 2005b. Overview: the languages of Central Maluku. In S. Musgrave (ed.), *Language Contact, Hybrids and New Varieties: Emergent Possessive Constructions*. Monash University Linguistic Papers 4 (1/2).

Florey, M. and van Engelenhoven, A.T.P.G. 2001. Language documentation and maintenance programs for Moluccan languages in the Netherlands. *International Journal of the Sociology of Language* 151:195–219.

Florey, M. and Bolton, R.A. 1997. Personal names, lexical replacement, and language shift in eastern Indonesia. *Cakalele: Maluku Research Journal* 8:27–58.

Florey, M., Healey, C.J., Wolff, X.Y. and Manakane, W. n.d. From the Edge of the Forest: Alune Ethnobiology in a Changing World. Unpublished ms.

Florey, M. and Wolff, X.Y. 1998. Incantations and herbal medicines: Alune ethnomedical knowledge in a context of change. *Journal of Ethnobiology* 18(1):39–67.

Ford, R.I. 1978. Ethnobotany: historical diversity and synthesis. In R.I. Ford (ed.), *The Nature and Status of Ethnobotany*, pp. 33–50. Ann Arbor, MI: Museum of Anthropology, University of Michigan.

Forth, G. 2001. *Dualism and Hierarchy: Processes of Binary Combination in Keo Society*. Oxford: Oxford University Press.

Fox, J.J. 1971. Sister's child as plant: metaphors in an idiom of consanguinity. In R. Needham (ed.), *Rethinking Kinship and Marriage*, pp. 219–52. London: Tavistock.

——— 1980. Models and metaphors: comparative research in eastern Indonesia. In J. Fox (ed.), *The Flow of Life: Essays on Eastern Indonesia*, pp. 327–33. Cambridge, MA: Harvard University Press.

——— 1989. Category and complement: binary ideologies and the organization of dualism in eastern Indonesia. In D. Maybury-Lewis and U. Almagor (eds), *The Attraction of Opposites*, pp. 33–56. Ann Arbor, MI: University of Michigan Press.

——— 1991. Managing the ecology of rice production in Indonesia. In J. Hardjono (ed.), *Indonesia: Resources, Ecology and Environment*, pp. 61–84. Singapore: Oxford University Press.

Frake, C.O. 1961. The diagnosis of disease among the Subanun of Mindanao. *American Anthropologist* 63:113–32.

——— 1962. The ethnographic study of cognitive systems. In T. Gladwin and W.C. Sturtevant (eds), *Anthropology and Human Behaviour*, pp. 72–85. Washington DC: Anthropological Society of Washington.

——— 1964. Notes on queries in ethnography. *American Anthropologist* 66(3, part 2):132–45.

——— 1980. *Language and Cultural Description: Essays by Charles O. Frake*, (ed.) A.S. Dil. Stanford, CA: Stanford University Press.

Friedberg, C. 1968. Les méthodes d'enquête en ethnobotanique. *Journal d'Agriculture Tropicale et de Botanique Appliquée* 15:297–324.

——— 1970. Analyse de quelques groupements de végétaux comme introduction à l'étude de la classification botanique Bunaq. In J. Pouillon and P. Maranda (eds), Échanges *et Communications: Mélanges Offerts à Claude Lévi-Strauss*, vol. 2:1092–131. The Hague: Mouton.

——— 1979. Socially significant plant species and their taxonomic position among the Bunaq of central Timor. In R.F. Ellen and D. Reason (eds), *Classifications in Their Social Context*, pp. 81–101. London: Academic Press.

——— 1990. *Le Savoir Botanique des Bunaq: Percevoir et Classer dans le Haut Lamaknen (Timor, Indonesie)*. Paris: Mémoires du Museum National d'Histoire Naturelle, Botanique 32.

Galvão, A. 1971 [1544]. *A Treatise on the Moluccas c.1544, Probably the Preliminary Version of the Lost Historia das Moluccas* (ed., annot. and trans. H.Th.Th.M. Jacobs). Rome: Jesuit Historical Institute.

Gaoue, O.G., Coe, M.A., Bond, M., Hart, G., Seyler, B.C. and McMillen, H. 2017. Theories and major hypotheses in ethnobotany. *Economic Botany* 71(3):269–87.

Garibaldi, A. and Turner, N.J. 2004. Cultural keystone species: implications for ecological conservation and restoration. *Ecology and Society* 9(3):1.

Geirnaert-Martin, D. 1992. *The Woven Land of Laboya: Socio-cosmic Ideas and Values in West Sumba, Eastern Indonesia*. Leiden: Centre for Non-Western Studies, Leiden University.

Gell, A. 1975. *Metamorphosis of the Cassowaries: Umeda Society, Language and Ritual*. London: Athlone Press.

——— 1999. *The Art of Anthropology: Essays and Diagrams*. London: Athlone Press.

Gelman, S. 2003. *The Essential Child: Origins of Essentialism in Everyday Thought*. Oxford: Oxford University Press.

Germeraad, J.H. 1946. Geology of central Seran. In L. Rutten and W. Holz (eds), *Geological, Petrographical and Palaeontological Results of Explorations, Carried Out From September 1917 till June 1919 in the Island of Ceram*. Amsterdam: J.H. de Bussy.

Gianno, R. 1990. *Semelai Culture and Resin Technology*. New Haven, CT: Connecticut Academy of Arts and Sciences Memoir 22.

Gibson, J.J. 1979. *The Ecological Approach to Visual Perception*. Boston, MA: Houghton Mifflin Harcourt.

Gibson, R.W. 2009. A review of perceptual distinctiveness in landraces including an analysis of how its roles have been overlooked in plant breeding for low-input farming systems. *Economic Botany* 63:242–55.

Giddens, A. 1986. *The Constitution of Society: Outline of the Theory of Structuration*. London: Polity.

Glatzel, G. 1992. *Report of a Fact Finding Mission on the Proposal for a Centre for Sustainable Land Use on Forested Islands of the Moluccas*. Vienna: Bundeskanzleramtes der Republik Österreich.

Glover, D.M 2018. Classification as narrative: a renewed perspective on a longstanding topic in ethnobiology. *Journal of Ethnobiology* 38(1):105–23.

Glover, I.C. and Ellen, R.F. 1975. Ethnographic and archaeological aspects of a flaked stone collection from Seram, eastern Indonesia. *Asian Perspectives* 18(1):51–61.

Glover, I.C. and Higham, C.F.W. 1996. New evidence for early rice agriculture in South, Southeast and East Asia. In D.R. Harris (ed.), *The Origins and Spread of Agriculture and Pastoralism in Eurasia*, pp. 413–41. Washington DC: Smithsonian Institution Press.

Goodwin, Z.A., Harris, D.J., Filer, D., Wood, J.R.I. and Scotland, R.W. 2015. Widespread mistaken identity in tropical plant collections. *Current Biology* 25(22):R1066–R1067.

Goody, J. 1977. *The Domestication of the Savage Mind.* Cambridge: Cambridge University Press.

——— 1993. *The Culture of Flowers.* Cambridge: Cambridge University Press.

——— 1982. *Cooking, Cuisine and Class: A Study in Comparative Sociology.* Cambridge: Cambridge University Press.

Groube, L. 1989. The taming of the rainforests: a model for the late Pleistocene rainforest exploitation in New Guinea. In D.R. Harris and G.C. Hillman (eds), *Foraging and Farming: The Evolution of Plant Exploitation,* pp. 292–304. London: Unwin and Hyman.

Guzman, C.C. de and Siemonsma, J.S. (eds) 1999. *Plant Resources of South-East Asia: Spices.* Bogor, Indonesia: PROSEA Foundation.

Hage, P., and W. Miller. 1976. 'Eagle' = 'bird': a note on the structure and evolution of Shoshoni ethnoornithological nomenclature. *American Ethnologist* 3:481–87.

Hapsari, L., Kennedy, J., Lestari, D.A., Masrum, A. and Lestarini, W. 2017. Ethnobotanical survey of bananas (Musaceae) in six districts of East Java, Indonesia. *Biodiversitas* 18(1):160–74.

Harris, D.R. 1966. Agricultural systems, ecosystems and the origins of agriculture. In P.J. Ucko and G.W. Dimbleby (eds), *The Domestication and Exploitation of Plants and Animals,* pp. 3–15. Chicago, IL: Aldine.

——— 1973. The prehistory of tropical agriculture: an ethnoecological model. In C. Renfrew (ed.), *The Explanation of Culture,* pp. 391–417. London: Duckworth.

——— 1977. Alternative pathways towards agriculture. In C.A. Reed (ed.), *Origins of Agriculture,* pp. 179–243. The Hague: Mouton.

——— 1989. An evolutionary continuum of people-plant interaction. In D.R Harris and G.C. Hillman (eds), *Foraging and Farming: The Evolution of Plant Exploitation,* pp. 11–26. London, Unwin Hyman.

——— 1996. Domesticatory relationships of people, plants and animals. In R. Ellen and K. Fukui (eds), *Redefining Nature: Ecology, Culture and Domestication,* pp. 437–63. Oxford: Berg.

Hastrup, K. 1990. The ethnographic present: a reinvention. *Cultural Anthropology* 5(1):45–61.

Hawthorne, W. and Harris, S. 2006. Plant names and botanical publication. In A. Lawrence and W. Hawthorne (eds), *Plant Identification: Creating User-friendly Field Guides for Biodiversity Management,* pp. 61–90. London: Earthscan.

Hays, T.E. 2005. Vernacular names for tubers in Irian Jaya: implications for agricultural prehistory. In A. Pawley, R. Attenborough, J. Golson and R. Hide (eds), *Papuan Pasts: Cultural, Linguistic and Biological Histories of Papuan-speaking Peoples,* pp. 625–70. Canberra: Pacific Linguistics.

Headland, T.N. and Bailey, R.C. 1991. Introduction: have hunter-gatherers ever lived in tropical rain forest independently of agriculture? *Human Ecology* 19(2):113–22.

Healey, C.J. 1978–9. Taxonomic rigidity in biological folk classification: some examples from the Maring of New Guinea. *Ethnomedizin* 3/4:361–84.

Heckler, S. and Zent, S. 2008. Piaroa manioc varietals: hyperdiversity or social currency. *Human Ecology* 36:679–97.

Heinsohn, T.E. 2009. Marsupials as introduced species: longterm anthropogenic expansion of the marsupial frontier and its implications for zoogeographic interpretation. In S. Haberle, J. Stevenson and M. Prebble (eds), *Altered Ecologies: Fire, Climate and Human Influence on Terrestrial Landscapes,* pp. 133–76. Canberra: ANU Press.

Heinz, H.J. and Maguire, B. 1974. *The Ethno-biology of the !Kõ Bushmen.* Gaborone: The Botswana Society.

Henrich, J., Heine, S. and Norenzayan, A. 2010. The weirdest people in the world? *Behavioral and Brain Sciences* 33(2–3):61–83.

Hermanto, C., Sutanto, A., Edison, H.S., Riska, A., Hosang, E., Daniells, J. and Hilman, Y. 2013. *Triangle Banana Exploration Report, Central Maluku and Lesser Sunda Islands, Indonesia 16 February – 6 March 2013.* Montpelier, France: MusaNet, Indonesian Tropical Fruit Research Institute (ITFRI), Queensland Department of Agriculture, Fisheries and Forestry (DAFF) and Bioversity International.

Heyne, K. 1913–17. *De Nuttige Planten van Nederlandsch-Indië.* Batavia: Ruygrok, Departement van Landbouw, Nijverheid en Handel.

Hill, R.D. 2004. Towards a model of the history of 'traditional' agriculture on Southeast Asia. In P. Boomgaard and D. Henley (eds), *Smallholders and Stockbreeders: History of Foodcrop and Livestock Farming in Southeast Asia,* pp. 19–45. Leiden: KITLV Press.

Hirschfeld, L.A. and Gelman, S.A. 1994. *Mapping the Mind: Domain Specificity in Cognition and Culture.* Cambridge: Cambridge University Press.

Holman, E.W. 2005. Domain-specific and general properties of folk classifications. *Journal of Ethnobiology* 25(1):71–91.

Hsu, E. 2010. Introduction: plants in medical practice and common sense. In E. Hsu and S. Harris (eds), *Plants, Health and Healing: On the Interface of Ethnobotany and Medical Anthropology,* pp. 1–48. Oxford: Berghahn.

Hull, D. 1992. Biological species: an inductivist's nightmare. In M. Douglas and D. Hull (eds), *How Classification Works: Nelson Goodman Among the Social Sciences,* pp. 42–68. Edinburgh: Edinburgh University Press.

Hunn, E. 1976. Toward a perceptual model of folk biological classification. *American Ethnologist* 3(3):508–24.

——— 1977. *Tzeltal Folk Zoology: The Classification of Discontinuities in Nature.* London: Academic Press.

——— 1982. The utilitarian factor in folk biological classification. *American Anthropologist* 84:830–47.

——— 1999. Size as limiting the recognition of biodiversity in folk biological classifications; one of four factors governing the cultural recognition of biological taxa. In D.L. Medin and S. Atran (eds), *Folkbiology,* pp. 47–69. Cambridge, MA: Harvard University Press.

——— 2002. Evidence for the precocious acquisition of plant knowledge by Zapotec children. In J.R. Stepp, F.S. Wyndham and R.K. Zarger (eds), *Ethnobiology and Biocultural Diversity*, pp. 604–13. Athens, GA: International Society of Ethnobiology.

——— 2008. *A Zapotec Natural History: Trees, Herbs, and Flowers, Birds, Beasts, and Bugs in the Life of San Juan Gbëë.* Tucson, AZ: University of Arizona Press.

Hunt, C.O and Premathilake, R. 2012. Early Holocene vegetation, human activity and climate from Sarawak, Malaysian Borneo. *Quaternary International* 249:105–19.

Hunt, C.O and Rabett, R.J. 2014. Holocene landscape intervention and plant food production strategies in island and mainland Southeast Asia. *Journal of Archaeological Science* 51:22–33.

Ingold, T. 1986. *The Appropriation of Nature: Essays on Human Ecology and Social Relations.* Manchester: Manchester University Press.

——— 2000. *The Perception of the Environment: Essays on Livelihood, Dwelling and Skill.* London: Routledge.

——— 2006. Rethinking the animate, re-animating thought. *Ethnos* 71:9–20.

——— 2011. *Being Alive: Essays on Movement, Knowledge and Description.* London: Routledge.

Iskandar, J. and Ellen, R. 1999. In situ conservation of rice landraces among the Baduy of West Java. *Journal of Ethnobiology* 19(1):97–125.

Jeffrey, C. 1982. *An Introduction to Plant Taxonomy.* Cambridge: Cambridge University Press.

Jensen, A.E. 1948. *Die Drei Ströme: Züge aus dem Geistigen und Religiösen Leben oder Wemale, einen Primitieven-Volk in den Molukken.* Leipzig: Otto Harrrassowitz.

Kato, M. 1990. The fern flora of Seram. In P. Baas, K. Kalkman and R. Geesink (eds), *The Plant Diversity of Malesia*, pp. 225–34. Dordrecht: Kluwer.

Kay, P. 1971. On taxonomy and semantic contrast. *Language* 47:886–7.

Keil, F. 1979. *Semantic and Conceptual Development: An Ontological Perspective.* Cambridge, MA: Harvard University Press.

Kennedy, J. 2008. Pacific bananas: complex origins, multiple dispersals? *Asian Perspectives* 47(1):75–94.

Kirksey, E. 2015. Species: a praxiographic study. *Journal of the Royal Anthropological Institute* 21(4):758–80.

Kocher-Schmid, C. 1991. *Of People and Plants: A Botanical Ethnography of Nokopo Village, Madang and Morobe provinces, Papua New Guinea.* Basel: Wepf and Co. AG Verlag

Kok, R. de 2008. The genus Vitex (Labiatae) in the Flora Malesiana region, excluding New Guinea. *Kew Bulletin* 63(1):17–40.

——— 2012. A revision of the genus Gmelina (Lamiaceae). *Kew Bulletin* 67(3):293–329.

Komaromi, R. 2009. Sharing Knowledge: Intra-cultural Variation of Ethnobotanical Knowledge and the Factors that Pattern it in a Mambila Community in the Cameroon-Nigeria Borderland. MSc thesis: University of Kent.

Lakoff, G. 1987. *Women, Fire and Dangerous Things: What Categories Reveal About the Mind.* Chicago, IL: University of Chicago Press.

Lalumera, E. 2014. Whorfian effects in color perception: deep or shallow. *The Baltic International Yearbook of Cognition, Logic and Communication* 9:1–13.

Latinis, D.K. 2000. The development of subsistence models for island Southeast Asia and near Oceania: the nature and role of arboriculture and arboreal-based economies. *World Archaeology* 32(1):41–67.

——— 2002. Protohistoric Archaeology and Settlement in Central Maluku, Eastern Indonesia. PhD thesis: National University of Singapore.

Latour, B. 1993. *We Have Never Been Modern* (trans. C. Porter). Cambridge, MA: Harvard University Press.

Lawrence, A. and Hawthorne, W. (eds) 2006. *Plant Identification: Creating User-friendly Field Guides for Biodiversity Management.* London: Earthscan.

Leach, E. 1964. Anthropological aspects of language: animal categories and verbal abuse. In E.H. Lenneberg (ed.), *New Directions in the Study of Language*, pp. 23–63. Cambridge, MA: MIT Press.

——— 1972. *Humanity and Animality.* London: South Place Ethical Society.

——— 2000 [1977]. Anthropos. In S. Hugh-Jones and J. Laidlaw (eds), *The Essential Edmund Leach*, pp. 324–80. New Haven, CT: Yale University Press.

Lende, D.H. and Downey, G. 2012. *The Encultured Brain: An Introduction to Neuroanthropology.* Cambridge, MA: MIT Press.

Lévi-Strauss, C. 1962. *Le Totémisme Aujourd'hui.* Paris: Presses Universitaires de France.

——— 1966. *The Savage Mind.* London: Weidenfeld and Nicolson.

Lewis, E.D. 1988. *People of the Source: The Social and Ceremonial Order of Tana Wai Brama on Flores.* Dordrecht: Foris.

Lewis-Jones, K.E. 2016. Introduction: people and plants. *Environment and Society: Advances in Research* 7:1–7.

Linnaeus, C. 1738. *Classes Plantarum.* Leiden: Conrad Wishoff.

Liswanti, N., Fripp, E., Silaya, T., Tjoa, M. and Laumonier, Y. 2013. *Socio-economic Considerations for Land Use Planning: The Case of Seram, Central Maluku.* Bogor, Indonesia: CIFOR.

Lloyd, G.E.R. 2007. *Cognitive Variations: Reflections on the Unity and Diversity of the Human Mind.* Oxford: Clarendon Press.

Long-Solis J. 1986. *Capsicum and Culture: The History of Chilli.* Mexico, DF: Fondo de Cultura Economica.

Ludwig, D. 2018. Revamping the metaphysics of ethnobiological classification. *Current Anthropology* 59(4):415–38.

Lugo, A.E. and Lowe, C. 1995. *Tropical Forests: Management and Ecology.* New York: Springer-Verlag.

MacKinnon, K., Hatta, G., Halim, H. and Mangalik, A. 1997. *The Ecology of Kalimantan.* Oxford: Oxford University Press.

Maffi, L. (ed.) 2001. *On Biocultural Diversity.* Washington DC: Smithsonian Institution Press.

Maloney, B.K. 1998 The long-term history of human activity and rainforest development. In B.K. Maloney (ed.), *Human Activities and the Tropical Rainforest*, pp. 66–85. Dordrecht: Kluwer.

Martin, G.J. 1995. *Ethnobotany: A Methods Manual.* London: Chapman and Hall.

Matallana, G., Wendt, T., De Araujo, D. and Scarano, F. 2005. High abundance of dioecious plants in a tropical coastal vegetation. *American Journal of Botany* 92 (9):1513–19.

Matoke, H. and Matoke, S. 2004. *Atuhu - Tanaman.* Ambon: Program Kerja Sama, Badan Pemberdayaan Masyarakat, Propinsi Maluku dan SIL [Summer Institute of Linguistics] International, Cabang Maluku.

Matthews, P.J. 2014. *On the Trail of Taro: An Exploration of Natural and Cultural History.* Osaka: National Museum of Ethnology.

Matthews, P.J. and Gosden, C. 1997. Plant remains from water-logged sites in the Arawe Islands, West New Britain Province, Papua New Guinea: implications for the history of plant use and domestication. *Economic Botany* 51:121–33.

Mauss, M. 1973 [1934]. Techniques of the body. (Trans. B. Brewster.) *Economy and Society* 2:70–88.

Mayr, E. 1963. *Animal Species and Evolution.* Oxford: Oxford University Press.

Merrill, E.D. 1913. The botanical exploration of Amboina by the Bureau of Science, Manila. *Science* (NS) 38:499–502.

——— 1917. *An Interpretation of Rumphius's Herbarium Amboinense.* Manila: Bureau of Printing.

——— 1946. Further notes on tobacco in New Guinea. *American Anthropologist* 48:22–30.

Michaux, B. 2010. Biogeology of Wallacea: geotectonic models, areas of endemism, and natural biogeographical units. *The Biological Journal of the Linnean Society.* 1473:193–212.

Miller, G. 1956. The magical number seven, plus or minus two: some limits on our capacity for processing information. *The Psychological Review* 63:81–97.

Miller, T. 2016. 'Living lists': coming to know indigenous Canela encounters with plants through ethnobotanical classification. *Journal of Ethnobiology* (special issue) 36(1):105–24.

Miller, T., Eoin, N., Daly, L. and French, K. (eds) 2016. *Botanical Ontologies. Journal of Ethnobiology* (special issue) 36 (1).

Mithen, S. 1996. *The Prehistory of the Mind: A Search for the Origins of Art, Religion and Science.* London: Thames and Hudson.

Mogea, J.P. 1991. Indonesia: palm utilization and conservation. In D. Johnson (ed.), *Palms for Human Needs in Asia: Palm Utilization and Conservation in India, Indonesia, Malaysia and the Philippines*, pp. 37–73. Rotterdam: A.A. Balkemma for WWF and IUCN.

Molnar, A.K. 2000. *Grandchildren of the Ga'e Ancestors: Social Organization and Cosmology Among the Hoga Sara of Flores.* Leiden: KITLV Press.

Monk, K.A., de Fretes, Y. and Reksodiharjo-Lilley, G. 1997. *The Ecology of Nusa Tenggara and Maluku.* Hong Kong: Periplus.

Morris, B. 1976. Whither the savage mind; notes on the natural taxonomies of a hunting and gathering people. *Man* (NS) 11:542–57.

Nabhan, G.P. and Rea, A. 1987. Plant domestication and folk-biological change: the upper Piman/Devil's Claw example. *American Anthropologist* 89:57–73.

Nabhan, G.P. and St Antoine, S. 1993. The loss of flora and faunal story: the extinction of experience. In S.R. Kellert and E. Wilson (eds), *The Biophilia Hypothesis*, pp. 229–50. Washington DC: Island Press.

Nesbitt, M., McBurney, R.P.H., Broin, M. and Beentje, H.J. 2010. Linking biodiversity, food and nutrition: the importance of plant identification and nomenclature. *Journal of Food Composition and Analysis* 23:486–98.

Nevenimo, T., Moxon, J., Wemin, J., Johnston, M., Bunt, C. and Leakey, R.R.B. 2007. Domestication potential and marketing of Canarium indicum nuts in the Pacific: 1. A literature review. *Agroforestry Systems* 69:117–34.

Niggemeyer, H. 1951–2. Alune-Sprache; Texte, Wörterverzeichnis und Grammatik einer Sprache West-Cerams. *Zeitschrift für Ethnologie* 76:50–69, 288–300; 77:116–132, 238–50.

Nolthenius, A.B.T. 1935. *Overzicht van de Literatuur Betreffende de Molukken (Exclusief Nieuw Guinea)*, vol. 2 (1922–1933). Amsterdam: Molukken-Instituut.

Ochse, J.J., Backer, C., Bakhuizen van den Brink, R.C. and Cornelis, R. 1931. *Fruits and Fruiticulture in the Dutch East Indies.* Batavia: Kolff, The Hague: Nijhoff.

Ochse, J.J. and Bakhuizen van den Brink, R.C. 1931. *Vegetables of the Dutch East Indies (Edible Tubers, Bulbs, Rhizomes and Spices Included).* Buitenzorg-Java: Archipel Drukkerij.

Oh Swee Cheng 2005. *Tuamane Nene Muaine: Makanlah Alam Untuk Tanah.* Cabang Maluku: Program Kerja Sama, Badan Pemberdayaan Masyarakat Propinsi Maluku dan SIL International.

Olson, D.R. 1994. *The World on Paper: The Conceptual and Cognitive Implications of Writing and Reading.* Cambridge: Cambridge University Press.

Ong, W.J. 1982. *Orality and Literacy: The Technologizing of the Word.* London: Methuen.

Oskarsson, M., Cornelya, C.R., Klütsch, F.C., Boonyaprakob, U., Wilton, A., Tanabe, Y. and Savolainen, P. 2012. Mitochondrial DNA data indicate an introduction through mainland Southeast Asia for Australian dingoes and Polynesian domestic dogs. *Proceedings of the Royal Society B* 279(1730):967–74.

Panoff, F. 1969. Some facets of Maenge horticulture. *Oceania* 40:20–31.

Parris, B.S. 1993. Studies of the phytogeography and altitudinal zonation of the Pteridophyte flora of Seram. In I.D. Edwards, A.A. MacDonald and J. Proctor (eds), *Natural History of Seram, Maluku, Indonesia*, pp. 75–89. Andover: Intercept.

Payton, R.W. 1993. Soils of the Manusela National Park. In I.D. Edwards, A.A. MacDonald and J. Proctor (eds), *Natural History of Seram, Maluku, Indonesia*, pp. 19–61. Andover: Intercept.

Peeters, A. 1979. Nomenclature and classification in Rumphius's 'Herbarium Amboinense'. In R.F. Ellen and D. Reason (eds), *Classifications in Their Social Context*, pp. 145–66. London: Academic Press.

Phillips, O. and Gentry, A.H. 1993. The useful plants of Tambopata, Peru: I. Statistical hypotheses tests with a new quantitative technique. *Economic Botany* 47:15–32.

Pina-Cabral, J. de 2000. The ethnographic present revisited. *Social Anthropology* 8(3):341–48.

Platenkamp, J.D.M. 1990. The severence of origin: a ritual of Tobelo of north Halmahera. [In Ritual: a socio-cosmic order in eastern Indonesian societies, Part 2, Maluku.] *Bijdragen tot de Taal-, Land- en Volkenkunde* 146(1):74–92.

Powell, J.M. 1976. Ethnobotany. In K. Paijmans (ed.), *New Guinea Vegetation*, pp. 106–83. Canberra: Commonwealth Scientific and Industrial Research Organization in association with Australian National University Press.

PROSEA 1992–2003. *Plant Resources of South-East Asia, Volumes 1–19*. Bogor: PROSEA Foundation.

Praet, I. 2014. *Animism and the Question of Life*. London: Routledge.

Prud'homme van Reine, W.F. and Trono, G.C. (eds) 2002. *Plant Resources of South-East Asia 15 (1) Cryptogams: Algae*. Bogor: PROSEA Foundation.

Puri, R.K. 2005. Post-abandonment ecology of Penan fruit camps: anthropological and ethnobiological approaches to the history of a rain-forested valley in East Kalimantan. In M.R. Dove, P.E. Sajise and A. Doolittle (eds), *Conserving Nature in Culture: Case Studies From Southeast Asia*, pp. 25–82. New Haven, CT: Yale University Council on Southeast Asia Studies.

Purwanto, Y. and Cosiaux, A. 2013. *Studi Sistem Pertanian Tradisional Masyarakat Negeri Saleman, Seram Utara, Kabupaten Maluku Tengah*. Montpellier, France: Laporan Penelitian COLUPSIA Project, CIRAD dan UNIEROPA.

Raharjo, S., Natori, Y., Yamasaki, A., Kitajima, A. and Yamamoto, M. 2018. Exploration and DNA analysis of local Citrus genetic resources on Ambon Island of Indonesia. *Tropical Agriculture and Development* 62(3):115–23.

Randall, R.A. 1976. How tall is a taxonomic tree?: some evidence for dwarfism. *American Ethnologist* 8:229–42.

Randall, R.A. and Hunn, E.S. 1984. Do life-forms evolve or do uses for life?: some doubts about Brown's universals hypothesis. *American Ethnologist* 11:329–49.

Ranlund, A. 2011. Structure and Tree Diversity of Lowlamd Limestone Forest on Seram Island, Indonesia, MSc Thesis: Department of Ecology, Swedish University of Agricultural Sciences, Uppsala.

Rant, A. 1934. Twee botanische reizen naar Ambon. *Natuurkundig Tijdschrift voor Nederlandsch-Indië* 94:100–33.

Rappaport, R.A. 1971. Ritual, sanctity and cybernetics. *American Anthropologist* 73(1):59–76.

Reed, E.S. 1988. The affordances of the animate environment: social science from the ecological point of view. In T. Ingold (ed.), *What Is an Animal?*, pp. 110–26. London: Unwin Hyman.

Revel, N. 1990. *Fleurs de Paroles: Histoire Naturelle Palawan, vol. 1: Les Dons de Nägsalad*. Paris: Editions Peeters.

Rindos, D. 1984. *The Origins of Agriculture: An Evolutionary Perspective*. Orlando: Academic Press.

Rival, L. (ed.) 1998. *The Social Life of Trees: Anthropological Perspectives on Tree Symbolism*. Oxford: Berg.

——— 2012. Animism and the meaning of life: towards an understanding of manioc domestication. In O. Ulturgasheva, M. Brightman and V.E. Grotti (eds), *Animism in Rainforest and Tundra: Personhood, Animals, Plants and Things in Contemporary Amazonia and Siberia*, pp. 119–41. Oxford: Berghahn.

Roger, A.J. and Simpson, A.G.B. 2009. Evolution: revisiting the root of the Eukaryote tree. *Current Biology* 19(4):R165–7.

Rosch, E. 1977. Human categorization. In N. Warren (ed.), *Studies in Cross-cultural Psychology, 1*, pp. 1–49. London: Academic Press.

Rosch, E., Mervis, C., Gray, W., Johnson, D. and Boyes-Braem, P. 1976. Basic objects in natural categories. *Cognitive Psychology* 8:382–439.

Roullier, C., Kambouo, R., Paofa, J., McKey, D. and Lebot, V. 2013. On the origin of sweet potato (*Ipomoea batatas* (L.) Lam.) genetic diversity in New Guinea, a secondary centre of diversity. *Heredity* 110(6):594–604.

Rubenkoning, J.A. 1959. Preface. In H.C.D. de Wit (ed.), *Rumphius Memorial Volume*, pp. i–v. Baarn: Hollandia NV.

Rumphius, G.E. 1741 [1705]. *D'Amboinische Rariteitkamer*. Amsterdam: Jan Roman de Jonge.

——— 1999 [1705]. *The Ambonese Curiosity Cabinet* (trans. E.M. Beekman). New Haven, CT: Yale University Press,

——— 1741-50. *Herbarium Amboinense*. 6 vols. Amsterdam: J. Burmannus.

——— 2011 [1741–50]. *The Ambonese Herbal*, vols 1–6. (trans. E.M. Beekman). New Haven, CT: Yale University Press.

Ryan, J.C. 2012 Passive flora?: reconsidering nature's agency through human-plant studies (HPS). *Societies* 2(3):101–21.

Sakamoto, S., Kimata, M. and Kawase, M. 1980. An analysis of cultivated millet collected in Halmahera. In N. Ishige (ed.), *The Galela of Halmahera: A Preliminary Survey*, pp. 181–90. Osaka: National Museum of Ethnology.

Samarin, W.J. 1967. *Field Linguistics: A Guide to Linguistic Fieldwork*. New York: Holt, Rinehart and Winston.

Sasaoka, M. and Laumonier, Y. 2012. Suitability of local resource management practices based on supernatural enforcement mechanisms in the local social-cultural context. *Ecology and Society* 17(4):6.

Sasaoka, M., Laumonier, Y. and Sugimura, K. 2014. Influence of indigenous sago-based agriculture on local forest landscapes in Maluku, east Indonesia. *Journal of Tropical Forest Science* 26(1):75–83.

Scaglion, R. and Soto, K.A. 1994. A prehistoric introduction of the sweet potato in New Guinea? In A.J. Strathern and G. Stürzenhofecker (eds), *Migration and Transformations: Regional Perspectives on New Guinea*, pp. 257–94. Pittsburgh, PA: University of Pittsburgh Press.

Sheil, D., Basuki, I., German, L., Kuyper, T.W., Limberg, G., Puri, R.K., Sellato, B., van Noordwijk, M. and Wollenberg, E. 2012. Do anthropogenic dark earths occur in the interior of Borneo?: some initial observations from East Kalimantan. *Forests* 3:207–29.

Sillitoe, P. 1980. Confusions in the classifications: how Wola name their plants. *Ethnos* 45(3–4):133–56.

——— 1981. The gender of crops in the Papua New Guinea highlands. *Ethnology* 20:1–14.

——— 1983. *Roots of the Earth: Crops in the Highlands of Papua New Guinea.* Manchester: Manchester University Press.

——— 1995. An ethnobotanical account of the plant resources of the Wola region, southern highlands province, Papua New Guinea. *Journal of Ethnobiology* 15(2):201–35.

Slooten, D.F. van 1959. Rumphius as an economic botanist. In H.C.D. de Wit (ed.), *Rumphius Memorial Volume*, pp. 295–338. Baarn: Hollandia NV.

SOED 1973. *The Shorter Oxford English Dictionary*, 2 vols. Oxford: Clarendon Press.

Sperber, D. 1994. The modularity of thought and the epidemiology of representations. In L.A. Hirschfeld and S.A. Gelman (eds), *Mapping the Mind: Domain Specificity in Cognition and Culture*, pp. 39–67. Cambridge: Cambridge University Press.

Spriggs, M. 1998. Research questions in Maluku archaeology. *Cakalele: Maluku Research Journal* 9(2):51–64.

Stark, K. and Latinis, D.K. 1996. The response of early Ambonese foragers to the Maluku spice trade: the archaeological evidence. *Cakalele: Maluku Research Journal* 7:51–67.

Stearn, W.T. 1973. *Botanical Latin: History, Grammar, Syntax, Terminology and Vocabulary.* Newton Abbot: David and Charles.

Steenis, C.G.G.J. van 1950a. The delimitation of Malaysia and its main plant geographical divisions. *Flora Malesiana* I(1):lxx–lxxv. National Herbarium of the Netherlands: Koeltz Scientific Books.

——— 1950b. Malaysian plant collectors and collections. *Flora Malesiana* I(1). National Herbarium of the Netherlands: Koeltz Scientific Books.

Sterly, J. 1997. *Simbu Plant-lore: Plants Used by the People in the Central Highlands of New Guinea*, vols 1–3. Berlin: Dietrich Reimer Verlag.

Stevens, P.F. 1974. A review of Calophyllum L. (Guttiferae) in Papuasia. *Australian Journal of Botany* 22:349–411.

——— 2001. Angiosperm Phylogeny Website. Version 12: www.mobot.org/MOBOT/research/APweb (accessed July 2012).

Stoller, P. 1989. *The Taste of Ethnographic Things: The Senses in Anthropology.* Philadelphia, PA: Pennsylvania University Press.

Stresemann, E. 1923. Religiöse Gebräuche auf Seran. *Tijdschrift voor Indische Taal-, Land- en Volkenkunde* 62(2):305–424.

Stross, B. 1973. Acquisition of botanical terminology by Tzeltal children. In M.S. Edmonson (ed.), *Meaning in Mayan Languages*, pp. 107–41. The Hague: Mouton.

Sugishima, T. 1994. Double descent, alliance, and botanical metaphors among the Lionese of central Flores. *Bijdragen Tot de Taal-, Land- en Volkenkunde* 150:146–70.

Suharno, D.M.W. 1997. Représentation de l'environment vegetal et pratiques agricoles chez les Alune de Lumoli, Seram de l'ouest (Moluques centrales, Indonésie de 'est). These de Doctorat de l'Université Paris VI. [Trans. in Indonesian as: 2013. *Pulau Seram: Pencitraan Lingkunan Alamya dan Perlaku Pertanian Orang Alune.* Jakarta: Kepustakaan Populer Gramedia.]

Suharno, D.M.W. and Friedberg, C. 1996. Gestion de la biodiversité dans les kwesie: froches, jachères des Alune de Seram de l'Ouest (Moluques centrales, Indonésie de l'est). *Journal d'Agriculture Tropicale et de Botanique Appliquée* 38(1):131–52.

Taiz, L. and Taiz, L. 2017. *Flora Unveiled: The Discovery and Denial of Sex in Plants.* Oxford: Oxford University Press.

Taylor, P.M. 1990. *The Folk Biology of the Tobelo People: A Study in Folk Classification.* Washington DC: Smithsonian Institution Press.

Teijsmann, J.E. 1877. Verslag eener botanische dienstreis naar de Molluken. *Natuurkundig Tijdschrift voor Nederlandsche-Indië* 37:76–148.

THE PLANT LIST 2013. Version 1.1: www.theplantlist.org.

Tichelman, G.J. 1918. Rubber-cultuur op eiland Ceram. *Algemeen Landbouwweekblad voor Nederlandsch-Indië* 3:288–90.

——— 1918–19. Iets over de klappercultuur en copra-bereiding aan de Taloetihbaai (Zuid Ceram). *Algemeen Landbouwweekblad voor Nederlandsch-Indië* 3:290–92; 10:323–4.

Traube, E.G. 1986. *Cosmology and Social Life: Ritual Among the Mambai of East Timor.* Chicago. IL: University of Chicago Press.

Tsing, A. 1999. Becoming a tribal elder, and other green development fantasies. In T. Li (ed.), *Transforming the Indonesian Uplands: Marginality, Power and Production*, pp. 159–202. Amsterdam: Harwood.

——— 2005. *Friction: An Ethnography of Global Connection.* Princeton, NJ: Princeton University Press.

Turland, N.J., Wiersema, J.H., Barrie, F.R., Greuter, W., Hawksworth, D.L., Herendeen, P.S., Knapp, S., Kusber, W.-H., Li, D.-Z., Marhold, K., May, T.W., McNeill, J., Monro, A.M., Prado, J., Price, M.J. and Smith, G.F. (eds), 2018: *International Code of Nomenclature for Algae, Fungi, and Plants (Shenzhen Code) Adopted by the Nineteenth International Botanical Congress, Shenzhen, China, July 2017.* Glashütten: Koeltz Botanical Books.

Valeri, V. 1989. Reciprocal centers: the Siwa-Lima system in the central Moluccas. In D. Maybury-Lewis and U. Almagor (eds), *The Attraction of Opposites*, pp. 117–41. Ann Arbor, MI: University of Michigan Press.

——— 2001. *Fragments from Forests and Libraries.* Durham, NC: Carolina Academic Press.

Valmayor, R.V., Jamaluddin, S.H., Silayoi, B., Kusumo, S., Danh, L.D., Pascua, O.C. and Espino, R.R.C. 2000. *Banana Cultivar Names and Synonyms in Southeast Asia.* Los Baños, Philippines: International Network for Improvement of Banana and Plantain – Asia and the Pacific Office.

Vaughan, J.G. and Geissler, C.A. 1997. *The New Oxford Book of Food Plants.* Oxford: Oxford University Press.

Vavilov, N.I. 1951. *The Origin, Variation, Immunity and Breeding of Cultivated Plants*: New York: Ronald.

Veen, M. van der 2014. The materiality of plants: plant–people entanglements. *World Archaeology* 46(5):799–812.

Verheiji, E.W.M. and Coronel, R.E. (eds) 1992. *Plant Resources of South-East Asia, 2: Edible Fruits and Nuts.* Bogor: PROSEA Foundation.

Verheijen, J.A.J. 1984. Plant names in Austronesian linguistics. *Nusa: Linguistic Studies of Indonesian and Other Languages of Indonesia* 20:0–98.

——— 1993. Glossary of taxonomic – Indonesian plant names. *Nusa: Linguistic Studies of Indonesian and Other Languages of Indonesia* 35:0–58.

Voegelin, E. 2004. *The Collected Works,* (eds) W. Petropulos and G. Weiss, vol. 33, pp. 290–1. Columbia, MO: University of Missouri Press.

Votava, E.J., Baral, J.B. and Bosland, P.W. 2005. Genetic diversity of Chile (*Capsicum annuum* var. annuum L.) landraces from northern New Mexico, Colorado, and Mexico. *Economic Botany* 59(1):8–17.

Wallace, A.R. 1863. On the physical geography of the Malay archipelago. *Journal of the Royal Geographical Society* 7:205–12.

Warburg, O. 1893 Vegetationsschilderungen aus Südost-Asien. *Botanische Jahrbuch* 17:169–76.

——— 1902. Die Botanische Erforschung der Molukken seit Rumpfs' Zeiten. In M. Greshoff (ed.), *Rumphius Gedenkboek 1702–1902*, pp. 63–78. Haarlem: Koloniaal Museum.

Waterton, C. 2003. Performing the classification of nature. In B. Szerszynski, W. Heim and C. Waterton (eds), *Nature Performed: Environment, Culture and Performance,* special issue *Sociological Review* 51(2):111–29.

Watt, K.E. 1966. *Systems Analysis in Ecology.* New York: Academic Press.

WCMC 1994. *Conservation Status Listing of Plants Database: Moluccas.* Cambridge: World Conservation Monitoring Centre.

Welzen, P.C. van, Parnell, J.A.N. and Slik, J.W.F. 2011. Wallace's Line and plant distributions: two or three phytogeographical areas and where to group Java? *Biological Journal of the Linnean Society* 103(3):531–45.

Werner, O. and Campbell, D.T. 1970. Translating, working through interpreters, and the problem of decentering. In R. Naroll and R. Cohen (eds), *A Handbook of Method in Cultural Anthropology*, pp. 398–420. New York: Columbia University Press.

White, C.M.N. 1975. The problem of the cassowary in Seran. *Bulletin of the British Ornithologist's Club* 95(4):165–70.

Whitehouse, C. 2014. Why do plant names change? *The Garden* 139(7):64.

Whitehouse, H. 2004. *Modes of Religiosity: A Cognitive Theory of Religious Transmission.* Walnut Creek, CA: Altamira Press.

Whitmore, T.C. 1981. Palaeoclimate and vegetation history. In T.C. Whitmore (ed.), *Wallace's Line and Plate Tectonics*, pp. 36–42. Oxford: Clarendon Press.

Whitmore, T.C., Tantra, I.G.M. and Sutisna, U. (eds) 1989. *Tree Flora of Indonesia: Check List for Maluku.* Bogor: Ministry of Forestry, Agency for Forestry Research and Development (Forest Research and Development Centre).

Wiebusch, T. 2009. Stability and Adaptation of Classification Systems in Cross-cultural Perspective. Paris: COST Action 31: Monitoring Progress Report.

Wigboldus, J.S. 1979. De oudste Indonesische maiscultuur. In F. van Anrooij, D.H.A. Kolff, J.T.M. van Laanen and G.J. Telkamp (eds), *Between People and Statistics: Essays on Modern Indonesian History Presented to P. Creutzberg*, pp. 19–31. The Hague: Martinus Nijhoff.

Wit, H.C.D. de 1959. A checklist to Rumphius's Herbarium Amboinense. In H.C.D. de Wit (ed.), *Rumphius Memorial Volume*, pp. 339–460. Baarn: Hollandia NV.

Witkowski, S.R. and Brown, C.H. 1983. Marking reversals and cultural importance. *Language* 59(3):569–82.

Witkowski, S.R., Brown, C.H. and Chase, P.K. 1981. Where do tree terms come from? *Man* (NS) 16:1–14.

Wolff, X.Y. and Florey, M.J. 1998. Foraging, agricultural, and culinary practices among the Alune of west Seram, with implications for the changing significance of cultivated plants as foodstuffs. In S. Pannell and F. von Benda-Beckmann (eds), *Old World Places, New World Problems: Exploring Resource Management Issues in Eastern Indonesia*, pp. 267–321. Canberra: Centre for Resource and Environmental Studies, Australian National University.

Wright, C.A. 2007. The medieval spice trade and the diffusion of the chile. *Gastronomica* 7:119–37.

Yen, D.E. 1974. *The Sweet Potato and Oceania: An Essay in Ethnobotany.* Honolulu: Bishop Museum Press.

——— 1989. The domestication of environment. In D.R. Harris and G.C. Hillman (eds), *Farming and Foraging: The Evolution of Plant Exploitation*, pp. 55–78. Boston, MA: Unwin Hyman.

——— 1995. The development of Sahul agriculture with Australia as a bystander. *Antiquity* 69:831–47.

Yesomkor, A. 2005. *Irahu Paha Msinae: Bercocok Tanam Bawang Merah.* Cabang Maluku: Program Kerja Sama, Badan Pemberdayaan Masyarakat Propinsi Maluku dan SIL Internatiobal.

Yoshida, S. 1980. Cultivated bananas amongst the Galela. In N. Ishege (ed.), *The Galela of Halmahera: A Preliminary Survey*, pp. 119–37. Osaka: National Museum of Ethnology.

Zent, S. 2014. Processual perspectives on traditional environmental knowledge: continuity, eroision, transformation, innovation. In R. Ellen, S.J. Lycett and S.E. Johns (eds), *Understanding Cultural Transmission in Anthropology: A Critical Synthesis*, pp. 213–65. New York: Berghahn.

INDEX

www.ingramcontent.com/pod-product-compliance
Lightning Source LLC
LaVergne TN
LVHW060412220926
848484LV00015B/2513
* 9 7 8 1 9 1 2 3 8 5 2 5 6 *